Advances in Critical Zone Science

Series Editor

Steven Banwart, The University of Sheffield, Sheffield, UK

Earth's critical zone (CZ) is the near-surface layer of the planet which determines the availability of life-sustaining resources - extending from the surface of unaltered bedrock to the atmospheric boundary layer. Critical zone science is the interdisciplinary study of the natural processes that shape the critical zone and determine its evolution and the effects of natural and human-induced change.

The rapid expansion of CZ science and its growing relevance across a range of disciplines means the time is ripe for a new series of books that will synthesise the current state of knowledge in key areas of CZ science, highlight new research results, and suggest exciting avenues for future investigation. Books in the series will be edited or authored by leading international experts at the forefront of critical zone science. Each volume will be relatively short, on the order of 150-300 pages in length, and have a target audience comprising advanced students, researchers, and applied scientists engaged with the research agenda.

The aim of the Series is to present the study of Earth's critical zone as a new field of integrating science. The scientific content draws on interdisciplinary research and subject expertise from the basic sciences of physics, chemistry, biology and mathematics, and applied sciences such as geological sciences, ecological sciences, environmental sciences, human and physical geography, engineering and expertise at the interface of applied research with policy and practice.

Timothy White · Antonello Provenzale
Editors

Critical Zone and Ecosystem Dynamics

Springer

Editors
Timothy White
Earth and Environmental Systems Institute
The Pennsylvania State University
University Park, PA, USA

Antonello Provenzale
National Research Council
Institute of Geosciences and Earth Resources
Pisa, Italy

ISSN 2731-5851 ISSN 2731-586X (electronic)
Advances in Critical Zone Science
ISBN 978-3-031-69075-4 ISBN 978-3-031-69076-1 (eBook)
https://doi.org/10.1007/978-3-031-69076-1

This Springer imprint is published by the registered company Springer Nature Switzerland AG
The registered company address is: Gewerbestrasse 11, 6330 Cham, Switzerland

We dedicate this book to Steve Banwart.
Devoted husband, father, son, friend, colleague.
Critical zone science ambassador to the world.
1959–2023

Acknowledgements

This volume derived substantial benefit from support provided by U.S. National Science Foundation grants: ICER-1445246 Crossing the Boundaries of Critical Zone Science with a Virtual Institute, and ICER-2129402 Abandonment and rebound: Societal views on landscape- and land-use change and their impacts on water and soils.

Contents

Introduction to the Critical Zone

Timothy White, Antonello Provenzale, and Steven Banwart

Abstract A diverse array of transdisciplinary Earth surface and environmental scientists are considering the Critical Zone (CZ)—the thin outer veneer of the near-surface Earth extending from the top of the vegetation canopy down to subsurface depths at which fresh groundwater freely circulates. The CZ is a major link in the Earth system, defining the mass and energy flux interface connecting the solid Earth (lithosphere) to the atmosphere, biosphere and hydrosphere—initially it forms and evolves through geological processes of the rock cycle: tectonic uplift, subsidence, weathering, erosion and sedimentation. A series of physical, chemical and biological processes transform CZ structure and function and is reflected in a vertical sequence from intact rock at the base of the CZ, to fully formed soils and vegetation cover at the land surface. CZ depth refers to the solid foundation beneath the land surface, but also to a temporal component, deep time, that holds the legacy of past geologic, tectonic, and paleoclimatic events, and even to future events, e.g. ongoing climate and land use change, that will change the form and function of the CZ that we know today. As the physical character of the CZ evolves it affects the storage and flux of water, solutes, sediments, gases, biota and energy—critical services to humanity. Humans have modified more than half of Earth's land surface and have wrought substantial changes in Earth's life support system, yet paradoxically the best projections suggest a continued accelerated trajectory of harmful human activities. Due to its vertical connectivity, CZ functions that provide benefits to people are thus vulnerable to the adverse impact of human-induced environmental change. However, due to its proximity to the land surface, the CZ is also accessible to intended human interventions. A primary tenet of CZ science is that to achieve environmental sustainability, society must first understand the CZ system, the natural processes and services of the CZ

T. White (✉)
Earth and Environmental Systems Institute, The Pennsylvania State University, University Park, PA, USA
e-mail: tsw113@psu.edu

A. Provenzale
Institute of Geosciences and Earth Resources, National Research Council, Pisa, Italy

S. Banwart
Global Food and Environment Institute, University of Leeds, Leeds, England

T. White and A. Provenzale (eds.), *Critical Zone and Ecosystem Dynamics*, Advances in Critical Zone Science, https://doi.org/10.1007/978-3-031-69076-1_1

that are of value to society, and how those processes operate with and without the presence of humanity. Critical Zone Observatories (CZOs) are natural laboratories for investigating Earth surface processes and they can be at the core of collaborative centers comprising field sites, samples, data and models, and people. Most CZOs share a similar approach to answering similar questions: 1) What controls CZ properties and processes? 2) What will be the response of CZ structure, and its stores and fluxes, to climate and land use change? and 3) How can improved understanding of the CZ be used to enhance resiliency and sustainability and restore function of the CZ? The environmental diversity and transdisciplinary approach of CZOs globally promises to more rapidly advance new understanding of key societally important issues in Earth surface and environmental science. Currently, a model of the whole CZ is still missing. Building such a model is not an easy task, not that different from modeling the full Earth System: a large number of dynamical variables and components, a vast range of spatial and temporal scales, and the need for parameterizing unknown or unresolved processes.

1 Humanity's Relationship with Earth

The landmark paper "Human Domination of Earth's Ecosystems" [72] well documented and quantified the degraded state of Earth environments, observations subsequently reproduced and emphasized in many other publications, e.g., [31, 36, 44, 52, 55, 58, 68]. While these synthesis publications and many other studies in the scientific literature primarily documented human impacts on ecological systems, Wilkinson et al. [78] demonstrated that humanity also dominates abiotic systems, moving more sediment annually than all natural processes combined. Humanity's dominating effect on Earth's geology, ecosystems and climate is so significant that a new geological epoch has been proposed to encompass it—the Anthropocene (e.g., [17, 59, 84], and many others).

Hooke [32] summed up the topics and tone of these reports concluding that humans have modified more than half of Earth's land surface, that the current rate of land transformation is unsustainable, and that human activities have wrought substantial changes in Earth's life support system. To many scientists and citizens, these threats to our life support system have reached an acute level, yet paradoxically the best projections suggest a continued accelerated trajectory of harmful human activities [44, 59]. Still the science of understanding and managing anthropogenic threats mostly remains embedded within individual disciplines and the ability to markedly slow or reverse environmental degradation appears beyond society's current capacity [8]. Humanity desperately needs the greater Earth surface and environmental science community to rapidly develop transdisciplinary approaches to studying and understanding natural processes, functions and services, and human impacts on them. Following the footsteps established by Alexander von Humboldt almost two centuries ago, science should adopt the systemic vision of natural processes and fully address the interlinked and interactive aspects of biosphere and geosphere dynamics. Along

Fig. 1.1 The "Nature Painting" (Naturgemälde) by Alexander von Humboldt and Aimée Bonpland (1805), a view of the Chimborazo volcano in terms of what will become Earth System Science, encompassing climate, rocks, living beings, soils, and water

these lines, Fig. 1.1 shows the von Humboldt and Bonpland "Naturgemälde", the picture of the Chimborazo volcano that unites geology, biology, hydrology, climate and soils. Transdisciplinary science can bypass the "silo" conundrum between scientific disciplines and help to overcome society's incapacity to achieve environmental sustainability, and to meet sustainable development goals in the face of burgeoning populations, intensifying land use and climate change.

2 Toward a System View

The history of humanity's relationship with varied landscapes and ecosystems has waxed and waned through the ages often in synch with the needs and mores of dominating cultures and industries. After the pioneering views of Alexander von Humboldt and Alfred Russel Wallace in the nineteenth century, during the twentieth century scientists began to more formally consider the complex relationships between life and the environment. Vladimir Vernadsky wrote a pioneering book on biogeochemistry in 1926, "Biosphere" (in Russian), translated in French in 1929 and in English for the first time only in 1979. [62] coined the term "ecosystem" and described the intimate link between organisms and their abiotic environment, while [9, 53] include wonderful reviews and consideration of other early scientists' work in understanding biotic-abiotic relationships. Unfortunately, the potential development of integrated approaches was historically stifled by the subdivision of terrestrial ecosystem science into aboveground and belowground realms, as well as by the divisions between disciplinary realms (e.g., biology, geology and geomorphology, atmospheric sciences). However, the recognition that terrestrial ecosystem function is governed by complex interactions *between* aboveground- and belowground-organisms and their abiotic environment, especially in response to environmental change, is now widely accepted [7, 9] and included references therein), and a diverse array of Earth surface and environmental scientists are considering the Critical Zone—where rock meets life—the next evolutionary step in defining humanity's relationship to and understanding of the terrestrial Earth, including the many facets of the interactions between geodiversity and biodiversity.

3 Critical Zone Science

The Critical Zone (CZ; [69]) encompasses the thin outer veneer of the near-surface Earth extending from the top of the vegetation canopy down to subsurface depths at which fresh groundwater freely circulates. It is an object [28]—the near-surface heterogeneous amalgamation of landscapes, ecosystems and subsurface environments—that can be probed and measured, monitored and assayed. Its most obvious and visible components are the land surface we live and work and walk on every day, the vegetation that lives on that surface, and the lowermost atmosphere surrounding the vegetation. It includes a central component of the CZ—soil, the living biomantle formed by the transformation of rock and organic matter. But most of the CZ extends to greater depths [81], unseen beneath us and difficult to explore [28]. The physical character of the CZ evolves and in doing so it affects the storage and flux of water, solutes, sediments, gases, biota and energy. Further, by mediating these stores and fluxes, the CZ provides critical services to humanity, e.g., carbon storage, water supply, nutrients, and functional forest and river ecosystems [46]. It is our life support system, sustaining most terrestrial life including humanity: a single entity across the terrestrial surface of the planet capable of nurturing life, supporting agriculture, cleansing water, buffering atmospheric gas levels, to name a few processes. CZ function is supported by complex coupled physical, biological and chemical processes manifested as incredibly complex spatial and temporal heterogeneity. As a consequence of this complexity, a transdisciplinary approach engaging myriad areas of scientific expertise is required to understand the zone and its processes [28]: geology, soil science, biology, ecology, geochemistry, geomorphology, hydrology, atmospheric science, and many more. In the words of Bruno Latour and Peter Weibel, Critical Zones are central to the science and politics of landing on Earth [39].

A primary tenet of CZ science is that to achieve environmental sustainability, society must first understand the CZ system, the natural processes and services of the CZ that are of value to society, and how those processes operate with and without the presence of humanity [74]. An immediate challenge is to develop a predictive ability—earthcasting [25]—for how the CZ may respond to burgeoning population pressures manifested most evidently by climate and land-use changes, pollution, disruption of biogeochemical cycles (e.g., nitrogen) and biodiversity loss, and to address quantitatively the value of CZ functions and services [8]. This predictive ability must be founded on broad knowledge of: (1) the CZ system and CZ processes, to describe the interactions of the varied climatic, ecologic and geologic factors that distinguish different geographic regions—a primary focus of the scientific and educational efforts at now defunct U.S. Critical Zone Observatories [77] and references therein [8], for some information on international observatories), (2) CZ resiliency, the ability of the CZ to continue key functions that provide ecosystem services in the presence of disturbance [2], and (3) a CZ perspective that expands the context of ecosystem services by considering that the geological framework constrains the provisioning of ecosystem services by setting the spatial and temporal template that directs the coevolution of physical and biological systems that ultimately results in societally beneficial services [8, 19, 59].

4 The Multiple Facets of Critical Zone Dynamics

If we consider the onset of the Anthropocene as a step-change in land surface inputs to the CZ, the rate of propagation of the effects depends principally on (1) the distribution of transport mechanisms, pathways and travel times from the surface into the critical zone and (2) the rates of material, energy and biotic transformations that occur along transport pathways. Response times to change the subsurface physical and chemical environment by changing infiltration water flow or chemistry depends on the hydraulic properties of the soil, vadose and groundwater zones with travel times ranging from e.g., a few hours ($\sim 10^{-4}$ years) through a thin, permeable soil, to 10^6 years within regional aquifers. Chemical process response times may be considered as a nominal half-life, e.g. for rapid (minutes) removal of dissolved Fe(II) from mildly alkaline groundwater by formation of Fe(III) hydroxide precipitate in response to $O_2(aq)$ transported with infiltrating water, to slow weathering transformation, e.g. conversion of silicate minerals in metamorphic rock to secondary clay phases. Biological transformation at organism and ecosystem level can range from the doubling time of a few hours for species within the soil microbiome to development of mature climax land surface ecosystems over centuries. The response times of these CZ process can be compared to the rate of human change in inputs to the CZ globally, e.g., changing atmospheric composition and the global climate system since the onset of the industrial revolution, or locally, e.g., immediate deforestation impacts on infiltration, litter input to soil, and the thermal balance of the land surface. A crucial role of CZ science is therefore to gain fundamental understanding of processes, their spatial extent and rates, and their dependencies on biophysical conditions throughout the vertical extent of the CZ. This basic research agenda allows analysis of process responses to natural and human forcing of the CZ, and the time scales and extent of impact on CZ functions. Application includes assessing the potential adverse impact on CZ structure and functions, or the potential for beneficial intervention to protect or restore CZ function.

5 Geological Framework

The CZ is a major link in the Earth system, defining the mass and energy flux interface connecting the solid Earth (lithosphere) to the atmosphere, biosphere and hydrosphere. This thin outermost layer of Earth's continental crust—ranging from a few tens of cm in the high mountains to ~120 m [81] to hundreds of meters and perhaps kilometers in well-developed sedimentary basins (Riebe et al. [54] and references therein)—forms and evolves its purely physical form through geological processes of the rock cycle: tectonic uplift, subsidence, weathering, erosion and sedimentation. The upper boundary of the CZ extends to the top of a vegetation canopy, and comprises ~20% of total CZ thickness globally [81], it is relatively easy to see but is heterogeneous and complex, and typically requires high frequency information to fully understand. On average, 80% of total CZ thickness globally lies below ground [81], comprised of minerals, water, gases, and living and dead organisms that exist interconnected between the surface and soils, and the deep geosphere. The lower boundary of the CZ equates to the base of the freely-circulating fresh groundwater zone—it is diffuse and variable, more difficult to define than the upper boundary, and the relative paucity of life and lack of atmosphere somewhat simplifies study and can require lower frequency data to understand ongoing processes.

On geological timescales, tectonic environments may be generally characterized by (1) active processes of uplift and/or subsidence wherein uplifted terrains are typically dominated by erosion, while sedimentation dominates in subsiding landscapes, and (2) tectonically quiescent, passive environments which characteristically may be dominated by weathering. Each of these broad categories of geological landscapes forms a suite of bedrock types and surface deposits with unique geochemical, mineralogical, textural, and structural characteristics that define the physical subsurface template and topography upon which the CZ forms.

Many geological processes influence and establish the solid framework and structure in which the CZ forms and evolves, and the CZ inherits the result of these processes. Studying these processes lies within the realm of the greater geosciences, but it is helpful to understand how they may relate to CZ function. Other geological processes, e.g., Earth surface processes, occur within the CZ and may reshape and further establish the solid framework and structure in which the CZ forms and evolves. Studying these processes also lies within the realm of the greater geosciences—but they are necessary to the science of the CZ when they link to processes occurring between the other "spheres" of the CZ.

Rock (and sediment) composition is inherited from the original geological environment of formation and has been demonstrated to have a profound impact on CZ development, including the biosphere on local to global scales. For example, [29] showed that bedrock geochemistry and mineralogy is on par with climate as a regulator of forest productivity in the U.S. Sierra Nevada, and likely in other granitic mountain ranges, while [33] determined that bedrock may be as large a source of terrestrial nitrogen as the atmosphere, representing 6–17% of total nitrogen inputs to the surface environment globally. Rock structure strongly controls the development of the CZ, as three-dimensional topography and tectonic stress regimes develop stress fields and bedrock fracture patterns beneath Earth's surface, that in turn produce spatial variations in the rates of surface processes and the development of non-uniform groundwater flow paths [45].

Earth surface processes in the critical zone may profoundly influence the development of bedrock fracture patterns. For example, climate effects have been well documented, as in the impacts of frost damage below mobile regolith where rock strength is diminished [5], and the widespread nature of frost cracking of bedrock and doubled erosion rates in unglaciated terrains during the Last Glacial Maximum [42]. And vegetation also strongly impacts shallow CZ development, especially at the tree-rock boundary. For example, arborturbation or tree throw, has long been known to influence soil turnover and downslope erosion rates (e.g., [21], and many others), though [43] concluded that "charismatic tree throw may matter less than belowground damage" by trees that detach rock through daily water uptake and wind-driven tree sway. Other Earth surface processes in the critical zone in turn influence erosion rates, for example West et al. [74] found that at lower erosion rates, mineral supply limits weathering rates, whereas at higher erosion rates there is abundant material but climatic factors limit weathering.

6 Geological Time of the Critical Zone

The depth of the CZ does not refer simply to the solid foundation beneath the land surface, but also to a temporal component, deep time, that holds the legacy of past geologic, tectonic, and paleoclimatic events, and even to future events, e.g. ongoing climate and land use change, that will change the form and function of the CZ that we know today. For example, the overall trajectory of landscape evolution (and CZ development) through geologic time evolved and changed from that of an abiotic Earth to one in which the evolution of plants provided the stabilizing impact of roots on the evolution of soils and fluvial morphology [23]. Another example can be observed in simple regolith evolution models wherein the depth of rock weathering increases from polar to equatorial latitudes as a function of climate [11]. Such a conceptual approach may be used to similarly classify the distribution of CZ thickness globally, yet the thickest CZ measurements are observed in mid latitudes [81]. This conundrum might be explained as a legacy of past environmental change: during well-documented ancient greenhouse episodes (e.g., [83]), more intense weathering environments extended further poleward (e.g., [55, 75]), in turn increasing the potential for developing a thicker CZ in mid to high latitudes. During icehouse episodes, the advance of continental glaciers from high latitudes, scraped and gouged underlying soil, surficial deposits, and bedrock [16], and carried it to and deposited it on mid latitude landscapes, that already had been subjected to intensive (and deeper) periglacial frost cracking [42].

7 Forming CZ Biophysical Structure and Function

CZ biophysical properties and processes are dominated by the sharp vertical gradient in physical characteristics that govern contrasting rates of fluid mass transport. For example, some sedimentary rocks and sediments exhibit a primary porosity and permeability readily amenable to fluid transmission, whereas in other rock types with very little primary porosity, stress loading and unloading create bedrock strain and fracturing that forms an initial porosity. The physical processes range from slow diffusion of aqueous solutes at grain boundaries within intact bedrock, hydrodynamic flow and advection within rock fractures and unconsolidated aquifers, recharge flow and gas diffusion within the vadose zone, diffusive gas exchange between soil and atmosphere, and the aerodynamic flow of the lower atmospheric boundary layer. Hence, the geological formation of the critical zone is considered as the spatial and temporal transition from unaltered bedrock as soil parent material at the base of the CZ, through formation of unconsolidated and erodible regolith, to soil and land surface biota supporting photosynthesis.

A generalized conceptual model of the CZ in regions of topographic steady state or net denudation (uplift balanced by erosion) holds that fluids are supplied by precipitation at the top of the critical zone reactor, and the downward advance of the weathering front brings unweathered rock into the reactor which is also stirred at the top by biological and physical processes. Thus, the vertical extent of the CZ—the size of the reactor—reflects the relative rates of downward weathering-front advance and erosion from the top [3]. Alternatively, [49] determined that slow downward drainage of groundwater from fresh bedrock creates a bottom-up control on the elevation and topography of the transition from weathered bedrock to underlying fresh bedrock. Brantley et al. [14] concluded though that lateral flow (on slopes) decreases downward flow thereby decreasing the weathering advance rate in underlying zones. Therefore, the rate of weathering advance is affected by the three-dimensional distribution of reaction zones that affect permeability across a watershed [14].

Hydraulic contrast between the bulk of intact rock and the relatively more hydraulically transmissive fractures permits penetration of meteoric water from the land surface into the rock mass. The primary minerals of bedrock are far from chemical equilibrium with respect to the composition of groundwater and undergo slow, irreversible chemical dissolution reactions. Porosity is further developed from the chemical action of water, solutes and biota on the dissolution of minerals, weakening of mechanical strength with greater fracturing and the development of internal porosity and increasing reactive mineral surface area. As the porosity and transmissivity fields change, there is increased water circulation and mass transport of reaction product solutes through groundwater discharge to surface waters, and the formation within the subsurface of secondary minerals that contribute to regolith composition. As bedrock transforms to regolith, the soil parent material becomes transportable as sediment with continued CZ evolution at the point of sediment deposition.

Following initial physical alteration to allow fluid and solute penetration into the bedrock mass, biogeochemical processes also contribute to development of porosity and CZ evolution. Colonization of bedrock by biota occurs as the physical scale of internal pores and their connection increases to the point that size exclusion of microbes no longer occurs, with transport of microbes into the rock mass with groundwater circulation. Many different strains of microbes can influence mineral transformation reactions, accelerating the development of porosity. Ecological selection of microbes is influenced by the physical and geochemical conditions within the subsurface. A gradient in redox conditions exists from between the oxic environment at the land surface and anoxic conditions that develop below the water table, due to the diffusion barrier of water-filled pores to O_2 mass transfer into groundwater. The diffusion coefficient for O_2 in water is on the order of 10^5 times smaller than that for diffusion in air.

Under oxic conditions with input of organic matter from photosynthetic organisms at the land surface (dead biomass derived from leaf litter), microbes utilize the carbon and energy source in the organic matter for aerobic respiration with O_2 as the terminal electron acceptor. Below the water table, aerobic respiration, even at a slow rate, consumes dissolved O_2 at a much faster rate than it can be replenished by diffusion, and anaerobic processes become dominant. Most importantly across this vertical redox cline in the critical zone, the biogeochemical transformations of Fe(II/III) minerals plays an important role in bedrock alteration. Anaerobic respiration by dissimilatory metal-reducing organisms use Fe(III) as the terminal electron acceptor. These organisms cause reduction of Fe(III) within minerals to soluble Fe(II)(aq) forms that results in mineral dissolution. Alternatively, in the presence of O_2, e.g. at shallower depths, any soluble Fe(II) or Fe(II)-bearing minerals can be transformed to Fe(III) hydroxide or oxide minerals. These minerals usually have a larger molar volume than the mineral source of Fe(II), creating mineral expansion and localized strain and micro fracturing of the surrounding rock, where the oxidized minerals form. This further weakens the rock, creates micro porosity that accelerates the diffusion of water and solutes into the interior of rock blocks, and contributes to the further chemical and physical alteration of the CZ.

Fracture development at the land surface, or the uplift of regolith, allows establishment of rooted plants, which is a major step in developing the vertical structure and functioning of the CZ. Release of soluble P and K during rock alteration and mineral dissolution, provides essential nutrient elements that create selective pressure to support plant colonization. Root penetration and growth in fractures creates strain and further fracturing of the shallow rock, which increases O_2 transport, precipitation infiltration, and groundwater recharge and circulation. Photosynthesis also directs chemical energy in the form of reactive organic compounds from plant biosynthesis into the subsurface through the root system. Exuded compounds provide carbon and energy for microbial respiration that supports the development of the soil microbiome. Many of these compounds are known to be reactive towards minerals and accelerate dissolution and formation of secondary phases such as clay minerals. Most land plants form symbiotic associations with mycorrhizal fungi that attach to the plant roots and gain carbon and energy for growth from the organic compounds produced by the plant. Mycorrhizal fungi form large mycelial networks that permeates the surrounding subsurface and through exudation of reactive compounds, solubilize inorganic nutrients (P, K, Fe, Ca) from minerals and organic nutrients (N, P) from dead organic matter and transport the nutrient elements back to the root systems to support plant growth. Hence, photosynthesis and the establishment of vegetation and canopy cover creates a transformative biological solar engine at the land surface that drives organic matter inputs to the sub-surface to support soil formation.

This series of physical, chemical and biological transformations of the CZ provides a conceptual model for the temporal evolution of CZ structure and function that is also reflected in the vertical structure. With continuous uplift of altered rock to the land surface, and removal by erosion, the spatial sequence of alteration is observed in a vertical sequence from intact rock at the base of the critical zone, to fully formed soils and vegetation cover at the land surface.

8 Vertical Connectivity of the Critical Zone in the Anthropocene

Soil formation is a central component of CZ development. Soil functions describe the sets of processes that transport and transform material, energy and biota within soil layers, as a component of Earth's life support system that create benefits for people. These functions include storing and transmitting water, transforming nutrients, providing nutrients for plant growth, storing carbon and nitrogen and preventing their release as greenhouse gases, filtering contaminants by sorption onto organic matter and minerals, and providing habitat and supporting biodiversity. Noting the processes of soil functions described here, many of these occur even at the earliest stages of bedrock alteration; formation of porosity to store and transmit water, mineral dissolution to provide essential nutrients for biota, formation of a sub-surface microbiome, microbial respiration and decomposition of dead biomass, and plant rooting in shallow rock fractures. Soil functions can therefore be considered as a subset of CZ functions that nourish the development of terrestrial ecosystems and support a sub-surface microbiome and the land surface ecology. Intrinsic to these CZ functions is the vertical connectivity of the movement of material, energy and genetic diversity, particularly through fluid flow and transport. This includes water penetration from the land surface, mineral element transport from weathering rock and the movement, colonization and selection of indigenous populations of biota through the full depth of the CZ, from bedrock to treetop.

Vertical connectivity is highly relevant to CZ function in the Anthropocene. Human impacts, particularly at the land surface, are propagated through the vertical extent of the CZ. These include land management that alters land cover such as transition of native ecosystems to agriculture and the effects of urbanization. Changing water use such as groundwater abstraction and surface irrigation affect groundwater circulation and recharge. Human impacts also occur from atmospheric pathways, such as combustion emissions that create anthropogenic nutrient and acid loads in precipitation and altered weather patterns due to greenhouse gas emissions and changing climate. Due to its vertical connectivity, CZ functions that provide benefits to people are thus vulnerable to the adverse impact of human-induced environmental change. However, due to its proximity to the land surface, the CZ is also accessible to intended human interventions, e.g., through land management practices, to protect, restore and enhance CZ functions. Examples of intervention include changing tillage practices on agricultural land to reduce physical erosion and loss of soil organic matter, either as sediment or through microbial GHG emissions of CO_2, NO_2 and CH_4; afforestation to sequester organic carbon, changing vegetation cover to increase precipitation infiltration and groundwater recharge while reducing runoff and flood peaks in surface waters.

9 Biodiversity, Geodiversity and Ecosystem Engineers

The interaction of geological and biological processes in the CZ is reflected in the interplay of biodiversity and geodiversity. Biodiversity is at the core of ecosystem functioning [41, 63] and it involves all types of organisms, from top vertebrate predators to trees and grasses to bacteria. Microbiota, fungi and invertebrate biodiversity in the soil, as well as the type of vegetation cover, are of special importance to CZ development and dynamics [7]. On the other hand, geodiversity is the variety of rocks, minerals, fossils, landforms, sediments and soils, together with the natural processes which form and alter them, and it is sometimes called the "silent partner of biodiversity" (www.geodiversityday.org/what-is-geodiversity). As such, geodiversity and the diversity of landscapes modulate and control biodiversity, favoring its development [65] and must inform nature conservation plans [4], possibly increasing ecosystem durability to droughts [82], even though the relationships between the two types of diversity can be quite complex [51].

On the other hand, biodiversity can favor geodiversity through modifications of the physical and chemical environment and landscape differentiation by the action of living organisms. This is especially true in the Critical Zone, where most of the weathering and soil formation processes are mediated and amplified by biotic processes. Even on a grand scale, the difference of landscapes is associated to different rock types, many of which have a direct biological origin. In this framework, the concept of "ecosystem engineers" plays a central role [38]. Organisms are capable of drastically modifying the environment in which they live, of course for their own benefit but, in doing so, also affecting the whole habitat and the selection pressures on other organisms [18]. Classical examples include beavers, which transform a valley floor into a lake, or humans, that are transforming the whole planet—undoubtedly affecting biodiversity and geodiversity on a large scale. A relevant example for the Critical Zone is the role played by shrubs and the bacterial crust in water-limited ecosystems [24]. The bacterial crust affects the permeability of the soil surface, slowing down water infiltration and consequently favoring surface water flow and then water infiltration in proximity of the shrubs, which break the crust. This effect becomes even more important as aridity increases, and it generates islands of moisture and fertility around the shrubs. In turn, this allows for the presence of a whole ensemble of organisms around the shrubs, including isopods which bioturbate the soil and further enhance local water infiltration. In these "patchy", spatially structured ecosystems, the properties of the Critical Zone are modulated by the inhomogeneous distribution of the organisms.

10 Droughts, Fires and the Critical Zone

Climate change affects the global water cycle, with a redistribution of precipitation across different regions [67] and a generalized tendency towards the intensification of the hydrological cycle [25]. In large areas of the world, including the Mediterranean basin, sub-Saharan Africa and the southwestern North America, there is a significant reduction of precipitation with an increased risk of longer and more extended drought episodes [36]. Given the crucial role played by water in Critical Zone processes, in coming years the assessment of drought impacts will become one of the great challenges in CZ dynamics, management and preservation. How droughts affect groundwater resources, processes in the unsaturated zone, microbiological and vegetation dynamics already is a central research and practical question in CZ science (e.g., [6]).

Similarly, temperature rise is intensifying wildfires in several areas of the world, generating extended fires in regions that traditionally were not affected by large fire events and leading to large emissions of carbon dioxide [85]. An apparently new phenomenon is the emergence of megafires, which can become extremely difficult to manage and generate large-scale damages, suggesting that our planet may have hit a tipping point [60]. The annual burned area is expected to increase in several regions, such as the Mediterranean basin [66], where the burned area is affected also by vegetation continuity and diversity [62]. Wildfire changes CZ processes, the flow of water, carbon and other elements, and the properties of the soil [56]. A detailed understanding of how the CZ responds to fires, and what CZ characteristics can help a fast recovery from fire events and/or reduce the risk of megafires, will probably become a central point in the agenda of CZ research.

11 Critical Zone Observatories

Critical Zone Observatories (CZOs) are natural laboratories for investigating Earth surface processes and they can be at the core of collaborative centers comprising field sites, samples, data and models, and people, all available to the greater scientific community for research, education and outreach. CZO research seeks to understand the many coupled processes in the CZ across all timescales using quantitative models parameterized from observations of meteorological variables, streams, and groundwater, and sampling and analyzing landforms, bedrock, soils, and ecosystems [77].

Intensive field measurements at CZOs provide data that can guide process understanding to develop the models that explain CZ evolution, forecast possible future states, and guide land use decisions.

In the U.S., teams of cross-disciplinary scientists at nine CZOs worked to further CZ science using field and theoretical approaches, education and outreach. The sites represented a wide range of geologic, climatic, and land use settings that provided an opportunity to develop a broad and general understanding of the evolution and function of the CZ [77]. While the US CZO network was terminated in 2020, the science achieved established a firm foundation from which the successor Critical Zone Collaborative Network works to develop systems-level models to predict how the CZ is responding to natural and human-altered processes. But the U.S. networks represent only a fraction of the formally funded networks internationally, in China, Australia, France, Germany, and United Kingdom, while still more exist as self-identified CZOs [8]. In Italy, three CZOs are coordinated by the National Research Council, and others are being established in the framework of the large-scale national infrastructure project ITINERIS. In the whole of Europe, an attempt to coordinate the different CZOs comes from the activities of the eLTER Research Infrastructure, while a pan-Arctic initiative on establishing a network of Arctic CZOs is currently ongoing in the Terrestrial Working Group of the International Arctic Science Committee (IASC).

Most of the CZOs share a similar approach to answering similar questions: (1) What controls CZ properties and processes? (2) What will be the response of CZ structure, and its stores and fluxes, to climate and land use change? and, (3) How can improved understanding of the CZ be used to enhance resiliency and sustainability and restore function of the CZ? The environmental diversity and transdisciplinary approach of CZOs globally promises to more rapidly advance new understanding of key societally important issues in Earth surface and environmental science. To address the above questions, however, several challenges have to be met. Most importantly, there is the need to identify a minimum set of common measurements, variables or indicators that provide concise but quantitative information on the various components of the Critical Zone [12], in the spirit of the "Essential Variables" already defined in other frameworks (e.g., [80]).

12 CZ Science and Earth System Modeling

After much efforts to implement measurements and monitoring, and data analysis, there is the need for developing models of Critical Zone processes and—hopefully—of the whole CZ system [19]. This endeavor helps both the testing of ideas and interpretations, allowing for the design of new observations and measurements, as well as the quantitative prediction of the CZ response to climate end environmental changes and/or of how the CZ can react to specific management and preservation actions.

A "model" can be conceptual, non-mathematical, and used to order ideas and relations between the different elements of a problem—or of a system. On the other hand, most "models", as they are intended now, are based on a set of mathematical descriptions that can be expressed in terms of numerical procedures to solve them. Along these lines, models can be either process-based ("deterministic"), based on differential equations, or data-driven correlation approaches (such as the various linear or nonlinear incarnations of regression models), or machine-learning techniques based on the various application of Artificial Intelligence (AI) methods. Or a blend of the different types.

Models in CZ science originate from various disciplines, such as soil dynamics, surface and groundwater flows, reactive transport and vegetation processes, but also soil–vegetation–atmosphere exchanges, geomorphological evolution, or ecosystem dynamics (including microbiota ecosystems). Some of the models have been developed ad hoc, for the specific problem at hand, while others are open-source "community models" that have been built in the course of time by a large group of researchers (see for example the case of the USGS MODFLOW model for groundwater dynamics, [34]). In all cases, often a model must be completed and extended in order to include new processes that emerged from the measurements or to tackle a given question.

Currently, a model of the whole CZ is still missing. Building such a model is not an easy task, and it is not that different from the construction of a full Earth System model: a large number of dynamical variables and components, a vast range of spatial and temporal scales, and the need for parameterizing unknown or unresolved processes. A serious effort that cannot be undertaken by a single research group but would require a community strategy, not dissimilar from that adopted in the development of a climate model. In such framework, it also becomes interesting to consider the current efforts on the development of "digital twins", which expand the notion of models to a new dimension [10]. A digital twin is a mixture of AI methods, process-based models, data assimilation and analysis, complemented by a continuous flow of data from ground (and below-ground) sensors and remote sensing. A new frontier would be to develop and implement a CZ digital twin for a specific (presumably small) watershed in a given CZO, and try to see whether the real and the digital siblings behave similarly.

The contributions to this volume are divided into two main parts. In Part I, chapters deal with the basic processes and approaches in Critical Zone science, addressing a set of interconnected themes and challenges. In Part II, some specific examples of the practical implementation of CZ studies are discussed, focusing on the activities at three selected CZ Observatories across the globe.

13 Part I

Chapter 2, by Billings et al., addresses the interplay of biotic and abiotic factors in Critical Zone functioning. It thus focuses understanding the Critical Zone as a problem that is, at its most fundamental level, an ecosystem ecology problem. The chapter describes in detail the interactions among landscape and soil constraints on CZ processes, and the functional roles of biota both physically and biologically, that result in CZ state and functioning.

Chapter 3, by Beierkuhnlein and Shrestha, explores the biogeographical facets of Critical Zone dynamics, focusing on CZ carbon cycling in the different biomes and on the interplay between biodiversity and ecosystem functioning. The relevance of coordinated field experiments is discussed, together with the need for the interplay of experiments, modelling and monitoring, also addressing the role of the unavoidable uncertainties in our knowledge.

Chapter 4, by Wymore et al., deals with the interplay of Critical Zone functioning and surface water chemistry, linking ecosystem ecology and Critical Zone science to develop conceptual models of the multiple environmental controls on surface water chemistry. The main point here is that the chemical characteristics of freshwaters integrate the multiple processes active in the watershed, both above, at and below ground, and thus a multi-scale, systemic approach is needed for a proper understanding of the whole system.

Chapter 5, by Sullivan et al., discusses the complexities of the interplay of groundwater and Critical Zone dynamics. Water—and especially groundwater—is one of the “pillars” of Critical Zone functioning. Groundwater, both in the saturated and unsaturated layers, controls the dynamics of many terrestrial ecosystems and it is crucial to humans as a primary source of freshwater. Here they explore how the structure of the CZ interacts with groundwater to regulate recharge, evapotranspiration, groundwater-surface water interactions, groundwater flow paths (even km deep), chemical weathering, interbasin groundwater flow, and finally coastal and submarine groundwater discharge dynamics. Altogether they provide a holistic understanding of how CZ processes and structure help to regulate one of Earth’s most vital resources.

Chapter 6, by Dere and Santini, introduces soils, one of the basic components of the Critical Zone. This chapter opens a series of following contributions devoted to different aspects of soil dynamics. Here, we learn the main soil-forming mechanisms and various terms applied to soil components and the parent materials that form soils. Descriptions of soil forming models are complemented by how to study soils practically and effectively in the field, and some of the methods used to measure and quantify soil characteristics and processes. The dimensionless mass transfer coefficient tau is discussed in detail as one approach to interpret biogeochemical profiles in soils. The chapter concludes with two case studies that apply soil forming frameworks and soil profile data to understand drivers of soil formation as a function of climate and time, respectively.

Chapter 7, by Moreland et al., continues the exploration of soils, dealing with the dynamics of organic matter, composed of a mixture of the decaying remains of once-living plants, microorganisms, insects, and animals. Soil organic matter typically accounts for only a small fraction of soil weight, but it has significant importance on Critical Zone functioning. This organic matter is the main energy and nutrient source for microorganisms and is key for soil's water holding capacity, tilth, and physical stability. Hence, the understanding of CZ dynamics cannot be achieved without considering soil organic matter and its links with ecosystem dynamics.

Chapter 8, by Lilli et al., delves into the processes of soil degradation, aggregation and restoration, focusing on soil structure and exploring how to reduce soil degradation and achieve restoration through the addition of organic matter and the use of proper agroecological practices. In this chapter, it is shown how soil organic matter amendments improve soil structure, sequester more carbon and nutrients and enhance the ecosystem services provided by soils and landscapes. The example of the Koiliaris CZO is used to illustrate the concepts, and the CAST (Carbon, Aggregation, and Structure Turnover) model is introduced.

Chapter 9, by Aronson et al., considers the role of microbial diversity, dynamics, patterns, and processes in the Critical Zone. It is demonstrated how microbial niches are influenced by environmental gradients, which in turn depend also on climatic factors. Again, the roles of soil aggregates and soil micro-habitat diversity are shown to be central to Critical Zone functioning. The role of microbial importance in rock weathering is emphasized. The chapter also discusses how microbial communities regulate biogeochemical cycling between pools, through multiple trophic strategies, and how the soil and atmosphere habitats are connected by multiple processes, including dust deposition.

Chapter 10, by Giamberini and Provenzale, deals with the fluxes of CO_2 in the CZ carbon cycle, showing how they are closely related to soil and vegetation dynamics through photosynthesis and respiration/decomposition. Carbon, the "globetrotter of life", is the central element of life and climate regulation, through its role in greenhouse gasses such as carbon dioxide and methane. After reporting some of the main quantitative estimates of the main carbon pools and fluxes, the chapter considers the control factors of CO_2 uptake and releases at the atmosphere—vegetation—soil interface and gives a short overview of the measurement methods of carbon dioxide fluxes at different scales, from point-wise flux chamber measurements, to eddy covariance, to larger-scale remote sensing observations, focusing on the specific example of Arctic environments.

Chapter 11, by Magnani and Provenzale, considers the complexity of Critical Zone modelling, mentioning the different approaches which can be, and have been, adopted in past studies. Data-driven empirical (correlative) models, machine-learning algorithms, and process-based models have all been used, and sometimes combined, in the attempt to describe individual Critical Zone processes. The next step is to try achieving a "whole-system" approach, capable of reproducing most processes at once, in the (perhaps visionary) desire of developing a "digital twin" of the real Critical Zone.

Chaper 12, by Lohse et al., closes the first part of the volume and shows how the Critical Zone supports many essential services, including to humanity—such processes develop over long periods of time and have been rarely quantified. The chapter thus addresses the issue of how the maintenance of ecosystem services through time requires process understanding at multiple time scales. In particular, it is discussed how intermediate scale measurements enhance understanding of climate, topography, and bedrock controls on ecosystem services, while measurements at deeper spatial and temporal scales inform management of baseline rates that are key to predicting responses to extreme events.

14 Part II

Chapter 13, by Chorover et al., considers the interactions among water, carbon and weathering fluxes in the Catalina-Jemez Critical Zone Observatory in the southwestern US. The authors view the CZ through the lens of an open system thermodynamic model where climatic forcing, quantified as "effective energy and mass transfer" (EEMT), predicts rates and trajectories of CZ evolution. EEMT model predictions are tested against an impressive set of field data, with special attention to the coupled carbon–water dynamics in surface waters and in the vadose and saturated zones. They show that the thermodynamic model can predict CZ structure. However, they also illustrate how resolving the mechanisms—whereby throughfluxes of water and carbon drive weathering reactions—requires probing the black box of the CZ with direct measurements of coupled processes. By implementing a sensor and sampler array that reports on coupled hydro-biogeochemical processes in real time, they show that moisture dynamics are linked to biological production of CO_2 that, in turn, drives a pulsed release of silicate weathering products into stream waters.

Chapter 14, by Baneschi et al., presents the Gran Paradiso Critical Zone Observatory, located in a high-altitude mountain environment in the northwestern Italian Alps. Carbon flux measurements and modelling are the focus of the chapter—and of this Observatory. After describing the typical characteristics of Alpine grasslands and of the Gran Paradiso National Park, the measuring methods (from portable flux chambers to eddy covariance towers) and the modelling results on the drivers of carbon dioxide fluxes are discussed.

Chapter 15, by Leopold et al., introduces the Avon River Critical Zone Observatory, a part of the Australian Critical Zone Observatories Network. One of the few CZ Observatories in the Southern Hemisphere, the AR-CZO is characterised by about 150 years of recent history of European land clearing and agricultural land use, superimposed on about 40,000 years of Aboriginal land management, notably through fire management. In the chapter, the various biogeochemical and hydrological processes of this area are discussed, together with ecosystem and agro-ecosystem dynamics, agricultural management and anthropogenic impacts, and CZ-services.

References

1. Agrawal A, Chhatre A, Hardin R (2008) Changing governance of the world's forests. Science 320(5882):1460–1462
2. Amundson R, Richter D, Humphreys G, Jobbagy E, Gaillardet J (2007) Coupling between biota and earth materials in the critical zone. Elements 3(5):327–332
3. Anderson S, von Blanckenburg F, White A (2007) Physical and chemical controls on the critical zone. Elements 3(5):315–319. https://doi.org/10.2113/gselements.3.5.315
4. Anderson MG, Comer PJ, Beier P, Lawler JJ, Schloss CA, Buttrick S, Albano CM, Faith DP (2015) Case studies of conservation plans that incorporate geodiversity. Conserv Biol 29:680–691. https://doi.org/10.1111/cobi.12503
5. Anderson R, Anderson S, Tucker G (2013) Rock damage and regolith transport by frost: an example of climate modulation of the geomorphology of the critical zone. Earth Surface Processes and Landforms 38(3). https://doi.org/10.1002/esp.3330
6. Bales RC, Dietrich WE (2020) Linking critical zone water storage and ecosystems. Eos 101. https://doi.org/10.1029/2020EO150459
7. Banwart SA, Nikolaidis NP, Zhu YG, Peacock CL, Sparks DL (2019) Soil functions: connecting Earth's critical zone. Annu Rev Earth Planet Sci 47:333–359
8. Banwart S, Chorover J, Gaillardet J, Sparks D, White T, Anderson S et al (2013) Sustaining earth's critical zone basic science and interdisciplinary solutions for global challenges, vol 47. University of Sheffield, UK. https://www.czen.org/
9. Bardgett R, Wardle D (2012) Aboveground-belowground linkages biotic interactions, ecosystem processes, and global change. Oxford series in ecology and evolution. Oxford University Press 1:1–14
10. Bauer P, Stevens B, Hazeleger W (2021) A digital twin of Earth for the green transition. Nat Clim Chang 11:80–83
11. Brady N, Weil R (2002) The nature and properties of soils. Prentice Hall, Upper Saddle River, NJ, USA, 13:56, Figure 2.23
12. Brantley SL et al (2016) Designing a suite of measurements to understand the critical zone. Earth Surf Dyn 4:211–235. https://doi.org/10.5194/esurf-4-211-2016
13. Brantley S, Buss H, Lebedeva M, Fletcher R, Ma L (2011) Investigating the complex interface where bedrock transforms to regolith. Appl Geochem 26:S12–S15. https://doi.org/10.1016/j.apgeochem.2011.03.017
14. Brantley SL, Lebedeva MI, Balashov VN, Singha K, Sullivan P, Stinchcomb G (2017) Toward aconceptual model relating chemical reaction fronts to water flow paths in hills. Geomorphology 277:100–117
15. Clark EA, Sheffield J, van Vliet M, Nijssen B, Lettenmaier DP (2015) Continental runoff into the oceans (1950–2008). J Hydrometeorol 16:1502–1520. https://doi.org/10.1175/JHM-D-14-0183.1
16. Clarke G (1987) A short history of scientific investigations on glaciers. J Glaciol 33(S1):4–24. https://doi.org/10.3189/S0022143000215785
17. Crutzen P, Stoermer E (2000) The 'Anthropocene'. Global Change Newsletter 41:17–18
18. Cuddington K et al (eds) (2007) Ecosystem engineers: plants to protists. Academic Press, USA
19. Duffy C et al (2014) Designing a suite of models to explore critical zone function. Proc Earth Planetary Sci 10:7–15
20. Field J, Breshears D, Law D, Villegas J, Hoffman L, Brooks P, Chorover J et al (2015) Critical zone services: expanding context constraints and currency beyond ecosystem services. Vadose Zone J 14(1)
21. Gallaway J, Martin Y, Johnson E (2009) Sediment transport due to tree root throw: integrating tree population dynamics wildfire and geomorphic response. Earth Surf Proc Land 34(9):1255–1269
22. Gettelman A, Rood RB (2016) Simulating terrestrial systems in demystifying climate models: a users guide to earth system models. Springer, Berlin 1:109–137. https://doi.org/10.1007/978-3-662-48959-8

23. Gibling M, Davies N (2012) Palaeozoic landscapes shaped by plant evolution. Nat Geosci 5:99–105
24. Gilad E, von Hardenberg J, Provenzale A, Shachak M, Meron E (2007) A mathematical model of plants as ecosystem engineers. J Theor Biol 244(4):680–691. https://doi.org/10.1016/j.jtbi.2006.08.006
25. Giorgi F, Im E-S, Coppola E et al (2011) Higher hydroclimatic intensity with global warming. J Clim 24:5309–5324. https://doi.org/10.1175/2011JCLI3979.1
26. Goddéris Y, Brantley SL (2013) Earthcasting the future critical zone. Elem Sci Anth (1):000019. https://doi.org/10.12952/journal.elementa.000019
27. Golub A, Hertel T, Lee H, Rose S, Sohngen B (2009) The opportunity cost of land use and the global potential for greenhouse gas mitigation in agriculture and forestry. Resour Energy Econ 31:299–319
28. Grant G, Dietrich W (2017) The frontier beneath our feet. Water Resourc Res Comment 53. https://doi.org/10.1002/2017WR020835
29. Hahm W, Riebe C, Lukens C, Araki S (2014) Bottom-up control of ecosystems and landscapes. Proc Natl Acad Sci 111(9):3338–3343. https://doi.org/10.1073/pnas.1315667111
30. Hertel T (2017) Land use in the 21st century: contributing to the global public good. Rev Dev Econ 21(2):213–236. https://doi.org/10.1111/rode.12295
31. Hoekstra J, Molnar J, Jennings M, Revenga C et al (2010) The atlas of global conservation: changes, challenges, and opportunities to make a difference, Molnar JL (ed). University of California Press, Berkeley
32. Hooke R, Martin-Duque J, Pedraza J (2012) Land transformations by humans: a review. GSA Today 22(12):4–10
33. Houlton B, Morford S, Dahlgren R (2018) Convergent evidence for widespread rock nitrogen sources in Earth's surface environment. Science 360:58–62. https://doi.org/10.1126/science.aan4399
34. Hughes W, Langevin CD, Banta ER (2017) Documentation for the MODFLOW 6 framework. USGS report. https://doi.org/10.3133/tm6A57. https://www.usgs.gov/publications/documentation-modflow-6-framework
35. Humboldt A (2010) Cosmos: sketch of a physical description of the universe (Cambridge library collection - physical sciences), Sabine E (ed). Cambridge University Press, Cambridge. https://doi.org/10.1017/CBO9780511708985
36. IPCC (2014) Climate change 2014: synthesis report, contribution of working groups I, II and III to the fifth assessment report of the intergovernmental panel on climate change [Core Writing Team, R.K. Pachauri and L.A. Meyer (eds)]. IPCC, Geneva, Switzerland 151
37. IPCC (2023) Climate Change 2023:AR6 Synthesis Report. https://www.ipcc.ch/report/ar6/syr/
38. Jones CG, Lawton JH, Shachak M (1994) Organisms as Ecosystem Engineers. Oikos 69:373–386
39. Klos P, Goulden M, Riebe C, Tague C, O'Geen A et al (2018) Subsurface plant-accessible water in mountain ecosystems with a Mediterranean climate. WIREs Water 5(3):e1277. https://doi.org/10.1002/wat2.1277
40. Latour B, Weibel P (2020) Critical zones. MIT Press, The science and politics of landing on earth
41. Malte J et al (2020) The results of biodiversity–ecosystem functioning experiments are realistic. Nature Ecol. Evol 4:1485–1494
42. Marshall J, Roering J, Bartlein P, Gavin D et al (2015) Frost for the trees: Did climate increase erosion in unglaciated landscapes during the late Pleistocene? Sci Adv 1(10):e1500715. https://doi.org/10.1126/sciadv.1500715
43. Marshall J (2018) From ice to trees, surprising insights into past and present processes that sculpt our earth. AGU Fall meeting 2018: abstract #EP44A-01
44. Millenium Ecosystem Assessment (2005) Ecosystems and human well-being. Synthesis. Island Press, Washington, DC
45. Moon S, Perron J, Martel S, Holbrook W, St. Clair J (2017) A model of three-dimensional topographic stresses with implications for bedrock fractures, surface processes, and landscape

evolution: Three-dimensional topographic stress. J Geophys Res: Earth Surface 122:823–846. https://doi.org/10.1002/2016JF004155
46. Nie W, Guo H and Banwart SA (2021) Economic valuation of earth's critical zone: framework, theory and methods. Environ Develop, Article Number 100654
47. Prinn R (2013) Development and application of earth system models. PNAS 110(1):3673–3680
48. Reibe CS, Hahm JW, Brantley SL (2017) Controls on deep critical zone architecture: a historical review and four testable hypotheses. Earth Surf Proc Land 42:128–156
49. Rempe D, Dietrich W (2014) A bottom-up control on fresh-bedrock topography under landscapes. Proc Natl Acad Sci 111(18):6576–6658. https://doi.org/10.1073/pnas.1404763111
50. Rempe D, Dietrich W (2018) Direct observations of rock moisture, a hidden component of the hydrologic cycle. PNAS 115(11):2664–2669
51. Ren Y, Lü Y, Hu J, Yin L (2021) Geodiversity underpins biodiversity but the relations can be complex: Implications from two biodiversity proxies. Global Ecology Conservat 31:e01830
52. Richardson K, Steffen W, Lucht W, Bendtsen J, Cornell, S, Donges J, Druke, M, Fetzer I, Bala G, Bloh W, Feulner G, Fiedler S, Gerten D, Gleeson T, Hofmann M, Huiskamp W, Kummu M, Mohan C, Nogues-Bravo D, Petri S, Porkka M, Rahmstorf S, Schaphoff S, Thonicke K, Tobian A, Virkki V, Wang-Erlandsson L, Weber L, Rockstrom J (2023) Earth beyond six of nine planetary boundaries. Sci Adv 9(37). https://doi.org/10.1126/sciadv.adh2458
53. Richter D, Billings S (2015) "One physical system": Tansley's ecosystem as Earth's critical zone. New Phytol 206:900–912
54. Riebe CS, Hahm WJ, Brantley SL (2017) Controls on deep critical zone architecture: a historical review and four testable hypotheses. Earth Surf Process Land 42(1):128–156
55. Ripple W, Wolf C, Gregg J, Rockstrom J, Newsome T, Law B, Marques L, Lenton T, Xu C, Huq S, Simons L, King D (2023) The 2023 state of the climate report: entering uncharted territory. Bioscience 73(12):841–850
56. Sanchez RA (2023) Forest fires in the critical zone: the impact of wildfires on solute fluxes stream flow hydrologic flowpat. Development and biogeochemical processes. PhD thesis, University of Arizona
57. Sigleo W, Reinhardt J (1988) Paleosols from some cretaceous environments in the southeastern United States. In: Reinhardt J, Sigleo WR (eds) Paleosols and weathering through geologic time: Principles and applications. GSA Spec Pap, vol 216, pp 123–142
58. Smith D (2012) The Penguin state of the world. In: Atlas 9th Ed Penguin books, New York, vol 144
59. Steffen W, Broadgate W, Deutsch L, Gaffney O, Ludwig C (2015) The trajectory of the anthropocene: the great acceleration. Anthropocene Rev 2(1):81–98
60. Struzik E (2020) The age of megafires: the world hits a climate tipping point. Yale Environ 360. https://e360.yale.edu/features/the-age-of-megafires-the-world-hits-a-climate-tipping-point
61. Sullivan P et al (2022) Embracing the dynamic nature of soil structure: a paradigm illuminating the role of life in critical zones of the Anthropocene. Earth-Sci Rev 225(103873)
62. Tansley A (1935) The use and abuse of vegetational concepts and terms. Ecology 16:284–307
63. Tilman D, Isbel F, Cowley JM (2014) Biodiversity and ecosystem functioning. Ann Rev Ecol Evolut Syst 45:471–493
64. Trucchia A et al (2023) Wildfire hazard mapping in the eastern mediterranean landscape. Int J Wildland Fire. https://doi.org/10.1071/WF22138
65. Tukiainen H, Toivanen M, Maliniemi T (2022) Geodiversity and biodiversity. In: Geological society, London, special publications visages of geodiversity and geoheritage, p 530 (1)
66. Turco M et al (2018) Exacerbated fires in Mediterranean Europe due to anthropogenic warming projected with nonstationary climate-fire models. Nat Commun 9:382. https://doi.org/10.1038/s41467-018-06358-z
67. Turco M, Palazzi E, von Hardenberg J, Provenzale A (2015) Observed climate change hotspots. Geophys Res Lett 42. https://doi.org/10.1002/2015GL063891
68. UNEP (2005) One planet many people: atlas of our changing environment. Division of early warning and assessment, United Nations Environment Programme, Nairobi, Kenya

69. USNRC (2001) Basic research opportunities in the earth sciences. U.S. National Research Council, National Academy Press, Washington, DC, pp 35–38
70. Unruh J (2008) Carbon sequestration in Africa: the land tenure problem. Glob Environ Chang 18:700–707
71. Vernadsky V (1926) Biosfera, Leningrad. Annotated English edition 2013, "The biosphere", Nevramount Publishing Company, New York
72. Vitousek P, Mooney H, Lubchenko J, Melillo J (1997) Human domination of Earth's ecosystems. Science New Series 277(5325):494–499
73. Wallace AR (1876) The geographical distribution of animals. Harper, New York
74. West A, Galy A, Bickle M (2005) Tectonic and climatic controls on silicate weathering. Earth Planetary Sci Lett 235(1–2):211–228. https://doi.org/10.1016/j.epsl.2005.03.020
75. White T, Bradley D, Haeussler P, Rowley D (2017) Early Eocene paleosols and a new measure of the transport distance of Alaska's Yakutat terrane. J Geol 125:113–123
76. White T, Wymore A, Dere A, Hoffman A, Washburne J, Conklin M (2017) Integrated interdisciplinary science of the Critical Zone as a foundational curriculum for addressing issues of sustainability. J Geosci Educ 65:136–145
77. White T, Brantley S, Banwart S, Chorover J, Dietrich W (2015) The role of critical zone observatories in critical zone science. In: Giardino R, Hauser C (eds) Principles and dynamics of the critical zone, Elsevier. Chap 2, Developments in earth surface processes, vol 19, pp 15–78
78. Wilkinson BH (2005) Humans as geologic agents: a deep-time perspective. Geology 33(3):161–164
79. World Meteorological Organization (WMO) (2018) Executive summary: scientific assessment of ozone depletion: 2018, World meteorological organization, global ozone research and monitoring project – report no. 58, Geneva, Switzerland, p 67
80. World Meteorological Organization (WMO) (2022) Essential climate variables. https://public.wmo.int/en/programmes/global-climate-observing-system/essential-climate-variables
81. Xu X, Liu W (2017) The global distribution of Earth's critical zone and its controlling factors. Geophys Res Lett 44. https://doi.org/10.1002/2017GL072760
82. Yizhaq H, Stavi I, Shachak M, Bel G (2017) Geodiversity increases ecosystem durability to prolonged droughts. Ecol Complex 31:96–103
83. Zachos J, Dickens G, Zeebe R (2008) An early Cenozoic perspective on greenhouse warming and carbon-cycle dynamics. Nature 451:279–283
84. Zalasiewicz J, Williams M, Steffen W, Crutzen P (2010) The New World of the Anthropocene. Environ Sci Technol 44(7):2228–3223
85. Zheng B et al (2023) Record-high CO2 emissions from boreal fires in 2021. Science 379:912–917. https://doi.org/10.1126/science.ade0805

The CZ as an Ecological Problem: How the Interplay of Biotic and Abiotic Actors Determines the Functioning of Earth's Living Skin

Sharon A. Billings, Pamela L. Sullivan, Daniel Hirmas, Jesse B. Nippert, Daniel D. Richter, Zachery Brecheisen, Charles W. Cook, and Emma Hauser

1 An Emerging Recognition of the Critical Zone

Long before the term 'critical zone' (CZ) was coined by Ashley [82] to encompass Earth's biological and geological features from the top of the vegetative canopy to the depths of circulating groundwater, many scientists have recognized that both biotic and abiotic actors are centrally important for understanding many of Earth's most fundamental processes [141]. Scientists such as Alexander von Humboldt (1769–1859), Charles Darwin (1809–1882), Jacques-Joseph Ébelmen (1814–1852), Vasily Dokuchaev (1846–1903), Vladimir Vernadsky (1863–1945), Arthur Tansley (1871–1955), Hans Jenny (1899–1992), Robert Berner (1925–2015), and James

S. A. Billings (✉) · E. Hauser
Department of Ecology and Evolutionary Biology, Kansas Biological Survey and Center for Ecological Research, University of Kansas, Lawrence, KS, USA
e-mail: sharon.billings@ku.edu

P. L. Sullivan
College of Earth, Ocean, and Atmospheric Science, Oregon State University, Corvallis, OR, USA

D. Hirmas
Department of Plant and Soil Science, Texas Tech University, Lubbock, TX, USA

J. B. Nippert
Division of Biology, Kansas State University, Manhattan, KS, USA

D. D. Richter · C. W. Cook
Nicholas School of the Environment, Duke University, Durham, NC, USA

Z. Brecheisen
Purdue University, West Lafayette, IN, USA

E. Hauser
Department of Forest and Wildlife Ecology, University of Wisconsin - Madison, Madison, WI, USA

T. White and A. Provenzale (eds.), *Critical Zone and Ecosystem Dynamics*, Advances in Critical Zone Science, https://doi.org/10.1007/978-3-031-69076-1_2

Lovelock (1919–2022) worked across centuries and continents to demonstrate how momentary- and molecular-scale biotic actions (e.g., photosynthesis, respiration, production of organic acids, root growth) generate powerful chemical and physical forces that when scaled up govern environmental conditions of the planet. Indeed, it was a botanist familiar with plants' consumption of CO_2—Eunice Newton Foote—who first developed and published the concept of greenhouse gases warming the planet [53, 151], an idea developed as well by John Tyndall [165]. Today, we can recognize how these individuals and their colleagues integrated a diversity of ideas from biotic and abiotic processes into their work, a hallmark of today's CZ science.

Tansley's introduction of the ecosystem concept [159] offered an intellectual home for those interested in the systemic interaction of biota with the abiotic [130]. Ecosystem ecology, a particular branch of the broader field of ecology, is defined as the study of life as it interacts with the abiotic world [146]. In the nearly 100 years since the coinage of 'ecosystem' [159], this discipline has developed a literature, conceptual and mathematical models, and subdisciplines that have enhanced our understanding of how the biotic and abiotic interact on Earth's surface (e.g., [20, 74, 97, 103, 112, 118, 120, 122].

Concomitant with these developments, scientific institutions such as departmentally-organized universities and scientific societies, journals, and funding agencies also evolved, promoting more disciplinary approaches to science [131]. As a result, scientists began to face strong pressures to self-identify with particular disciplines. This development countered the ecosystem approach, which is defined by its interdisciplinarity. Perhaps because ecosystem ecology was linked by its name to other, purely biological subdisciplines of ecology (e.g., population and community ecology), it has often been assumed to be a fundamentally biological pursuit. Many contemporary ecosystem ecologists have been trained in and remain focused on biologically-focused areas such as photosynthesis, plant respiration, microbial cycling of organic matter, and plant–microbe interactions, and often lack training in the chemical, physical, and geologic dimensions of ecosystems. One manifestation of this problem is the superficial coverage of the belowground component of ecosystems [131, 109]. This feature was noted by Binkley [16], who highlighted that reviews of the ecological literature omit mention of soil, the very medium from which terrestrial ecosystems derive most of their required resources. Most ecological studies of soil focus on soil microbes and fauna interacting with each other or their organo-substrates, or on roots and their interactions with microbes—all important but largely biologically-focused endeavors (e.g., [5]).

This biological focus of many ecosystem-trained scientists and the institutions in which many function can limit our understanding of the interactive nature of many Earth surface processes. Multiple examples demonstrate this assertion. First, rock- and mineral-derived nutrients are critical factors driving biological activity in terrestrial systems [145]; with the exception of most N, even nutrients derived from organic matter recycling originally are released from mineral weathering. Second, the inherited geologic structures of bedrock strongly controls ecosystem water storage and transmission to groundwater and streams [93]. Third, landscape position and geomorphological characteristics constrain light availability [12]. These well-studied phenomena demonstrate the degree to which an ecosystem's ability to capture nutrients, water, and light—and thus its productivity—is dictated by chemical and physical site characteristics.

A fourth phenomenon exemplifies an especially rapidly expanding branch of CZ science that, like the first three, clearly links biotic and abiotic processes. Multiple investigations reveal the importance of dust nutrient inputs to ecosystems as key promoters of ecosystem productivity. This feature of terrestrial ecosystem nutrition has been known for years (e.g., [32, 96, 150]), but recent CZ studies help expand this concept to more fully reveal the importance of dust-derived inputs to many ecosystems. Where highly weathered soils would otherwise promote relatively low productivity, dust can serve as an especially critical nutrient source [32, 59]. In addition to inoculating recipient sites with dust-bound microbial populations [101], dust can travel great distances [47, 184] to provide nutrients like P to biota [102, 170] at supply rates that can outpace that of bedrock itself in some systems [1]. Dust also can influence soil structure [42] and generate meaningful material thickness within soil profiles [95, 102]. Dust inputs are not an important source of nutrition in all systems (e.g., [166]), but the nutrient subsidies provided by dust in many may be sufficient to mitigate rock and mineral weathering that would otherwise supply nutrients. This echoes the hypothesis suggesting that OM-derived nutrients can mitigate rock and mineral weathering [25]. More broadly, evolution of dust-focused CZ literature rests within a concept central to ecosystem, and thus CZ, science—that nutrients are either obtained internally through de novo weathering or OM recycling, or from external sources, and that nutrient sources thus dictate weathering rates.

The interactive nature of many of Earth's CZ processes is further demonstrated by biological phenomena influencing multiple physical and chemical attributes of the CZ, going far beyond the role of vegetation as a key sink for atmospheric CO_2 and nutrients. Roots, for example, both help prevent erosion [153] on timescales of seasons to decades and help promote soil creep over longer timescales [22, 124]. Roots and the soil microbial communities they nurture generate organic acids and CO_2 that contribute to the weathering of soil minerals and bedrock [90, 104]. Roots and microbes are also capable of recycling organic matter-bound nutrients, mitigating subsequent weathering processes as nutrient demand is met partially by organic matter decay [69]. Holistic terrestrial ecosystem studies thus must embrace some combination of physical and chemical sciences (e.g., lithology, mineralogical weathering, physical geography, geomorphology, hydrology) along with the biological phenomena that interact with these features.

The emergence of the CZ paradigm [23, 82] promoted an interdisciplinarity already evident in the works of scientists such as Humboldt and Darwin [141, 132]. The explicit linkage of the bio- and geosciences in CZ studies provides the scaffolding needed to address environmental puzzles at a diversity of scales [15, 156, 185]. Realization of the CZ approach is hardly trivial: Implementing studies that integrate the bio- and geosciences requires collaboration among practitioners who represent disciplinary siloes, speak distinct vocabularies, and view scientific puzzles through separate lenses ([131]; Fig. 2.1). These traits can impede even well-intentioned collaborations [174]. However, many lessons have emerged by bridging the bio- and geo-focused disciplines to examine the 'critical zone ecosystem' [130].

Here, we describe findings that emphasize how life, emphasizing vegetation and microbes, responds to and shapes the physical environment in which it persists, yielding feedbacks for Earth's climate, primarily through modifications to hydrologic functioning. We focus on the interactions of biota and the physical and chemical features of soil pedons and landscapes as they drive ecosystem-scale hydrologic fluxes. We acknowledge that the flowpaths for soil water are also the same conduits through which soil gases flow [83] and are thus key to understanding ecosystem functions (e.g., [67]). In the current work, we emphasize hydrologic flows due to their importance for vegetative water uptake and thus productivity, flows of soil organic C and nutrients, and soil weathering processes. These processes all reflect diverse disciplines that coalesce in Earth's soils (Fig. 2.1). We focus on hydrologically-relevant features because of the long history of individual disciplines telling us about the large-scale importance of these processes, and because of emerging research highlighting the importance of the intersection of these disciplines for projecting future

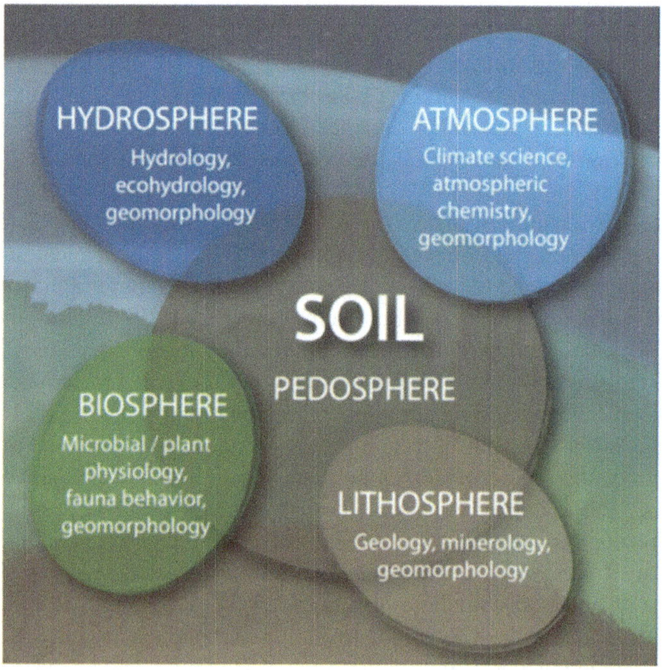

Fig. 2.1 Depiction of soil as representing the Earth science realms of the atmosphere, lithosphere, biosphere, and hydrosphere. Examples of scientific disciplines often associated with each realm are indicated. Note that all realms intersect with other realms as pairs (e.g., atmospheric chemistry interacts with plants to represent biosphere–atmosphere interactions), but soil is the salient feature of Earth's surface where all four realms intersect. Geomorphology is presented as a feature in multiple locations due to its influence on, and its capacity to be influenced by, multiple realms

ecosystem functioning on a rapidly changing Earth. The knowledge we spotlight reveals Earth's CZ as a fundamentally ecological problem.

2 Soil and Landscape Constraints on CZ Hydrologic Functioning

Within a given climate regime, hydrologic function is strongly influenced by the properties of and interactions between biotic and abiotic components of the CZ ([7, 39], Fig. 2.2). Here we adopt a physically-oriented, catchment-scale definition for hydrologic function as the partitioning, storage, and discharge of water [171, 181]. Although different locations within the CZ contain both biotic and abiotic components, it is the interaction between these components in the soil environment that exerts a strong control on hydrologic function through its effect on soil hydraulic properties. This near-surface soil environment is subject to steep and dynamic energy and soil moisture gradients that drive the relatively rapid flux of water and the temporal variations in the direction and velocity of this flow [176]. As soil biota (e.g., microorganisms, plant roots, and macrofauna) respond to these variable near-surfaces fluxes of water, they shape the organization of soil particles and pores [156] and thus influence the development of soil horizons especially within shallow depths where the soil biological activity is concentrated. This modification happens primarily through the production of soil organic C (SOC), the formation and stabilization of soil aggregates, and the direct and indirect creation of macropores that perforate the soil. However, the ability of biota to modify the soil environment is constrained by the physical and rheological properties of the material, which are largely controlled by soil texture.

Soil texture is important to the hydrologic functioning of the CZ because it directly affects the relative surface area of the soil and the soil pore-size distribution. Pore size (i.e., effective pore diameter) and relative surface area control the affinity of water to the soil matrix and impart a potential energy to the soil water (i.e., matric potential; [71]). Matric potential, in combination with components such as gravitational and hydrostatic pressure potential, governs the total potential energy of the soil water. Differences in the total potential energy state drive infiltration and the movement of water through the CZ. Thus, texture strongly affects water flux through soil by governing potential energy gradients. Soil texture also affects the ability of the soil to conduct water because of its control over surface area and the porosity of the soil matrix (especially, pore size and tortuosity). That is, as the soil particle-size distribution becomes finer, so do the pores between those particles, reducing the hydraulic conductivity of the soil. As the soil dries, air is introduced, bounding the soil water between the air–water interface and surfaces of the soil particles, this modifies the lengths and tortuosity of the paths that the water must follow to move through the soil, further reducing the hydraulic conductivity [83]. Therefore, soil texture—both on its own and in conjunction with degree of saturation of the soil—is

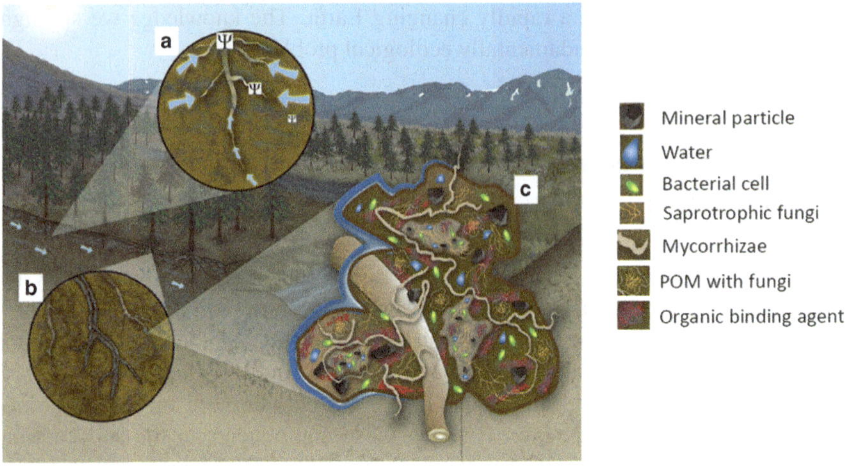

Fig. 2.2 Schematic describing examples of belowground features of the critical zone (CZ). (1) At the landscape, catena, and pedon scale, water flows down potential energy gradients (arrows) and through relatively large soil pores (water depicted near roots in **a**). Larger ψ symbols indicate more negative water potentials. Where soil profiles have distinct boundaries, water flows can change abruptly. (2) Roots are opportunistic drivers of hydrologic functioning. Roots take up water primarily from surficial horizons and to a lesser extent from deeper in the soil profile (**a**). Where roots can reach the water table, productivity tends to be higher. (3) Because roots and microbes influence soil structure, within a given climate region they shape their own water availability. Roots must perforate the soil to grow, creating channels for flow (**b**). When they die, shrink, and eventually decay (larger channel in **b**), these channels become more open, becoming reinforced as preferential flow paths. Microbes and roots also govern soil structure by exuding soil binding agents. When microbes die, their necromass appears to have an especially outsized role in serving as "sticky" binding agents. Binding agents, no matter their source, help to form soil aggregates (**c**), and it is those aggregates that dictate where the voids are in soil through which water can flow and be stored. The relatively large pores surrounding macroaggregates are especially important for water flow. The mineralogy, particle texture, and capacity of clay minerals to bind with organic matter all influence aggregate characteristics. Within macroaggregates (**c**), microaggregates harbor their own microbial communities, stores of organic matter and minerals, water, and gases (visible as outlined shapes containing these features, within the macroaggregate). POM, particulate organic matter

critical for understanding the retention and flux of water in the CZ due to its influence on both the total potential energy gradient driving flow and the conductivity of the soil through which the water moves.

Macroporosity of the soil is another important determinant of fluxes of water through the CZ. Although there are different classifications of macropores that depend on their origin and shape, here we restrict our discussion to macropores that are large (often greater than a few hundred micrometers), planar or tubular, and continuous; these macropores function as potential pathways for the preferential transmission of water ([64, 78], Fig. 2.2). Macropores of this type result from either biotic or abiotic processes in the soil. Abiotic processes include the drying and concomitant shrinkage of the soil matrix especially in fine-textured materials, the formation of aggregated structural units (e.g., clay tactoids, clusters of clay and silt domains, and subsurface peds like angular blocks and prisms) due to the cohesion of the soil particles or precipitation of inorganic cements (e.g., $CaCO_3$, opaline Si, or Fe-oxides), and the reinforcement of preferential pathways by the development of surface coatings [78, 161]. This latter process is largely due to the illuviation of clay (e.g., argillans, expanded on below) or the reorientation of the soil fabric at the pore surface from mechanical pressures (e.g., fabric hypocoatings, [154]). Biotic processes responsible for the formation of macropores include the activity of fossorial fauna [126], the growth and subsequent decay of roots [64], and the formation of stable aggregates through, for instance, the egestion of soil particles by earthworms [81], enmeshing of particles by mycorrhizal fungi and fine roots [79, 135], and secretion by soil microorganisms of extracellular polymeric substances (EPS) that glue soil particles together [36].

Macropores created through the processes outlined above typically make up only a minor fraction of the soil volume, but have an inordinate effect on water flux through the soil in saturated and even unsaturated conditions [113, 175]. The reason for this phenomenon is that volumetric flow through a pore is proportional to the fourth power of its radius (i.e., Poiseuille's law). Watson and Luxmoore [175] demonstrated this principle by reporting that, under ponded conditions, macropores may be responsible for 96% of the water flow though only accounting for 0.32% of the soil's volume. Where these macropores intersect the land surface (e.g., open biopores or aggregated surface soil horizons), they are important for the infiltration of water into the soil and the potential for runoff generation, and, thus, the loss of sediments and nutrients through erosion [10, 92, 149]. Recent findings have shown that soil macropores are sensitive to shifts in climate, likely driven by soil biota [30, 73]. Given the large control over water flow that macropores exhibit, even small changes to these pores in response to climate is likely to affect the water cycle and the hydrologic function of the CZ via changes to infiltration, aquifer recharge, and the lateral redistribution of water by runoff [73].

As soils form in response to the addition, loss, transformation, and translocation of mass and energy, soil horizons develop that are distinguished from each other, in part, on the basis of macromorphological properties exhibited as color, consistence, texture, structure, and porosity [140, 164]. Differences between vertically adjacent soil horizons in the expression of these latter properties create layered conditions that are important for water flow through the vadose zone. This is because the boundaries between horizons, especially if they are abrupt, can represent significant discontinuities in soil hydraulic conductivity and water content [71]. The discontinuities—for example, a coarse-textured A horizon above a clay-enriched Bt horizon—act to retard the rate of soil water infiltration due to the reduced saturated hydraulic conductivity of the lower compared to the upper layer [127]. However, even in cases where fine-textured layers or horizons with concentrated organic matter overlie more conductive, coarse-textured materials (e.g., an O horizon directly over an E horizon, or a clay-rich horizon over a sandy horizon), the boundary between these horizons forms an obstacle (often temporary) to the wetting front that reduces infiltration. This is due to the differences in potential energy of the soil water across the boundary of the two layers with the fine-textured material imparting a sufficiently lower total potential energy compared to the underlying layer [127]. The water hangs at this boundary until the saturated zone behind the leading edge of the wetting front increases sufficiently for the added hydrostatic pressure to overcome the matric forces holding the water. In addition to effects on infiltration, differences in soil hydraulic properties through and across horizons strongly influence the generation of preferential flow [46].

At a broader scale, ecosystems, soils, and landforms coevolve to shape the soil-geomorphic properties of the landscape [182]. For example, slope curvature affects soil thickness and the concentration/dispersion of surface water (e.g., [123]), and elevation controls snow accumulation, rainfall, and evapotranspiration (e.g., [178]). These features and the soil properties discussed above all constrain the hydrologic functioning of the CZ by driving the lateral redistribution of water and sediments along topographic gradients from ridges to valleys and by controlling the radiative conditions, and, thus, the soil temperature, effective moisture, and ecological properties corresponding to land-surface aspect [48]. The lateral fluxes of water along topographic gradients occur both at the land surface and within the subsurface, controlling the vertical distribution of plant-available water and, therefore, plant rooting depth across the landscape [50]. Milne's [106] concept of the soil catena that was used in part to explain different drainage classes along a hillslope [182] is especially useful for understanding this relationship. A catena is a chain of adjacent and geomorphologically-related soils along a hillslope, from summit to base, perpendicularly transecting the topographic contours [140]. Within this framework, soil morphological differences at different hillslope positions are explained by the lateral and vertical fluxes of surface water, groundwater, and sediments. Indeed, catenas can be conceptualized as hillsheds that undergo soil creep [26]. The movement of soil material and redistribution of water from steeply sloping positions to the base of hillslopes give rise to textural differences on the landscape that control CZ hydrologic functioning such as soil water retention (Fig. 2.3).

All of these soil and landscape physical constraints on CZ hydrologic functioning are both the product of, and influence, biotic functioning. Soil geomorphic properties of landforms, and the soil properties of soil profiles, strongly affect the lateral and vertical distribution of water and, thus, the distribution of vegetation across the landscape and the distribution of roots within a profile. However, the influence of vegetation on both surface and subsurface hydrology significantly affects landform evolution and soil formation as well (e.g., [33, 179]). Indeed, rooting depths appear sensitive to the different vertical distributions of soil moisture that result from hillslope-driven fluxes [48] and the properties of a given profile. In well-drained summit positions, for instance, the pool of plant-available water is largely contained at relatively shallow depths (i.e., from the infiltration of rainwater) with the water table too deep for plant roots to access; this is in contrast to poorly-drained toeslope positions that favor wetland species due to the creation of permanent water-logged conditions from a water table that is near or above the land surface at the base of the hillslope [48]. Between these two ends of the hillslope, plant-rooting distributions are affected by the relative positions of infiltration-controlled soil moisture and the capillary fringe above the water table, with plant roots sensitive to the

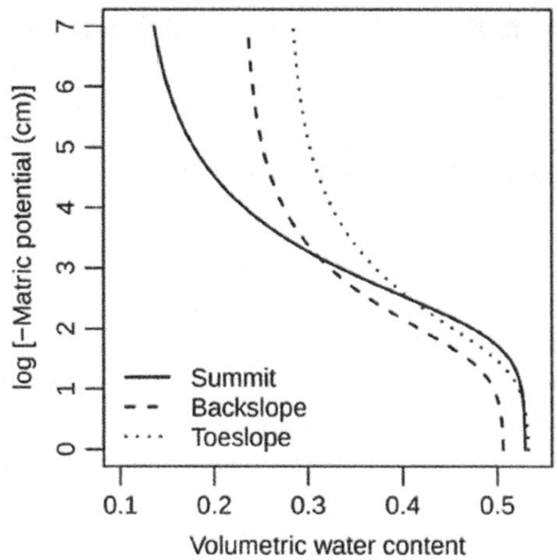

Fig. 2.3 Soil water retention curves relating soil volumetric water content to the energy at which that water is held in the soil (matric potential) plotted for three surface A horizons representing different hillslope positions along a catena in the Fitch Natural History Reservation at the University of Kansas Field Station near Lawrence, KS, USA depicting the soil and landscape constraints on hydrologic functioning of the CZ. Soils on the broad, relatively-level summit position of this hillslope have developed within the fine-textured loess parent material without significant lateral additions or losses of soil material. The texture of the A horizon has coarsened naturally in this position through the vertical translocation of fines (silt and clay) to deeper subsurface (B) horizons, yielding a surface horizon water retention curve with a steeper slope (i.e., greater change in water content with unit change in matric potential) in the plant-available water range (i.e., y-axis approximately between 2.5 and 4.2). The steeper slope translates to soil water that would be released relatively easily in response to increasing root-water demand. By contrast, soils in the steeply sloping backslope position have developed under conditions of frequent erosion of material to lower-lying toeslope positions where surface A horizons inherit properties of the exhumed clay-rich subsurface layers. The higher clay content of these surface horizons in both the backslope and the more distal toeslope positions that have received inputs from movement along the catena create a greater proportion of very fine pores that retain more water in the soil even under low matric potentials (i.e., a relatively shallow slope in the plant-available water range) and a greater amount of residual water content which can be seen as high water content even under extremely low matric potentials (6–7). Material moved to the surface horizon of the backslope is likely to have been derived immediately upslope from a higher position along the backslope. The curves were assumed to follow a [168] water retention function with the parameters of the function predicted from sand, silt, clay, bulk density, and the volumetric water contents at 33 and 1,500 kPa using the Rosetta3 pedotransfer function [186]. Although the water retention curves in this figure are plot with matric potential on the y-axis and water content on the x-axis, they are often plot with these axes switched. We note that these two equally-accepted conventions of plotting the water retention curve stems from the lack of matric potential and water content in being a clear response variable

seasonal changes in water table depth in upper backslope and footslope positions [48]. Thus, without a transdisciplinary knowledge base embracing geomorphology, soil science, and vegetation water uptake patterns, CZ water balance would remain poorly understood.

3 Roots as Direct, Biotic Drivers of CZ Hydrologic Functioning

3.1 Ecohydrological Considerations

Discerning root water uptake patterns must be informed by root biology and the soil and landscape constraints on soil physical properties described above. Historically, root water uptake has been the purview of ecophysiologists and ecosystem ecologists with training in ecophysiology, and to a certain extent soil physicists. However, plant water availability and root water uptake impose cascading influences on whole-CZ structure and function. It has long been recognized that the physical and biogeochemical changes that occur with changes in species composition and root architecture co-evolve with topography, geology, soil, and climate to control the trajectory of catchment hydrology (e.g., [65]). Understanding such connections and non-linear feedbacks is particularly important as we confront the need to forecast Earth system evolution in the face of accelerating climate change and human perturbations [155]. Thus, in addition to the role of landscape position influencing system hydrology discussed above, we must also emphasize that vegetation plays a role in governing water flows as well.

With ample nutrient availability, vegetative growth tends to be optimized where water availability is balanced by evaporative demand [136]. Though at the largest scale this is governed by the overall climate, within each climate zone, the local CZ (e.g., the landscape features discussed above) exerts strong governance over water abundance and depth to the water table. If rooting networks are able to access this groundwater, plants can flourish in what might otherwise be sub-optimal conditions [49]. This action can result in an altered position of the water table, even in low landscape positions, as plants meet their evaporative demands [157]. The depth of this drawdown is a function of the soil's water holding capacity, which can be impacted by plant-induced clogging of pores, or perhaps pore generation [158]. Thus, the relationship between vegetation and groundwater or water table position, and factors that possibly alter water availability and position, can have significant impacts on how the CZ-ecosystem functions.

An additional layer of complexity in ecohydrologic processes has been hinted at for decades and is emphasized in recent, explicitly CZ-focused work. In multiple ecosystems, rock water extraction by vegetation has been observed [2, 21, 152]. More recently, neutron probe data from the mountainous western U.S. demonstrate a meaningful reliance of some vegetation on deep rock moisture—water retained in the weathered rock vadose zone [108, 129]. Given that rates of root growth into the subsurface can outpace rates of soil loss from erosion, roots thus not only influence water storage in these zones, but also govern the detachment of bedrock and the initiation of soil formation [137], factors that feedback to govern the generation of porosity and thus water storage. In the Anthropocene, rates of forest harvesting may be outpacing the rate at which root growth by trees impart their ability to fracture or detach bedrock in working forests [137].

The human influence in the Anthropocene is also evident via shifting rooting depth and/or rooting function at a diversity of scales. Given the role of roots as soil architects, human activities that modify root activities are likely altering the physical structure of the subsurface. Within a given biome, changes in environmental conditions that shift the amount and timing of plant water use (e.g., via changes in plant water use efficiency, the depth to which water is extracted, or the frequency with which soils may undergo wetting and drying) may alter aggregate formation and stability, and thus soil structure [27, 35, 40, 41, 121]. Where permafrost thaws or alpine tundra warms, more deeply rooted, woody vegetation can develop [180]. Woody encroachment of deeply-rooted shrubs into grasslands in warmer climates is also accelerating [139]. Globally, estimates suggest that an additional ~19,200 km^3 of soil have become rooted in regions experiencing root deepening in the Anthropocene ([68]; Fig. 2.4). These root-deepening phenomena have the potential to sculpt the subsurface in ways that influence water flow through the subsurface [156]. Root shallowing also is an important Anthropocene phenomenon; roots are absent from an estimated 30,100 km^3 of soil today compared to soil volumes rooted by potential vegetation [68]. Roots have been removed from an estimated 13,700 km^3 of this volume where perennial systems have been replaced with annual agriculture, a landscape feature expected to expand in the future ([68]; Fig. 2.4).

The net effect of human activities in the Anthropocene thus has been a shallowing of roots, and a decline in the soil volume directly mined by roots for water and nutrients [68]. This feature prompts the question of how soil formation processes are transformed as root shallowing limits the production of weathering agents, particularly given the long time periods over which root regeneration occurs [13]. More generally, this phenomenon highlights the fundamental role of an ecological consideration—rooting depth—as an agent of soil engineering at depth and thus of CZ functioning.

3.2 *Ecophysiological Considerations*

The water uptake patterns permitted by local climate and physical landscape- and pedon-scale features, described above, are the scaled-up result of processes occurring at the momentary and cellular temporal and spatial scales, respectively. The dynamics of plant water uptake and loss are rapid. Indeed, while water accounts for up to 95% of the fresh weight of herbaceous plants, the majority of water absorbed by most plant species (~98%) is lost to the atmosphere via transpiration within minutes of being absorbed from the soil. Thus, plants need enormous amounts of water to offset transpiration losses and facilitate C uptake (and photosynthesis). While some species can absorb a significant portion of water through leaf pores (as dew or vapor) [9, 70], the vast majority of plant species rely on root uptake to meet water demands. Roots provide several key functions for plants, including anchorage to the terrestrial surface, but the functional consequences of water and nutrient absorption from the

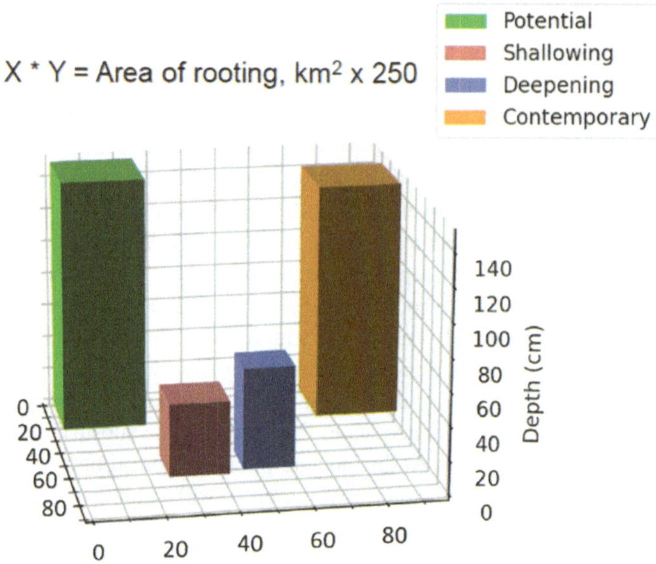

Fig. 2.4 Scaled depiction of estimates of the volume of Earth's rooted soil rooted with potential vegetation (green) versus actual contemporary vegetative cover (orange), and the two competing changes driving that difference. The red bar represents the volume of soil where roots are absent now that contemporary vegetation has replaced potential vegetation (~30,100 km3), largely due to agricultural clearing. The blue bar represents the volume of soil now experiencing rooting that, prior to the Anthropocene, was not rooted (~19,200 km3); this is largely due to woody encroachment. The top of each bar is scaled to reflect the area represented by the data. Hence, the top of the green and orange bars are of equal area to reflect all of Earth's vegetated surface; the red bar's top area is larger than the blue bar to reflect the larger area over which roots are shallower in the Anthropocene compared to potential vegetation. The depth axis reflects the average depth to which roots extended globally prior to the Anthropocene (green) and today (orange), and the average depth to which roots in regions experiencing shallowing (red) and deepening (blue) extend. In all cases, 'depth' refers to the depth of 99% of root biomass. Data derived from maps of potential vegetation, remote sensing-derived estimates of current vegetation, and biome-specific rooting depth distributions in the literature; for details see Hauser et al. [68]

soil are key determinants of nutrient, C, and water recycling within ecosystems and serve as a link between belowground and aboveground physiological processes [77].

Plant roots absorb water from the soil based on the hydrostatic pressure gradients established from soil matric to leaf water potentials. Water flows down this potential energy gradient as long as the potential energy in the leaf that is established by water lost via transpiration is lower than that in the soil. As the soil dries, the potential energy gradient from soil to leaf declines, resulting in less plant-available water. In this situation, plants: (1) close leaf stomata to reduce the water potential gradient, (2) lower the leaf water potential (by increasing transpiration) to re-establish a larger pressure gradient from soil to leaf allowing for greater potential extraction of water from soil spaces, or (3) utilize water from potentially wetter zones of the soil. The degree of leaf stomatal control used to regulate transpiration flux has been commonly described as isohydric or anisohydric [105, 160]. Isohydry refers to a strategy of stomatal regulation to maintain a constant water potential prior to wilting and prior to the establishment of air embolisms and cavitation in the xylem vasculature. Thus, isohydric strategies reduce transpiration to minimize the chance of wilting and hydraulic failure, but with a cost of reduced C assimilation and growth. Plant species that utilize anisohydric strategies regulate transpiration at the expense of potential hydraulic failure to maintain the hydrostatic pressure gradient as soils dry and water is less available. Thus, anisohydric species are able to maintain C assimilation but at greater risk of desiccation. Strategies of isohydry or anisohydry vary by species and genotypes according to changes in soil water availability, which in turn vary across catenas and landscapes (Fig. 2.5). While these strategies present dichotomous endpoints of acclimation to low water availability, most plant species exhibit aspects of each, with examples of isohydric behavior during specific periods of growth or portions of a season and anisohydric behavior during others [87].

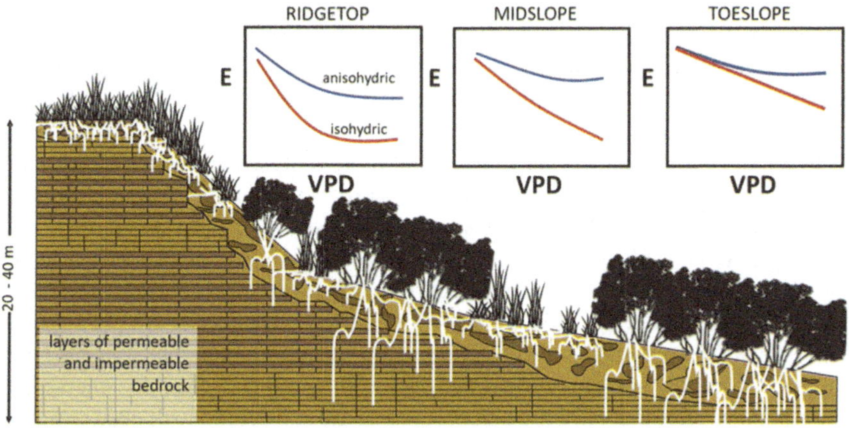

Fig. 2.5 Plant functional types and rooting depths change across complex landscapes with varying soil depths and varying layers of permeable and impermeable bedrock. For example, in regions where grasses can proliferate, fibrous-rooted grasses dominate in shallow soils associated with ridgetops, with surficial bedrock penetration. These soils often exhibit periods of high and low soil moisture, with low soil moisture often accompanying seasonally high vapor pressure deficit (VPD). Isohydric species show high transpiration (E) when surface moisture is high (and VPD is low) but illustrate greater stomatal regulation in dry soils during periods when atmospheric demand (i.e., VPD) is high; this response to VPD is similar but muted for anisohydric species. At midslope locations, deeper soil layers facilitate species with fibrous and tap-roots that utilize soil moisture across relatively large depth intervals. Penetration of bedrock by tap-rooted species is possible. Anisohydric species at midslope positions exhibit markedly reduced declines in E across a range of VPD, supported by decreased stomatal regulation and the larger soil moisture storage capacity of the soil and rock layers. Soil depth tends to be greatest at toeslope locations, resulting in relatively few differences between isohydric and anisohydric species in E across a range of VPD. In all locations, isohydric species employ drought tolerance strategies (anatomical, physiological) to maximize E across a range of VPD

Root traits have long been examined to try to understand plant, and thus ecosystem, functions. The root trait most frequently reported as a predictor of ecosystem function has been maximum rooting depth, which has been reported for a diversity of species and varies with biogeography [142, 143, 153, 163]. However, if we want to improve our ability to predict how species coexist within ecosystems or how global environmental changes may impact communities in space and time, other characteristics of root systems beyond maximum rooting depth require consideration [88, 114]. For example, not all roots are created equal. Angiosperm species can be tap-rooted or fibrous, with branching tap-roots more common in eudicots, and dense fibrous root systems more common in monocots [17]. Fibrous and tap-rooted species have inherently different morphologies, with different root densities within a given layer of the soil, varying depth distributions of roots, and varying root diameters within depth distributions of the soil [76, 107]. Species with tap roots tend to have roots of larger diameter, with a deeper maximum rooting depth but reduced fine root density at any particular depth compared to fibrous root systems [76]. Species with fibrous root systems tend to be more efficient at absorbing water, can typically resist lower water potentials [37], and have higher root turnover (lower longevity). Even within species (and genotypes), these root morphological traits have high variability [34, 89], highlighting the need for greater investigation of what drives this variability in root morphological traits.

In addition to differences in root system morphology, the functional attributes (i.e., physiology, including water absorption) of roots vary among species, locations, and temporal periods of the growing season. Using stable isotopes as natural tracers of resource uptake, plant scientists have discovered that root presence at a particular soil depth does not equate to water uptake (root function) from roots that exist within a particular depth [31, 38, 115]. This result implies that just because a plant species has roots within a particular zone of the soil it does not mean that water is being absorbed from that soil zone. Interestingly, this outcome has been documented for many plant species that have roots in portions of the soil profile with plant-available water. For many herbaceous species, deep roots typically contribute very little to the overall plant water budget despite roots being present at depth. It remains unclear why plants possessing roots in zones of the soil with available water may not use that reservoir. As previously described, water moves primarily by mass flow to roots and root hairs down a pressure potential gradient. Following absorption, water can move towards the vascular cylinder via apoplastic (between cells) and symplastic (from cell to cell across membranes) movement pathways. However, once water reaches the endodermis, only symplastic transport is possible across the Casparian Strip (CS) [62]. Plant biochemists still have a limited understanding of how water movement is regulated across the CS, but transport across this barrier can be turned on and off [6, 62]. Thus, vascular plants have the ability to regulate water transport through cellular mechanisms associated with transport across the endodermis. Finally, the microanatomical features of roots vary within an individual based on root order, depth in the soil, and phenological stage [116]. For herbaceous species, this typically means reduced hydraulic conductivity (and reduced rates of water transport) from deeper soil depths [116, 119]. To better understand how, when, and where root functionality occurs requires a greater understanding of the variability in root micro-anatomical features.

Microanatomy is a historically overlooked aspect of plant roots that links physiology and whole-root system morphology—and thus to CZ-ecosystem functioning. The microanatomical characteristics provide insights into the larger functional contexts of root systems. Root traits like the ratio of cortex to stele, vessel size, vessel number, xylem wall thickness, and hydraulic conductance, all provide clues towards understanding the investment in safety versus efficiency of water transport through root systems [51]. In the seminal work by [172], the authors used root microanatomical traits (e.g., cross sectional area, number of xylem, xylem wall thickness) with aboveground processes like relative growth rate. This work illustrated tradeoffs in safety versus efficiency of water transport in root microanatomy that predicted whole plant growth traits for perennial grasses [172]. Based on these observations, subsequent studies have highlighted how the internal root structure sets a foundation for the whole root system function [116, 173, 119]. Relationships between root microanatomy and aboveground physiology / growth vary by within communities [119], plant functional type [173] and phylogeny [167] and are an active topic of investigation. Because microanatomical images of roots require more preparation time than whole root systems, they have been utilized much less than other traits. For this reason, there is much remaining to discover with regard to linking the internal anatomy of roots to the broader water and nutrient uptake patterns of plant species and communities [56].

As described so far, plant roots vary in both morphology (whole-plant and microanatomical) and physiology among plant types, species, and ecosystems, and the biological focus of this knowledge base is clear. Unsurprisingly, these root differences translate to variability in resource uptake, subsurface porosity, and weathering within the soil profile in space and time. For these reasons, inferring root functional processes (e.g., water uptake) is rarely as intuitive as measuring maximum rooting depth, or root biomass and yet has great importance for projecting whole CZ-ecosystem functioning. Improving linkages between root physiology and CZ-ecosystem properties requires better measurements of root longevity, associating roots to species (in mixed-species environments), quantifying growth rates and biomass by depth through time, and linking these features to the dynamic physical and chemical properties of the soil. Resolving these biologically-focused frontiers in the context of the abiotic constraints on CZ hydrology will undoubtedly assist with linking root presence and activities with ecosystem modification, and further our understanding of how root structure and function regulate pools of water and nutrients and fluxes of water, C, and nutrients in the CZ [22, 39]. Specifically, such efforts are necessary for understanding how ecological processes feedback to CZ hydrologic functioning.

4 Roots and Microbes as Indirect Biotic Drivers of CZ Hydrologic Functioning

In addition to the direct effect of roots on water uptake throughout soil profiles and across landscapes, roots are active sculptors of the subsurface (Fig. 2.6). As such, they represent key agents linking the biotic to the abiotic realms within a CZ. Roots have significant impacts on the depth to which water can infiltrate and the position of the water table. Specifically, roots promote flow. Macropores, acknowledged above as a soil constraint on hydrologic functioning, are often formed via root growth [3, 99, 185] and reinforced via deposition and accumulation of clay films [147]. After root shrinkage, saturated films on root surfaces or along pore walls provide evidence of flow in these pores [16]. Where roots persist at depth, development of preferential flow paths is thought to trigger a positive feedback by enhancing the water drainage to depth [44, 125, 169]. Not only do roots generate pores, but their growth can both build and destroy soil aggregates through enmeshing small particles or cleaving apart big aggregates [123], a process that alters the size and distribution of macropores. Because roots control soil-hydraulic properties and the generation of preferential flow, small changes in rooting depth distributions can alter water flow significantly [11, 117]. Indeed, given that roots transform far more soil volume than rhizosphere dimensions might initially suggest [134], extant aggregate size distributions and pore networks likely represent the legacy of past generations of roots.

Root growth also imparts an important impact on the translocation of clay minerals and clay-sized particles, a process that influences both soil water flows and storage. Though clay-rich horizons can develop as clay forms within the horizon, in many locations clay accumulates in a horizon with illuviation from upper horizons [29,

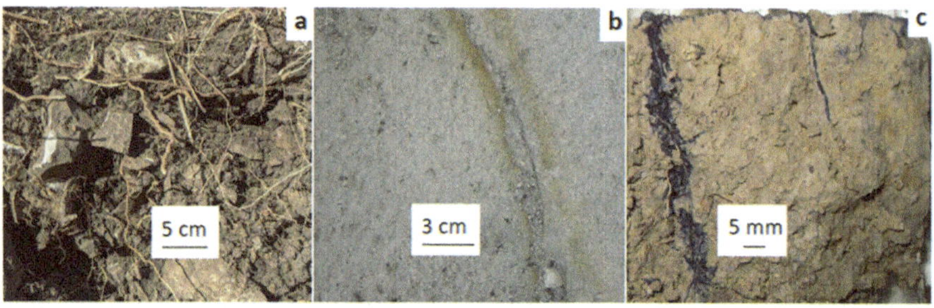

Fig. 2.6 Visual evidence of the role of roots as soil structural architects and their interaction with rocks and minerals is offered by **a** roots in surface horizons of Mollisols in Kansas, U.S., in intimate contact with limestone cobbles which experience dissolution upon exposure to the increased acidity imposed by root and microbial activities; **b** pine roots penetrating granite underlying well-developed Ultisolsin the southeastern U.S., resulting in enhanced porosity and thus oxygen flow, oxidation of organic reductants in the rhizosphere, and subsequent reduction of Fe^{3+} which diffuses away from the rhizosphere and undergoes reoxidization; and **c** methylene blue dye revealing a large, tubular pore likely formed via root growth and now serving as a preferential flow path in a Mollisol from the University of Kansas Field Station, Kansas, U.S. Photo credits: **a** J. Nippert; **b** D. Richter; **c** D. Hirmas

176]. The downward movement of clay-sized particles and clay minerals is linked to climate, occurring where effective precipitation is sufficient to promote downward movement of particles through the profile [57, 58, 91] and likely facilitated by seasonal wetting and drying. However, the downward transport of clay across horizons also is governed by the generation of flow paths through which clay illuviation can proceed [128]. The resulting clay-rich horizons influence profile water dynamics via lower saturated hydraulic conductivity and thus enhanced water storage, and depth-dependent hydraulic conductivity due to the textural discontinuities with neighboring soil horizons [133], Fig. 2.7). Clay-rich horizons also can result in a layer of soil difficult for roots to penetrate [61]. In this indirect way, then, roots govern their own hydrologic environment.

Microbes, too, sculpt the soil. Soil microbes exude compounds that can serve as binding agents [162]. The complex mix of exudation compounds (e.g., polysaccharides, organic acids, enzymes, diverse waste products, often referred to collectively as EPS (see above)) is composed of materials critical for soil aggregate formation and preservation [79]. Microbial necromass is emerging as an important feature for

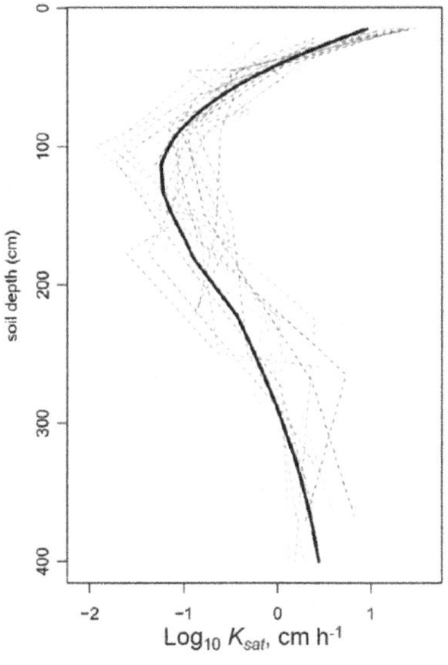

Fig. 2.7 Depth dependence of log10-transformed saturated hydraulic conductivity (K_{sat}) in cm h^{-1}. Data derived from measurements made at 21 locations within the Calhoun Critical Zone Observatory from 2017 to 2019. Each dashed line corresponds to one of the 21 profiles. For each profile, repeated K_{sat} measurements at the same soil depths were averaged within the 21 locations and a rolling mean of K_{sat} was calculated for each profile (window width = 3) across depths. The solid bold line is a LOESS fit (Local Polynomial Regression Fitting) of all 21 soil profiles for which K_{sat} was measured. All sites exhibit a Bt horizon, typically between ~40 cm and ~150 cm depth, that results from climate and biota promoting vertical clay movement via infiltrating soil moisture (i.e., lessivage). The accumulation and concentration of fine clay particles in these Bt horizons results in diminished K_{sat} relative to the coarser textured soils above and below

aggregate formation and preservation. The remnants of dead soil microbes appear to persist in soil far longer than some plant-derived compounds [94], potentially serving as glue and adhering mineral and organic particles together into aggregates [28]; the observation that microaggregates tend to harbor relatively old organic C [138] suggests that smaller aggregates may retain microbial necromass to a greater extent than in larger aggregates. Thus, necromass and the exudates of living microbes govern soil void geometries by influencing soil aggregate formation and durability. Like all soil microbial activities, these processes are especially prevalent in and around the rhizosphere, but any such microbial actions whether in bulk or rhizosphere soil can be important for structural change.

Additional microbial actions further can drive soil structure. Soil microbes induce soil organic matter (SOM) decay via exo-enzymes, and transform some of the C into CO_2 and released nutrients and C into biomass. As SOM is transformed from a solid into a solute, and then into either a gaseous or microbial form, soil aggregates can collapse [27]. The fate of the non-decayed SOM that had resided within collapsed aggregates is unclear, though it is presumed that the removal of aggregate structural protection [148] means that any remaining SOM experiences a greater probability of undergoing decay. Whether aggregate collapse associated with SOM decay promotes a loss or a gain in soil porosity is a matter of some debate. Over broad spatial scales, higher concentrations of SOM generally correlate with greater porosity [54]. However, the transformation of a solid material into a gaseous or solute form results in the presence of small pores in the space where that solid material previously existed. The net effect of SOM mineralization within an aggregate on soil aggregate stability and size distribution is unclear, though some studies suggest that pore opening dominates (X. Zhang, P. Sullivan, S. Billings, et al., unpublished data). Discerning the conditions under which SOM decay induces porosity gains or losses is a current research focus.

Another mechanism by which both soil microbes and roots can influence soil structure is via the exudation of organic acids and CO_2. Organic acids release nutrients from minerals [4, 84, 100], often via ligand exchange [60, 66, 84]. Carbon dioxide, once dissolved into the carbonic acid system, can also induce acid-promoted rock and mineral transformations (e.g., [85]). Acid transformations of rock are a key feature of long-term soil development [24, 43, 154]. Indeed, over long timescales, such momentary-scale processes can enhance soil porosity in the regolith and bedrock itself, and ultimately promote soil production from rock. Thus, microbial and root acid losses can be important determinants of soil structure over diverse timescales.

Because roots and microbes influence soil structure, and because soil microbes tend to proliferate near roots, any change in rooting depth distributions has the potential to alter soil aggregate and void arrangement, and thus CZ functioning (Fig. 2.2). Recent work highlighting the role of humans during the Anthropocene in regionally-dependent deepening or shallowing of roots indicates the massive scale at which these phenomena are occurring [68]. Ecosystem process models, reactive transport models, and continental-scale models of biosphere–atmosphere exchanges of energy and water offer the opportunity to test hypotheses probing the effects of modified abundances of deep root influences on soil structure [156]. Modeling efforts like these will further strengthen the intellectual ties among ecosystem ecologists and the diversity of investigators working on CZ problems.

5 Conclusion

Understanding the CZ as an ecological problem and designing research that invokes the CZ paradigm offers us a way forward as we attempt to project future environmental processes in the Anthropocene. In this chapter, we provide examples of biotic-abiotic processes across a diversity of scales that govern CZ functioning and that are impossible to understand and predict without the use of the CZ paradigm. We specifically focus on hydrologic functioning given water's importance in both the biotic and abiotic realms.

By describing the influence of soil texture and macroporosity on hydrologic flow patterns, we highlight how fundamental CZ constraints ([52]; here, the particle surface area present in a soil profile and the capacity of that soil to move water with relatively little tension) can govern the storage and fluxes of a critical resource—water. In turn, water fluxes through a soil profile influence the extent and pace of soil horizonization, which can result in varied capacity of soil at a given depth to store and release water (Fig. 2.7). Roots also contribute to soil development in numerous ways (Fig. 2.6), among them their capacity for perforating soil and thus generating pores through which fluids can flow. Of course, roots also drive water loss from soils. Combined, these actions—root growth and water uptake—promote movement of soil particles at small (~nm to cm) scales, further modifying the soil pore networks through which liquids and gases flow. Widespread changes in rooting depth in the Anthropocene (Fig. 2.4) likely have modified pore networks in meaningful but unquantified ways. Microbes further contribute to these dynamics, in multiple ways. One salient role of soil microbes is their mineralization of soil organic C into CO_2. This mineralization removes a structural agent from the soil, and likely can induce the collapse of soil structural features that may protect soil organic matter from microbial attack (Fig. 2.2). Microbes and roots also generate acids that impose structural change to the regolith and bedrock via rock and mineral transformations. These root and microbial actions, and associated flows of water and soil particles, only sometimes scale up in a predictable way. However, at a landscape scale, we can expect spatial variability in soil capacity to retain water (Fig. 2.3) and plant adaptations to varied water availability (Fig. 2.5) that reflect water's tendency to flow towards more negative water potentials. Scaling the flows of fluids and solids, and patterns of root water uptake and soil microbial mineralization, across watersheds and landscapes remains a challenge important to address given spatially-varied responses of water availability to a changing climate. This challenge cannot be met without appreciating both the biotic and abiotic characteristics of the soil profile or landscape in question.

A multitude of biotic-abiotic interactions not addressed in this work but that govern CZ functioning also require a transdisciplinary, CZ approach to develop a predictive understanding of the processes at large. Consider, for example, how within-canopy air currents govern system energy and water dynamics throughout the CZ [19, 111]. Linkages among soil structural attributes and hydrologic flows and resulting soil C and nutrient pools and fluxes are well-established [176, 133]. Other examples of biotic-abiotic interactions include the generation of channels by soil fauna through which gases and solutes flow [126], and how SOM flows across landscapes with erosion [8, 43] result in a redistribution of this key agent of soil structure and a source of CO_2 and organically-bound nutrients [14]. All of these processes participate in the complex interplay of the biota with the abiotic that structure the physical and chemical conditions in which life persists.

We emphasize that the biotic responses to land use, temperature, precipitation patterns, and atmospheric CO_2 concentrations—often explored in ecosystem ecology—can be rapid. Because biota influence soil structure, rapid responses of biota to environmental conditions may prompt rapid changes in the structure and function of the CZ, and thus impart meaningful alterations to CZ-climate feedbacks [156]. It is no longer possible to examine these environmental puzzles effectively from the perspective of a single or even a pair of disciplines. The knowledge bases required to understand these processes emerge from a diversity of disciplines, including geomorphology, soil science, ecohydrology, plant physiology, and ecosystem ecology. It is no coincidence that biotic-abiotic interactions all revolve around soil; soil is where the lithosphere, biosphere, atmosphere, and hydrosphere interact to form Earth's living skin, and is perhaps the most intuitive location within the CZ where biotic and abiotic processes come to govern the functioning of the CZ ecosystem (Fig. 2.1). Integrating knowledge from these realms has offered CZ scientists the opportunity to gain new insights about the functioning at Earth's surface, which in turn allows the community to project Earth's future functioning via both empirical and modeling studies. Taking our cues from the likes of Humboldt, Foote, Tansley, and Berner, we can continue to contribute to the rich history of the science of the CZ ecosystem, and move forward with greater confidence in our ability to project future CZ functioning.

Acknowledgements The authors have been supported by National Science Foundation (NSF) grants 13331846 (SAB, DDR), 2026874 (PLS, SAB, DRH), 2034232 (PLS, SAB, DRH), 2012633 (SAB), 2012796 (PLS), 2025849 (JN), 2121694 (PLS, SAB, DRH, JN), and 2024388 (PLS, JN), and by the U.S. Department of Energy grant DE-SC0019037 (JN). This work was also supported by the USDA National Institute of Food and Agriculture through Hatch funds (CA-R-ENS-5195-H, project accession no. 1022418, DRH) and a Signals in the Soil grant (2021-67019-34341 (DRH) and 2021-67019-34338 (SAB)). This material is also based on work supported by the NSF under OIA-1656006 with matching support from the State of Kansas through the Kansas Board of Regents (SAB and PLS).

References

1. Aciego SM, Riebe CS, Hart SC et al (2017) Dust outpaces bedrock in nutrient supply to montane forest ecosystems. Nat Comm Med. https://doi.org/10.1038/ncomms14800

2. Anderson MA, Graham RC, Alyanakian GJ et al (1995) Late summer water status of soils and weathered bedrock in a giant sequoia grove. Soil Sci 160:415–422
3. Angers DA, Caron J (1998) Plant-induced changes in soil structure: processes and feedbacks. Biogeochemistry 42:55–72
4. Aoki M, Fujii K, Kitayama K (2012) Environmental control of root exudation of low-molecular weight organic acids in tropical rainforests. Ecosystems. http://www.jstor.org/stable/23328067
5. Baldrian P (2019) The known and the unknown in soil microbial ecology. FEMS Microbiol Ecol. https://doi.org/10.1093/femsec/fiz005
6. Barberon M, Vermeer JEM, De Bellis D et al (2016) Adaptation of root function by nutrient-induced plasticity of endodermal differentiation. Cell 164(3):447–459. https://doi.org/10.1016/j.cell.2015.12.021
7. Bennett JA, Klironomos J (2018) Mechanisms of plant-soil feedback: Interactions among biotic and abiotic drivers. New Phytol 222:91–96
8. Berhe AA, Barnes RT, Six J et al (2018) Role of soil erosion in biogeochemical cycling of essential elements: C, nitrogen, and phosphorus. Annu Rev Earth Planet Sci. https://doi.org/10.1146/annurev-earth-082517-010018
9. Berkelhammer M, Hu J, Bailey A et al (2013) The nocturnal water cycle in an open-canopy forest. J Geophys Res Atmos 118:10225–10242. https://doi.org/10.1002/jgrd.50701
10. Beven K, Germann P (1982) Macropores in water flow in soils. Water Resour Res 18(1311–1325):655
11. Beven K, Germann P (2013) Macropores and water flow in soils revisited. Water Resour Res 49:3071–3092
12. Bilir TE, Fung I, Dawson TE (2021) Slope-aspect induced climate differences influence how water is exchanged between the land and atmosphere. J Geophys Res: Biogeosci. https://doi.org/10.1029/2020JG006027
13. Billings SA, Hirmas D, Sullivan PL et al (2018) Loss of deep roots limits biogenic agents of soil development that are only partially restored by decades of forest regeneration. Elementa: Science of the Anthropocene 6:34 https://doi.org/10.1525/elementa.287
14. Billings SA, Richter DdeB, Ziegler SE et al (2019) Distinct contributions of eroding and depositional profiles to land-atmosphere CO2 exchange in two contrasting forests. Front Earth Sci 7 https://doi.org/10.3389/feart.2019.00036
15. Billings SA, Sullivan PL (2022) Working across scales to project soil biogeochemical responses to climate
16. Binkley D (2006) Soils in ecology and ecology in soils. In: Warkentin BP (ed) Footprints in the soil. Amsterdam, the Netherlands: Elsevier, 259–278
17. Blair J, Nippert J, Briggs J (2014) Grassland ecology 14. Ecol Environ 389:389–423
18. Bogner C, Gaul D, Kolb A et al (2010) Investigating flow mechanisms in a forest soil by mixed-effects modelling. Eur J Soil Sci 61:1079–1090
19. Bonan GB (2016) Ecological climatology: concepts and applications, 2nd edn. Cambridge University Press, New York, p 681
20. Bormann FH, Likens GE (1967) Nutrient cycling: small watersheds can provide invaluable information about terrestrial ecosystems. Science 155:424–429
21. Bornyasz M, Graham R, Allen M (2005) Ectomycorrhizae in a soil-weathered granitic bedrock regolith: linking matrix resources to plants. Geoderma 126:141–160
22. Brantley SL, Eissenstat DM, Marshall JA et al (2017) Reviews and syntheses: on the roles trees play in building and plumbing the CZ. Biogeosciences 14(5115–5142):690
23. Brantley S, Goldhaber MB, Ragnarsdottir KV (2007) Crossing disciplines and scales to understand the CZ. Elements 3:307–314
24. Brantley, SL, Lebedeva, M, Hausrath, EM (2012) A geobiological view of weathering and erosion. Fundament Geobiol 205–227. https://doi.org/10.1002/9781118280874.ch12
25. Brantley SL, Megonigal JP, Scatena FN (2011) Twelve testable hypotheses on the geobiology of weathering. Geobiology 9:140–165

26. Brecheisen ZS, Richter DD, Moon S et al (2021) Quantitative analysis of hillshed geomorphology and critical zone function: raising the hillshed to watershed status. GSA Bull 134:2007–2021
27. Bronick CJ, Lal R (2005) Soil structure and management: a review. Geoderma 124(3–22):704
28. Buckeridge KM, La Rosa AF, Mason KE et al (2020) Sticky dead microbes: rapid abiotic retention of microbial necromass in soil. Soil Biol Biochem 149:107929
29. Calabrese S, Richter DD, Porporato A (2018) The formation of clay-enriched horizons by lessivage. Geophys Res Lett. https://doi.org/10.1029/2018GL078778
30. Caplan JS, Giménez D, Hirmas DR et al (2019) Decadal-scale shifts in soil hydraulic properties as induced by altered precipitation. Sci Adv. https://doi.org/10.1126/sciadv.aau6635
31. Case MF, Nippert JB, Holdo RM et al (2020) Root-niche separation between savanna trees and grasses is greater on sandier soils. J Ecol 108:2298–2308
32. Chadwick OA, Derry LA, Vitousek PM et al (1999) Changing sources of nutrients during four million years of ecosystem development. Nature 397:491–497
33. Chamberlin CA, Bianchi TS, Brown AL et al (2019) Mass balance implies Holocene development of a low-relief karst patterned landscape. Chem Geol 527:118782
34. Chen W, Zeng H, Eissenstat DM et al (2013) Root trait variation and evolution. Glob Ecol Biogeogr 22:846–856
35. Cosentino D, Chenu C, Le Bissonnais Y (2006) Aggregate stability and microbial community dynamics under drying-wetting cycles in a silt loam soil. Soil Biol Biochem 38(2053–2062):734
36. Costa OYA, Raaijmakers JM, Kuramae EE (2018) Microbial extracellular polymeric substances: ecological function and impact on soil aggregation. Front Microbiol 9(1636):737
37. Craine JM, Ocheltree TW, Nippert JB et al (2013) Global diversity of drought tolerance and grassland climate-change resilience. Nat Clim Chang 3:63–67
38. Dawson TE, Ehleringer JR (1991) Streamside trees that do not use stream water. Nature 350(335–337):742
39. Dawson TE, Hahm WJ, Crutchfield-Peters K (2020) Digging deeper: What the CZ perspective adds to the study of plant ecophysiology. New Phytol 226:666–671
40. Degens BP, Sparling GP (1995) Repeated wet-dry cycles do not accelerate the mineralization of organic C involved in the macro-aggregation of a sandy loam soil. Plant Soil 175:197–203
41. Denef K, Six J, Bossuyt H et al (2001) Influence of dry-wet cycles on the interrelationship between aggregate, particulate organic matter, and microbial community dynamics. Soil Biol Biochem 33:1599–1611
42. Derry LA, Chadwick OA (2007) Contributions from Earth's atmosphere to soil. Elements 3:333–338
43. Dialynas YG, Bastola S, Bras RL et al (2016) Topographic variability and the influence of soil erosion on the C cycle. Glob Biogeochem Cycles 30. https://doi.org/10.1002/2015GB005302
44. D'Odorico P, Okin GS, Bestelmeyer BT (2012) A synthetic review of feedbacks and drivers of shrub encroachment in arid grasslands. Ecohydrology 5:520–530
45. Dontsova K, Balogh-Brunstad Z, Chorover J (2020) Plants as drivers of rock weathering. In: Dontsova Z, Balogh-Brunstad Z, Le Roux G (eds) Biogeochemical cycles American geophysical union geophysical monograph series. https://doi.org/10.1002/9781119413332.ch2
46. Dyck MF, Kachanoski RG (2010) Spatial scale-dependence of preferred flow domains during infiltration in a layered field soil. Vadose Zone J 9:385–396
47. Eger A, Almond PC, Condron LM (2013) Phosphorus fertilization by active dust deposition in a super-humid, temperate environment – soil phosphorus fractionation and accession processes. Glob Biogeochem Cycles 27:108–118
48. Fan Y, Clark M, Lawrence DM et al (2019) Hillslope hydrology in global change research and earth system modeling. Water Resour Res 55. https://doi.org/10.1029/2018WR023903
49. Fan Y (2015) Groundwater in the Earth's CZ: relevance to large-scale patterns and processes. Water Resour Res 51:3052–3069

50. Fan Y, Miguez-Macho G, Jobbagy EG et al (2017) Hydrologic regulation of plant rooting depth. Proc Natl Acad Sci 114(10572–10577):775
51. Field JP, Breshears DD, Law DJ et al (2015) Critical zone services: expanding context, constraints, and currency beyond ecosystem services. Vadose Zone J. https://doi.org/10.2136/vzj2014.10.0142
52. Field JP, Breshears DD, Law DJ, Villegas JC, Lopez-Hoffman L, Brooks PD, Chorover J, Barron-Gafford GA, Gallery RE, Litvak mE, Lybrand RA, McIntosh JC, Meixner T, Niu G-Y, Papuga SA, Pelletier JD, Rasmussen CR, Troch PA (2015) Critical zone services: Expanding context, constraints, and currency beyond ecosystem services. Vadose Zone J https://doi.org/10.2136/vzj2014.10.0142
53. Foote E (1856) Circumstances affecting the heat of the sun's rays. American Journal of Science and Arts 22:382-383.
54. Franzluebbers AJ (2011) Stratification of soil porosity and organic matter. In: Gliński J, Horabik J, Lipiec J (eds) Encyclopedia of Agrophysics. Encyclopedia of Earth Sciences Series. Springer, Dordrecht. https://doi.org/10.1007/978-90-481-3585-1_225
55. Franzenleubbers AJ (2014) Stratification of soil porosity and organic matter. In: Glinski J, Horabik J, Lipiec J (eds) Encyclopedia of agrophysics. https://doi.org/10.1007/978-90-481-3585-1_225
56. Freschet GT, Roumet C, Comas LH et al (2021) Root traits as drivers of plant and ecosystem functioning: current understanding, pitfalls and future research needs. New Phytol 232:1123–1158. https://doi.org/10.1111/nph.17072
57. Goddard TM, Runge ECA, Ray BW (1973) The relationship 3+ between rainfall frequency and amount to the formation and profile distribution of clay particles. Soil Sci Soc Am J https://doi.org/10.2136/sssaj1973.03615995003700020037x
58. Honeycutt CW, Heil RD, Cole CV (1990) Climatic and topographic relations of three Great Plains soils: I. Soil morphology. Soil Sci Soc Am J https://doi.org/10.2136/sssaj1990.03615995005400020030x
59. Gallardo A, Fernandez-Palacios JM, Bermudez A et al (2020) The pedogenic Walker and Syers model under high atmospheric P deposition rates. Biogeochemistry 148:237-253.
60. Ganor J, Reznik IJ, Rosenberg YO (2009) Organics in Waterrock interactions. Rev Mineral Geochem 70:259–369
61. Gao W, Hodgkinson L, Jin K et al. (2016) Deep roots and soil structure. Plant, Cell and Environment 39:1662-1668.
62. Geldner N (2013) The endodermis. Annu Rev Plant Biol 64:531–558
63. Gilbert GK (1877) Report on the geology of the Henry mountains. US geographical and geological survey of the Rocky Mountain Region. U.S. Gov. Print. Office, Washington, DC
64. Gimenez D, Hirmas DR (2017) Macroporosity. In: Lal R (ed.) Encyclopedia of Soil Science, Vol. 2, 3rd Ed. CRC Press, Boca Raton, FL
65. Harman C, Troch PA (2014) What makes Darwinian hydrology "Darwinian"? Asking a different kind of question about landscapes. Hydrol Earth Syst Sci 18:417–433. https://doi.org/10.5194/hess-18-417-2014.
66. Hasegawa S, MacDonald CA, Power SA (2015) Elevated carbon dioxide increases soil nitrogen and phosphorus availability in a phosphorus-limited Eucalyptus woodland. Glob Change Biol 22:1628–1643
67. Hasenmueller EA, Jin L, Stinchcomb GE et al. (2015) Topographic controls on the depth distribution of soil CO2 in a small temperate watershed. Applied Geochemistry 63:58-69.
68. Hauser E, Sullivan PL, Flores L et al (2022) Global-scale shifts in Anthropocene rooting depths pose unexamined consequences in critical zone functioning. Earth's Future. https://doi.org/10.1029/2022EF002897
69. Hauser E, Richter, DdeB, Markewitz D, Brecheisen Z, Billings SA (2020) Persistent anthropogenic legacies structure the depth dependence of regenerating rooting systems and their functions. Biogeochemistry https://doi.org/10.1007/s10533-020-00641-2
70. Hill AJ, Dawson TE, Dody A (2021) Dew water-uptake pathways in Negev desert plants: a study using stable isotope tracers. Oecologia 196:353-361.

71. Hillel D (1998) Environmental soil physics. Academic Press, California
72. Hillel D (2003) Introduction to environmental soil physics. Elsevier, New York
73. Hirmas DR, Gimenez D, Nemes A et al. (2018) Climate-induced changes in continental-scale soil macroporosity may intensify water cycle. Nature 561:100-103.
74. Hutchinson GE (1940) Bio-ecology by F.E. Clements and V.E. Shelford. Review. Ecology 21:267-268.
75. Jackson R (2020) Eunice Foote, John Tyndall and a question of priority. Notes Records: R Soc J Hist Sci 74(105–118):845
76. Jackson RB, Canadell J, Ehleringer JR et al. (1996) A global analysis of root distributions for terrestrial biomes. Oecologia 108:389-411.
77. Jackson RB, Sperry JS, Dawson TE (2000) Root water uptake and transport: using physiological processes in global predictions. Trends Plant Sci 5:482–488
78. Jarvis N, Larsson M (2001) Modeling macropore flow in soils: field validation and use for management purposes. In: Council NR (ed) Conceptual models of flow and transport in the fractured vadose zone. National Academy Press, Washington, DC, pp 189–215
79. Jastrow JD, Miller RM, Lussenhop J (1998) Contributions of interacting biological mechanisms to soil aggregate stabilization in restored prairie. Soil Biol Biochem 30(905–916):858
80. Jenny H (1946) Arrangement of soil series and types according to functions of soil-forming factors. Soil Sci 61:375–392
81. Johnson-Maynard JL, Strawn DG (2016) Linking physical and biogeochemical properties and processes in the drilosphere. Soil Science 181:126-132.
82. Jordan T, Ashley GM, Barton MD et al (2001) Basic research opportunities in Earth science, vol 867. National Academy Press, Washington, DC
83. Jury WA, Horton R (2004) Soil physics, 6th edn. Wiley, Hoboken
84. Keiluweit M, Bourgoure JJ, Nico PS (2015) Mineral protection of soil carbon counteracted by root exudates. Nat Clim Chang 5:588–595
85. Keller K (2019) Carbon exports from terrestrial ecosystems: a critical-zone framework. Ecosystems 22:1691–1705
86. Klaus J, Jackson CR (2018) Interflow is not binary: a continuous shallow perched layer does not imply continuous connectivity. Water Resour Res 54(5921–5932):878
87. Klein T (2014) The variability of stomatal sensitivity to leaf water potential across tree species indicates a continuum between isohydric and anisohydric behaviours. Functional Ecology 28:1313-1320.
88. Kong D, Ma C, Zhang Q et al. (2014) Leading dimensions in absorptive root trait variation across 96 subtropical forest species. New Phytologist 203:863-872.
89. Kumordzi BB, Aubin I, Cardou F et al (2019) Geographic scale and disturbance influence intraspecific trait variability in leaves and roots of North American understorey plants. Funct Ecol 33:1771–1784
90. Landeweert R, Hoffland E, Finlay RD et al (2001) Linking plants to rocks: ectomycorrhizal fungi mobilize nutrients from minerals. Trends Ecol Evol 16:248–254
91. Lawrence CR, Schulz MS, Masiello CA, Chadwick OA, Harden JW (2021) The trajectory of soil development and its relationship to soil carbon dynamics. Geoderma 403:115378
92. Léonard J, Perrier E, Rajot JL (2004) Biological macropores effect on runoff and infiltration: a combined experimental and modelling approach. Agr Ecosyst Environ 104:277–285
93. Leone JD, Holbrook WS, Riebe CS et al (2020) Strong slope-aspect control of regolith thickness by bedrock foliation. Earth Surf Proc Land 45:2998–3010
94. Liang C, Schimel JP, Jastrow JD (2017) The importance of anabolism in microbial control over soil C storage. Nat Microbiol 2:17105. https://doi.org/10.1038/nmicrobiol.2017.105
95. Lin Y-C, Feng J-L (2015) Aeolian dust contribution to the formation of alpine soils at Amdo (Northern Tibetan Plateau) Geoderma 259–260:104–115
96. Lovett GM (1994) Atmosphere deposition of nutrients and pollutants to North America: an ecological perspective. Ecol Appl 4:629–650
97. Lovett GM, Cole JJ, Pace ML (2006) Is net ecosystem productivity equal to ecosystem carbon accumulation? Ecosystems 9:1–4

98. Lu J, Zhang Q, Werner AD et al (2020) Root-induced changes of soil hydraulic properties–A review. J Hydrol 125203
99. Lucas M, Schlüter S, Vogel H-J et al (2019) Roots compact the surrounding soil depending on the structures they encounter. Sci Rep 9:16236. https://doi.org/10.1038/s41598-019-52665-w
100. Lugli L, Andersen KM, Aragao LEOC et al (2019) Multiple phosphorus acquisition strategies adopted by fine roots in low-fertility soils in Central Amazonia. Plant Soil. https://doi.org/10.1007/s11104-019-03963-9
101. Maltz MR, Carey CJ, Freund HL et al (2022) Landscape topography and regional drought alters dust microbiomes in the Sierra Nevada of California. Front Microbiol. https://doi.org/10.3389/fmicb.2022.856454
102. Marcon V, Hoagland B, Gu X et al (2021) How the capacity of bedrock to collect dust and produce soil affects phosphorus bioavailability in the northern Appalachian Mountains of Pennsylvania. Earth Surf Proc Land. https://doi.org/10.1002/esp.5209
103. Markewitz D, Richter DD (1998) The bio in aluminum and silicon geochemistry. Biogeochemistry 42:235–252
104. Marschner P, Rengel Z (2007) Nutrient cycling in terrestrial ecosystems. Springer, New York
105. McDowell N, Pockman WT, Allen CD et al (2008) Mechanisms of plant survival and mortality during drought: why do some plants survive while others succumb to drought? New Phytol 178(719–739):934
106. Milne G (1936) Normal erosion as a factor in soil profile development. Nature 138:548–549
107. McCormack ML, Dickie IA, Eissenstat DM et al (2015) Redefining fine roots improves understanding of below-ground contributions to terrestrial biosphere processes. New Phytol 207:505–518
108. McCormick EL, Dralle DN, Hahm WJ et al (2021) Widespread woody plant use of water stored in bedrock. Nature 597:225–229
109. Mobley ML, Lajtha K, Kramer MG, Bacon AR, Heine PR, Richter DdeB (2015) Surficial gains and subsoil losses of soil carbon and nitrogen during secondary forest development 21:986–996
110. Molles MC, Sher AA (2019) Ecology: concepts and applications. McGraw Hill, New York
111. Monson R, Baldocchi D (2014) Terrestrial biosphere-atmosphere fluxes. Cambridge University Press, Cambridge
112. Morford SL, Houlton BZ, Dahlgren RA (2016) Geochemical and tectonic uplift controls on rock nitrogen inputs across terrestrial ecosystems. Glob Biogeochem Cycles 30. https://doi.org/10.1002/2015GB005283
113. Nimmo JR (2012) Preferential flow occurs in unsaturated conditions. Hydrological Processes 26:786-789.
114. Nippert JB, Holdo RM (2015) Challenging the maximum rooting depth paradigm in grasslands and savannas. Funct Ecol 29:739–745
115. Nippert JB, Knapp AK (2007) Linking water uptake with rooting patterns in grassland species. Oecologia 153:261–272
116. Nippert JB, Wieme RA, Ocheltree TW et al (2012) Root characteristics of C 4 grasses limit reliance on deep soil water in tallgrass prairie. Plant Soil 355:385–394
117. Noguchi S, Tsuboyama Y, Sidle RC et al (1997) Spatially distributed morphological characteristics of macropores in forest soils of Hitachi Ohta Experimental Watershed, Japan. J For Res 2:207–215
118. Odum EP (1968) Energy flow in ecosystems: a historical review. Am Zool 8:11–18
119. O'Keefe K, Bachle S, Keen R, Tooley EG, Nippert JB (2022) Root traits reveal safety and efficiency differences in grasses and shrubs exposed to different fire regimes. Funct Ecol 36:368–379
120. O'Neill RV (2001) Is it time to bury the ecosystem concept? (with full military honors, of course!). Ecology 82:3275–3284
121. Park EJ, Sul WJ, Smucker AJM (2007) Glucose additions to aggregates subjected to drying/wetting cycles promote C sequestration and aggregate stability. Soil Biol Biochem 39:2758–2768

122. Pataki DE, Ehleringer JR, Flanagan LB (2003) The application and interpretation of Keeling plots in terrestrial carbon cycle research. Global Biogeochem Cycles 17:1022
123. Patton NR, Lohse KA, Godsey SE (2018) Predicting soil thickness on soil mantled hillslopes. Nat Commun 9:3329
124. Pawlik Ł (2013) The role of trees in the geomorphic system of forested hillslopes—a review. Earth Sci Rev 126:250–265. https://doi.org/10.1016/j.earscirev.2013.08.007
125. Pawlik Ł, Phillips JD, Šamonil P (2016) Roots, rock, and regolith: biomechanical and biochemical weathering by trees and its impact on hillslopes—a critical literature review. Earth Sci Rev 159:142–159
126. Platt BF, Kolb DJ, Kunhardt CG et al (2016) Burrowing through the literature: the impact of soil-disturbing vertebrates on physical and chemical properties of soil. Soil Sci 181:175–191
127. Radcliffe DE, Šimůnek J (2010) Soil physics with HYDRUS: modeling and applications. CRC Press, Boca Raton
128. Rebertus RA, Buol SW (1985) Iron distribution in a developmental sequence of soils from mica gneiss and schist. Soil Sci Soc Am J https://doi.org/10.2136/sssaj1985.03615995004900030037x
129. Rempe DM, Dietrich WE (2018) Direct observations of rock moisture, a hidden component of the hydrologic cycle. Proc Natl Acad Sci 115:2664–2669
130. Richter DD, Billings SA (2015) 'One physical system': Tansley's ecosystem as Earth's critical zone. New Phytol Tansley Rev. https://doi.org/10.1111/nph.13338
131. Richter DD, Billings SA, Groffman PM et al (2018) Ideas and perspectives: Strengthening the biogeosciences in environmental research networks. Biogeosciences 15:4815–4832
132. Richter DdeB, Billings SA (2020) Ansichten der Calzone: Views of the Calhoun Critical Zone Observatory. In: Latour B, Weibel P (eds) Critical Zones: The Science and Politics of Landing on Earth. MIT Press, Cambridge, MA
133. Richter DD, Markewitz D (2001) Understanding Soil Change: Soil Sustainability of Millennia, Centuries, and Decades. Cambridge University Press, New York.
134. Richter DD, Oh NH, Fimmen R et al. (2007) The rhizosphere and soil formation. pp. 179-200 In: Whitbeck JL, Cardon ZG (eds.) The Rhizosphere. Academic Press, NY
135. Rillig MC, Mummey DL (2006) Mycorrhizas and soil structure. New Phytol 171:41–53
136. Roebroek CTJ, Melsen LA, Hoek van Dijke AJ et al (2020) Global distribution of hydrologic controls on forest growth. Hydrol Earth Syst Sci 24:4625–4639
137. Roering J, Marshall J, Booth AM et al (2010) Evidence for biotic controls on topography and soil production. Earth Planet Sci Lett 298:183–190
138. Romkens PFAM, Hassink J, van der Plicht J (1998) Soil organic 14C dynamics: effects of pasture installation on arable land. Radiocarbon 40:1023–1031
139. Saintilan N, Rogers K (2015) Woody plant encroachment of grasslands: a comparison of terrestrial and wetland settings. New Phytol 205:1062–1070
140. Schaetzl RJ, Anderson S (2005) Soils: genesis and geomorphology. Cambridge University Press, Cambridge
141. Schaffer S (2020) Beware of precursors: How not to trace the history of the critical zone. In: Latour B, Weibel P (eds.) Critical zones: the science and politics of landing on earth. ZKM Karlsruhe, Germany and MIT Press, Cambridge
142. Schenk HJ, Jackson RB (2002) The global biogeography of roots. Ecol Monogr 72:311–328
143. Schenk HJ, Jackson RB (2005) Mapping the global distribution of deep roots in relation to climate and soil characteristics. Geoderma 126:129–140
144. Schenk HJ, Jackson RB (2002) Rooting depths, lateral root spreads and below-ground/above-ground allometries of plants in water-limited ecosystems. J Ecol 90:480–494
145. Schlesinger W, bernhardt E (2020) Biogeochemistry: An Analysis of Global Change. 4th ed. Academic Press, New York.
146. Sher A, Molles M (2022) Ecology: concepts and applications. McGraw Hill, New York
147. Sigen C, Franklin RE, Johnson AD (1997) Clay film effects on ion transport in soil. Soil Sci 162:91–96

148. Six J, Paustian K (2014) Aggregate-associated soil organic matter as an ecosystem property and a measurement tool. Soil Biol Biochem 68:A4–A9
149. Smettem KRJ, Chittleborough DJ, Richards BG et al (1991) The influence of macropores on runoff generation from a hillslope soil with a contrasting textural class. J Hydrol 122:235–252
150. Soderberg K, Compton JS (2007) Dust as a nutrient source for fynbos ecosystems, South Africa. Ecosystems 10:550–561
151. Sorenson RP (2011) Eunice Foote's pioneering research on CO2 and climate warming. Search and Discovery #70092
152. Sternberg P, Anderson M, Graham R et al (1996) Root distribution and seasonal water status in weathered granitic bedrock under chaparral. Geoderma 72:89–98
153. Stone EL, Kalisz PJ (1991) On the maximum extent of tree roots. For Ecol Manage 46:59–102
154. Stoops G (2003) Guidelines for the analysis and description of soil and regolith thin sections, 2nd edn. Soil Science Society of America, Madison, Wisconsin
155. Sullivan PL, Godderis Y, Shi Y, Gu X, Schott J, Hasenmueller EA, Kaye J, Duffy C, Jin L, Brantley SL (2019) Exploring the effect of aspent to inform future earthcasts of climate-driven changes in weathering of shale. J Geophy Res Earth Surf https://doi.org/10.1029/2017JF004556
156. Sullivan PL, Billings S, Hirmas D et al (2022) Embracing the dynamic nature of soil structure: a key paradigm shift for illuminating the role of life in the critical zones of the Anthropocene. Earth Sci Rev 225:103873
157. Sullivan PL, Price RM, Ross MS et al (2011) Hydrologic processes on tree islands in the Everglades (Florida, USA): tracking the effects of tree establishment and growth. Hydrogeol J 19:367–378
158. Sullivan PL, Price RM, Ross MS et al (2016) Trees: a powerful geomorphic agent governing the landscape evolution of a subtropical wetland. Biogeochemistry 128:369–384
159. Tansley AG (1935) The use and abuse of vegetational concepts and terms. Ecology 16:284–307
160. Tardieu F, Simonneau T (1998) Variability among species of stomatal control under fluctuating soil water status and evaporative demand: modelling isohydric and anisohydric behaviours. J Exp Bot 49:419–432
161. Thoma SG, Gallegos DP, Smith DM (1992) Impact of fracture coatings on fracture/matrix flow interactions in unsaturated, porous media. Water Resour Res 28:1357–1367
162. Tisdall JM, Oades JM (1982) Organic matter and water-stable aggregates in soils. J Soil Sci 33:141–163
163. Tumber-Dávila SJ, Schenk HJ, Du E et al (2022) Plant sizes and shapes above and belowground and their interactions with climate. New Phytol 235:1032–1056
164. Turk JK, Chadwick OA, Graham RC (2012) Pedogenic processes. In: Huang PM et al (eds.) Handbook of soil sciences: Properties and processes. CRC Press, Boca Raton, pp 30-1–30-29
165. Tyndall J (1861) The Bakerian Lecture: on the absorption and radiation of heat by gases and vapours, and on the physical connexion of radiation, absorption, and conduction. Philos Trans R Soc 151:28–29
166. Uhlig D, Schuessler JA, Bouchez J et al (2017) Quantifying nutrient uptake as a driver of rock weathering in forest ecosystems by magnesium stable isotopes. Biogeosciences 14:3111–3128
167. Valverde-Barrantes OJ, Authier L, Schimann H et al (2021) Root anatomy helps to reconcile observed root trait syndromes in tropical tree species. Am J Bot 108:744–755
168. van Genuchten MTh (1980) A closed-form equation for predicting the hydraulic conductivity of unsaturated soils. Soil Science Society of America Journal 44:892-898.
169. Viglizzo EF, Nosetto MD, Jobbágy EG et al (2015) The ecohydrology of ecosystem transitions: a meta- analysis. Ecohydrology 8:911–921
170. Vogel C, Helfenstein J, Massey MS et al (2021) Microspectroscopy reveals dust-derived apatite grains in acidic, highly-weathered Hawaiian soils. Geoderma 381:114681
171. Wagener T, Sivapalan M, Troch P et al (2007) Catchment classification and hydrologic similarity. Geogr Compass 1:901–931
172. Wahl S, Ryser P (2000) Root tissue structure is linked to ecological strategies of grasses. New Phytol 148:459–471

173. Wargowsky IK, NeSmith JE, Holdo RM (2021) Root vascular traits differ systematically between African savanna tree and grass species, with implications for water use. Am J Bot 108:83–90
174. Watson P (2017) Convergence: the idea at the heart of science. Simon and Schuster, New York
175. Watson KW, Luxmoore RJ (1986) Estimating macroporosity in a forest watershed by use of a tension infiltrometer. Soil Sci Soc Am J 50:578–582
176. Weil RR, Brady NC (2017) The nature and properties of soils, 15th edn. Pearson Publishing, Boston, Massachusetts
177. Wen H, Sullivan PL, Billings SA, Ajami H, Cueva A, Flores A et al. (2022) From soils to streams: connecting terrestrial carbon transformation, chemical weathering, and solute export across hydrological regimes. Water Resour Res 58. https://doi.org/10.1029/2022WR032314
178. Wigmosta MS, Vail LW, Lettenmaier DP (1994) A distributed hydrology-vegetation model for complex terrain. Water Resources Research 30:1665-1679.
179. Williamson TN, Gessler PN, Shouse PJ et al (2006) Pedogeneis-terrain links in zero-order watersheds after chaparral to grass vegetation conversion. Soil Sci Soc Am J 70:2065–2074
180. Wilmking M, Harden J, Tape K (2006) Effect of tree line advance on C storage in NW Alaska. J Geophys Res 111:G02023. https://doi.org/10.1029/2005JG000074
181. Wlostowski AN, Molotch N, Anderson SP et al (2020) Signatures of hydrologic function across the CZ observatory network. Water Resour Res 57:e2019WR026635
182. Wysocki DA, Schoeneberger PJ, Hirmas DR et al (2012) Geomorphology of soil landscapes. In: Huang PM et al (eds) Handbook of soil sciences: properties and processes. CRC Press, Florida, pp 29-1–29-26
183. Yang Y, Keiluweit M, Senesi N, Xing B (eds) (2022) In: Multi-scale biogeochemical processes in soil ecosystems: critical reactions and resilience to climate changes, vol. 5 in IUPAC series on biophysico-chemical processes in environmental systems. Wiley
184. Yu H, Chin M, Yuan T et al (2015) The fertilizing role of African dust in the Amazon rainforest: a first multiyear assessment based on data from cloud-aerosols Lidar and infrared pathfinder satellite observations. Geophys Res Lett 42:1984–1991
185. Zhang Y, Niu J, Yu X et al (2015) Effects of fine root length density and root biomass on soil preferentialflow in forest ecosystems. Forest Syst 24. https://doi.org/10.5424/fs/2015241-06048
186. Zhang Y, Schaap MG (2017) Weighted recalibration of the Rosetta pedotransfer model with improved estimates of hydraulic parameter distributions and summary statistics (Rosetta3). J Hydrol 547:39–53

The Biogeography of the Earth Critical Zone

Carl Beierkuhnlein and Samip Narayan Shrestha

1 Introduction

The active skin of the Earth, the Earth Critical Zone (ECZ), is a complex layer with diffuse limits towards the atmosphere and the lithosphere resulting from interacting physical, chemical, and biological processes. Biodiversity across all biological realms, including bacteria, archaea, and eukaryotes, is a substantial component and contribution to the ECZ. Without the activity of organisms, this zone and its ecosystems would not be functional or even exist.

Organisms control the physical structure, both above- and below-ground, and the chemical composition of substrates and the atmosphere. Flows and storage of matter, energy, and information are regulated by the activity of organisms. This regulation results in emerging specific traits such as soil types and processes like carbon sequestration [4]. Both historic and current processes have been or are acting with different dynamics. The recording of the fluxes alone cannot satisfy without the understanding of the drivers and processes.

Relevant time scales for the ecological processes in the ECZ range from within-days to centuries and millennia, and beyond. They include diurnal rhythms of biological activity (e.g., photosynthesis), seasonal fluctuations of populations (e.g., regeneration cycles, migration), accumulation of biomass (storage of carbon in non-active tissues), organization of communities and ecosystems (e.g., emerging species assemblages through biotic interactions and respective feedback on fluxes of energy and matter), formation of soil organic matter (e.g., accumulation of peat in wetlands) as well as selection and speciation of organisms (e.g., development of genetically different biota in separated locations but in comparable ecological niches). Understanding the temporal scaling of such processes is crucial when scrutinizing the functioning of the ECZ.

C. Beierkuhnlein (✉) · S. N. Shrestha
Department of Biogeography, University of Bayreuth, Bayreuth, Germany
e-mail: carl.beierkuhnlein@uni-bayreuth.de

T. White and A. Provenzale (eds.), *Critical Zone and Ecosystem Dynamics*, Advances in Critical Zone Science, https://doi.org/10.1007/978-3-031-69076-1_3

The major reflection of biological activity in the ECZ is the accumulation of carbon in living organisms and soils [51]. The global perspective illustrates that climatically favored regions and biomes with high productivity and developed biomass do not store carbon in soils due to the high biological activity of decomposing soil organisms [36, 38]. Surprisingly, the biggest stocks of Soil Organic Carbon (SOC) are detected in parts of the polar and boreal biomes that were completely glaciated until the beginning of the Holocene (Fig. 3.1). This illustrates that low-productivy ecosystems can accumulate large amounts of carbon within relatively short periods, and also that large current biomass does not necessarily indicate carbon sequestration under conditions of rapid turnover.

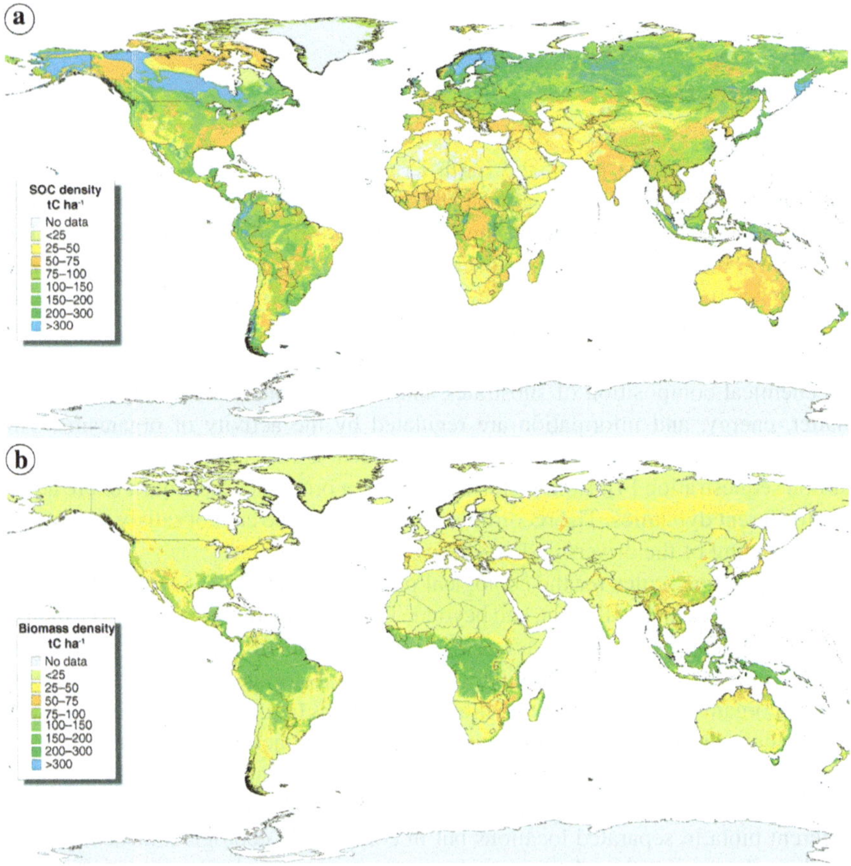

Fig. 3.1 Comparison between the Carbon expressed as **a)** Carbon density (tC ha -1) stored in soils as Soil Organic Carbon (SOC) within 1 m soil depth and **b)** in (above and belowground) vegetation biomass. Highly productive regions in the tropics and subtropics that exhibit high decomposition rates and carbon turnover show only low amounts of carbon in soils, and low productivy regions with limited biological activity in soils exhibit substantial accumulation of carbon in soils. Maps are based on [49], soil data from [29], and biomass data from [48]

Carbon turnover in the ECZ is governed by the interplay between the storage of carbon in living organisms and their decay and decomposition. In addition to photosynthesis and decomposition in soils, short-term events such as bark-beetle outbreaks or wildfires in boreal forest ecosystems can rapidly modify the ECZ. Important carbon sinks can turn into substantial sources through natural disturbances [40]. Canada has used such findings as a negotiation tool to exclude natural insect-calamities in forests from global carbon accounting treaties.

Spatial scales of biotic processes range from the microscale of bacteria, archaea, fungi and other organisms - that are found on the surface of soil particles, through the depth of soil profiles and the height of vegetation structures - to the extent of ecosystems and biomes. The diversity of organisms, their assemblages, and their functional contribution to the mechanisms that govern the ECZ is substantially modified on biogeographical scales [7]. Understanding the linkages between spatial scales and the limits of up- and downscaling between these levels is a second prerequisite to avoid pitfalls in data use and interpretation as well as unrealistic modelling approaches.

The primary spatial differentiation of the terrestrial surface is in terms of its bedrock. Petrography and the age of the parent material differ considerably. Tectonic mechanisms and glacial processes have shaped diverse landscapes. The geochemical composition of the bedrock is the major driver of nutrient availability. Physical processes have created minerals and rock types that are prone or resistant to erosion and thus soil formation.

Historic and current climatic conditions have modified this geo-ecological setting. The legacy of the Pleistocene, as an example, is still highly relevant in large areas of the Northern Hemisphere. Today, the functioning of the ECZ is reflecting latitudinal gradients of annual solar energy input and its seasonal distribution. Longitudinal gradients of continentality explain precipitation regimes, cloudiness, and radiative emissions (nighttime cooling).

Climatic conditions drive above-ground biomass production and below-ground biological activity. Fungal and bacterial biomass carbon in topsoil is less compared to phytomass and SOC but it is a crucial component of carbon cycling and biotic interactions (e.g., for nutrient cycling and uptake). The spatial distribution is less well scrutinized compared to plant biomass, but clear global patterns emerge showing the high importance of fungi in polar soils and the strong dependence of bacterial biomass on moisture (Fig. 3.2).

This uncertainty about belowground biological activity also includes roots and other plant organs. Ecosystems and biomes that seem to build up less aboveground biomass, such as the tundra (polar) biome, can exhibit large amounts of root biomass [32], and a substantially longer seasonal activity [12] compared to the apparent aboveground vegetation. Such findings support the ECZ approach that is not just "superficial".

To evaluate the drivers and the sensitivity of the ECZ, continental to global scales need to be taken into consideration. The energy exchange between landmasses and oceans has a strong influence on atmospheric circulation patterns. Large mountain ranges add climatic complexity through elevational gradients connected with increasing UV radiation, decreasing average temperature, and partial pressure of

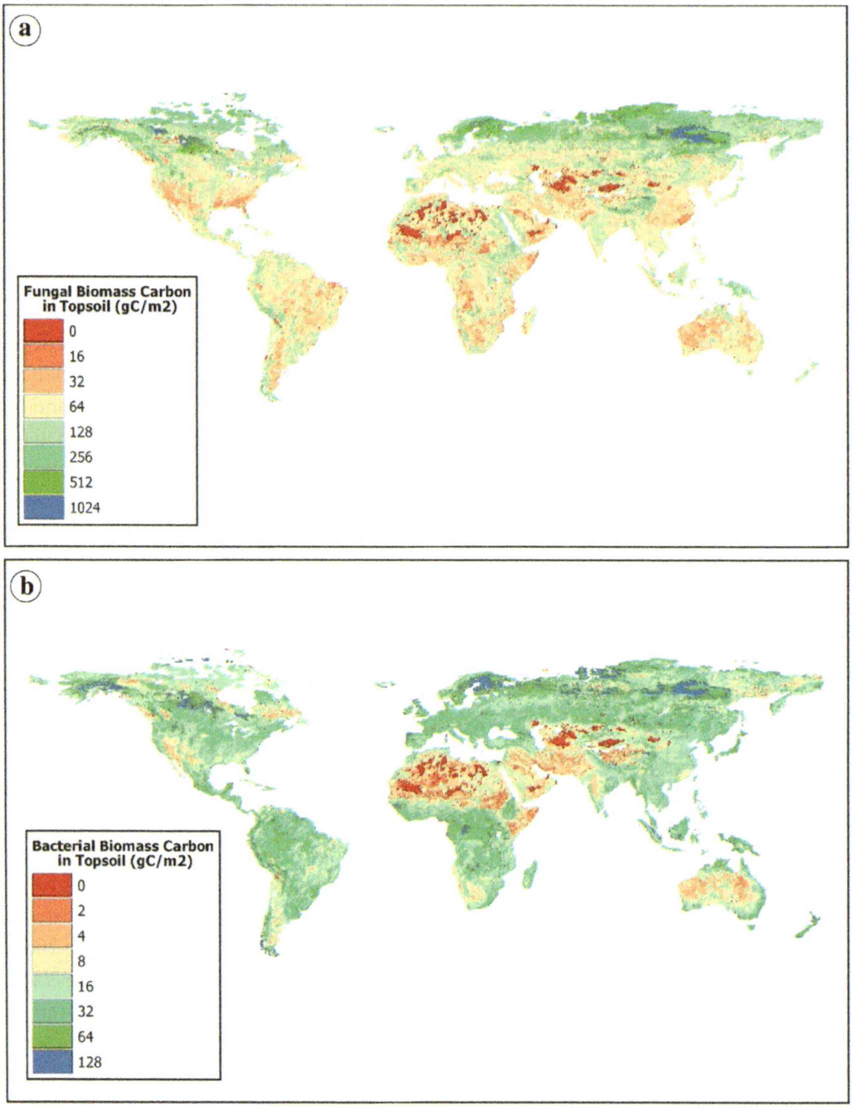

Fig. 3.2 **a)** Fungal and **b)** bacterial biomass carbon in topsoil (10 kgC ha-1). Although in both cases, arid climate is limiting microbial biomass, spatial patterns are substantially different in tropical and subtropical regions. Data sourced from He et al. [28]

CO_2. Besides the availability of solar energy and water, spatial aspects related to topography, elevation, aspect, or slope steepness, include important dynamics within the ECZ for instance in terms of erosion or mass movements. Physical disturbance regimes are strongly linked to the availability of kinetic energy and are responsible not only for the destruction and restarting of successional trajectories but also for providing fresh parent material and nutrients for organismic activity [60]. It is evident that a purely physical perspective does not apply to the ECZ. Vegetation cover and ecosystem functioning are crucial components that are responding to the physio-chemical background and controlling it at the same time.

Surface structures are linked with geographical patterns of spatial units (e.g., islands, high mountains, lakes, rivers, rocks) where biotic components (organisms, ecosystems) of the ECZ are modified by spatial filters. This applies to dispersal and accessibility but also to the required space available for maintaining viable population sizes. If certain Earth surfaces cannot be reached by functional guilds or key species in trophic food chains, or if these have become regionally extinct during Earth history, then their specific functional contribution to the functioning of the ECZ is not provided. If isolated habitats cannot be reached by certain functional types of organisms, then their contribution to ecosystem functioning is not found.

Islands, for instance, are often lacking mammal herbivores and predators with fundamental consequences for trophic cascades and the turnover of matter (in addition to missing adaptations in island biota to these functional groups). If isolated high mountains were not reached by specific tree types (e.g., conifers) that are adapted to low temperatures, the treeline would be lower than climatically expected with consequences for soil formation and erosion processes in the extended alpine elevational zone. This is the case for instance in the highest mountains of the Hawaiian Archipelago.

Physical processes and conditions exhibit highly differentiated effects on the performance of organisms. Plants, fungi, animals, protozoa, archaea, and bacteria are responding in specific ways to the availability of abiotic resources and stressors. Species have evolved into these ecological niches. Adaptive radiation and the selection of biota that are best equipped with traits to optimize the use of diverse resources are the main causes of biodiversity. And vice versa, the loss of these biotic components of the ECZ would translate into repercussions in soil organic matter and its turnover.

Besides physical factors, it is the chemical composition of near-surface layers of parent material that translates into the availability of nutrients or the presence of toxic compounds under certain climatic conditions. Again, species-specific, and ecosystem-related requirements and restrictions need to be considered. Extreme (rare) chemical site conditions, for instance on ultramafic serpentine outcrops with high concentrations of heavy metals, can be toxic for many species but favor those that can tolerate these characteristics [14].

The geodiversity of rock types is a reflection of Earth's history. Suprisingly, very young terrestrial ecosystems are located on the oceanic crust. Most oceanic (volcanic) islands are not older than a few million years and still dominated by repeated eruptions of pyroclastic material resulting in basalt, tephra, and respective initial soil types. Biogeochemical processes on these geologically quite uniform islands are modified by the geographic location and the respective climate. Old continents, in contrast, can harbor a high diversity of rocks and corresponding soils, and ecosystems, which emerge as complex adaptive systems [24] through the long-term interaction of the biotic realm with chemical and physical features. The composition of the continental crust is affected by the biosphere. It was irrevocably altered by the emergence of land plants in the Palaeozoic era, which caused changes in patterns of weathering, erosion and deposition [52] as well as the storage of large amounts of carbon. In all cases, organisms have been selected through evolutionary processes, and ecosystems emerged that are adapted to the availability and restrictions of resources under specific climatic conditions (Fig. 3.3).

2 The Biosphere and ECZ

No "sphere" can be seen in isolation. The ECZ and the complexity of its boundary conditions are the parts of the biosphere, where the biotic activity and the exchange processes with the abiotic setting of the environment take place. The biosphere reaches beyond the ECZ. It reaches high into the atmosphere, where spores and microorganisms are dispersed, and down to the lithosphere [1], where archaea are active even hundreds of meters below the surface. Atmosphere, pedosphere, and hydrosphere provide the basics for life. Walter and Breckle [57] introduced the term "geobiosphere" to hint at the close linkage between the biotic realm and its spatial, abiotic, and ecologic context. So, the overlap is an intrinsic quality of these spatial volumes. Ecological processes that are regulated by organisms are an integral part of the ECZ. Human existence, however, is dependent on these spheres. The term biosphere and the human influence on it is discussed since the geo-ecological studies by Vernadsky (in English translation published in 1945).

It seems quite practical to identify and investigate many organisms such as the population of a tree species. Nevertheless, the detection of individual specimens can be limited in cases of clonal organisms with subterranean organs. Characterizing biotic units at higher levels of organization becomes more of a challenge, even if these are ecologically more relevant for the ECZ. The increasing organizational complexity with the respective levels and scales is resulting in less distinct units or types. Gradual transitions are predominating at large scales.

The organismic contribution to the ECZ can be first subdivided into living and dead organic material and additionally into solid matter, solute compounds, such as Dissolved Organic Carbon (DOC), and gases, such as methane and carbon dioxide. Biologically inactive remnants of organisms, however, are the habitat of a highly diverse and important cascade of decomposition resulting in the cycling of carbon and nutrients. Even if there is no true equilibrium in ecological systems, the turnover of organic matter is an interplay between the growth and formation of living biomass of all kinds, and its decay and finally mineralization. Similar to how metabolism is the result of anabolism and katabolism, the functioning of ecosystems is a constant cycle of creation and destruction of life, producing complex organic molecules.

These cycles include many different components and mechanisms at different temporal and spatial scales and cannot be quantified and measured completely. We have good estimates about Above-ground Net Primary Production (ANPP), but there are fundamental restrictions in assessing the role of below-ground production because

a
b

◂**Fig. 3.3** Contrasting ecosystems on parent material of very different age. **a)** Highly diverse mixed boreal forest based on a large continental species pool. This vegetation is adapted to intense weathering and leaching of nutrients during Holocene soil formation on the Laurentian shield (Proterozoic Period) in landscapes shaped by glacial processes during the Pleistocene. **b)** Remnant species-poor vegetation on volcanic ashes half a year after the Cumbre Vieja volcanic eruption in autumn 2021 dominated by endemic plant species based on a small species pool of an oceanic island. This vegetation is adapted to frequent natural hazards and initial soil formation with high mineral nutrient availability but limited nitrogen supply (Location: La Palma, April 2022). (photos C. Beierkuhnlein)

of limited accessibility and the fact that investigations are mostly destructive. Our understanding of the behavior of carbon dioxide and methane in the atmosphere has improved in the face of climate change, but the complex processes of surface-near mixing and circulation are difficult to assess at large scales.

Another fundamental aspect of biotic components in the ECZ is the fact that they cannot be seen in isolation. The Biosphere 2 facility in the desert of Arizona was an eye-opening experiment, illustrating unintentionally the fact that no ecosystem can be isolated [16]. To understand the interactions between the Biosphere and the Earth's Critical Zone, it is simpler to divide the components affecting the Earth's Critical Zone into vertical layers. The topmost layer begins with the surface organisms and birds, with vegetation as the next layer; followed by the rhizosphere layer, which is concentrated in the upper 20–30 cm of the topsoil, and finally the subsurface layer [39]. This zonation considers differences in the transport vectors of the biotic components.

3 Biomes

The most integral organization of life on Earth emerges in global biomes (ecozones). These ecological units of continental extent are characterized by a mosaic of characteristic ecosystems and landscapes. Structural differences in biological complexity are apparent as well in the gradient from three-dimensional multilayered forest canopies to two-dimensional vegetation in the steppe (prairie) and the arctic biomes (tundra), and from complete vegetation cover in all forested biomes to sparse or lacking vegetation in desert landscapes [10].

In classic approaches (e.g., [58]) the climatic conditions of biomes are often characterized by average annual temperature and precipitation. However, the duration of biotic activity ranges from a few weeks in the polar zones to the entire year. Additionally, seasonality performs gradients in terms of temperature and precipitation, translating into specific cycles of nutrient and water uptake, and rhythms of carbon sequestration. The biological heartbeat of ecosystems in different biomes follows the abiotic framework of fluctuations in energy and moisture. The conditions and processes of the sub-surface ECZ are regulated by these rhythms.

On a global basis, large-scale spatio-ecological units serve for orientation about the natural background of ecosystems. However, from various disciplinary perspectives

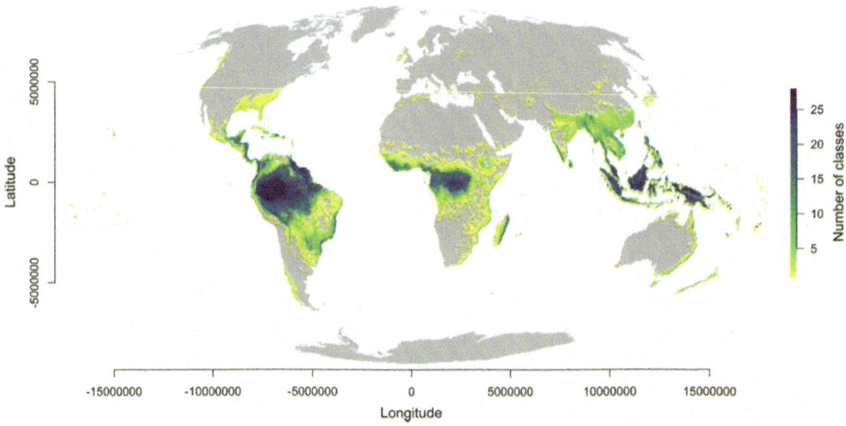

Fig. 3.4 Global pattern of tropical rain forest biome clusters identified through AI/ML approaches based on 31 published biome and ecozone classifications. The color scheme indicates the degree of overlap and agreement (dark blue signifies high agreement). Interestingly, no pixel showed complete agreement of all 31 considered biome classification [19]

(climatology, vegetation science, landscape ecology, physical geography), a series of global approaches have been developed, based on expert opinion, big data, modelling, and remote sensing resulting in different numbers of units. Even if certain regions are uniformly categorized (e.g., hot deserts lacking vegetation cover, per-humid tropical rain forest), a specific disciplinary background and methodology is resulting in deviating classifications or types across a range of approaches [19] (Fig. 3.4).

Besides the 2-dimensional organization of biomes, ecoregions, and ecozones [3, 21, 51, 59], the third dimension, topography, mountains, and bathymetry, needs to be considered. In the terrestrial realm elevational gradients in high mountain regions result in a close neighborhood of ecosystems that would span large latitudinal gradients (Fig. 3.5). From a global perspective, this fine-grained three-dimensional organization cannot be reflected. Therefore, large high mountain regions are categorized as orobiomes [56] and often shaded out in global biome maps.

The moisture regime of biomes is an important driver of biological processes in addition to the temperature regime. Biomes and climate zones (e.g., referred to in IPCC reports) are closely linked. This is reflected very directly in vegetation structures and bacterial biomass, diversity, and activity [15]. Also, the accumulation of carbon in different strata of the ECZ is predominantly controlled by moisture [49] (Fig. 3.6).

Characteristic types of vegetation and ecosystems are considered main traits of biomes and are known to be closely correlated to climate [61]. Hitherto, however, all maps on the organization of life on Earth in terms of biomes and ecozones were focused on above-ground emerging biota (e.g., [18, 44, 56, 58]). This does not mean that below-ground life is considered to be less important. It is well known that in the polar and boreal biome the vegetation, which is dominated by members of the

◂**Fig. 3.5** Close neighborhood of substantially different ecosystems along an elevational gradient at Mount Lemmon close to Tuscon, Arizona, representing biomes that are distributed over hundreds of kilometers in latitudinal gradients. **a)** Succulent scrub with Saguaro cactus (*Carnegia gigantea*) in low elevation (1000 m asl); **b)** Mediterranean deciduous forest with oak trees (*Quercus emoryi*) in mid-elevation (1400 m asl); **c)** Boreal conifer forest with pine, spruce, and fir species (*Pinus ponderosa, Picea engelmannii, Abies lasiocarpa, Abies concolor*) at high elevation (2500 m asl) on the same mountain slope. The interaction of climate, bedrock, slope, and vegetation is reflected in highly diverse soil conditions within a horizontal distance of less than 10 km. (photos C. Beierkuhnlein)

Fig. 3.5 (continued)

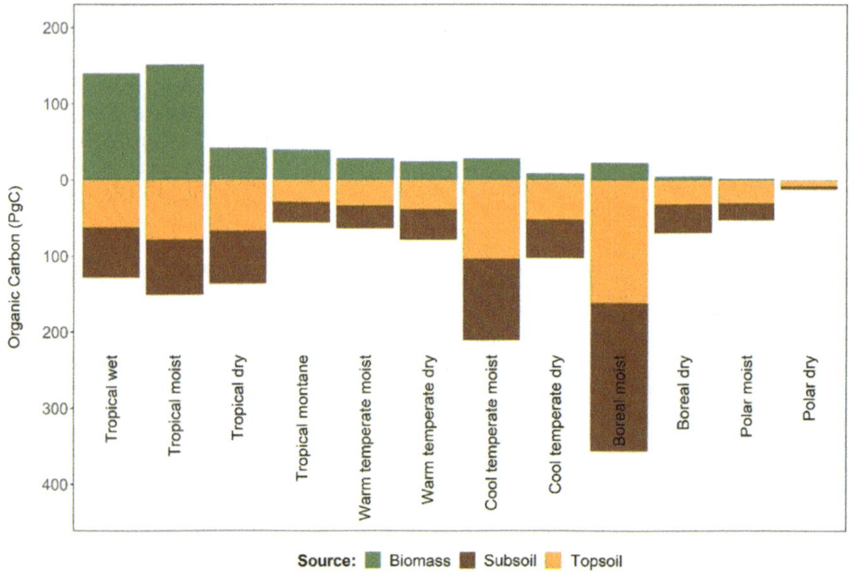

Fig. 3.6 Distribution of organic carbon stocks across IPCC climatic regions subdivided in sub-soil, top-soil and above (following [29]) and belowground biomass [48]. Compilation is based on data from [49]. Data: New IPCC Tier-1 Global Biomass Carbon Map for the Year 2000 (lbl.gov)

Ericaceae and Pinaceae family, is directly dependent on mycorrhiza, just to give one example. But there is still a considerable lack of knowledge and missing spatial coverage of data on the belowground functioning of ecosystems.

Even though humans have already modified biomes since millennia, the anthropogenic impact has increased in recent decades tremendously [42]. As a result, most spatial biome products must be understood as the current ecological potential of the terrestrial Earth surface, and not as a reflection of the current state. And in consequence, this potential spatial organization will be modified considerably, if the climatic conditions are changing [26].

4 Biodiversity and Ecosystem Functioning

Organisms contribute substantially to the processes, mechanisms and functioning of the ECZ. This contribution differs quantitatively and qualitatively between ecosystems and biomes. In many cases the major contribution to biogeochemical cycles comes from key species or ecosystem engineers, who are regulating fluxes of matter or energy (Fig. 3.7).

Ecological systems are composed of different species whose co-existence can be seen as a reflection of the fulfilment of different functional roles (ecological niches) without which they would outcompete each other. If this applies, then the loss of

Fig. 3.7 Ecosystem engineers are species that contribute substantially to key processes in ecosystems. Examples are **a)** beavers with their landscape-modifying contribution to water cycles and flooding regimes (Quebec, Canada), and **b)** termites with their crucial contribution to the carbon cycle in savanna landscapes (Coral Bay, North-Western Australia). (photos C. Beierkuhnlein)

species would be coupled with the loss of a functional contribution. Thus, it has been postulated that the loss of species, or biodiversity, translates into a decline of ecosystem functioning, including functions that are relevant for humankind, known as ecosystem services. Some species groups, such as plants, are well recorded in large parts of the world (e.g., [5]). However, only 20% of the assumed species diversity of the planet is known to science with large uncertainties. This uncertainty largely applies to soil biodiversity, where the simplistic understanding to determine, list, and count species at the organismal level is limited. Many soil microorganisms cannot be determined to the species level since they cannot be isolated and cultivated outside of their natural environment. Then, we just get a genetic signal as a proxy for soil biodiversity.

The preservation of biodiversity has been an established international policy goal since the early 1990s with the advent of the UN Conference on Environment and Development (UNCED), the "Earth Summit" in Rio de Janeiro, and the ratification of the Convention on Biological Diversity (CBD). Yet, the term "biodiversity" is still too loosely defined, and often referred to as "life on Earth". Biodiversity means more than just a number of species. Wilson [59] emphasized that "it is important to recognize that biological diversity is about the variety of living organisms at all levels—from genetic variants belonging to the same species, through arrays of species, families and genera, and through population, community, habitat and even ecosystem levels." This is very close to the UNCED definition: "'Biological Diversity' means the variability among living organisms from all sources, including, inter alia, terrestrial, marine and other aquatic ecosystems and the ecological complexes of which they are part, this includes diversity within species, between species and of ecosystems."

This definition addresses important aspects of biodiversity such as type of organism (including Eukaryots, Archaea, Bacteria), different levels of organization (organisms, communities, ecosystems), dissimilarity between biotic objects (genetic, morphological, functional), number of classified types (species, functional types,

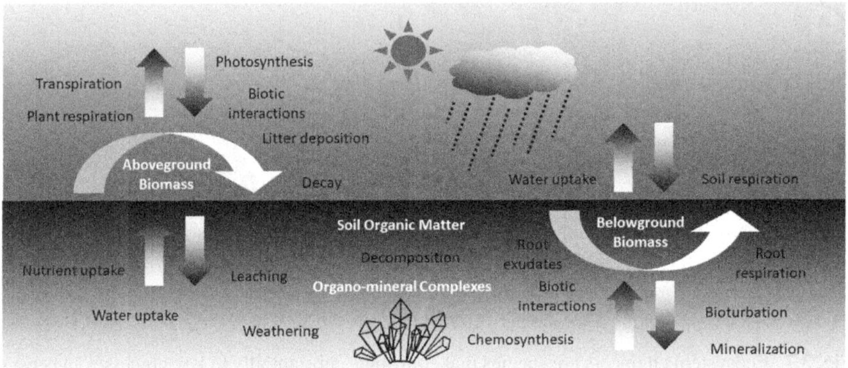

Fig. 3.8 Selected processes in the ECZ interface that are controlled by organisms. The ECZ biological cycle includes major pathways of matter and energy regulation. Soil organic matter as well as aboveground and belowground biomass are highlighted as three main compartments of the ECZ. Abiotic frame conditions such as solar energy, precipitation, and parent material are illustrated by symbols. The contribution of the broad spectrum of organisms including fungi, bacteria, insects etc. is indicated by "biotic interactions"

life forms), and the interactive complexity between biotic objects. We can conclude: "Biodiversity characterizes qualitative, quantitative and functional aspects of biotic units at various levels of organization in a concrete or abstract context, and at a given temporal or/and spatial scale."

With higher levels of organization, the challenge of segregating biotic units, objects, and types becomes increasingly prominent. Interactions, such as the exchange of genetic information, energy flow, and fluxes of nutrients, increase. Structural boundaries are diffuse and transitions between ecosystems are a common phenomenon. Gaston and Spicer [23] note: "whilst it is helpful … to label different habitat types, it is not always obvious precisely where one ends and another begins." This is especially important in natural systems with low human impact where ecotones between systems are common.

Various aspects of biodiversity control the processes within the ECZ (Fig. 3.8). The richness of species is simply a basic indicator of the diversity of processes, which support the stability, resilience, and insurance of functioning under fluctuating environmental conditions [31, 63].

5 Biogeographical Scales

Fundamental aspects of the variability of the biotic components contributing to the ECZ can be related to space and time. Grain and resolution of information (and knowledge) is dependent on the spatial and temporal extent of the subject. Natural units have different sizes and lifetimes (life cycles, turnover) depending on the degree of integration they are covering. In consequence, to compare them and to scrutinize their variability, the size of an investigation plot or volume and the duration of investigation must be adequate. But what is adequate?

Implicitly visual homogeneity concerning certain — in most cases physiognomic — criteria, becomes the decisive factor for defining units. Even more difficult is the evaluation of time scales. We know almost nothing about lifetimes of natural units of higher levels (communities, ecosystems) although we notice that they are changing. Astonishingly perhaps, this is also true for species in natural systems. What age does *Gentiana germanica* normally attain? How long does it take for the microbiome of fungi and bacteria to support plant nutrient uptake on young pyroclastic volcanic ashes? And, how long does one have to investigate a chalk grassland to find out whether the occurring species are a vital constant part of the system and not only aging without efficient reproduction?

To a large portion, the questions of science that are worked on intensively depend on the technical facilities for their analysis. Due to the restrictions of representing multiple dimensions in graphs, one-dimensional analyses (gradient analysis, transects) and two-dimensional graphs (as maps) prevail. Natural systems, biotic units, are organized in three dimensions in space (plus time). For highly integrative, complicated systems, two dimensional maps are not appropriate. Innovative computer techniques can offer new methods to visualize three-dimensional variability in space, their spatial biodiversity, and dynamics.

The scale of investigation periods is given less thought. Here too, the temporal resolution is strongly connected to the objects that are analyzed. A short observation period might not include all the subunits and their properties, whereas making the sampling too long might lead to a loss of distinction between the objects in time.

There are obvious natural rhythms that influence the performance of biotic units including short term responses to meteorological events, diurnal and seasonal rhythms, and long-term trends over several years. They are all relevant to biodiversity research because these trends and fluctuations control the presence or absence of certain units (e.g., species) or their quantitative or qualitative properties. However, biotic units of different levels react with different degrees of resilience to temporal dynamics.

Historical periods or even eras on the geologic scale are important for comparison, for the evaluation of evolutionary processes and base lines, but are not of direct relevance for actual problems related to biodiversity. In any case, we are not able to collect comparable and really complete data from former times. Even very powerful approaches such as pollen analysis only informs about fractions of species compositions (anemophilous plants) for singular units of landscapes (lakes, raised bogs).

6 Increasing Knowledge Through Experiments, Modelling, and Monitoring

The diversity, heterogeneity, and complexity of the ECZ requires multi-disciplinary access to its patterns, processes, and functioning. Individual case studies approached from a disciplinary perspective deliver important pieces to this mosaic, but they need to be understood as a contribution to a coherent framework. In consequence, the toolkit for ECZ studies must include a spectrum not only of disciplines, but also of methodologies. Observational, experimental, and modelling studies on the ECZ should be designed in a way that can relate to each other and enable temporal and spatial extrapolation and generalization (Fig. 3.9).

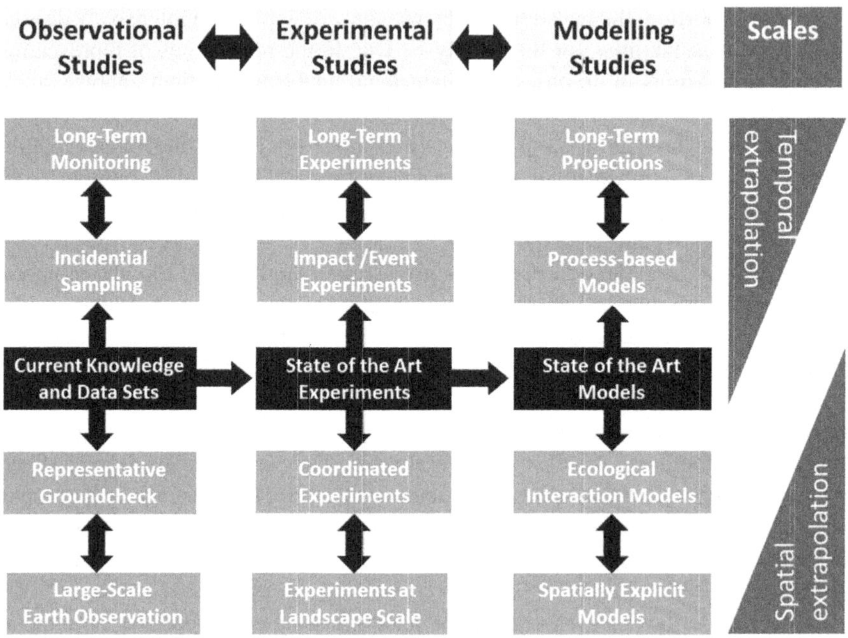

Fig. 3.9 Conceptual framework for ECZ research allowing for large-scale and long-term conclusions for ecological responses. As case studies are limited in area, duration and in ecological complexity, it is mandatory to link studies and to connect perspectives at spatial and temporal scales

7 Coordinated Experiments

Experimental studies enable testing hypotheses and identifying drivers and processes in the ECZ. However, single site approaches can also be just singularities. Generality cannot be concluded from local findings. Therefore, spatially distributed experiments are needed to understand which processes are just local or of a wider relevance.

Fraser et al. [21] made a claim for Coordinated Distributed Experiments (CDEs) to overcome local singularities that are restricting the significance and explanatory power of single-site experiments. This approach may result in a trade-off towards less elaborate and complex experiments, but it can possibly show whether experiments will yield comparable responses in different places and ecosystems. As it is already a challenge to convince partners to participate in coordinated action, where the trade-off in research output requires an equal share and the integration of many colleagues, it is likely that such kind of approaches are implemented mainly at sites where high performing research groups are established.

The geographic coverage across large spatial scales is often a clear ambition in CDEs, but it is not the major constraint or rule. Gradients, transition zones and ecotones can be used to install coordinated experiments. Besides the spatial distribution of experimental sites, the spatial provenance of specimen, seeds, or populations can also matter [8]. Common Garden experiments are experimental approaches that focus on the comparison of different genotypes (or species) under the same conditions that are given at a certain location.

Here, biogeography is important and the selection of locations which are considered should follow defined protocol and criteria, but practicability decides in most cases. Often, the availability of seed or plant material is limited, and it is rarely possible to investigate those populations that would be most appropriate to test a hypothesis. More likely, it is the mere accessibility and availability of biotic material. In consequence specific sampling of the most appropriately distributed populations should be done, but the workload and risk (germination rates are unclear) of such approaches guides against it.

The search for generality in species and ecosystem responses to the major impacts and distortions such as climate change, land use change, or biodiversity loss requires intense research which goes beyond single stands and sites. However, there are principal restrictions to field research and monitoring that have shifted the focus in ecology away from field research to experimental approaches. The quality of field research, to be close to reality, is seen as being constrained by its low heuristic value and the inability to separate coincidence from cause when independent controls are missing. However, the capability to really test hypotheses via experiments is also seen as being constrained by their artificiality and singularity because ambitious experiments are almost necessarily limited to a single site (e.g., Jena Experiment, Cedar Creek, EVENT).

Reviews, data mining and meta-analyses are aiming to put scattered knowledge together (e.g., [31]). Nevertheless, their limits in identifying general principles and patterns are obvious and mostly related to non-standardized sampling protocols. In consequence, there is an emerging need to combine both perspectives. The disadvantages and limitations of both approaches cannot be neglected, but a proper concept that neither ignores biogeographical gradients nor the requirements of sound experimental design show a great promise to tackle uncertainty and novel challenges.

CDEs are aiming at a better understanding of large-scale biogeographical patterns and adjusting biogeographical and experimental concepts. Biogeographically Designed Experiments (BDEs) are a special type of Coordinated Distributed Experiments (CDEs) considering the coverage of biogeographically defined areas for the placing of experimental sites. However, such conceptual approaches are still very rare. Relevant spatial and temporal scales of the experimentally manipulated system and its processes need to be understood a priori, which is often ignored [6]. However, the three-dimensional heterogeneity of the ECZ requires an in-depth consideration of corresponding organization.

8 Biogeographical Modelling

Biogeographical models are, by definition, a simplification of natural mechanisms. Simplification with a focus on key traits and processes is helpful to understand complex systems. However, there is a broad technical spectrum of modelling approaches ranging from qualitative to quantitative and from process-based to correlative models. The aim in all cases is to explore the interplay between the biogeographical patterns (e.g., in biomass, vegetation structure, carbon storage, species richness) and the processes that are shaping them.

Simple models are a way to mathematically formulate our knowledge. However, based on selected processes they can show whether expected system behavior is likely to be realistic or not. Hence, models can be seen as explorative tools. If simplifying biotic processes does yield reasonable results, then the considered processes are likely to be important. Limitations in process-based models, however, are increasing with the biodiversity and complexity of the system under consideration.

Limitations of correlative modelling approaches such as SDMs (Species Distribution Models) and ENMs (Environmental Niche Models) are in data availability. Even with an increasing global accumulation and availability of biotic data (e.g., GBIF, TRY, Catalogue of Life) accomplished by regional and specific data bases, Big Data are mostly restricted to certain groups of organisms such as plants or birds. Species interactions and traits are increasingly integrated into correlative models (e.g., Joint Species Distribution Models) as well as dispersal and mobility distances [33], but still there is a gap to bridge between mechanistic and correlative modelling approaches.

It is a logical and intrinsic restriction that models are based on existing knowledge and evolve due to the progress in algorithms and technology. The development of knowledge and technology happens in parallel to biotic and abiotic changes in the environment and the respective repercussions or lagged responses in ecosystems. Models on future ecosystem response must consider changes in the dominance patterns of key organisms as well as in biotic interactions and inform about the consequences in the performance of ecosystem functioning. Therefore, a functional perspective rather than a taxonomic approach is required. Species identities can only indirectly inform about their functional traits. Even if the data in this field are growing rapidly [37], there are still enormous knowledge gaps. Nevertheless, models can help to better target specific mechanisms in experiments, and they can be compared with evidence from observation and monitoring.

9 Biogeographical Monitoring

The recording and analysis of spatial biogeographical data provides the necessary matrix for nested designs and upscaling of case studies of the ECZ. Biogeographical monitoring of species, diversity, ecosystems, and biomes enables us to link biotic processes with the biogeochemical fluxes at the Earth surface. However, monitoring the entire pool of organisms and correlated organismic processes within the ECZ is almost impossible, even at a selected location.

Increasingly, recorded in-situ data are made available in large data bases such as GBIF, but there are different kinds of bias and complete quality control is not feasible. The Group on Earth Observation (GEO) is making geoinformation, including biogeographic data, openly available. GEO BON is sharing data with precise spatial geographical locations and compiling indicators of biodiversity such as Essential Biodiversity Variables (EBVs.). However, the knowledge basis related to the ECZ is still scarce when it comes to remote areas or rare species.

The heterogeneity and scaling issues of the ECZ call for spatially and temporally nested approaches in observational monitoring. Replicability and representativity are key. Remote sensing offers fascinating perspectives but is only providing a two-dimensional projection which must be linked with point data from the field that explores the vertical structure of the ECZ and with the temporal development. Earth Observation (EO) integrates both in-situ observational monitoring and remote sensing.

A series of modern sensors for environmental monitoring are open access and can be used for biodiversity monitoring [2] (Fig. 3.10). The type of use is dependent on the spectral and spatial resolution. The technological progress is enormous but the use of such data in ecology is still scarce. Engineers rarely have a solid ecological background, and ecologists rarely have the technological skills to efficiently exploit the growing amount of data. Interface tools are needed as provided for instance in the "Protected Areas from Space" toolkit of the EU ECOPOTENTIAL project.

The "Spectral Species Concept" [47] is a pragmatic approach for supporting biodiversity monitoring under the auspice of the new generation of spaceborne and airborne sensors. Spectral diversity in signals is linked with the diversity in functional traits, species assemblages, and ecosystems. However, the conclusion on the ECZ is

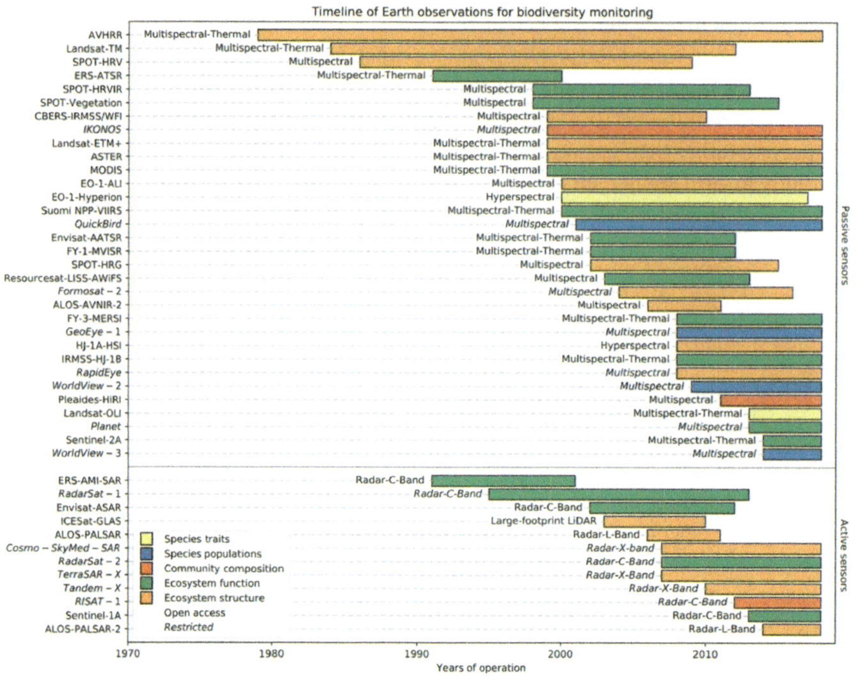

Fig. 3.10 Temporal extents of current and historic satellite sensors, colored by biodiversity pattern type. Temporally coincident measurements can be leveraged for analysis in a multi-sensor fusion framework to increase data dimensionality [2]

still an indirect one. High temporal and spatial resolution of ecosystem responses can be linked with soil moisture, nutrients, and biotic activity in target-oriented ground truthing.

Long-Term Ecological Research (LTER) sites enable the detection of processes that cannot be covered within the short time frame of most research projects. This approach was first prominent in the United States and funded by the National Science Foundation. The aim was to establish a network of sites that collect ecological data over long periods of time across major biotic regions in North America [13].

In recent years, the National Ecological Observatory Network (NEON) has added new dimensions to long-term ecological research. NEON, which has 81 freshwater and terrestrial sites across the United States, is a network which fosters understanding of ecology and ecosystem fluxes by expanding monitoring of the sites through sensors and experiments that can provide data of greater temporal and spatial scales [41]. There has been a global development of such networks across regional and global scales. The model developed by LTERs has been expanded to form a global network also called International LTER (ILTER) [54].

10 Uncertainty and Lack of Information

It is unlikely that the complex structure of the ECZ and the diversity of all its components can be completely measured and understood. Substantial knowledge gaps, for instance in terms of soil microorganisms, will persist. Many soil microorganisms are just known as genetic signals. Knowledge gaps on interactions between plants and fungi or bacteria are pervasive. There is missing information on belowground biotic organization in large parts of the world [32]. We have seen that it is difficult to define spatial limits of the ECZ. However, even if these were postulated, the uncertainty about the vertical extent of the ECZ is profound (Fig. 3.11).

Knowledge gaps translate into uncertainty, which limits the quality of assessments on the current and future functioning of the ECZ. Such knowledge gaps apply to spatial information and mapping of biota but are often ignored [46]. The temporal resolution of species records is often categorized to sections of 50 years and the spatial resolution is often very coarse. Changes that are going on, such as the decline

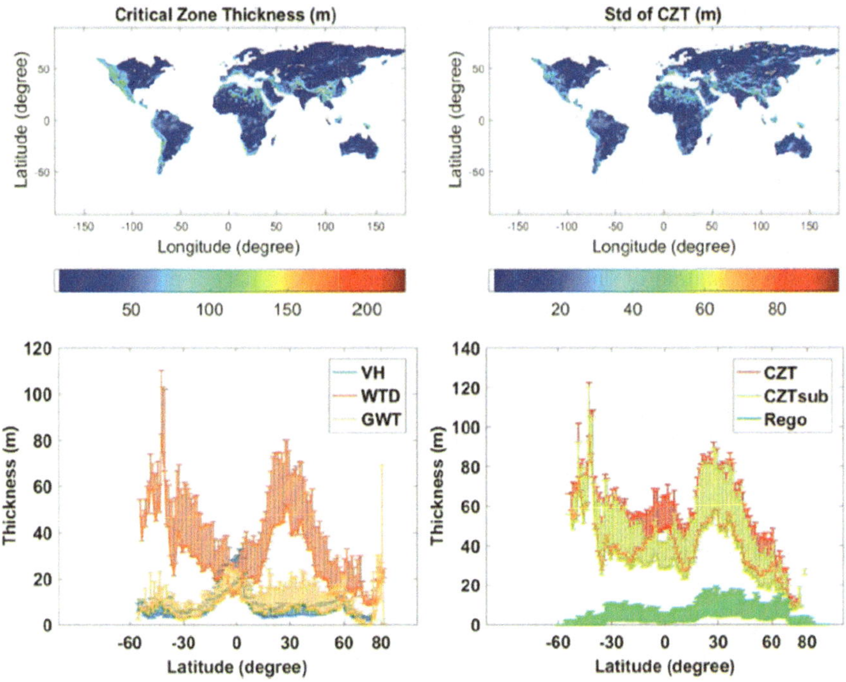

Fig. 3.11 Spatial distribution of critical zone (CZ) thickness and its uncertainty. VH, vegetation height; WTD, water table depth; GWT, groundwater thickness; CZTsub, the underground part of CZ (WTD + GWT); CZT, critical zone thickness (VH + WTD + GWT); Rego, regolith thickness; and SD, standard deviation. Note that the latitude profile of different components is presented with mean ± one standard deviation (the error bar) [62]

of flowering plants and pollinators, cannot be seen in such data structures. Biodiversity loss may take place but does not become visible in data as monitoring is primarily focused on recording occurrences but not on precisely recording the loss of individual populations. There are many black boxes of biotic uncertainties. They relate to changes in species assemblages, biotic interactions, local extinctions and loss of functions, pathogens, and invasive species.

The projection of future climatic conditions is mostly dependent on the decision for a certain scenario (RCP) and thus a consequence of political decisions and of societal and technological development. In addition to uncertainty about climatic drivers, there are uncertainties in climate models. However, it is likely that uncertainties in ecosystem models are even larger. Global Vegetation Models (GVMs) can be used to identify if the choice of a specific Global Circulation Model (GCM) creates singularities in the response of vegetation.

Climate change affects the functioning of the ECZ in many ways. Long-term trends in temperature and precipitation regimes, shifted seasonality patterns, and increasingly extreme conditions of different duration are impacting the biogeochemical cycles. The growing mismatch between previously developed adaptations and selected biota and a changing environment relates to the speed of change and the capacity to respond to these changes. Furthermore, the term "extreme" can cause misunderstanding if it is just defined by probability of occurrence. Hitherto "extreme" (rare) or unexperienced climatic conditions will, by re-occurrence, become less statistically extreme, but more ecologically extreme if more frequent formerly unexperienced events can no longer be compensated in ecological systems. This opens a large research arena for ecosystem research and for the ECZ [34].

Extreme events of pulses may drive the systems towards novel states and support even the formation of "novel ecosystems" that may not have a precursor in Earth history [30]. However, only parts of the system are likely to respond to these changes, others may just tolerate such an impact or exhibit inertia with survival strategies that would enable them to become active or to reproduce again when the impact has come to an end. Former ecosystem performance does not permit us to conclude on the future performance under these conditions.

11 Perspectives

The biogeographical perspective of the ECZ informs us about global patterns in ecological processes that are governing the structure and functioning of this sensitive space. The formation of soils is a main component in the functioning of the ECZ, but this formation is driven by a combination of abiotic and biotic processes. The ECZ is characterized by the overlap of the pedosphere, the hydrosphere, the lithosphere, the atmosphere, and the biosphere. This overlap translates into the necessity to integrate a broad spectrum of scientific disciplines into ECZ research.

Even if ECZ research is strongly related to ecological research, the focus in ecology is more on the biotic and in the ECZ more on the abiotic compartments and mechanisms. However, there is no way to separate these two perspectives. There has been an acknowledgement, over the past decade, of this type of interactions which couple the biosphere with other abiotic components [23, 46]. In this chapter, the biotic and biogeographical view is elaborated, but this would be incomplete without considering abiotic conditions for life.

Ecology is subject to rapid paradigm shifts. The historic focus on field research is being increasingly overtaken by modelling and experiments. Data mining, ecoinformatics, and meta-analyses are attracting the attention of big audiences and uncovering a larger picture (e.g., [11]). Technological progress in Earth observation is providing new insights towards a better understanding of large-scale spatial patterns. This is nothing less than the detection of spatial patterns of the ECZ.

The advantages of such new perspectives are evident. Experiments allow hypothesis testing, which is hardly possible in nature where a multitude of factors coincide and affect vegetation responses to the variables under consideration [9]. New research fields are opened and adequately addressed with sophisticated experimental approaches creating a legacy of subsequent detail studies and leading to new insights (e.g., [35]).

The spatial validity of ecology has been extended to the continental and global scale in the era of big data [28]. The findings from single sites can be tested across biomes bringing academic debates with previously irreconcilable views closer to consent (e.g., [22]).

With ecological models, the effects of processes can be detected and, based on existing data species distribution, models can project future trends and patterns [53]. Limitations and shortcomings of these approaches are intensely addressed, and methods are increasingly improved.

However, in the face of these powerful advances, there is a risk of neglecting research that is focused on the local plot scale and on real ecosystems. Increasingly, expertise in field research is getting scarce. Yet more importantly, specialization can hamper the integration of various techniques in vegetation studies. We make a plea for explicitly addressing this area of methodological tension. Qualities and restrictions of different approaches should be integrated in a conceptual way to reach a clearer and less biased understanding.

Providing a better picture of the ECZ requires the detection of the temporal dynamics of three-dimensional conditions at the Earth surface, and the assessment and investigation of cross-sectional vertical gradients from the overlapping near-surface atmosphere, through biosphere, and pedosphere to the lithosphere, while processes in the hydrosphere are linking all these sections of the ECZ. Intelligent monitoring concepts should integrate remote sensing, be flexible in the face of responses to short-term events, provide a high spatial resolution of information, and aim for repeatability.

Research on the future development of the ECZ needs to consider different qualities of uncertainty. There is uncertainty in our current state of knowledge, data, and approaches about the functioning of the ECZ. Much of this uncertainty refers to the biological aspects of the ECZ as we have limited knowledge of the soil microorganisms that are contributing substantially to the functioning of the ECZ.

Ecosystem services ("nature's contribution to people") are an important motivation for a better understanding of the ECZ and to communicate the societal importance of ecosystems [17]. Society is dependent on and, at the same time, negatively influencing the functioning of the ECZ. These interactions need to be considered by integrating the human perspective into studies on the ECZ. The ecological integrity of this zone and the threats that are leading to its degradation are of vital importance for humankind. Therefore, the approach of eLTER and NEON, networks of long-term socio-ecological research sites, is timely. It is not sufficient to focus only on the spectrum and diversity in the functioning of ecosystems. Various aspects of human activity, dependence, impacts and threats to the ECZ must be included into ECZ research.

Integrating the broad spectrum of complexity that applies to the ECZ requires a unified theoretical framework. The ECZ and its ecological systems are governed by processes and cannot just be condensed to mere spatial entities, cross-sections, or volumes. It can be questioned if current approaches and concepts are adequate or just too sectorial. Old paradigms may have to be overcome to make the next big leap in ECZ research. Multi-scale studies should involve a spectrum of mechanisms and methodologies. This necessitates multi-disciplinarity, multi-scaling, multi-methodology, and multi-factoriality.

12 Conclusion

The biogeographical context of ECZ research is dependent on the research question. In one case, structural units such as biomes (e.g., savanna, boreal forest) that are defined by predominant plant growth forms will be appropriate. For other questions, the phylogenetic relatedness of the flora or the bio-climatically driven ecosystems will be more adequate. Finally, the context of biomes at continental to global scales needs to be kept in mind.

As research questions are becoming increasingly globalized, this must be reflected in ECZ research as well. However, with an increasing extent of study area, grain size and spatial representation of vegetation, regional heterogeneity is likely to decrease. Nested spatial design can help to tackle this challenge.

Clearly, a sequential approach is recommended: analyze the gradient of interest first and decide on experimental site or sampling location in a second step. If the design of data or sample collection is adequate to the required distribution for a specific research question, then subsequent upscaling of results and interpolation between places can be justified.

Such a step-wise procedure may be time-consuming. It may interfere with the common practice of avoiding pre-studies and hierarchical structure during the progress of a study due to time constraints. However, seeing the tremendous amount of workload that has already been invested in previous studies, this seems to be justifiable.

References

1. Akob DM, Küsel K (2011) Where microorganisms meet rocks in the Earth's critical zone. Biogeosciences 8(12):3531–3543. https://doi.org/10.5194/bg-8-3531-2011
2. Anderson CB (2018) Biodiversity monitoring earth observations and the ecology of scale. Ecol Lett 21(10):1572–1585
3. Bailey RG (1998) Ecoregions: the ecosystem geography of oceans and continents. Springer, New York
4. Banwart SA, Nikolaidis NP, Zhu YG, Peacock CL, Sparks DL (2019) Soil functions: connecting Earth's critical zone. Annu Rev Earth Planet Sci 47(1):333–359. https://doi.org/10.1146/annurev-earth-063016-020544
5. Barthlott W, Hostert A, Kier G, Küper W, Kreft H, Mutke J, Rafiqpoor MD, Sommer H (2007) Geographic patterns of vascular plant diversity at continental to global scales. Erdkunde 61:305–315
6. Beier C, Beierkuhnlein C, Wohlgemuth T, Penuelas J, Emmett B, Körner C, Hde B, Hesselbjerg CJ, Leuzinger S, Janssens IA, Hansen K (2012) Precipitation manipulation experiments – challenges and recommendations for the future. Ecol Lett 15:899–911
7. Beierkuhnlein C (2007) Biogeographie – Die räumliche Organisation des Lebens in einer sich verändernden Welt. Eugen Ulmer Verlag Stuttgart
8. Beierkuhnlein C, Jentsch A, Thiel D, Willner E, Kreyling J (2011) Ecotypes of European grass. Species respond specifically to warming and extreme drought. J Ecol 99(3):703–713
9. Beierkuhnlein C, Neßhöver C (2006) Biodiversity experiments – artificial constructions or heuristic tools? Progress Botany 67:486–535
10. Beierkuhnlein C, Fischer JC (2021) Global biomes and ecozones – Conceptual and spatial communalities and discrepancies. Erdkunde 75(4):249–270
11. Bekker RM, van der Maarel E, Bruelheide H et al (2007) Long-term datasets: from descriptive to predictive data using ecoinformatics. J Veg Sci 18:458–462
12. Blume-Werry G, Wilson SD, Kreyling J, Milbau A (2016) The hidden season: Growing season is 50% longer below than above ground along an arctic elevation gradient. New Phytol 209(3):978–986
13. Brantley SL, McDowell WH, Dietrich WE, White TS, Kumar P, Anderson SP et al (2017) Designing a network of critical zone observatories to explore the living skin of the terrestrial Earth. Earth Surface Dyn 5(4):841–860. https://doi.org/10.5194/esurf-5-841-2017
14. Chiarucci A, Robinson BH, Bonini I et al (1998) Vegetation of tuscan ultramafic soils in relation to edaphic and physical factors. Folia Geobot 33:113–131.
15. Delgado-Baqzerizo M, Eldrige DJ (2019) Cross-biome drivers of soil bacterial alpha diversity on a worldwide scale. Ecosystems 22:1220–1231. https://doi.org/10.1007/s10021-018-0333-2
16. Dempster WF (2008) Tightly closed ecological systems reveal atmospheric subtleties – experience from Biosphere 2. Adv Space Res 42(12):1951–1956
17. Díaz S, Pascual U, Stenseke M, Martín-López B, Watson RT, Molnár Z et al (2018) Assessing nature's contributions to people. Science 359(6373):270–272. https://doi.org/10.1126/science.aap8826
18. Dinerstein E, Olson D, Joshi A, Vynne C, Burgess ND, Wikramanayake E et al (2017) An ecoregion-based approach to protecting half the terrestrial realm. Bioscience 67(6):534–545. https://doi.org/10.1093/biosci/bix014
19. Fischer JC, Walentowitz A, Beierkuhnlein C (2022) The biome inventory – standardizing global biogeographical land units. Glob Ecol Biogeogr 31:2172–2183. https://doi.org/10.1111/geb.13574
20. FAO - Food and Agriculture Organization of the United Nations. (2015) Global ecological zones for FAO forest reporting: 2010 update. www.fao.org/3/ap861e/ap861e00.pdf
21. Fraser LH, Henry HAL, Carlyle CN, White SR et al (2013) Coordinated distributed experiments: an emerging tool for testing global hypotheses in ecology and environmental science. Front Ecol Environ 11:147–155

22. Fraser LH, Pither J, Jentsch A, Sternberg M et al (2015) Worldwide evidence of a unimodal relationship between productivity and plant species richness. Science 349(6245):302–305
23. Gaston KJ, Spicer JI (1998) Biodiversity: an Introduction. Blackwell Science Ltd. ISBN 1-4051-1857-1
24. Gell-Mann M (1994) The quark and the jaguar: adventures in the simple and the complex. Freeman, New York
25. Goddéris Y, Brantley SL and Blum JD (2013) Earthcasting the future critical zone. Elementa: Sci Anthropocene 1(000019)
26. Guiot J, Cramer W (2016) Climate change: the 2015 Paris agreement thresholds and Mediterranean basin ecosystems. Science 354(6311):465–468
27. He L et al (2020) Global biogeography of fungal and bacterial biomass carbon in topsoil. Soil Biol Biochem 151(108024)
28. Heffernan JB et al (2014) Macrosystems ecology: understanding ecological patterns and processes at continental scales. Front Ecol Environ 12:5–14
29. Hiederer R, Köchy M (2011) Global soil organic carbon estimates and the harmonized world soil database. Publication office of the European Union
30. Hobbs RJ, Higgs E, Harris JA (2009) Novel ecosystems: implications for conservation and restoration. Trends Ecol Evol 24(11):599–605
31. Isbell F et al (2015) Biodiversity increases the resistance of ecosystem productivity to climate extremes. Nature 526:574–577
32. Iversen CM et al (2015) The unseen iceberg: plant roots in arctic tundra. New Phytol 205:34–58
33. Jaeschke A, Bittner T, Reineking B, Beierkuhnlein C (2013) Can they keep up with climate change? - Integrating specific dispersal abilities of protected Odonata in species distribution modelling. Insect Conservat Divers 6:93–103
34. Jentsch A, Beierkuhnlein C (2008) Research frontiers in climate change: effects of extreme meteorological events on ecosystems. CR Geosci 340:621–628
35. Jentsch A, Kreyling J, Beierkuhnlein C (2007) A new generation of climate change experiments: events not trends. Front Ecol Environ 5:365–374
36. Jobbágy EG, Jackson RB (2000) The vertical distribution of soil organic carbon and its relation to climate and vegetation. Ecol Appl 10(2):423–436
37. Kattge J et al (2020) TRY plant trait database - enhanced coverage and open access. Glob Change Biol 26:119–188
38. Kirschbaum MUF (1995) The temperature dependence of soil organic matter decomposition and the effect of global warming on soil organic C storage. Soil Biol Biochem 27(6):753–760. https://doi.org/10.1016/0038-0717(94)00242-S
39. Küsel K, Totsche KU, Trumbore SE et al (2016) How deep can surface signals be traced in the critical zone? Merging biodiversity with biogeochemistry research in a central German Muschelkalk landscape. Front Earth Sci 4(32). https://doi.org/10.3389/feart.2016.00032
40. Kurz W, Dymond C, Stinson G et al (2008) Mountain pine beetle and forest carbon feedback to climate change. Nature 452:987–990. https://doi.org/10.1038/nature06777
41. Loescher HW, Kelly EF, Lea R, Chabbi A (2017) National ecological observatory network: beginnings programmatic and scientific challenges and ecological forecasting. In: Terrestrial ecosystem research infrastructures: Challenges new developments and perspectives. CRS Press, Boca Raton
42. Moncrieff GR, Bond WJ, Higgins SI (2016) Revising the biome concept for understanding and predicting global change impacts. J Biogeogr 43:863–873
43. Smith LA, Stern N (2011) Uncertainty in science and its role in climate policy. Phil Trans Soc 369:1–24. https://doi.org/10.1098/rsta.2011.0149
44. Olson DM, Dinerstein E, Wikramanayake ED, Burgess ND et al (2001) Terrestrial ecoregions of the world: a new map of life on Earth. Bioscience 51(11):933–938
45. Rasmussen C, Pelletier JD, Troch PA, Swetnam TL, Chorover J (2015) Quantifying topographic and vegetation effects on the transfer of energy and mass to the critical zone. Vadose Zone J 14(11):1–16

46. Rocchini D, Hortal J, Lengyel S, Lobo JM, Jimenez-Valverde A, Ricotta C et al (2011) Accounting for uncertainty when mapping species distributions: the need for maps of ignorance. Prog Phys Geogr 35(2):211–226
47. Rocchini D, Santos MJ, Ustin SL, Feret JB, Asner GP, Beierkuhnlein C et al (2022) The spectral species concept in living color. J Geophys Res Biogeosci 127:e2022JG007026
48. Ruesch A, Gibbs H (2008) New IPCC Tier-1 global biomass carbon map for the year 2000. Carbon Dioxide Information Analysis Center, Oak Ridge National Laboratory, Tennessee. http://cdiac.ess-dive.lbl.gov
49. Scharlemann JP, Tanner EVJ, Hiederer R, Kapos V (2014) Global soil carbon: understanding and managing the largest terrestrial carbon pool. Carbon Manag 5(1):81–91. https://doi.org/10.4155/cmt.13.77
50. Schultz J (1995) The ecozones of the world: the ecological divisions of the geosphere. Springer, Berlin
51. Spawn-Lee S, Sullivan C, Lark T, Gibbs H (2020) Harmonized global maps of above and belowground biomass carbon density in the year 2010. Sci Data 7(112). https://doi.org/10.1038/s41597-020-0444-4
52. Spencer CJ, Davies NS, Gernon TM, Wang X, McMahon WJ, Morrell TRI, Lu GM (2022) Composition of continental crust altered by the emergence of land plants. Nat Geosci 1–6
53. Thuiller W, Albert C, Araujo MB et al (2008) Predicting global change impacts on plant species' distributions: future challenges. Perspect Plant Ecol Evolut Systemat 9:137–152
54. Vanderbilt K, Gaiser E (2017) The international long term ecological research network: a platform for collaboration. Ecosphere 8(2):e01697
55. Vernadsky W (1945) The biosphere and the noosphere. Am Sci 33:1–12
56. Walter H (1973) Vegetation of the earth in relation to climate and eco-physical conditions. Springer, New York
57. Walter H, Breckle S (1985) Ecological systems of the geobiosphere. Ecological principles in global perspective 1. Springer, New York
58. Whittaker RH (1975) Communities and ecosystems. 2nd (revised ed) MacMillan Publishing Company Inc
59. Wilson DS (1992) Complex interactions in metacommunities with implications for biodiversity and higher levels of selection. Ecology 73(6):1984–2000
60. Wohlgemuth T, Jentsch A, Seidl R (2022) Disturbance ecology. Springer, Switzerland
61. Woodward I, Lomas MR, Kelly CK (2004) Global climate and the distribution of plant biomes. Philos Trans R Soc: Biol Sci 359:1465–1476
62. Xu X, Liu W (2017) The global distribution of Earth's critical zone and its controlling factors. Geophys Res Lett 44(7):3201–3208. https://doi.org/10.1002/2017gl072760
63. Yachi S, Loreau M (1999) Biodiversity and ecosystem productivity in a fluctuating environment: the insurance hypothesis. Proc Natl Acad Sci USA 96:1463–1468

Surface Water Chemistry as an Integrated Signal of Ecosystem Vectors and Critical Zone Reactors

Adam S. Wymore, Hannah M. Fazekas, Desneiges S. Murray, and Arial J. Shogren

Abstract Lying at the valley of nearly all critical zones is a freshwater ecosystem. Due to precipitation, gravity, and hydrological connectivity, the chemistry of these aquatic habitats provides an integrated signal of geologic, meteorologic, and biologic processes occurring within the landscape. Here we explore the intersection between ecosystem and critical zone science to develop a holistic multidimensional perspective of surface water chemistry. We leverage both foundational concepts from ecosystem ecology and geochemistry and discuss interactions between vectors of material transport throughout the critical zone and the ramifications for spatial and temporal variation in surface water chemistry. We then evaluate methods that can be applied at the watershed, ecosystem, and critical zone scale to characterize these integrated signals. We end by highlighting a series of open questions that are central to understanding the controls of surface water chemistry and water quality in a changing world.

Keywords Water quality · Surface water chemistry · Critical zone · Ecosystem ecology · Feed-through reactor · Watersheds · Concentration-discharge relationships

A. S. Wymore (✉) · H. M. Fazekas · D. S. Murray
Department of Natural Resources and the Environment, University of New Hampshire, Durham, NH, USA
e-mail: adam.wymore@unh.edu

H. M. Fazekas
Biology Department, Saginaw Valley State University, University Center, MI, USA

A. J. Shogren
Department of Biological Sciences, The University of Alabama, Tuscaloosa, AL, USA
e-mail: ashogren@ua.edu

T. White and A. Provenzale (eds.), *Critical Zone and Ecosystem Dynamics*, Advances in Critical Zone Science, https://doi.org/10.1007/978-3-031-69076-1_4

1 Introduction

It has been long recognized that streams are intrinsically linked to the valleys in which they drain. In his seminal Edgardo Baldi Memorial Lecture, HBN Hynes [70] declared "*...in every respect the valley rules the stream. Its rock determines the availability of ions, its soil, its clay, even its slope. The soil and climate determine the vegetation and the vegetation rules the supply of organic matter. The organic matter reacts with the soil to control the release of ions, and the ions, particularly nitrate and phosphate, control the decay of the litter, and hence lie right at the root of the food cycle*". This is emphasized by the central observation (and tenet of stream ecology) that most streams and rivers on the earth's surface require allochthonous inputs (i.e., from outside the system) to support their largely heterotrophic metabolism [47, 75, 172]. While allochthony is classically applied to energy subsidies of organic matter [165, 179], allochthonous inputs can be diverse, revealing the multiple ways in which the chemistry of surface water is connected to the terrestrial landscape. For example, silica, with inputs associated with both atmospheric deposition and weathering, is an essential nutrient for the growth of freshwater diatoms a community of algae which directly controls concentrations of nitrogen across global rivers [171, 187]. Surface water concentrations of nitrogen are also primarily sourced from the landscape including wet deposition, the decomposition of organic matter, and groundwater [6]. This connectivity across the terrestrial-aquatic interface points to the link between surface water chemistry and lateral, vertical, and longitudinal dimensions of a watershed.

The critical zone (CZ) is defined as Earth's thin skin from the weathering front to vegetative canopy at the lower atmospheric boundary [128]. This vertical profile is inclusive of physical characteristics including lithology, topography and aspect, surface and groundwater flow paths and reservoirs of water and solutes, the structure of regolith and soil, and the distribution of porosity. These physical attributes determine the rate and magnitude in which water and solutes move across the terrestrial-aquatic interface. The CZ determines rates of multiple Earth surface processes and biogeochemical cycles that support and sustain life [9, 28, 144] actualized as variability in surface water chemistry. The CZ framework is unique from other multidimensional concepts (e.g., Jenny's state factors of soil development [73]) in that the CZ is a physical entity which exists on the surface of the Earth, fully integrating processes and structural characteristics across its entire depth [21]. The CZ is recognized as a single open system of energy and matter exchange [120, 144] capable of functioning as both a transporter and transformer of material [9, 46, 82, 176].

Like the CZ, both terrestrial and aquatic ecosystems are defined as open systems that exchange and transform energy and material [48, 133, 147, 166] with process rates determined by ecological and evolutionary constraints, and interactions between biota and the physical environment. Ecosystems are thus more than biological systems [94] and are recognized as the basic units of nature on the face of the earth (Hutchinson 1940). There are many parallels between the CZ and ecosystem concepts, and like the CZ, processes occurring within and across ecosystems are expressed as variability in surface water chemistry [149, 177]. For example, phenological shifts in biological processes influence oscillations in solute concentrations including carbon and nitrogen [5, 106]. Borne from the early recognition that inputs into the terrestrial landscape had consequences for the aquatic ecosystem, was the development of the "whole-ecosystem approach" [17, 91] which conceptualized the watershed-ecosystem complex as an array of vectors that drove material flux across the terrestrial-aquatic interface (TAI). These vectors, identified as *geologic*, *meteorologic*, and *biologic*, could be uniquely studied through examining stream chemistry of lotic ecosystems that drain small watersheds and their accompanying valleys [91].

Here we coalesce concepts from the fields of ecosystem ecology and CZ science to demonstrate the multiple controls on surface water chemistry highlighting the need for holistic multidimensional models that incorporate multiple process and spatiotemporal scales (e.g., [45]). We center our work on the three vectors of material inputs and fluxes identified by Likens and Bormann [91]. We build out this initial framework to also describe the CZ as a feed through reactor (sensu lato [9] which modulates vectors of material and energy inputs expressed over time as the integrated signal of surface water chemistry. We examine methods that provide an integrative signal of the ecosystem-CZ complex (e.g., concentration-discharge relationships) to understand the critical zone-ecosystem complex. We highlight methods that support a mechanistic interpretation of the various signals contained with surface water chemistry including how to parse the contribution of terrestrial versus aquatic processes. We intentionally did not overly focus on any single measurement of stream water chemistry, or element, to promote a holistic perspective on terrestrial-aquatic linkages in the broadest sense. And while we acknowledge the role of terrestrial ecosystems and the CZ to impact ponds and lakes (e.g., [37] we primarily focus on lotic systems (streams and rivers) because of the major role discharge plays in impacting surface water chemistry [29]. We do not intend this review to be exhaustive but rather as an opportunity to bring together two fundamental concepts that describe and characterize the Earth's near surface and its influence on a vital resource—*freshwater*.

2 The Whole-Ecosystem Approach and Vectors of Material Flux

The whole-ecosystem approach, as originally articulated in Borman and Likens [17], proposed a conceptual model of three vectors of material transport and flux that linked the terrestrial and aquatic environments (Fig. 4.1). These include the geologic, meteorologic, and biologic vectors. This work was largely a product of early work at the Hubbard Brook Experimental Forest (New Hampshire, USA) in response to the feedbacks between industrial emissions and the acidification of ecosystems via the wet deposition of NO_x and sulfate, with measurable consequences on the chemistry of streams draining these hardwood forests including decreasing pH and increasing concentrations of calcium (Ca^{2+}), aluminum (monomeric Al), and nitrate (NO_3^-).

These changes in surface water chemistry had significant negative effects on aquatic life including native brook trout [178]. While the chemistry of precipitation can directly contribute to surface water chemistry, especially in lakes [37], much of the observed changes in Hubbard Brook streams were due to the stripping and leaching of base cations and the diminution of the acid neutralizing capacity of soils. It was these observations that led to the conceptualization of the meteorologic vector. This early work points to early perceptions of the CZ described as interactions among processes operating at various time scales including event (precipitation), seasonal (vegetation and nutrient and biomass accrual), and millennia (weathering and soil development) that drive variability in surface water chemistry.

The whole-ecosystem framework also led to an appreciation of how forests and their changing rates of biomass accrual alter nutrient concentrations in streams. While Likens and Bormann [91] originally viewed animals as the biologic vector, a more contemporary definition of this vector includes, among other organisms, plants and microorganisms. For example, the complete denuding of the landscape

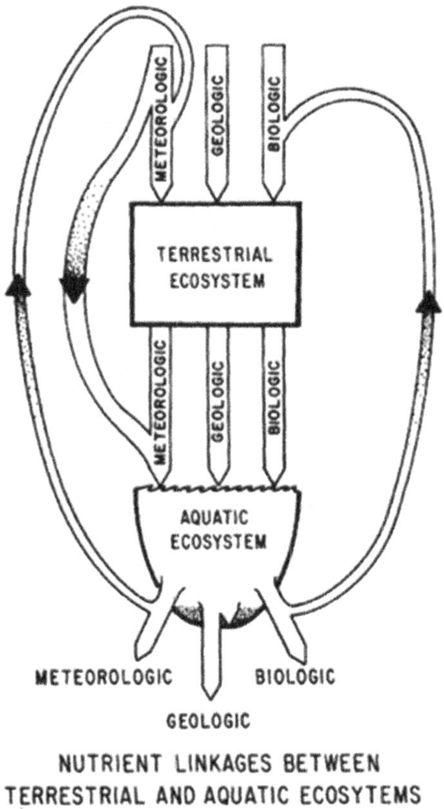

Fig. 4.1 The three ecosystem vectors as originally conceptualized by Likens and Bormann [91] that linked nutrients between terrestrial and aquatic ecosystems

through experimental whole-watershed tree harvesting resulted in NO_3^- concentrations approximately 50-fold higher than background, well exceeding levels recommended for human consumption and health [92]. Other significant increases occurred for calcium (Ca^{2+}), magnesium (Mg^{2+}), potassium (K^+), and sodium (Na^+) with net losses of all major cations and anions far greater than losses from the undisturbed control watershed [92]. Similar patterns are observed in "natural experiments" where NO_3^- and K^+ levels remain elevated for years following hurricane disturbance in streams draining tropical watersheds [153]. Differences among experimental harvests, which often include herbicide treatment to inhibit regrowth, and natural experiments such as wind, ice, and rain events, is likely attributable to activity in the rooting zone as trees take up nutrients from the soil post-disturbance. On ecosystem successional time scales, streams draining older forests are associated with higher NO_3^- concentrations while streams draining younger successional forests have lower concentrations of NO_3^- as rapidly growing trees retain more nitrogen via higher rates of biomass accrual [177]. In these younger, actively growing forests, the signal of photosynthesis is clear in surface water concentrations of NO_3^- as the forest goes through seasonal cycles of primary productivity actively incorporating nutrients into biomass during the growing season.

In the context of the whole ecosystem concept and the three vectors of material transport, the geologic vector primarily focuses on toxic substances, particulate matter, sediment, and dissolved nutrients [91]. Here the geologic vector is focused heavily on pollution via heavy metals, pesticides, and excessive nutrient inputs from agriculture and industry highlighting the role of humans to modify fluxes of material across the terrestrial-aquatic interface. Notably absent from the initial framing of the geologic vector is the role of weathering and the production of solutes which can influence biological processes (and thus the biologic vector) and surface water chemistry including the role of Ca^{2+} with critical thresholds for aquatic organisms [182] and silica which influence rates of diatomaceous primary productivity in streams and rivers [53]. Subsequent experiments in Hubbard Brook documented that the large-scale additions of calcium (as wollastonite, $CaSiO_3$) to watersheds could in fact replace the lost calcium due to acidification (via the meteorologic vector) and increase aboveground net primary productivity [14]. These experimental results demonstrate a role of mineral dissolution and weathering in the deep CZ to impact the physiology of terrestrial vegetation and by extension solute exports to surface waters. Calcium reaching the stream channel has a direct impact on other surface water properties including pH [138]. Here we view the geologic vector to reflect deeper and often slower processes occurring within the CZ that introduce essential elements primarily through chemical and geomicrobial weathering which are then transported vertically and laterally to the subsurface which can drive biological and ecological processes, and, ultimately, the chemistry of surface waters.

3 The Feed Through Reactor

A deeper integration of the "bottom up" perspectives of the CZ and the role of lithology with the "top-down" perspectives of ecosystem ecology provide a meaningful opportunity to integrate multiple processes to understand how interactions among vectors impact the export of material to streams and rivers. Material inputs via the three identified vectors are filtered through the CZ as they move into the stream channel. The critical zone has been conceptualized as a feed-through reactor [9] where weathered material enters the critical zone from the lowest boundary and is digested, mixed, and transported by chemical, physical and biological processes and erosion and infiltrated precipitation laterally moves material downslope to the river channel (Fig. 4.2). The balance of downward advancement of the weathering front and erosion from the top determines the vertical extent of the CZ and the volume of the reactor. The size of the reactor, or the depth of the CZ, is highly variable across space [34, 148] and the transformation and transportation of the material can be temporally dynamic given seasonality and disturbance regimes. Nested reaction fronts [164] can create vertical diversity in weathering products and determine the advection of solutes and water [22].

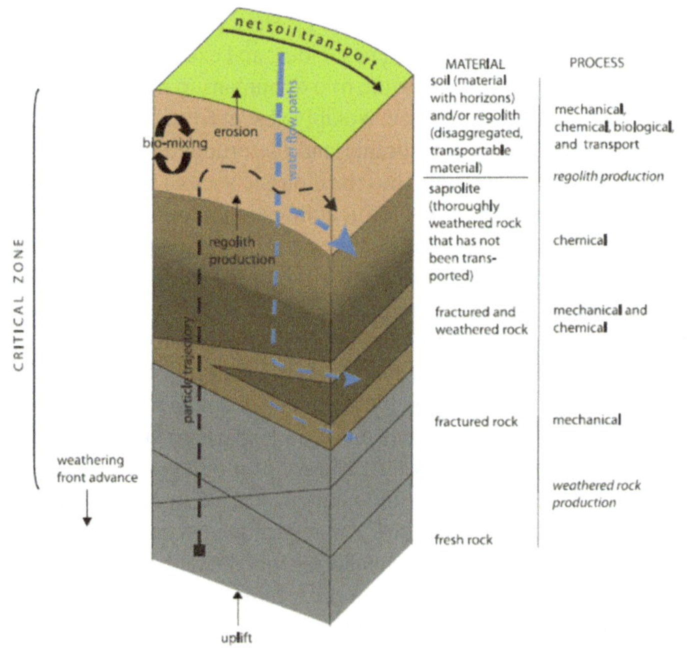

Fig. 4.2 The critical zone's feed-through reactor as originally conceptualized by Anderson et al. [9] where weathering products (particles) enter the reactor through the lower boundary of the critical zone. As particles move vertically they encounter water flow paths and are subjected to mixing via biological processes

Multiple 'geologic' processes control chemical weathering rates within the feed-through reactor. These processes vary across spatial scales from deep tectonic processes, hillslope creep (e.g., via freeze–thaw cycles), and mineral transformation, all which create variability in chemical weathering and affect water and material residence time within the reactor [9]. The framework of the feed-through reactor allows us to develop a more complete geologic vector and acknowledge the production of critical life-supporting solutes via weathering reactions with CO_2 (acid–base) and O_2 (redox), the biological mobilization and extraction and uplift of these nutrients which maintain high soil cation concentrations [74, 143], and the meteorically driven flux of this material to surface waters [164]. The structural complexity of the feed-through reactor in combination with the topography of the land surface [188], the structure of riparian zones and hillslopes [114], the diversity of biogeochemical reaction surfaces [28], and flow system heterogeneity [79] results in the non-linear filtration of both water and material [79, 115] giving rise to spatial and temporal variability in surface water chemistry.

While the feed-through reactor concept is highly useful to conceptualize the geologic vector, it may not be broadly applicable across space. For example, in regions such as the Midwestern United States, continuous glacial sediments predominate much of the geochemical characteristics of the CZ and those processes that drive variability in surface water chemistry have little connectivity to the underlying parent material [8]. This vertical discontinuity can also exist in alluvial, lacustrine, and coastal landscapes, as well as in ecosystems where aeolian sediments contribute disproportionately to geochemical properties [8]. In such environments, the geologic vector will have to be recontextualized to include these dominant geologic signatures which will imprint onto both ground- and surface-water chemistry. Here we apply the feed-through reactor concept broadly as a reconceptualization of the geologic vector originally proposed by Likens and Bormann [91] recognizing that the specific bio-geochemical properties of the reactor will vary as a function of CZ structure and history—both past and present [28, 82, 146].

It is also important to recognize that the feed-through reactor receives inputs of material (and water) from the surface and higher boundaries of the critical zone. Thus, we argue that the feed-through reactor has two primary vectors. One that represents the input of material from the weathering front and another vector that represents inputs from the meteorologic and biologic vectors (Fig. 4.3). While the former vector is more associated with weathering products and regolith and saprolite production, the latter can contribute nutrients and carbon that result from biological processes such as soil respiration, bacterial and fungal metabolism, organic acid production via photosynthate and organic matter decomposition. The production of carbonic and organic acids can be consequential for the reactor as they contribute to chemical weathering throughout the critical zone [7, 86, 147]. These top-down and bottom-up vectors can be distributed differently throughout the feed-through reactor due to variation in permeability near the top of the reactor (high) compared to lower and near the bottom of the reactor (low) [9]. This biologically induced weathering within the reactor directly contributes to spatial and temporal variation in dissolved organic and inorganic carbon in surface water [76]. Organic acids that

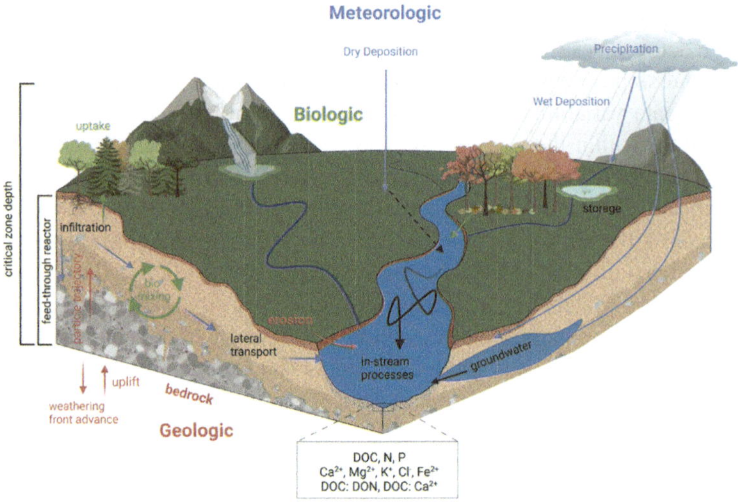

Fig. 4.3 Conceptual model illustrating how the three ecosystem vectors (meteorologic, biologic, geologic) interact with the critical zone feed-through reactor to determine variability in surface water chemistry. The model accounts for in-stream physical and biological processes that modulate surface water chemistry

enter the reactor from biological processes complex with other products of chemical weathering (aluminum, iron) increasing their solubility which increases chemical loads across global rivers [51].

4 Geologic

The structure and function of the geologic vector and the feed-through reactor is ultimately set by the composition of local parent material and tectonic history. The ramifications of lithology, uplift history, and landscape evolution have multiple consequences for surface water chemistry from the abundance and stoichiometry of solutes via weathering reactions to the development of relief that results from tectonic activity (i.e., elevation gradients) and flow-path architecture [30, 31, 188]. For example, in the tropical rainforest ecosystem of the Luquillo Mountains in Puerto Rico, variation in parent material (granitic vs. volcaniclastic) drives variability in surface water chemistry in response to major hurricane disturbance including the rate in which NO_3^- and K^+ return to background concentrations [112]. In the granitic terrain, NO_3^- took over a decade to return to long-term background concentrations while in the volcaniclastic landscape, concentrations were back to pre-disturbance levels after 1–2 years. The primary mechanism attributed to these spatial differences in the trajectories of the response of NO_3^- are coarse soils with deeper groundwater flow paths associated with the granitic terrain. Here, the deeper flow paths in the granitic feed-through reactor likely store solutes for slower release into the stream channel. This hypothesis is supported by variation in NO_3^- concentrations from well water samples in both terrains. In the granitic watersheds, NO_3^- concentrations are an order of magnitude higher than from wells in the volcaniclastic watershed [112].

Variation in hydrologic flow path architecture and zones of preferential flow within the reactor ultimately determine the magnitude by which the vectors are expressed as surface water chemistry. Tools such as hydrograph separation and mass balance improved conceptualizations of runoff sources to the stream channel during high flows [142] providing insight into what portion of the reactor was providing solutes and thus how the different vectors where interacting and imprinting on surface water chemistry. At low flows, most stream water originates from deeper flow paths (i.e., groundwater sources; [114]). In contrast, and under wet conditions, such as during snowmelt and intense storm events, precipitation infiltrates the hillslope soils, increasing hydraulic connectivity up the relatively shallower subsurface to the stream channel [85, 163]. High flow conditions are generally considered the result of the combination of (1) faster flows through saturated subsurface layers [72, 113, 114], (2) a rising groundwater table and translator flow [109, 159], (3) increased infiltration of precipitation through soil layers and preferential flow [122, 124], (4) displacement of riparian or hyporheic water [114], and/or (5) surface runoff via overland flow [40, 68]. Each of these distinct water sources carries a unique chemistry which is related to the geochemical characteristics of the reactor and transit and reaction times that are determined by the architecture of the CZ [88, 102]. For example, groundwater from deeper zones is often enriched in cations and weathering products [116], given longer transit times and high interaction with subsurface geology. Groundwater can also be relatively low in organic matter and high in reactive nutrients, given the lack of biotic processes and leaching from decaying matter. In contrast, precipitation that contributes to high flows often infiltrates the top layers of the soil profile, such that stream water from these sources are enriched in organics relative to deeper flowpaths [44, 71, 129, 169]. While greatly simplified here, stream water at any given moment contains an integrated signal of multiple water and solute sources that contribute to surface water chemistry. This notion is often referred to as the "shallow and deep" hypothesis [163, 198, 199] and provides vital clues to the role of the different vectors. It is now well-recognized that stream water is largely a mixture of water derived from flowpaths spanning various depths in the CZ [20, 109, 129].

Chemical weathering rates which determine the magnitude of solute generation along the geologic vector, and thus flux of material to surface waters, can have *transport-* or *surface reaction*-controlled weathering [100]. Briefly, solute generation and flux can be determined by both fluid residence time and the time required to achieve chemical equilibrium [100]. In the former, (dis)-equilibrium is controlled by the aqueous removal and transport of weathering products and the rate of chemical weathering is controlled by fluid residence time. Surface-reaction controlled weathering occurs when thresholds of aqueous transport are realized, and the limiting rate step is controlled by reaction kinetics on the mineral/weathering surface. Both mechanisms can be influenced by preferential flow (as mentioned above) that can drive (dis)-equilibria depending on the mineral surfaces available to react. And while peak residence times and hydraulic conductivity are thought to be a function of geology, the distribution of fluid residence times are related to aridity pointing to the role of climate to play a regulatory role (e.g., water table depth) in these chemical weathering processes [107]. When fluid residence times exert a dominant control of fluxes from the geologic vector (e.g., production of weathering-derived solutes), surface water chemistry results from the geometry of subsurface flow paths and the structure of the CZ including relief, runoff, and permeability [16, 101]. These physical attributes of the CZ may be more important than basin size in contributing to solute fluxes to the river network. Evidence exists for both mechanisms which likely vary as a function of spatial and temporal scale and collectively set the maximum flux and concentrations possible within a catchment (e.g., [188]). Reactive transport models are increasingly valuable to understand mechanisms which control solute fluxes to surface waters [88], however, many of these models have been developed in smaller catchments and how these processes and model assumptions scale with basin size remains an open question [101].

The weathering reactions and processes described above are also directly impacted by clay formation and chemistry (e.g., 1:1 or 2:1 silica-based clays). The role of clays in weathering can rival that of aqueous transport by controlling reaction affinities of dissolving minerals [103]. The formation of clay and their silica to aluminum ratios affects cation exchange capacity and nutrient availability, elemental fluxes, and both the quality and quantity of water that can be transmitted through porous media and exported to surface waters [55]. In tropical mountainous rainforest ecosystems, the prevalence, and densities of clay (which vary with lithology) influence infiltration forcing water either below or through the rooting zone of riparian vegetation. These environments vary in redox conditions and thus the quantity and form of nitrogen (organic versus inorganic N) exported to surface waters [111]. At larger scales clays can decrease rates of erosion and geochemical denudation by increasing the cohesive strength of soils.

While we have greatly generalized mechanisms here, stream water at any given time contains an integrated signal of multiple water sources that contribute to the water visible in the stream channel. As water passes through the reactor it carries with it tracers that can identify both the origins of the water as well as the solutes whether they be conservative or reactive. The resulting chemical signals can reflect layers of the CZ including subsurface soil, regolith and weather and fractured bedrock (Fig. 4.4). For example, elemental stoichiometry [21] can vary along a depth profile providing a means by which to identify sub-compartments of the CZ and the geologic vector. The conceptualization of the CZ as an integrator of multiple dynamic subsurface vectors suggests geochemical or solute compartments within the CZ whose contribution to surface water chemistry may vary over space and time. The lateral (land to water) and longitudinal (upstream to downstream) movement of water connects and integrates these resulting chemistries providing clues to the role of the geologic vector to input material into the reactor that can interact with the other vectors enroute to the point of sampling and the watershed outlet [36]. For example, groundwater from deeper zones is often enriched in cations and weathering products [116], given longer transit times and high interaction with subsurface geology.

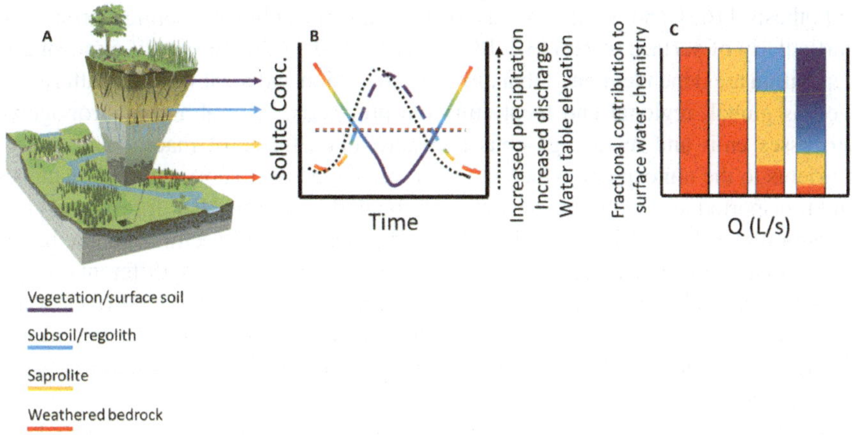

Fig. 4.4 Illustration of how spatially distinct areas within the critical zone contribute to solute signatures in surface waters. **a** Critical zone profile with colored arrows representing different depths and solute reservoirs. **b** Geochemical signatures can vary over time as a function of increasing precipitation, discharge, and water table elevation which engage different solute reservoirs and flow paths across the depth profile. Multicolored solid line represent the common response of weathering products (e.g., calcium, silica). Multicolored dashed line represents the response of common soil subsurface solutes (e.g., dissolved organic carbon). Multicolored dotted line (horizontal line) represents a solute with no response to flow and potentially variable or mixed contribution from the different depths. Dashed blue line is the response of precipitation, discharge, and water table elevation as a function of time. **c** The fractional contribution of different solute reservoirs as a function of discharge (Q). Intervals of Q are arbitrary as are the fractional contributions

Groundwater can also be relatively low in organic matter and high in reactive nutrients, given the lack of biotic processes and leaching from decaying matter. In contrast, precipitation that contributes to high flows are commonly infiltrating the top layers of the soil profile, such that stream water from these sources are enriched in organics relative to deeper flowpaths [44, 71, 129, 169]. The observations that shallow versus deeper water will carry unique signatures has been especially useful for identifying source areas and flow generation zones under different hydrologic conditions (e.g., [18, 43]). While much has yet to be resolved relating complex subsurface hydrology, an integration of the multi-dimensional movement of different water sources—and the materials they carry—through the CZ will be required to move hydrology beyond inference to process-based understanding [110].

5 Meteorologic

Meteorologic contributions of water and dissolved and particulate matter represent an important vector of material input to the reactor. The presence or absence of precipitation determines the hydrologic dynamics of a watershed (e.g., shallow vs. deep hypothesis; [163], and wet and dry deposition can impact biogeochemical processes, particularly of surface waters [3, 118, 125, 126, 130, 167]. Patterns of atmospheric deposition are dependent on global airmass circulation and local weather patterns, as well as global, regional and local emission processes. Natural and anthropogenic processes emit and release gaseous and particulate forms of macro- and micronutrients to the atmosphere which are then distributed over short regional distances and transported at a global scale such as through dust storms [10], forest fires [11], transpiration [59], and decomposition [58]. The meteorologic vector is unique in this regard in that it is the one vector that can relocate material between different ecosystems and critical zones. Such large-scale processes are translated to surface water chemistry via the feed-through reactor, which acts as a non-linear filter that moves precipitation, and dissolved solutes, from the atmosphere to the stream network [78, 79, 88].

Deposition plays a crucial role in sustaining the productivity and biodiversity of terrestrial and aquatic ecosystems by providing a source of bioavailable, and often limiting, nutrients. Inputs via deposition, however, can fundamentally change the geochemical properties of the feed-through reactor and the expression of other vectors and processes. Dry deposition of phosphorous (P), for example, contributes to the limiting supply of P available for trees, often outpacing the supply from weathering [4, 27, 108, 140]. Wet deposition represents the largest input of biologically available nitrogen (N) to undeveloped ecosystems [52], while also transporting other elements including carbon (C), Si, Ca^{2+}, and K^+ [9, 95, 195]. An acute example of the influence that the meteorologic vector has on the CZ, is the seven-fold increase in the deposition of reactive (Nr) to the biosphere from 1890–1990 (17 Tg N/year to 64 Tg/N/year, [52] attributable to the increased combustion of fossil fuels [67], which among greenhouse gases, also emit NO_x into the atmosphere. Following a suite of complex atmospheric reactions, anthropogenic activities have caused marked increases in the acidic composition of rain, resulting in consequences for terrestrial ecosystems, the biologic vector, and surface water chemistry.

The impacts of atmospheric N deposition on ecosystems are complex and varied, with impacts to surface water mediated by both the geologic and biologic vectors [3, 118, 125, 126, 130, 167]. For example, soil acidification is a significant consequence of chronic N and sulfate deposition in forested regions with low levels of Ca^{2+} and Mg^{2+} concentrations, thus acid-altered soils are less able to neutralize strong acids [19]. As a result, there are well-recognized consequences for surface water chemistry that reflect interactions among the vectors, including altered rates of plant productivity [130] and lower soil drainage capacity [93]. Intense acidification of ecosystems has caused increases in stream NO_3^- concentrations, particularly in headwater streams, except during the growing season when NO_3^- export to surface waters is attenuated by biomass and nutrient uptake [125]. In this way the meteorologic influence on surface water chemistry is mediated by the biologic vector. In regions where increased N deposition co-occurs with increased biomass accumulation (e.g., as observed through NDVI remote sensing) there is an increase in terrestrial N retention and a concomitant decrease in stream N loading [130]. Acid deposition also effects carbon cycling by suppressing leaf litter decomposition leading to a retention of organic matter in the upper soil horizons [134]. However, as atmospheric N and sulfate deposition concentrations have declined, there has been an observed increase in DOC leaching to streams due to changes in soil ionic strength [69, 135], which varies based on the parent material (i.e., the geologic vector) of a given catchment [81, 119]. For example, weathering rates in granitic catchments are unable to buffer against acid deposition due to low rates of base cations production including Ca^{2+} and Mg^{2+}. In contrast, ultramafic serpentinite catchments are associated with high weathering rates and elevated concentrations of Mg^{2+} which provides buffering capacity to the effects of acid deposition [81].

The theory of N saturation [3, 98] suggests that when the rate of atmospheric N deposition, or any element deposited on the landscape, exceeds an ecosystem's ability to retain it, the excess will be lost from the system resulting in elevated concentrations within receiving water bodies. Recovery from acid deposition, and the disproportionate influence of the meteorologic vector (as historically experienced in temperate forests of New England and Europe) on the biologic vector, can be highly variable due to interactions with other aspects of the CZ including the physical–chemical properties of soils [119] which have their own developmental history. Thus, interactions among the vectors and the properties and processes occurring within the feed-through rectors determine the hysteretic pattern of system recovery [54] and return to baseline stream chemistry [130].

The role of wet deposition, particularly to impact surface water chemistry remains an actively ongoing trajectory of ecosystem-CZ research. Much of this research now focuses on how changes in rain chemistry, that have resulted from the Clean Air Act (CAA) and associated amendments, interact with the biologic and geologic vectors within the CZ reactor [50, 118, 126, 149]. Understanding the contemporary role of the meteorologic vector to impact surface water chemistry will need to characterize how the vector, and the material load carried, interacts with other forcings wrought by global change including shifting seasonality in temperate ecosystems [127]. While historic increases in atmospheric N deposition have been attributed to increases in N exports across watersheds [23, 125] climate has taken on a stronger role in controlling surface water chemistry as N deposition declines [167, 184]. Simultaneously, while concentrations of NO_x deposition are declining due to the CAA [38, 89, 96] this policy focused on oxidized rather than reduced forms of N, leading to temporal patterns of atmospheric deposition differing among N forms. Other elements within deposition are also changing, including increases in ammonium [54], dissolved organic C and N [127, 175], lead [183], and phosphorous [136]. The effects of such trends on surface water chemistry are currently unresolved, with research suggesting that DON and DOC wet deposition loads can influence stream inorganic N concentrations [126]. Anthropogenic perturbations to global biogeochemical cycles have resulted in opportunities for research to explore the role of deposition and its interactions with competing vectors and the role of the reactor to modulate surface water chemistry.

6 Biologic

The biologic vector is nestled in the middle of the reactor receiving and integrating outputs from processes that occur throughout the CZ. Material carried through this vector is, in turn, subject to the ecological processes carried out by biota resulting in outputs that can be measured as an integrated signal in surface water chemistry. The productivity and reactive capacity of the biologic vector depends on the balance between the input, throughput, and output of water, directly connecting the outputs of this vector to the meteorologic vector. While in transit, considerable changes to carbon (C), nitrogen (N), and phosphorus (P), occur from biologic regulation via microorganisms and terrestrial vegetation. Many of the rates of biological transformation depend on water availability by limiting or changing dominant biogeochemical pathways. Saturated conditions restrict aerobic soil respiration by inducing anoxic conditions while overly dry conditions limit microbial metabolism through desiccation and restrict solute transport through soils [196]. The outputs of these metabolic processes to the stream channel are ultimately determined by hydrometeorological and climate regimes which often overwhelm the influence of other CZ attributes [45, 90]. Interactions with climate by the different vectors to determine the chemistry of freshwater varies across spatial scales. For example, at continental scales, the role of climate in determining stream water concentrations of carbon is mediated by saturated hydraulic conductivity. This indirect control of climate and hydrometeorology is lost while soil characteristics become the dominant driver of stream water concentrations of carbon at regional scales [45]. In contrast, solutes from deep within the CZ reactor and which originate from weathering are more stable inter-annually with concentrations less responsive to climate variability [200].

Source pools and transport affinities resulting from biological processing are reflected and discernable in stream chemistry signals. Organic C is often found in high quantities in shallow soils originating via primary productivity and microbially mediated organic matter decomposition and relatively easily transports to surface waters along flow paths [33, 141, 197]. Instream inorganic C concentrations are a function of subsurface CO_2 distribution over depth resulting from weathering and soil respiration [154, 163]. Under wet conditions, soil water with high DOC and CO_2 concentrations leach more into the deeper subsurface, enhancing weathering and lateral fluxes of C [181]. N, often in solution as nitrate (NO_3^-), has little affinity for the soil matrix and moves freely through soils or is retained in biomass via uptake by terrestrial vegetation [66, 117]. In contrast, P in soils (phosphate, PO_4^{-3}) adsorbs to soil particles and accumulates. Consequently, N often enters streams with groundwater baseflow [160] while P is transported via surface and subsurface flow paths in soluble form [41] or in particulate form with eroded soils [131]. This lateral transport of reactive elements results in soil and groundwater chemistries that are mirrored by surface water chemistry expressing the combined influence of deep and shallow processes occurring throughout the reactor [163].

Alteration to source pool availability via anthropogenic disturbance can saturate the biotic processing capacity of the terrestrial ecosystem and feed through reactor resulting in excess lateral flux of reactive solutes. For example, removal of natural vegetation for agriculture increases both the quantity of dissolved N and P entering surface water and the proportion of it that is inorganic [155] as the reactor becomes saturated and unable to process the excess nutrients. N-saturation in agricultural systems often results in the loss of substantial quantities of N to surface water enhanced by the replumbing of the landscape via tile-drains preventing infiltration of excess nutrients into the reactor where retention and reaction time can be enhanced. In contrast, most intact forests respond to N deposition by increasing carbon sequestration [99] and comparatively undisturbed and unpolluted ecosystems lose relatively little N. And that N that is lost, is often in the form of dissolved organic nitrogen [137, 191]. Biological reactions ("bio-mixing") at the surface of the CZ and within the reactor ultimately serve as a final filter of inputs from the geologic and meteorologic vectors and the influx of energy and nutrients to surface waters.

The chemistry of streams and rivers is influenced by their dual functionality as both a transporter and transformer of material [46, 82]. Streams and rivers are remarkably proficient at transformation of biologically reactive elements [133] determining the fate of elements through biogeochemical pathways. Within fluvial surface waters, carbon and nutrient fluxes are linked with the ecology of aquatic organisms, serving as a main control of solute concentrations in addition to rates of discharge and residence time [13, 15]. Excess supply of nutrients can accelerate the flux or loss of carbon from fluvial systems [152] and the ratio of C to N in flowing waters can determine the rate of N uptake [189], while metabolism is correlated with ambient concentrations of phosphorus [121]. This cycling of nutrients during longitudinal transport within the stream is termed nutrient spiraling and is a unique aspect of fluvial ecosystems whereby nutrients are transformed during the continual movement downstream [42, 186]. The capacity for in-stream processes to remove nutrients, however, is finite and saturation of uptake has both physical and biological constraints [123, 139]. When surface waters act as a transporter or a biological reactor is largely a function of residence time [15, 24, 26]. For example, under the forcings of negative precipitation anomalies, biogeochemical processes have greater control on solute surface water solute concentrations relative to discharge [46]. The role of the biologic vector to influence surface water chemistry is a product of the vertical extent of the CZ (from geologic to meteorologic vectors) and reflects the lateral (through each vector) and the longitudinal (within surface waters) transport of material as well as the bio-mixing that occurs within the feed-through reactor.

7 Concentration-Discharge Relationships as an Integrative Measure of Ecosystem Vectors and Critical Zone Reactors

Surface water chemistry can be conceptualized as a response "sensor" of the ecosystem vectors and the CZ reactor which collectively represent the sum of inputs and outputs, and transformations across the terrestrial-aquatic interface. The terrestrial and aquatic portions of the CZ are ultimately connected through complex flow-path architecture and storage-discharge relationships [79] where transit time can be highly variable [64] all of which is expressed as spatial and temporal variability in surface water chemistry. Characterizing a watershed's concentration-discharge (C-Q) behavior has been used as an integrated measure of critical zone structure and function providing a signal of how ecosystem vectors and CZ reactors impact solute generation and export to the stream and river network (Fig. 4.5, [29, 162, 168, 188]). Evaluating stream C-Q patterns has recently become a mainstay metric in the hydro-biogeochemical literature, largely given the relative ease of use and interpretation [77, 80, 157, 161, 162, 190, 193] and the growing global repositories of stream chemistry and discharge data [63].

C-Q relationships have been described for a wide variety of landscapes including among [197] and within biomes where certain factors can be held constant including in tropical [188], temperate [39, 77, 145], montane [80], and Arctic watersheds [157, 180]. Concentration-discharge relationships have also been applied to highly modified landscapes where anthropogenic forcing can supersede the influence of the CZ often by removing or modulating certain ecosystem vectors in both urban [39, 60] and agricultural [105, 161], settings. And more recently, C-Q responses have been applied to infer the size, location, and potential availability of material source areas [151, 173].

While the C-Q response can be described by several mathematical models [61, 62], modern C-Q analyses often rely on simple power-law equations (i.e., $C = aQ^b$) to assess how solute concentrations change with discharge [56, 57, 193]. Power-law based regression analyses assume a log-linear relationship between dependent (C) and independent (Q) variables, lumping together multiple hydro-biogeochemical processes that vary over space and time [170] and containing multiple assumptions including those of constant load, landscape homogeneity, a well-mixed control volume, and no preferential flows [61, 62, 193]. Recognized across this body of work are three simplified relationships between solute concentration and discharge represented by the slope (b) of the power law (i.e., source-limited, chemostasis, transport-limited). *Source limitation* (or dilution) indicates that the abundance, location, and production of material in the catchment cannot maintain supply to the river network at the same timescales as the storm event [12]. Diluting C-Q relationships, particularly for weathering products, also indicate mean fluid residence times are shorter than the time required to reach chemical equilibrium [100]. Chemostasis can reflect no strong relationship between flow and a constituent, or mixing of many tributaries [36, 105]. In contrast to diluting relationships, chemostatic C-Q patterns

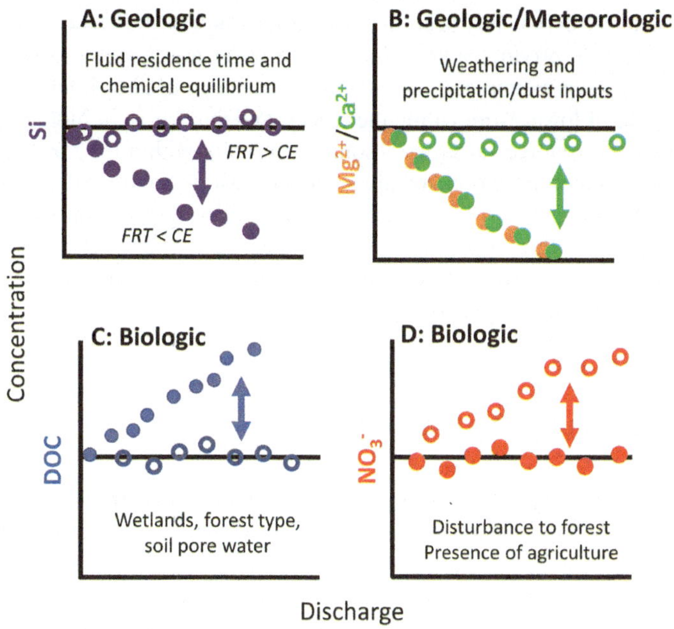

Fig. 4.5 Some examples and hypothesized mechanisms of how concentration-discharge (C-Q) relationships change with interactions among ecosystem vectors: geologic, meteorologic, biologic. Note: C-Q slopes are arbitrary with the emphasis on the directionality of change. Solid black lines represent a zero C-Q slope. **a** Variation in the silica (Si) C-Q slope is often interpreted to reflect the time required to reach chemical equilibrium (CE; closed circles) versus fluid residence time (FRT; open circles). **b** Magnesium (Mg^{2+}) and calcium (Ca^{2+}) C-Q relationships are often similar. Inputs via deposition can introduce new sources of Ca^{2+} resulting in a decoupling of Mg^{2+} and Ca^{2+} (open circles). **c** Dissolved organic carbon (DOC) C-Q relationships are often positive (closed circles). The presence of wetlands and variation in forest type can affect the slope (open circles). **d** Nitrate (NO_3^-) C-Q relationships are often chemostatic (closed circles). Forest disturbance can alter NO_3^- availability through reduced plant uptake (open circles). This meta-figure is not exhaustive but provides examples of interaction among vectors that influence variation in C-Q relationships. Associated literature: **a** [35, 57, 100, 101] **b** [57, 65, 188] **c** [32, 65, 156, 188, 197] **d** [32, 76, 77, 95, 188]

indicate fluid residence times that exceed the time needed to achieve chemical equilibrium and that the maximum concentrations of weathering-derived solutes have been reached [100]. Conversely, *transport-limitation* (or enrichment) indicates when the ecosystem is "leakier" to export, where material is available in excess and delivery to the stream network is controlled by the spatial extent and transport capacity of the terrestrial aquatic nexus [87, 201].

Concentration-discharge relationships temporally dynamic behaviors [162], such as event- and seasonal-scale hysteretic relationships, where solute concentrations vary at the same interval of discharge across the rising and falling limbs of the hydrograph [43, 97, 151]. C-Q relationships are now widely used to describe watershed responses as an integrated land to water signal, where the relationship between flow and material export has been used as an indicator of how ecosystem vectors and the CZ reactor produce and mobilize both water and material across the TAI [151, 157, 161]. For example, C-Q responses have been used to infer the export of material higher or lower from within the reactor and the degree of biological control [56], identify the relative location of solute reservoirs within the CZ [174], document whether solutes are mobilized along slow or fast flow paths [192], and deduce retention mechanisms [1, 49]. However, when stream C-Q responses are applied without quantifying the underlying CZ, ecosystem, and hydrologic processes (i.e., the vectors), C-Q responses are limited to inference of material sourcing [24, 185] and often left to be purely phenomenological.

8 Parsing Signals Within C-Q

Concentration-discharge relationships are a powerful tool that allows for the integration of multiple processes that vary across spatial and temporal scales from microbial respiration to slower weathering processes. At the same time, the drivers of C-Q relationships can be challenging to disentangle due to the vary integrative nature of the measurements. Here we developed a conceptual framework of the critical zone that incorporates fundamental concepts from ecosystem ecology (ecosystem vectors) and critical zone science (feed-through reactor) to help characterize how different processes at the Earth's near surface interact to influence the spatial–temporal variability in surface water chemistry. In this section we briefly highlight select methods and measurements (Fig. 4.6) that can be used to help parse various signals with surface water chemistry.

The measurement of surface water chemistry at a single sample point at some point in time provides a value that is integrative of all upstream and lateral fluxes. Given the spatial heterogeneity of the critical zone certain loci across the landscape may have disproportionate effects on downstream surface water chemistry. Temporal variability at multiple time steps may also cause different portions of the landscape to exert control on the variability in surface water chemistry observed downstream. Sampling protocols that leverage nested network design (Fig. 4.5) and high-frequency data, can provide the needed spatial and temporal coverage to characterize sub-catchment leverage [2, 158]. Such data can be used to quantify hot-spots and hot moments in solute production and removal in heterogenous landscapes as well as the stability in space and time of solute concentration [158, 193]. Synoptic data collected over time can thus be used to understand how processes at finer-spatial scales drive source-sink dynamics at larger spatial scales. Characterizing sub-catchment leverage has immediate management implications as it can help inform when and where mitigation efforts should be focused.

Combining measures of concentration and discharge with nutrient addition experiments would assist in determining the relative contribution of in-stream processes versus landscape contributions to the variability observed in long-term measurements of surface water chemistry. It is generally unclear how much of a C-Q pattern can be

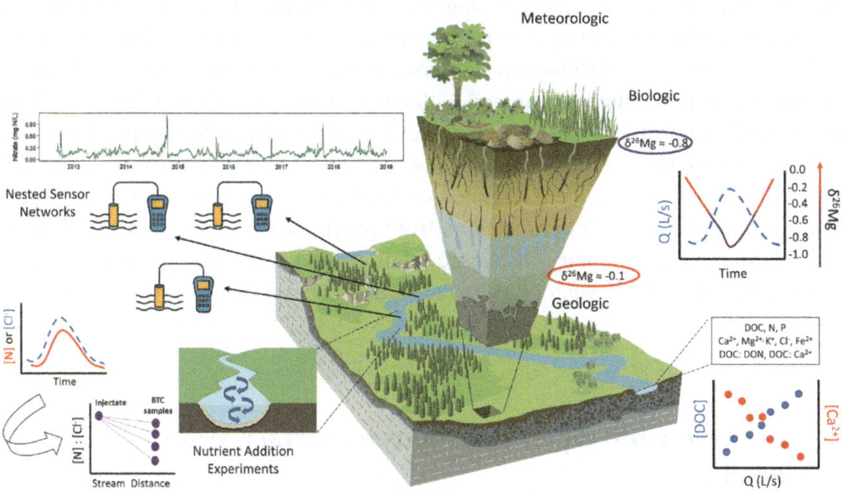

Fig. 4.6 Opportunities exist to synergistically collect data and perform experiments to understand how different ecosystem vectors (meteorologic, biologic, geologic) and the CZ reactor interact to influence surface water chemistry. Nested high-frequency sensor networks (top left) can be used to understand how various portions of the landscape influence downstream water quality (e.g., subcatchment leverage; [158]); nutrient addition experiment (bottom left) can be used to calculate nutrient uptake kinetics to understand the role of the in-stream biologic vector to control solute concentrations [150]; and magnesium (Mg^{2+}) isotopes (top right) and the stoichiometry of biological: weathering solutes (bottom right) can be quantified along intervals of time and discharge to characterize shallow and deep flow paths and associated solute reservoirs that contribute to surface water chemistry. Nitrate time series (top left) from [194], magnesium isotope data (top right) adapted from [83, 84], dissolved organic carbon (DOC)—calcium (Ca^{2+}) figure (bottom right) adapted from [163]. Central CZ figure from [28]

attributed to both biological and physical processes within the stream channel [185]. Nutrient addition experiments leverage the nutrient spiraling concept [42, 186] which quantifies both the unidirectional advection (via a conservative tracer, e.g., Cl^-) of solutes through the river channel and the physical and biological processes that collectively contribute to uptake of a reactive solute (e.g., nitrogen [N]). Nutrient additional experiments often occur at the reach scale and when replicated through a network would allow for the identification of in-stream hot-spots that disproportionately modulate solute concentrations. When combined with isotopes (e.g., ^{15}N) certain biogeochemical processes (e.g., nitrification; [139]) can be estimated. The ability for in stream production and retention of solutes is a key to understanding shifting controls on surface water chemistry. The channel-source hypothesis [185], for example, in one such model that considers the role of discrete source areas (McDonnell 2003) and solute generation from the stream itself during storm events.

Multi-solute perspectives are also vital to partition source areas from within the reactor and examine interactions among vectors. For example, concentrations of Mg and the fractionation of Mg isotopes ($\delta^{26}Mg$) can reveal the contribution of multiple vectors and processes occurring throughout the feed-through reactor including wet deposition and soil sorption–desorption processes to influence stream water Mg profiles [83]. When distinct isotopic patterns or source areas patterns are recognized, these measurements can be combined with other solutes to understand their distribution through the profile of the CZ. The co-examination of weathering products (e.g., Ca^{2+}) and biologically reactive solutes (DOC, NO_3^-), can demonstrate when deep and shallow source areas are decoupled or coupled [163] or when other sources of solutes are entering the system such as from precipitation or decomposition. Here variation in the stoichiometry-discharge relationship can be used to understand the contribution of multiple vectors to surface water chemistry and downstream solute fluxes.

9 Conclusion and Open Questions in Surface Water Chemistry

Here we developed a conceptual model that pulled from ecosystem ecology and critical zone science to offer a holistic multi-scale framework of the controls on the variability of surface water chemistry. The chemical composition of freshwaters is naturally integrative, and models must consider interactions of multiple vectors that contribute to these profiles. While we do not consider this framework exhaustive, it offers a template by which to consider meteorologic, biologic, and geologic vectors of material transport to the stream network and processes occurring within the feed-through reactor that are responsible for many of the transformations that occur across the terrestrial-aquatic interface. In our review we have also attempted to highlight the multiple ways in which the vectors interact, necessitating the need for data collection efforts and experiments that can parse the multiple signals the exist within stream water chemistry.

Variability in stream water chemistry reflects both processes and changes in the watershed and the strength of these signals varies by watershed size [36, 105]. Many outstanding questions remain, however, to understand the controls on spatial and temporal variability of surface water chemistry. Examples include: how will the chemical profiles of streams and river respond to global change? How do elements interact as they are co-exported and are these interactions constructive or destructive with respect to concentrations and fluxes? How will water quality respond to the continued intensification of land use land cover change? These questions emphasize understanding the interaction of ecosystem vectors and critical zone reactors with additional external forcings which may amplify or suppress their contributions to surface water chemistry.

Acknowledgements The authors acknowledge support from the US Department of Energy (DOE) EPSCoR Program and the project Integrating catchment expansion-contraction dynamics into cross-continental hydrobiogeochemical predictions (DE-SC0024636). ASW was supported by the National Science Foundation (NSF) EPSCoR project Canary in the Watershed (NSF EPS-1929148) and the NSF-funded Belmont Forum project ABRESO (NSF-2129383). Partial funding was also provided by the New Hampshire Agricultural Experiment Station, and the USDA National Institute of Food and Agriculture Hatch Multi-State Project 1022291. This is scientific contribution number 2972. DSM received support from the NASA FINESST program (Grant No 80NSSC1441). AJS was supported by the NSF-funded ConFines project (NSF DEB 2106076). The authors also acknowledge book editor Dr. Tim White and an anonymous reviewer whose feedback has significantly improved the manuscript.

References

1. Abbott BW, Baranov V, Mendoza-Lera C, Nikolakopoulou M, Harjung A, Kolbe T et al (2016) Using multi-tracer inference to move beyond single-catchment ecohydrology. Earth-Sci Rev 160:19–42
2. Abbott BW, Gruau G, Zarnetske JP, Moatar F, Barbe L, Thomas Z et al (2018) Unexpected spatial stability of water chemistry in headwater stream networks. Ecol Lett 21:296–308
3. Aber JD, Goodale CL, Ollinger SV, Smith ML, Magill AH, Martin ME et al (2003) Is nitrogen deposition altering the nitrogen status of northeastern forests? Bioscience 53:375–389
4. Aciego SM, Riebe CS, Hart SC, Blakowski MA, Carey CJ, Aarons SM, Dove NC, Botthoff JK, Sims KWW, Aronson EL (2017) Dust outpaces bedrock in nutrient supply to montane forest ecosystems. Nat Commun 8:1–10
5. Alexander RB, Böhlke JK, Boyer EW, David MB, Harvey JW, Mulholland PJ, Seitzinger SP, Tobias CR, Tonitto C, Wollheim WM (2009) Dynamic modeling of nitrogen losses in river networks unravels the coupled effects of hydrological and biogeochemical processes. Biogeochemistry 93:91–116
6. Alexander R, Smith R, Schwarz G (2000) Effect of stream channel size on the delivery of nitrogen to the Gulf of Mexico. Nature 403:758–776
7. Amundson R, Richter DD, Humphreys GS, Jobbágy EG, Gaillardet J (2007) Coupling between biota and earth materials in the critical zone. Elements 3:327–332
8. Anders AM, Bettis EA, Grimley DA, Stumpf AJ, Kumar P (2018) Impacts of Quaternary history on critical zone structure and processes: examples and a conceptual model from the Intensively Managed Landscapes Critical Zone Observatory. Frontiers in Earth Science 6
9. Anderson SP, von Blanckenburg F, White AF (2007) Physical and chemical controls on the critical zone. Elements 3:315–319
10. Arvin LJ, Riebe CS, Aciego SM, Blakowski MA (2017) Global patterns of dust and bedrock nutrient supply to montane ecosystems. Science Advances 3:eaao1588
11. Barrett TE, Ponette-González AG, Rindy JE, Weathers KC (2019) Wet deposition of black carbon: a synthesis. Atmos Environ 213:558–567
12. Basu NB, Rao PS, Thompson SE, Loukinova NV, Donner SD, Ye S, Sivapalan M (2011) Spatiotemporal averaging of in-stream solute removal dynamics. Water Resour Res 47
13. Battin TJ, Lauerwald R, Bernhardt ES, Bertuzzo E, Gener LG, Hall RO Jr et al (2023) River ecosystem metabolism and carbon biogeochemistry in a changing world. Nature 613:449–459
14. Battles JJ, Fahey TJ, Driscoll CT, Blum JD, Johnson CE (2014) Restoring soil calcium reverses forest decline. Environ Sci Technol 1:15–19
15. Benettin P, Fovet O, Li L (2020) Nitrate removal and young stream water fractions at the catchment scale. Hydrol Process 34:2725–2738
16. Beven K (1987) Towards the use of catchment geomorphology in flood frequency predictions. Earth Surf Proc Land 12:69–82
17. Bormann FH, Likens GE (1967) Nutrient cycling. Science 155:424–429
18. Botter M, Li L, Hartmann J, Burlando P, Fatichi S (2020) Depth of solute generation is a dominant control on concentration-discharge relations. Water Resour Res 56:e2019WR026695

19. Bowman WD, Cleveland CC, Halada Ĺ, Hreško J, Baron JS (2008) Negative impact of nitrogen deposition on soil buffering capacity. Nat Geosci 1:767–770
20. Boyer EW, Hornberger GM, Bencala KE, McKnight DM (1997) Response characteristics of DOC flushing in an alpine catchment. Hydrol Process 11:1635–1647
21. Brantley SL, Goldhaber MB, Ragnarsdottir KV (2007) Crossing disciplines and scales to understand the critical zone. Elements 3:307–314
22. Brantley SL, Lebedeva MI, Balashov VN, Singha K, Sullivan PL, Stinchcomb G (2017) Toward a conceptual model relating chemical reaction fronts to water flow paths in hills. Geomorphology 277:100–117
23. Brookshire ENJ, Valett HM, Thomas SA, Webster JR (2007) Atmospheric N deposition increases organic N loss from temperate forests. Ecosystems 10:252–262
24. Burns DA, Pellerin BA, Miller MP, Capel PD, Tesoriero AJ, Duncan JM (2019) Monitoring the riverine pulse: applying high-frequency nitrate data to advance integrative understanding of biogeochemical and hydrological processes. Wiley Interdiscip Rev: Water 6:e1348
25. Campeau A, Bishop K, Amvrosiadi N, Billett MF, Garnett MH, Laudon H et al (2019) Current forest carbon fixation fuels stream CO2 emissions. Nat Commun 10:1876
26. Catalàn N, Campo RD, Talluto M, Mendoza-Lera C, Grandi G, Bernal S, Schiller DV, Singer G, Bertuzzo E (2022) Pulse, Shunt and Storage: Hydrological contraction shapes processing and export of particulate organic matter in river networks. Ecosystems 12:1–20
27. Chadwick OA, Derry LA, Vitousek PM, Huebert BJ, Hedin LO (1999) Changing sources of nutrients during four million years of ecosystem development. Nature 397:491
28. Chorover J, Kretzschmar R, Garcia-Pichel F, Sparks DL (2007) Soil biogeochemical processes within the critical zone. Elements 3:321–326
29. Chorover J, Derry LA, McDowell WH (2017) Concentration–discharge relations in the critical zone: Implications for resolving critical zone structure, function, and evolution. Water Resour Res 53:8654–8659
30. Clow DW, Mast MA, Campbell DH (1996) Controls on surface water chemistry in the upper Merced River basin, Yosemite National Park, California. Hydrol Process 10:727–746
31. Clow DW, Sueker JK (2000) Relations between basin characteristics and stream water chemistry in alpine/subalpine basins in Rocky Mountain National Park, Colorado. Water Resour Res 36:49–61
32. Coble AA, Wymore AS, Shattuck MD, Potter JD, McDowell WH (2018) Multiyear trends in solute concentrations and fluxes from a suburban watershed: evaluating effects of 100-year flood events. J Geophys Res Biogeosci 123:3072–3087
33. Cole JJ, Prairie YT, Caraco NF, McDowell WH, Tranvik LJ, Striegl RG, Duarte CM, Kortelainen P, Downing JA, Middelburg JJ, Melack J (2007) Plumbing the global carbon cycle: integrating inland waters into the terrestrial carbon budget. Ecosystems 10:72–185
34. Condon LE, Markovich KH, Kelleher CA, McDonnell JJ, Ferguson G, McIntosh JC (2020) Where is the bottom of a watershed? Water Resourc Res 56:e2019WR026010
35. Conroy NA, Dann JB, Newman BD, Heikoop JM, Arendt C, Busey B, Wilson CJ, Wullschleger SD (2022) Chemostatic concentration–discharge behaviour observed in a headwater catchment underlain with discontinuous permafrost. Hydrol Process 36:e14591
36. Creed IF, McKnight DM, Pellerin BA, Green MB et al (2015) The river as a chemostat: fresh perspectives on dissolved organic matter flowing down the river continuum. Can J Fish Aquat Sci 72:1272–1285
37. Driscoll CT, Driscoll KM, Roy KM, Mitchell MJ (2003) Chemical response of lakes in the Adirondack region of New York to declines in acidic deposition. Environ Sci Technol 37:2036–2042
38. Du E, De Vries W, Galloway JN, Hu X, Fang J (2014) Changes in wet nitrogen deposition in the United States between 1985 and 2012. Environ Res Lett 9
39. Duncan JM, Band LE, Groffman PM (2017) Variable nitrate concentration–discharge relationships in a forested watershed. Hydrol Process 31:1817–1824
40. Dunn T, Black RD (1970) Partial area contributions to storm runoff in a small New England watershed. Water Resour Res 6:1296–1311

41. Dupas R, Gascuel-Odoux C, Gilliet N, Grimaldi C, Gruau G (2015) Distinct export dynamics for dissolved and particulate phosphorus reveal independent transport mechanisms in an arable headwater catchment. Hydrol Process 29:3162
42. Ensign SH, Doyle MW (2006) Nutrient spiraling in streams and river networks. J Geophys Res Biogeosci 111
43. Evans C, Davies TD (1998) Causes of concentration/discharge hysteresis and its potential as a tool for analysis of episode hydrochemistry. Water Resour Res 34:129–137
44. Fan Y, Clark M, Lawrence DM, Swenson S, Band LE, Brantley SL, Brooks PD, Dietrich WE, Flores A, Grant G, Kirchner JW (2019) Hillslope hydrology in global change research and earth system modeling. Water Resour Res 55:1737–1772
45. Fazekas HM, Brun J, Wymore AS (2023) Direct and indirect drivers of energy and nutrient availability in freshwater ecosystems across spatial scales. Global Biogeochem Cycles. https://doi.org/10.1029/2022GB007580
46. Fazekas HM, McDowell WH, Shanley JB, Wymore AS (2021) Climate variability drives watersheds along a transporter-transformer continuum. Geophysical Research Letters 48:e2021GL094050
47. Fisher SG, Likens GE (1973) Energy flow in bear brook, New Hampshire: an integrative approach to stream ecosystem metabolism. Ecol Monogr 43:421–439
48. Forbes SA (1887) The Lake as a Microcosm. Bull Peoria Sei Assoc
49. Frei RJ, Abbott BW, Dupas R, Gu S, Gruau G, Thomas Z et al (2020) Predicting nutrient incontinence in the Anthropocene at watershed scales. Front Environ Sci 7:200
50. Fuss CB, Driscoll CT, Campbell JL (2015) Recovery from chronic and snowmelt acidification: Long-term trends in stream and soil water chemistry at the Hubbard Brook Experimental Forest, New Hampshire, USA. J Geophys Res Biogeosci 120:2360–2374
51. Gaillardet J, Viers J, Dupré B (2004) Trace elements in river waters. In: Drever JI (ed) Surface and ground water, weathering, and soils. Treatise on geochemistry, 5 edn. Elsevier, California, pp 225–272
52. Galloway JN, Townsend AR, Erisman JW, Bekunda M, Cai Z, Freney JR et al (2008) Transformation of the nitrogen cycle. Science 320:889–892
53. Garnier J, Billen G, Coste M (1995) Seasonal succession of diatoms and Chlorophyceae in the drainage network of the Seine River: observation and modeling. Limnol Oceanogr 40:750–765
54. Gilliam FS, Burns DA, Driscoll CT, Frey SD, Lovett GM, Watmough SA (2019) Decreased atmospheric nitrogen deposition in eastern North America: predicted responses of forest ecosystems. Environ Pollut 244:560–574
55. Goddéris Y, François LM, Probst A, Schott J, Moncoulon D, Labat D, Viville D (2006) Modelling weathering processes at the catchment scale: the WITCH numerical model. Geochim Cosmochim Acta 70:1128–1147
56. Godsey SE, Hartmann J, Kirchner JW (2019) Catchment chemostasis revisited: water quality responds differently to variations in weather and climate. Hydrol Process 33:3056–3069
57. Godsey SE, Kirchner JW, Clow DW (2009) Concentration–discharge relationships reflect chemostatic characteristics of US catchments. Hydrol Process 23:1844–1864
58. Gray CM, Monson RK, Fierer N (2010) Emissions of volatile organic compounds during the decomposition of plant litter. J Geophys Res Biogeosci 115
59. Guenther A, Karl T, Harley P et al (2006) Estimates of global terrestrial isoprene emissions using MEGAN (model of emissions of gases and aerosols from nature). Atmos Chem Phys 6:3181–3210
60. Gwenzi W, Chinyama SR, Togarepi S (2017) Concentration-discharge patterns in a small urban headwater stream in a seasonally dry water-limited tropical environment. J Hydrol 550:12–25
61. Hall FR (1970) Dissolved solids-discharge relationships: 1. Mixing models. Water Resour Res 6:845–850
62. Hall FR (1971) Dissolved solids-discharge relationships: 2. Applications to field data. Water Resour Res 7:591–601

63. Hartmann J, Lauerwald R, Moosdorf N (2014) A brief overview of the GLObal RIver CHemistry Database, GLORICH. Procedia Earth and Planetary Science 10:23–27
64. Heidbuchel I, Troch PA, Lyon SW, Weiler M (2012) The master transit time distribution of variable flow systems. Water Resour Res 48
65. Herndon EM, Dere AL, Sullivan PL, Norris D, Reynolds B, Brantley SL (2015) Landscape heterogeneity drives contrasting concentration–discharge relationships in shale headwater catchments. Hydrol Earth Syst Sci 19:3333–3347
66. Hobbie SE, Finlay JC, Janke BD, Nidzgorski DA, Millet DB, Baker LA (2017) Contrasting nitrogen and phosphorus budgets in urban watersheds and implications for managing urban water pollution. Proc Natl Acad Sci 114:4177–4182
67. Holtgrieve GW, Schindler DE, Hobbs WO et al (2011) A coherent signature of anthropogenic nitrogen deposition to remote watersheds of the northern hemisphere. Science 334:1545–1548
68. Horton RE (1933) The role of infiltration in the hydrologic cycle. Eos Transc AGU 14:446–460
69. Hruška J, Krám P, McDowell WH, Oulehle F (2009) Increased dissolved organic carbon (DOC) in Central European streams is driven by reductions in ionic strength rather than climate change or decreasing acidity. Environ Sci Technol 43:4320–4326
70. Hynes HB (1975) The stream and its valley. Internationale Vereinigung für theoretische und angewandte Limnologie: Verhandlungen 19(1):1–5
71. Jasechko S, Kirchner JW, Welker JM, McDonnell JJ (2016) Substantial proportion of global streamflow less than three months old. Nat Geosci 9:126–129
72. Jencso KG, McGlynn BL (2011) Hierarchical controls on runoff generation: Topographically driven hydrologic connectivity, geology, and vegetation. Water Resour Res 47
73. Jenny H (1941) Factors of soil formation. McGraw-Hill, New York, 281 pp
74. Jobbágy EG, Jackson RB (2004) The uplift of soil nutrients by plants: biogeochemical consequences across scales. Ecology 85:2380–2389
75. Kaushik NK, Hynes HBN (1971) The fate of dead leaves that fall into streams. Arch Hydrobiol 68:465–515
76. Keller CK (2019) Carbon exports from terrestrial ecosystems: A critical-zone framework. Ecosystems 22:1691–1705
77. Khamis K, Blaen PJ, Comer-Warner S, Hannah DM, MacKenzie AR, Krause S (2021) High-frequency monitoring reveals multiple frequencies of nitrogen and carbon mass balance dynamics in a headwater stream. Front Water 3:668924
78. Kirchner JW, Feng X, Neal C (2001) Catchment-scale advection and dispersion as a mechanism for fractal scaling in stream tracer concentrations. J Hydrol 254:82–100
79. Kirchner JW (2009) Catchments as simple dynamical systems: catchment characterization, rainfall-runoff modeling, and doing hydrology backward. Water Resour Res 45
80. Knapp JLA, von Freyberg J, Studer B, Kiewiet L, Kirchner JW (2020) Concentration–discharge relationships vary among hydrological events, reflecting differences in event characteristics. Hydrol Earth Syst Sci 24:2561–2576
81. Krám P, Hruška J, Wenner BS, Discoll CT, Johnson CE (1997) The biogeochemistry of basic cations in two forest catchments with contrasting lithology in the Czech Republic. Biogeochemistry 37:173–202
82. Kumar P, Le PV, Papanicolaou AT, Rhoads BL, Anders AM, Stumpf A, Wilson CG et al (2018) Critical transition in critical zone of intensively managed landscapes. Anthropocene 22:10–19
83. Lara MC, Buss HL, von Strandmann PA, Dessert C, Gaillardet J (2014) Controls on the Mg cycle in the tropics: insights from a case study at the Luquillo Critical Zone observatory. Proc Earth Planetary Sci 10:200–203
84. Lara MC, Buss HL, von Strandmann PA, Schuessler JA, Moore OW (2017) The influence of critical zone processes on the Mg isotope budget in a tropical, highly weathered andesitic catchment. Geochim Cosmochim Acta 202:77–100
85. Laudon H, Seibert J, Köhler S, Bishop K (2004) Hydrological flow paths during snowmelt: congruence between hydrometric measurements and oxygen 18 in meltwater, soil water, and runoff. Water Resour Res 40

86. Lawrence C, Harden J, Maher K (2014) Modeling the influence of organic acids on soil weathering. Geochim Cosmochim Acta 139:487–507
87. Ledesma JL, Grabs T, Bishop KH, Schiff SL, Köhler SJ (2015) Potential for long-term transfer of dissolved organic carbon from riparian zones to streams in boreal catchments. Glob Change Biol 21:2963–79
88. Li L, Maher K, Navarre-Sitchler A, Druhan J et al (2017) Expanding the role of reactive transport models in critical zone processes. Earth Sci Rev 165:280–301
89. Li Y, Schichtel BA, Walker JT, Schwede DB et al (2016) Increasing importance of deposition of reduced nitrogen in the United States. Proc Natl Acad Sci USA 113:5874–5879
90. Li L, Stewart B, Zhi W, Sadayappan K, Ramesh S, Kerins D, Sterle G, Harpold A, Perdrial J (2022) Climate controls on river chemistry. Earth's Future 10:e2021EF002603
91. Likens GE, Bormann FH (1974) Linkages between terrestrial and aquatic ecosystems. Bioscience 24:447–456
92. Likens GE, Bormann FH, Johnson NM, Fisher DW, Pierce RS (1970) Effects of forest cutting and herbicide treatment on nutrient budgets in the Hubbard Brook watershed-ecosystem. Ecol Monogr 40:23–47
93. Likens GE, Driscoll CT, Buso DC (1996) Long-term effects of acid rain: response and recovery of a forest ecosystem. Science 272:244–246
94. Lindeman RL (1942) The trophic-dynamic aspect of ecology. Ecology 23:399–417
95. Liptzin D, Boy J, Campbell JL, Clarke N, Laclau J-P, Godoy R et al (2022) Spatial and temporal patterns in atmospheric deposition of dissolved organic carbon. Glob Biogeochem Cycles 36:e2022GB007393
96. Lloret J, Valiela I (2016) Unprecedented decrease in deposition of nitrogen oxides over North America: the relative effects of emission controls and prevailing air-mass trajectories. Biogeochemistry 129:165–180
97. Lloyd CEM, Freer JE, Johnes PJ, Collins AL (2016) Technical note: testing an improved index for analysing storm discharge–concentration hysteresis. Hydrol Earth Syst Sci 20:625–632
98. Lovett GM, Weathers KC, Sobczak WV (2000) Nitrogen saturation and retention in forested watersheds of the Catskill Mountains, New York. Ecol Appl 10:73–84
99. Magnani F, Mencuccini M, Borghetti M, Berbigier P, Berninger F, Delzon S et al (2007) The human footprint in the carbon cycle of temperate and boreal forests. Nature 447:849–851
100. Maher K (2010) The dependence of chemical weathering rates on fluid residence time. Earth Planet Sci Lett 294:101–110
101. Maher K (2011) The role of fluid residence time and topographic scales in determining chemical fluxes from landscapes. Earth Planet Sci Lett 312:48–58
102. Maher K, Chamberlain CP (2014) Hydrologic regulation of chemical weathering and the geologic carbon cycle. Science 343:1502–1504
103. Maher K, Steefel CI, White AF, Stonestrom DA (2009) The role of reaction affinity and secondary minerals in regulating chemical weathering rates at the Santa Cruz Soil Chronosequence, California. Geochim Cosmochim Acta 73:2804–2831
104. Marinos RE, Campbell JL, Driscoll CT, Likens GE, McDowell WH, Rosi EJ, Rustad LE, Bernhardt ES (2018) Give and take: a watershed acid rain mitigation experiment increases baseflow nitrogen retention but increases stormflow nitrogen export. Environ Sci Technol 52:13155–13165
105. Marinos RE, Van Meter KJ, Basu NB (2020) Is the river a chemostat?: scale versus land use controls on nitrate concentration-discharge dynamics in the upper Mississippi river basin. Geophys Res Lett 47:e2020GL087051
106. Marti E, Sabater F (1996) High variability in temporal and spatial nutrient retention in Mediterranean streams. Ecology 77:854–869
107. Maxwell RM, Condon LE, Kollet SJ, Maher K, Haggerty R, Forrester MM (2016) The imprint of climate and geology on the residence times of groundwater. Geophys Res Lett 43:701–708
108. McClintock MA, Brocard G, Willenbring J, Tamayo C, Porder S, Pett-Ridge JC (2015). Spatial variability of African dust in soils in a montane tropical landscape in Puerto Rico. Chem Geol

109. McDonnell JJ (2013) Are all runoff processes the same? Hydrol Process 27:4103–4111. https://doi.org/10.1002/hyp.10076
110. McDonnell JJ, Sivapalan M, Vaché K, Dunn S et al (2007) Moving beyond heterogeneity and process complexity: a new vision for watershed hydrology. Water Resour Res 43
111. McDowell WH, Bowden WB, Asbury CE (1992) Riparian nitrogen dynamics in two geomorphologically distinct tropical rain forest watersheds: subsurface solute patterns. Biogeochemistry 18:53–75
112. McDowell WH, Brereton RL, Scatena FN, Shanley JB, Brokaw NV, Lugo AE (2013) Interactions between lithology and biology drive the long-term response of stream chemistry to major hurricanes in a tropical landscape. Biogeochemistry 116:175–186
113. McGlynn BL, Seibert J (2003) Distributed assessment of contributing area and riparian buffering along stream networks. Water Resour Res 39
114. McGlynn BL, McDonnell JJ (2003) Role of discrete landscape units in controlling catchment dissolved organic carbon dynamics. Water Resour Res 39
115. McGuire KJ, McDonnell JJ (2010) Hydrological connectivity of hillslopes and streams: characteristic time scales and nonlinearities. Water Resour Res 46
116. McIntosh JC, Schaumberg C, Perdrial J, Harpold A, Vázquez-Ortega A, Rasmussen C, Vinson D, Zapata-Rios X, Brooks PD, Meixner T, Pelletier J (2017) Geochemical evolution of the Critical Zone across variable time scales informs concentration-discharge relationships: Jemez River Basin Critical Zone observatory. Water Resour Res 53:4169–4196
117. Van Meter KJ, Basu NB, Veenstra JJ, Burras CL (2016) The nitrogen legacy: emerging evidence of nitrogen accumulation in anthropogenic landscapes. Environ Res Lett 11(3):035014
118. Monteith DT, Stoddard JL, Evans CD, de Wit HA, Forsius M, Høgåsen T et al (2007) Dissolved organic carbon trends resulting from changes in atmospheric deposition chemistry. Nature 450:537–540
119. Monteith DT, Henrys PA, Hruška J, de Wit HA, Krám P, Moldan F et al (2023) Long-term rise in riverine dissolved organic carbon concentration is predicted by electrolyte solubility theory. Sci Advan 9:eade3491
120. Moravec B, Chorover J (2020) Critical zone biogeochemistry: linking structure and function. In: Dontsova K, Balogh-Brunstad Z, Le Roux G (eds) Biogeochemical cycles
121. Mulholland PJ, Fellows CS, Tank JL, Grimm NB, Webster JR, Hamilton SK et al (2001) Inter-biome comparison of factors controlling stream metabolism. Freshw Biol 46:1503–1517
122. Mulholland PJ, Wilson GV, Jardine PM (1990) Hydrogeochemical response of a forested watershed to storms: effects of preferential flow along shallow and deep pathways. Water Resour Res 26:3021–36
123. Mulholland PJ, Helton AM, Poole GC, Hall RO, Hamilton SK, Peterson BJ et al (2008) Stream denitrification across biomes and its response to anthropogenic nitrate loading. Nature 13:202–5
124. Mulholland PJ (1993) Hydrometric and stream chemistry evidence of three storm flowpaths in Walker Branch Watershed. J Hydrol 1:291–316
125. Murdoch PS, Stoddard JL (1992) The role of nitrate in the acidification of streams in the Catskill Mountains of New York. Water Resour Res 28:2707–2720
126. Murray DS, Moges E, Larsen L, Shattuck M, McDowell WH, Wymore AS (2023) Synchrony of nitrogen wet deposition inputs and watershed nitrogen outputs using information theory. Water Resour Res. https://doi.org/10.1029/2023WR034794
127. Murray DS, Shattuck MD, McDowell WH, Wymore AS (2022) Nitrogen wet deposition stoichiometry: the role of organic nitrogen, seasonality, and snow. Biogeochemistry 160:301–314
128. National Research Council (2001) Basic research opportunities in earth science. National. Academic Press, Washington DC
129. Neal C, Reynolds B, Rowland P, Norris D, Kirchner JW, Neal M, Sleep D, Lawlor A, Woods C, Thacker S, Guyatt H (2012) High-frequency water quality time series in precipitation and streamflow: From fragmentary signals to scientific challenge. Sci Total Environ 15(434):3–12

130. Newcomer ME, Bouskill NJ, Wainwright H, Maavara T, Arora B, Siirila-Woodburn ER et al (2021) Hysteresis patterns of watershed nitrogen retention and loss over the past 50 years in United States hydrological basins. Global Biogeochem Cycles 35:1–28
131. Ockenden MC, Deasy CE, Benskin CMH, Beven KJ, Burke S, Collins AL et al (2016) Changing climate and nutrient transfers: evidence from high temporal resolution concentration-flow dynamics in headwater catchments. Sci Total Environ 548–549:325–339
132. Oda T, Green MB, Urakawa R, Scanlon TM, Sebestyen SD, McGuire KJ, Katsuyama M, Fukuzawa K, Adams MB, Ohte N (2018) Stream runoff and nitrate recovery times after forest disturbance in the USA and Japan. Water Resour Res 54:6042–6054
133. Odum HT (1956) Primary production in flowing waters. Limnol Oceanogr 1:102–117
134. Oulehle F, Evans CD, Hofmeister J, Krejci R, Tahovska K, Persson T et al (2011) Major changes in forest carbon and nitrogen cycling caused by declining sulphur deposition. Glob Change Biol 17:3115–3129
135. Oulehle F, Hruška J (2009) Rising trends of dissolved organic matter in drinking-water reservoirs as a result of recovery from acidification in the Ore Mts. Czech Republic. Environ Pollut 157:3433–3439
136. Pan Y, Liu B, Cao J, Liu J, Tian S, Du E (2021) Enhanced atmospheric phosphorus deposition in Asia and Europe in the past two decades. Atmosp Ocean Sci Lett 14:100051
137. Perakis SS, Hedin LO (2002) Nitrogen loss from unpolluted South American forests mainly via dissolved organic compounds. Nature 415:416–419
138. Peters SC, Blum JD, Driscoll CT, Likens GE (2004) Dissolution of wollastonite during the experimental manipulation of Hubbard Brook watershed 1. Biogeochemistry 67:309–329
139. Peterson BJ, Wollheim WM, Mulholland PJ, Webster JR, Meyer JL, Tank JL et al (2001) Control of nitrogen export from watersheds by headwater streams. Science 292:86–90
140. Pett-Ridge JC (2009) Contributions of dust to phosphorus cycling in tropical forests of the Luquillo Mountains, Puerto Rico. Biogeochemistry 94:63–80
141. Pinay G, Peiffer S, De Dreuzy JR, Krause S, Hannah DM, Fleckenstein JH, Sebilo M, Bishop K, Hubert-Moy L (2015) Upscaling nitrogen removal capacity from local hotspots to low stream orders' drainage basins. Ecosystems 18:1101–1120
142. Pinder GF, Jones JF (1969) Determination of the ground-water component of peak discharge from the chemistry of total runoff. Water Resour Res 5:438–445
143. Porder S, Chadwick OA (2009) Climate and soil-age constraints on nutrient uplift and retention by plants. Ecology 3:623–636
144. Rasmussen C, Troch PA, Chorover J, Brooks P, Pelletier J, Huxman TE (2011) An open system energy-based framework for predicting critical zone structure and function. Biogeochemistry 102:15–29
145. Raymond PA, Saiers JE (2010) Event controlled DOC export from forested watersheds. Biogeochemistry 100:197–209
146. Richardson M, Kumar P (2017) Critical zone services as environmental assessment criteria in intensively managed landscapes. Earth's Future 5:617–632
147. Richeter DB, Billings SA (2015) 'One physical system': Tansley's ecosystem as Earth's critical zone. New Phytol 206:900–912
148. Riebe CS, Hahm WJ, Brantley SL (2017) Controls on deep critical zone architecture: a historical review and four testable hypotheses. Earth Surf Proc Land 42:128–156
149. Rodríguez-Cardona BM, Wymore AS, Argerich A, Barnes RT, Bernal S, Brookshire ENJ et al (2022) Shifting stoichiometry: Long-term trends in stream-dissolved organic matter reveal altered C: N ratios due to history of atmospheric acid deposition. Glob Change Biol 28:98–114
150. Rodríguez-Cardona B, Wymore AS, McDowell WH (2015) DOC:NO_3^- ratios control NO_3^- uptake in forested headwater streams. J Geophys Res – Biogeosci 121. https://doi.org/10.1002/2015JG003146
151. Rose LA, Karwan DL, Godsey SE (2018) Concentration–discharge relationships describe solute and sediment mobilization, reaction, and transport at event and longer timescales. Hydrol Process 32:2829–2844

152. Rosemond AD et al (2015) Experimental nutrient additions accelerate terrestrial carbon loss from stream ecosystems. Science 347:1142–1145
153. Schaefer DA, McDowell WH, Scatena FN, Asbury CE (2000) Effects of hurricane disturbance on stream water concentrations of fluxes in eight tropical forest watersheds of the Luquillo Experimental Forest, Puerto Rico. J Trop Ecol 16:189–207
154. Schlesinger WH, Bernhardt ES (2013) Biogeochemistry: an analysis of global change. Academic Press
155. Seitzinger SP, Harrison JA, Dumont E, Beusen AH, Bouwman AF (2005) Sources and delivery of carbon, nitrogen, and phosphorus to the coastal zone: an overview of Global Nutrient Export from Watersheds (NEWS) models and their application. Global Biogeochem Cycles 19
156. Shanley JB, McDowell WH, Stallard RF (2011) Long-term patterns and short-term dynamics of stream solutes and suspended sediment in a rapidly weathering tropical watershed. Water Resour Res 47
157. Shogren AJ, Zarnetske JP, Abbott BW, Iannucci F, Medvedeff A, Cairns S et al (2021) Arctic concentration–discharge relationships for dissolved organic carbon and nitrate vary with landscape and season. Limnol Oceanogr 66:S197–S215
158. Shogren AJ, Zarnetske JP, Abbott BW, Iannucci F, Frei RJ, Griffin NA, Bowden WB (2019) Revealing biogeochemical signatures of Arctic landscapes with river chemistry. Sci Reports 9:12894
159. Sklash MG, Farvolden RN (1979) The role of groundwater in storm runoff. J Hydrol 43:45–65
160. Song JH, Her Y, Guo T (2022) Quantifying the contribution of direct runoff and baseflow to nitrogen loading in the Western Lake Erie Basins. Sci Rep 12:9216
161. Speir SL, Tank JL, Bieroza M, Mahl UH, Royer TV (2021) Storm size and hydrologic modification influence nitrate mobilization and transport in agricultural watersheds. Biogeochemistry 156:319–334
162. Speir SL, Rose LA, Blaszczak JR, Kincaid DW, Fazekas HM, Webster AJ, Wolford MA, Shogren AJ, Wymore AS (2024) Catchment concentration–discharge relationships across temporal scales: a review. Wiley Interdisciplinary Rev Water 11(2):e1702
163. Stewart B, Shanley JB, Kirchner JW, Norris D, Adler T, Bristol C, et al (2022) Streams as mirrors: Reading subsurface water chemistry from stream chemistry. Water Resources Research 58:e2021WR029931. https://doi.org/10.1029/2021WR029931
164. Sullivan PL, Ma L, West N, Jin L, Karwan DL, Noireaux J et al (2016) CZ-tope at Susquehanna Shale Hills CZO: Synthesizing multiple isotope proxies to elucidate Critical Zone processes across timescales in a temperate forested landscape. Chem Geol 445:103–119
165. Tank JL, Rosi-Marshall EJ, Griffiths NA, Entrekin SA, Stephen ML (2010) A review of allochthonous organic matter dynamics and metabolism in streams. J N Am Benthol Soc 1:118–146
166. Tansley AG (1935) The use and abuse of vegetational concepts and terms. Ecology 16:284–307
167. Templer PH, Harrison JL, Pilotto F, Flores-Díaz A, Haase P, McDowell WH et al (2022) Atmospheric deposition and precipitation are important predictors of inorganic nitrogen export to streams from forest and grassland watersheds: a large-scale data synthesis. Biogeochemistry 160:219–241
168. Torres MA, Baronas J, Clark K, Feakins S, West A (2017) Mixing as a driver of temporal variations in river hydrochemistry:1. Insights from conservative tracers in the Andes-Amazon transition. Water Resour Res 53:3102–3119
169. Tunaley C, Tetzlaff D, Lessels J, Soulsby C (2016) Linking high-frequency DOC dynamics to the age of connected water sources. Water Resour Res 52:5232–5247
170. Tunqui Neira JM, Andréassian V, Tallec G, Mouchel JM (2021) Multi-objective fitting of concentration-discharge relationships. Hydrol Process 35:e14428
171. Turner RE, Rabalais NN, Justic D, Dortch Q (2003) Global patterns of dissolved N, P and Si in large rivers. Biogeochemistry 64:297–317
172. Vannote RL, Minshall GW, Cummins KW, Sedell JR, Cushing CE (1980) The river continuum concept. Canad J Fish Aquat Sci 37:130–137

173. Vaughan MCH, Bowden WB, Shanley JB, Vermilyea A, Wemple B, Schroth AW (2018) Using in situ UV-Visible spectrophotometer sensors to quantify riverine phosphorus partitioning and concentration at a high frequency. Limnol Oceanograp 16:840–855
174. Vaughan MCH, Bowden WB, Shanley JB, Vermilyea A, Sleeper R, Gold AJ et al (2017) High-frequency dissolved organic carbon and nitrate measurements reveal differences in storm hysteresis and loading in relation to land cover and seasonality. Water Resour Res 53:5345–5363
175. Verstraeten AP, Verchelde P, DeVos B, Neirynk J, Cools N, Roskams P, DeNeve S (2016) Increasing trends in dissolved organic nitrogen (DON) in temperate forest under recovery from acidification in Flanders, Belgium. Sci Total Environ 553:107–119
176. Vidon P, Allan C, Burns D, Duval TP, Gurwick N, Inamdar S et al (2010) Hot spots and hot moments in riparian zones: potential for improved water quality management. J Am Water Resour Assoc 46:278–298
177. Vitousek PM, Reiners WA (1975) Ecosystem succession and nutrient retention: a hypothesis. BioScience 25:376–381
178. Warren DR, Likens GE, Buso DC, Kraft CE (2008) Status and distribution of fish in an acid-impacted watershed of the northeastern United States (Hubbard Brook, NH). Northeast Nat 15:375–390
179. Webster JR, Benfield EF, Ehrman TP, Schaeffer MA, Tank JL, Hutchens JJ, D'angelo DJ (1999) What happens to allochthonous material that falls into streams? A synthesis of new and published information from Coweeta. Freshw Biol 4:687–705
180. Webster AJ, Douglas DA, Regier P, Scheuerell MD, Harms TK (2021) Temporal patterns in stream biogeochemistry indicate linked permafrost and ecological dynamics of boreal catchments. Ecosystems 25:1189–1206
181. Wen H, Sullivan PL, Macpherson GL, Billings SA, Li L (2021) Deepening roots can enhance carbonate weathering by amplifying CO 2-rich recharge. Biogeosciences 18:55–75
182. Weyhenmeyer GA, Hartmann J, Hessen DO, Kopáček J, Hejzlar J et al (2019) Widespread diminishing anthropogenic effects on calcium in freshwaters. Sci Rep 9:1
183. Winkler R, Rosner G (2000) Seasonal and long-term variation of 210Pb concentration in air, atmospheric deposition rate and total deposition velocity in south Germany. Sci Total Environ 263:57–68
184. de Wit HA, Stoddard JL, Monteith DT, Sample JE, Austnes K, Couture S et al (2021) Cleaner air reveals growing influence of climate on dissolved organic carbon trends in northern headwaters. Environ Res Lett 16:104009
185. Wondzell SM, Ward AS (2022) The channel-source hypothesis: empirical evidence for in-channel sourcing of dissolved organic carbon to explain hysteresis in a headwater mountain stream. Hydrol Process 36:e14570
186. Workshop SS (1990) Concepts and methods for assessing solute dynamics in stream ecosystems. J North Am Benthol Soc 9(2):95–119
187. Wurtsbaugh WA, Paerl HW, Dodds WK (2019) Nutrients, eutrophication and harmful algal blooms along the freshwater to marine continuum. Wiley Interdiscip Rev Water 6:e1373
188. Wymore AS, Brereton RL, Ibarra DE, Maher K, McDowell WH (2017) Critical zone structure controls concentration-discharge relationships and solute generation in forested tropical montane watersheds. Water Resour Res 53:6279–6295
189. Wymore AS, Coble AA, Rodríguez-Cardona B, McDowell WH (2016) Nitrate uptake across biomes and the influence of elemental stoichiometry: a new look at LINX II. Global Biogeochem Cycles 30:1183–1191
190. Wymore AS, Fazekas HM, McDowell WH (2021) Quantifying the frequency of synchronous carbon and nitrogen export to the river network. Biogeochemistry 152:1–2
191. Wymore AS, Johnes PJ, Bernal S, Brookshire EJ, Fazekas HM, Helton AM et al (2021) Gradients of anthropogenic nutrient enrichment alter N composition and DOM stoichiometry in freshwater ecosystems. Global Biogeochem Cycles 35:e2021GB006953
192. Wymore AS, Leon MC, Shanley JB, McDowell WH (2019) Hysteretic response of solutes and turbidity at the event scale across forested tropical montane watersheds. Front Earth Sci https://doi.org/10.3389/feart.2019.00126

193. Wymore AS, Larsen W, Kincaid DW, Underwood KL, Fazekas HM, McDowell WH et al (2023) Revisiting the origins of the power-law analysis for the assessment of concentration-discharge relationships. Water Resour Res 59:e2023WR034910. https://doi.org/10.1029/2023WR034910
194. Wymore AS, Shattuck MD, Potter JD, Snyder L, McDowell WH (2021) The Lamprey River hydrological observatory: suburbanization and changing seasonality. Hydrol Process 35(4):e14131
195. Xu-Yang Y, Dessert C, Losno R (2022) Atmospheric deposition over the Caribbean region: sea salt and Saharan Dust are sources of essential elements on the Island of Guadeloupe. Journal of Geophysical Research Atmospheres 127
196. Yan Z, Bond-Lamberty B, Todd-Brown KE, Bailey VL, Li S, Liu C, Liu C (2018) A moisture function of soil heterotrophic respiration that incorporates microscale processes. Nat Commun 9:2562
197. Zarnetske JP, Bouda M, Abbott BW, Saiers J, Raymond PA (2018) Generality of hydrologic transport limitation of watershed organic carbon flux across ecoregions of the United States. Geophys Res Lett 45:11–702
198. Zhi W, Li L (2020) The shallow and deep hypothesis: Subsurface vertical chemical contrasts shape nitrate export patterns from different land uses. Environ Sci Technol 54:11915–11928. https://doi.org/10.1021/acs.est.0c01340
199. Zhi W, Li L, Dong W, Brown W, Kaye J, Steefel C, Williams KH (2019) Distinct source water chemistry shapes contrasting concentration-discharge patterns. Water Resour Res 55:4233–4251. https://doi.org/10.1029/2018wr024257
200. Zhi W, Williams KH, Carroll RW, Brown W, Dong W, Kerins D, Li L (2020) Significant stream chemistry response to temperature variations in a high-elevation mountain watershed. Commun Earth Environ 1:43
201. Zimmer MA, Pellerin B, Burns DA, Petrochenkov G (2019) Temporal variability in nitrate–discharge relationships in large rivers as revealed by high-frequency data. Water Resour Res 55:973–989

Groundwater—The Dynamic Base of the CZ

Pamela L. Sullivan, David P. Genereux, Andrea E. Brookfield, Daniella Rempe, Randy L. Stotler, Holly R. Barnard, Leif Karlstrom, Jennifer C. McIntosh, Holly A. Michael, Karla Jarecke, Rene M. Price, Steven P. Loheide, Alissa White, and Kamini Singha

1 Introduction

Addressing the multiple and diverse threats to groundwater requires the holistic approach inherent to critical zone (CZ) science—the discipline that unites researchers across fields of earth and environmental science to study the interactions from the top of the canopy down to the depths of groundwater over timescales that span seconds to millennia [6, 24, 146]. Groundwater, the water in the saturated earth below the water table, is an important natural resource for humans and ecosystems around the world. It is the largest source of available freshwater [278], supporting over 2 billion

P. L. Sullivan (✉) · K. Jarecke
College of Earth, Ocean and Atmospheric Science, Oregon State University, Corvallis, OR, USA
e-mail: pamela.sullivan@oregonstate.edu

D. P. Genereux
Department of Marine, Earth, and Atmospheric Sciences, North Carolina State University, Raleigh, NC, USA

A. E. Brookfield · R. L. Stotler
Department of Earth and Environmental Sciences, University of Waterloo, Waterloo, ON, Canada

D. Rempe
Department of Geological Sciences, University of Texas at Austin, Austin, TX, USA

H. R. Barnard
Institute of Arctic and Alpine Research, Department of Geography, University of CO–Boulder, Boulder, CO, USA

L. Karlstrom
Department of Earth Sciences, University of Oregon, Eugene, OR, USA

J. C. McIntosh
Department of Hydrology and Atmospheric Sciences, University of Arizona, Tucson, AZ, USA

H. A. Michael
Department of Earth Sciences, Department of Civil and Environmental Engineering, University of Delaware, Newark, DE, USA

T. White and A. Provenzale (eds.), *Critical Zone and Ecosystem Dynamics*, Advances in Critical Zone Science, https://doi.org/10.1007/978-3-031-69076-1_5

people on Earth [141], however, another near 2 billion people live in areas where groundwater resources are under threat [102]. Approximately 70% of groundwater withdrawals are used for irrigation [59], representing about half of all water used for irrigation purposes [70, 257]. Groundwater is important for ecosystems globally: roughly 1/3 of terrestrial vegetation relies on groundwater [69], it is the source of baseflow to surface-water systems [279, 310], and through its reactions with the aquifer matrix, it plays a key role in controlling elemental cycling and fluxes [289]. Similarly, through baseflow contributions, groundwater plays a critical role in regulating heat in many surface-water bodies [121], often supporting specific aquatic fauna habitats (e.g., spawning ground for salmon; see [50] and references within). In addition to regulating energy fluxes, both the discharge of groundwater and the recharge of surface water through groundwater-surface water interaction controls the fate and transport of contaminants and nutrients in aquatic environments [28, 261]. Thus, perturbations to groundwater recharge or extraction induced by anthropogenic activities and climate change can threaten these CZ services [78].

At the most basic level, groundwater can be recharged naturally (e.g., precipitation, groundwater-surface water interactions) or artificially (e.g., irrigation, injection wells, infiltration ponds, and leaky urban infrastructure). Inputs to groundwater may change as a result of changing climatic conditions, such as the timing, amount, and phase of precipitation. Warmer temperatures can also increase evapotranspiration, which can have cascading effects on groundwater levels and fluxes. It has been projected that roughly half of global groundwater fluxes could be sensitive to shifts in climate and human activities within 100 years, potentially limiting available resources and their capacity for climate buffering due to altered recharge patterns [57]. Because many ecosystems are dependent upon using groundwater directly or the discharge of groundwater (e.g., wetlands, streams, and lakes), changes in climate may create key tipping points that drive ecosystems into alternative stable states [51, 54, 103]. For example, changes in the delivery of groundwater to streams may shift a stream from perennial to intermittent, flowing only part of the year. Given large-enough declines in the water table, an intermittent stream may transition to ephemeral, where the stream flows for a portion of the year, typically in response to precipitation events, and may also be disconnected from the groundwater [314]. Not only does stream intermittency impact the wetted length of streams and duration of surface water flow, but the degree and duration of groundwater-surface water

R. M. Price
Department of Earth and Environment, Florida International University, Miami, FL, USA

S. P. Loheide
Department of Civil and Environmental Engineering, University of Wisconsin-Madison, Madison, WI, USA

A. White
Department of Geosciences, University of Montana, Missoula, MT, USA

K. Singha
Hydrologic Science and Engineering Program, Colorado School of Mines, Golden, CO, USA

interactions, including those that thermally regulate stream-water temperature and support aquifer recharge [143]. In environments where the rooting zone of vegetation interacts and draws water from the water table, declines in the water table could lead to drastic shifts in the vegetation productivity [47] and shifts in community composition [51, 54, 103, 174]. The effects of climate change on water table position can be exacerbated by groundwater pumping [53, 140], causing water tables to lower at rapid rates. In many regions the water extracted is considered ancient, over pumping of groundwater—old or new—means water managers are likely to face growing scarcity issues in the future [75].

Solutions to these current and future challenges rely on understanding how much water is stored within the CZ. Yet the lower boundary, hidden well below our feet, is unclear; the base of the CZ has been defined as the bottom of the groundwater [106] or deepest depth of circulating groundwater [5], which may extend to depths greater than 2–3 km in some settings (e.g., [36, 196]). Entangled in these definitions are two questions: (1) What marks the bottom? and (2) What constitutes circulating water? This debate of where to draw the hydro-biogeochemical bottom of a watershed, or the CZ, was explored in [51, 54], where the authors provided three general definitions: (1) the depth to the low-conductivity boundary or "no-flow boundary", (2) the active circulation depth, which is "tunable" to the questions at hand–what flow paths are thought to contribute to the process or pattern under investigation, and (3) the depth to saline water. Our goal here is not to declare that any one of these is the best definition but to highlight that groundwater makes up an important component and driver of CZ processes, and that groundwater systems are complex and vary in the factors that threaten their resources.

Beyond groundwater quantity, groundwater scarcity is also controlled by water quality. Two spatially extensive processes threaten groundwater quality; the first is agriculture, which occupies roughly 40% of ice-free land [85] and is often associated with the contamination of groundwater by fertilizers and pesticides (e.g., [32, 249]). Contamination is particularly problematic in environments underlain by karstified aquifer systems having large conduits that allow for rapid connectivity between surface processes and groundwater through fast flow [116]. The second factor limiting access to potable groundwater occurs in coastal environments where seawater intrudes into coastal aquifers. It only takes about 2–3% seawater to yield non-potable water. Large-scale analysis of wells in the coastline of the contiguous USA shows groundwater levels are below sea level across 15% of the area [142], supporting landward hydraulic gradients that drive salt-water intrusion into coastal aquifers. Continued increases in sea levels create added pressure, as does coastal pumping, which can lower the water table and artificially create landward hydraulic gradients [15]. On a more regional scale, there are also millions of contaminated sites worldwide affected by a range of contaminants from heavy metals to pharmaceuticals to "forever" chemicals.

Below, we delve into the many ways in which groundwater dynamics and other subsurface-water stores (i.e., vadose zone water or "rock moisture") interact and feedback to control CZ functions, such as evapotranspiration, carbon cycling, and solute generation and export to surface waters and the ocean, and how these may be impacted under a changing climate. We also explore the ways in which CZ processes in turn influence the flow paths, transit times, and quality of groundwater, affecting water resources.

2 CZ Structure Controls Vadose Zone Thickness and Therefore Groundwater Recharge

Before we explore the role of groundwater as a dynamic boundary of the CZ, we highlight that the depth to the water table and the rates of groundwater recharge are dependent on the properties of the material above the water table (Fig. 1). We can conceptualize the subsurface as a three-phase system comprised of solids, liquid, and gases. The solid components include soil, saprolite, weathered rock, unweathered bedrock, and unconsolidated sediments, where liquid (water) and gases (e.g., oxygen (O_2), carbon dioxide (CO_2), and water vapor) fill the pore space created within the solid components. Collectively, these three components determine the heterogeneous CZ structure and its consequent function. Land surface, land cover, and atmospheric conditions interact with the underlying structure to determine the timing and amount of recharge to groundwater storage that feeds streams and springs.

Typically, the pathway of water transport from the surface to the water table requires the movement of water through some thickness of unsaturated subsurface known as the vadose zone. Fluid and nutrient fluxes in the vadose zone are driven by competing forces, with upward flow driven by evapotranspiration and capillarity, and gravity driving downward flow. In interbasin arid and semi-arid environments, wetting fronts are rarely deeper than the root zone [247]. In these environments, matric potentials are often lowest near the land surface, indicating that upward fluid

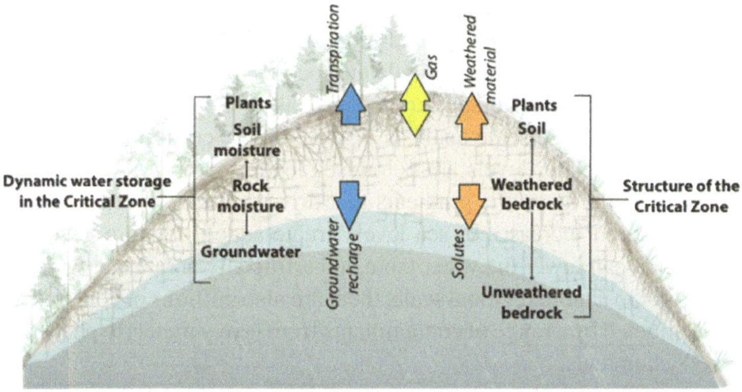

Fig. 1 A conceptual cross section through a hillslope shows the structure of the CZ (right) and water storage reservoirs within the critical zone (left). Arrows depict the fluxes of water (blue), gases (e.g., CO_2 in yellow), and products of the weathering of solid material (orange)

flows dominate [247, 295–298]. In more humid environments, matric potentials favor downward flow for longer periods of time, and water tables are closer to the surface. Downward flow dominates when water contents exceed the field capacity, where water draining beyond the root zone is potentially available for groundwater recharge. The balance of upward and downward fluxes drives the distribution and transport of nutrients and solutes in the subsurface. As external forcings change, such as shifts in the seasonal amount and intensity of precipitation or land cover/use changes, matric potential gradients in the subsurface will be altered and can change the distribution and flushing of solutes and nutrients [117, 198, 221, 222, 247, 287, 295].

The structure of the land surface, land cover, and vadose zone constrain infiltration, drainage, and fluid and solute movement in the vadose zone (Fig. 2). Drainage and recharge rates are not consistent across a landscape. Focused recharge is more likely to occur in depressions; areas where ponding frequently occurs; intermittent, ephemeral, and perennial stream beds with higher hydraulic potential than nearby aquifers; irrigated fields; or anywhere soil is wet for longer periods of time (e.g., [148, 198, 250, 291]). Gravity-induced drainage, defined by the specific yield, may also be aided by capillary flow, which acts in all directions. Ponded soils provide areas where downward pressure gradients are greater in duration and therefore able to drive fluids downward. This process also provides more opportunity to drive air out of pores, which increases the hydraulic conductivity. Infiltration and recharge are further constrained by biota, land cover and land use (e.g., [1, 240, 248, 308]). Worm, insect, and animal burrows, desiccation cracks, decayed root tubes, joints, and fractures provide pathways for rapid fluid flow past the root zone [124]. These preferential pathways essentially create a dual-porosity or dual-permeability medium, with water and solutes traveling to the water table at different rates [109, 239, 269].

In many locations, particularly in montane terrains, the vadose zone can consist of meters to tens of meters of weathered bedrock. This weathered bedrock can store exchangeable water, analogous to soil moisture, that has a distinctly different response to precipitation and drought than soil moisture [231]. Rock moisture storage may be greater than soil moisture storage, and continually expanding in active weathering environments through increases in secondary porosity (Fig. 1, [231, 294]). The influx, storage, and removal of rock moisture results in additional weathering, nutrient cycling, and geochemical mixing, likely supporting diverse and dynamic microbial populations [197, 231]. Importantly, rock moisture appears to represent water storage that mediates the rate of decline in plant-available water in droughts [231]. Consequently, plants with root systems that extend into fractured rock may be more resilient to seasonal droughts. Hahm and Rempe [111] highlighted the importance of rock moisture storage in controlling the regional distribution of plant communities, and evidence increasingly indicates that forest water use and productivity may be more closely related to rock moisture storage and topography than annual precipitation inputs [74, 276, 285]. Studies of soil moisture depletion have shown that many montane soils, particularly those on southern aspects, are thin and quick to dry, suggesting seasonally dry forests may rely on water from deeper reservoirs [11, 74, 126].

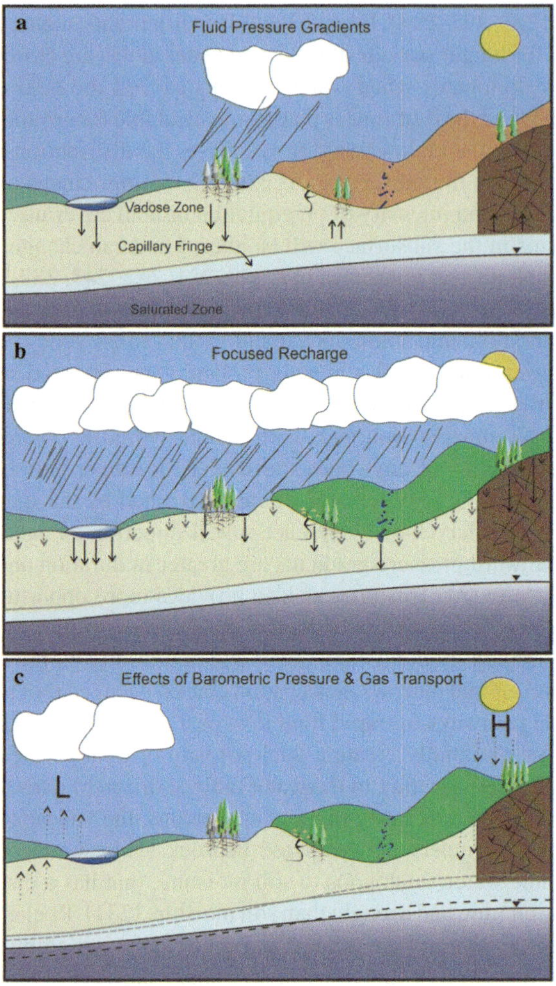

Fig. 2 Effects of changing conditions of fluid flow in the vadose zone. Note in this conceptualization surface-water bodies are disconnected from the water table and are losing water to the subsurface. **a** Fluid pressure gradients are downward beneath ponded water and during recharge events with gravity drainage, but upward during droughts and under actively transpiring plants. **b** Focused recharge pathways provide faster conduits for more water to the water table, and include ponded areas, depressions, and intermittent streams filled with water during and after a rain event, decayed root tubes, animal burrows, and joints and fractures. **c** Barometric pressure changes affect gas transport, as well as the position of both the capillary fringe and the water table. The upper dashed line, near the top of the capillary fringe, represents the location of the capillary fringe if barometric pressure conditions are constant across the watershed. Likewise, the lower dashed line represents the location of the water table under constant barometric pressure conditions. Reductions in barometric pressure result in upwards gas transport, while increases in barometric pressure result in downward gas transport

In recent decades, our understanding of how this unsaturated, weathered-bedrock vadose zone mediates fluxes to deeper groundwater has expanded rapidly as a result of innovations in sensing. Vadose-zone monitoring systems now allow monitoring of matric potential through advanced tensiometers, water content with multiple tools including flexible time-domain reflectometry, and soil and rock moisture and gas samples through sampling ports at multiple depths many meters below ground surface [65, 128, 168, 235, 236]. Near-surface geophysical tools also provide the ability to image the subsurface structure in minimally invasive ways and monitor changes across multiple temporal and spatial scales [218]. For example, shallow seismic refraction and electrical resistivity tomography have been used to map spatial variability in weathered and unweathered rock properties, and nuclear magnetic resonance documents moisture stored within the voids of weathered bedrock [81, 218, 253], which can be directly related to the structure of the vadose zone that overlies groundwater.

It is worth emphasizing that water moves in both fluid and vapor phases in the vadose zone (Fig. 2) and that reactive gases influence chemical reactions, nutrient availability, and solute transport (e.g., [158]). Gas transport is a complex process affected by both diffusion and pressure gradients of the gases, as well as a dissolved component in fluid [189]. The pneumatic diffusivity of gas in the subsurface is analogous to the hydraulic conductivity, and in the vadose zone, tends to be greater in matrices with large pore spaces (e.g., sands and gravels), with transport more likely in large capillaries and at low water contents [101, 301]. Because pressure gradients provide an important driving force for gas transport, changes in barometric pressure are integral to movement of gases in the vadose zone, even affecting the thickness of the capillary fringe and the position of the water table, which can affect groundwater flow directions and velocity [189, 264, 301]. Complex subsurface composition and pore structure can result in insertion and venting of gases in response to barometric pressure changes to great depths and long distances (i.e., 100 s m) [67, 169, 210]. Daily, seasonal, and climatic changes also affect connections and transport of gases, with increased water content, snowfall, and seasonal soil freezing reducing the prevalence and changing the locations in which barometric pressure gradients between the atmosphere and subsurface cause gas to flow into or out of the vadose zone (barometric pumping; [127, 188, 302]).

Recent investigations have expanded our understanding of vadose zone water storage, flow, and partitioning through the hydrologic cycle at multiple scales, and the effects on the connections of groundwater to the critical zone. Remote estimations collected over very large regions by the Gravity Recovery and Climate Experiment (GRACE) have revealed increases in total water storage in some areas despite significant on-going declines in water table elevation; these increases are driven by rising storage in the vadose zone [29]. Even as groundwater levels fall, the pores in the newly acquired vadose-zone storage do not completely drain [148]. Total water-storage estimates in conjunction with other ground-based observations indicate that groundwater removal in the Central Valley, CA, USA resulted in more uplift of the Sierra Nevada than tectonic uplift over a four-year drought period as an elastic response to the loss of water mass [8]. Yet ~70% of the uplift over this same four-year period, and 40–60% over an eleven-year period in which seven years are drought, is a result of drought-driven water loss deep in the vadose zone, alluvial aquifers, and crystalline basement [8]. Meanwhile, groundwater pumping in the Central Valley is resulting in localized subsidence of the land surface as decreasing fluid pressures increase effective stress, resulting in compaction of soils and permanent loss in aquifer and vadose-zone storage [73]. Groundwater-induced uplift and subsidence also reorganizes surface drainage and affects overland flow and sources of water for evapotranspiration [252]. The role of groundwater in frozen CZ systems, which might limit depth of flow profoundly, also remains an area ripe for additional research [232]. Work to date emphasizes the profound connection between subsurface water and the structure and functioning of the critical zone, and yet leaves us needing to fill several key knowledge gaps:

- How does the combined surface and subsurface topography drive spatial heterogeneity in water storage and flux patterns in and through the vadose zone to groundwater?
- To what degree can rock moisture support forest productivity under more variable weather conditions in the future (e.g., more severe drought)?
- How does an increasing depth to the water table interact with barometric pressure changes to affect fluid and gas fluxes and recharge rates through the vadose zone?

3 Groundwater and Plant Relationships

Plants rely on vadose-zone water stores for meeting transpiration demands that in turn have cascading impacts on groundwater recharge by influencing matric potentials. More directly, groundwater is responsible for an average of 10% of transpired water and up to 47% in lowlands [21, 49, 190, 201]. Thus, understanding the interconnections between transpiration—an inevitable consequence of photosynthesis and plant growth—and groundwater fluxes is thought to be one of the most important and emerging challenges in hydrology [207]. Transpiration is the largest water flux from the terrestrial surface and can be representative of vegetation productivity and how resilient it is to climate change [30]. A growing number of studies have shown that land–atmosphere feedbacks depend on regional groundwater [7, 77, 157]. For example, Maxwell and Kollet [191] found that the depth to the water table determines the relative sensitivity of areas to changes in temperature and precipitation. Similarly, Tromp-van Meerveld [285] indicate that hillslope-scale transpiration and tree basal area are more strongly related to subsurface storage than surface-water supply, highlighting the need for a larger-scale focus on subsurface structure to predict multiple sources of plant-available water and net carbon accumulation in forests.

Complex topography drives the hydrologic and microclimate dynamics of catchments and consequently the spatial and temporal variation of forest carbon accumulation and growth. Because surface topography is not always representative of subsurface processes, it has been difficult to identify mechanisms that determine the spatial distribution of aboveground biomass and its temporal feedbacks with groundwater fluxes. Dominant controls on subsurface moisture patterns often show substantial variability [220, 303], with spatial patterns of soil moisture controlled by lateral subsurface flow patterns that follow subsurface geologic features [147]. While the importance of soil properties in controlling plant-available water is well known (Billings et al. 22, Chap. 2 in this book), recent work has highlighted the importance of terrain and deeper subsurface physical structure in controlling plant water availability [134, 276]. The carbon accumulated in forests is a direct input into natural-resource dependent economies, consequently, quantifying the relationship between transpiration, groundwater, and accumulated carbon is essential to understanding the potential economic impacts to forestry resulting from climate change as well as the potential impacts on downslope agricultural water availability and agriculture-related economics [20, 30, 41]. Furthermore, understanding the relationship between transpiration, productivity, groundwater, and soils is essential for quantifying the magnitude and likelihood of success of so-called natural climate solutions that aim to sequester atmospheric carbon by enhancing forest primary productivity and increase in subsurface carbon stocks [133, 138, 164]. Consequently, it is expected that the terrain complexity will shape forest resilience and related downstream economic impacts with changing climate conditions.

While terrain complexity drives spatial heterogeneity in water availability, it is a species' ability to obtain resources—including nutrients, light, and water—in a given environment that controls its likelihood to thrive, or at least survive. In the face of competition with other species, water availability is one of the primary limits on plant productivity. Thus, a plant's strategy for obtaining water can give it an advantage under specific sets of environmental conditions (e.g., [259]). In environments where groundwater is deep relative to the root zone, all root water uptake occurs in the vadose zone and contributes to the unsaturated subsurface component of evapotranspiration (Sect. 2). However, some roots can access the water table and take up water from the saturated zone. This direct use of groundwater has long been recognized as a source of water for wetland species and phreatophytes [237, 270, 272, 273], and often, these are groundwater dependent ecosystems (GDEs), see review by [214] where their composition and function is controlled by the presence of and depth to groundwater. This connection is clearly demonstrated where diel water-table fluctuations are a diagnostic indicator of groundwater use by vegetation. When evaporative demand is high during the daylight hours, the water table falls, but as root water uptake slows in the evening, water is replenished due to groundwater flow from depth or from areas higher in the landscape [35, 115, 175]. Since the 1930s, diel fluctuations in the water table signal has been occasionally used to quantify evapotranspiration from groundwater [200, 305], but the use of diel signals to explore plant-water interactions has gained much more frequent use in recent decades [13, 35, 114, 187] and several methods have been introduced that allow for sub-daily estimates of evapotranspiration from groundwater [107, 145, 176]. In addition, the groundwater component of evapotranspiration has been quantified with isotopic methods, whereby the isotopic composition of xylem water can be portioned into end-member sources from both the vadose and saturated zones [12] (Fig. 3).

Fig. 3 Shallow groundwater subsidizes tree transpiration and growth in temperate forests when and where the water table is in the vicinity of plant roots. Trees growing where groundwater is shallow (left), experience more stable soil moisture regimes, sustained transpiration, and less year-to-year variability in growth whereas trees growing in areas of deep groundwater (right) experience greater climate-driven variability in moisture status and reduced growth, particularly in drier years. Figure modified from Cirruzi and Loheide [47]

Groundwater-dependent ecosystems in arid regions stand out as green oases within dry landscapes and have been used as indicators in prospecting for groundwater [237]. If shallow groundwater did not exist in these drylands, these ecosystems would not exist. However, this focus on groundwater-dependent ecosystems neglects other ecosystems that are not completely reliant on groundwater, but occasionally benefit from it. For example, Miller et al. [202] observed groundwater use in an oak savanna in a Mediterranean climate of California, [47] documented a groundwater component of evapotranspiration in temperate forests of northern Wisconsin, and Chimner and Resh [42] determined that Bur oak in Nebraska were obtaining 70–88% of their water from groundwater. Furthermore, Soylu et al. [262] and Zipper et al. [315] found that even maize uses shallow groundwater, which subsidizes both transpiration and crop yield. With this growing recognition that a much broader array of ecosystems use groundwater when and where it is available, even if the ecosystems' form and function may not rely on the presence of groundwater, it is time that we start to acknowledge groundwater-*influenced* ecosystems and better understand the interplay between groundwater and these ecosystems in the CZ.

The influence of groundwater extends beyond providing an additional source of water to support transpiration and subsidize growth [47, 178, 315], it is also a primary control on the distribution of ecosystems across the landscape and the spatial patterning and composition of vegetation within a given ecosystem. The range of hydrologic conditions that a plant can tolerate is its hydrologic niche. Individual species realize their hydrologic niches on the landscape based on their water requirements and tolerance for anaerobic conditions that inhibit root respiration. Shallow groundwater can help subsidize root water uptake by minimizing water stress while simultaneously increasing oxygen stress or water logging, which can lead to reduced productivity and even plant mortality. Individual species have evolved to have differing ability to benefit from increased water availability through adaptations such as deep tap roots and dimorphic root structures while other species better tolerate low-oxygen conditions through development of arenchema (a spongy tissue in the stem capable of transporting oxygen into the subsurface) and adventitious roots. It is these, and other, adaptations that increase the likelihood of an individual species to dominate under specific groundwater regimes by outcompeting others. The hydrologic niche of a species or vegetation community can be represented at a basic level by summary statistics of the groundwater regime that supports that vegetation, such as the mean water-table depth and range [3]. However, it is often extremes that drive plant mortality, and Henzey et al. [125] found that the 7-day high water level was the strongest predictor of riparian vegetation composition, likely because shallow groundwater of this duration can induce oxygen stress in unadapted plants, diminishing their competitiveness under these conditions. Furthermore, approaches such as those of Silvertown et al. [258] and Lowry et al. [179] recognize that the magnitude and duration of both wet and dry conditions control the ability of vegetation to thrive in shallow groundwater environments and have used water table hydrographs to integrate the area above and below thresholds of oxygen and water stress, respectively, as descriptors of the hydrologic regime and predictors of vegetation composition. Lastly, approaches such as that of the UK Environmental Agency [290] and vegetation threshold hydrographs of Loheide and Gorelick [174] incorporate the seasonality of groundwater levels in determining the suitability of a specific groundwater regime to support a vegetation type. These advances in our ability to quantitatively articulate the hydrologic niche of different vegetation types has enabled coupling of hydrologic and ecologic models to predict the two-way interactions and trajectory of ecohydrologic systems in response to restoration, management, climatic/land use change, and geomorphic evolution [46, 112, 173, 174, 229].

Another mechanism by which plants connect groundwater to the vadose zone is through hydraulic redistribution, the passive movement of water by roots from areas of high to low moisture content in the subsurface [37]. Root access to a continuous water source like groundwater may be particularly influential on strong upward hydraulic redistribution when the partial recovery of soil moisture in the upper soil layers is advantageous for plants to maintain shallow root function and sustain transpiration [105, 161, 167], such as during seasonal drought [61]. Hydraulic redistribution has been documented for a wide range of species, both woody and herbaceous, spanning arid to cool temperate regions as well as seasonally dry Mediterranean and tropical climates [139, 211]. This injection of water, however fleeting (typically wetting and drying daily), can increase dry-season transpiration and gross ecosystem productivity [60], stimulate microbial decomposition of organic carbon [89] and the production of CO_2 [110], and enhance the mobility of nutrients for root uptake [243]. Furthermore, the connectivity between groundwater and the vadose zone, as facilitated by hydraulic redistribution, can alter the water status of neighboring plants. For example, hydraulic redistribution of groundwater by deeply rooted sugar maple supplied understory species with up to 60% of their water [58] and hydraulic redistribution of groundwater by mesquite trees supported 14% of transpiration in shallow-rooted grasses [167]. These results highlight that the movement of groundwater through the process of hydraulic redistribution can impact both short-term drought resilience of plant communities as well as long-term above- and below-ground exchange of water, carbon, and nutrients as neighboring plant communities establish. Although still in a research stage, geophysical methods like self-potential, which measure naturally occurring electrical potentials in the ground and are thus sensitive to any source that creates a voltage (e.g., water movement and redox gradients), are being developed to characterize subsurface water flow around roots [293].

Finally, it is important to note that where roots mine water they also influence biogeochemical conditions directly through root exudation and solute uptake and indirectly via root-associated microbial activity [129]. Specifically, the root systems (roots and their microbes) respire CO_2, exude organic acids, and produce enzymes, all of which help plants meet their nutritional demands for growth [38, 120, 281]. Regardless of the substrate being mined, be it organic matter or mineral, the decomposition and dissolution of material can both generate and destroy pore space depending on the degree of expansion or compaction (e.g., Bazilevskaya et al. [14]). Over timescales of decades to millennia, these changes alter the subsurface hydraulic properties and therefore infiltration and groundwater recharge rates (Sullivan et al. [274] review and references therein).

Decades of research clearly demonstrate the importance of groundwater on supporting vegetation productivity, the ability of the CZ to accumulate carbon, and the ability of vegetation to control where groundwater flows. However, we are still left with several major knowledge gaps:

- When and where do plants rely on groundwater?
- How is the spatial distribution of aboveground biomass linked to patterns of groundwater uptake by vegetation?
- How will changes in land cover and plant function alter belowground biogeochemical functioning and thus terrestrial fluxes of carbon and nutrients to aquatic environments?
- Does hydraulic redistribution measurably alter soil carbon and weathering fluxes at the watershed scale?

4 Groundwater-Surface Water Interactions

Groundwater's contributions to streams, termed baseflow, sustain streams during dry seasons and drought, and provide valuable nutrients and thermal diversity to support healthy instream habitats [45, 155]. This exchange of water between the surface and subsurface is bi-directional, with conditions allowing for stream water to also recharge groundwater systems, which then often may return to surface flow down-gradient, defining the hyporheic zone (Boano et al. 2014). This flux of surface water into groundwater brings with it solutes such as nitrate (NO_3^-), dissolved organic carbon (DOC) and O_2, which enhance microbial activity in the subsurface (e.g., [108, 312, 313]). Consequently, groundwater-surface water interactions give rise to biogeochemical hotspots beneath streams (e.g., [17, 119]). Subsurface flow paths through the hyporheic zone are a major contributor to ecosystem health and stability, yet the controls on these flow paths are an unsolved problem in hydrology due, in part, to their spatial and temporal complexity [23]. One such example of complexity emerges from our understanding of oxygen consumption in such systems. In surface water, O_2 is thermodynamically favored for respiration, and we expect that hyporheic pore waters transition from bulk oxic to bulk anoxic conditions as the O_2 is respired (e.g., [39]). However, recent theoretical and experimental studies indicate the presence of anaerobic denitrification products, namely nitrous oxide (N_2O)—a potent greenhouse gas and an intermediary product in the microbially mediated reduction of NO_3^- to N_2 gas [66, 284] —and dinitrogen (N_2) in bulk oxic sediment [25, 118, 312]. Bulk oxic conditions should inhibit anaerobic processes, and yet products of anaerobic processes exist.

To explain this apparent paradox, the hypothesis of anoxic 'microzones' in groundwater was developed (e.g., [132]). These locations act as anaerobic reaction sites and thus, important sinks of N [79, 118]. Microzone formation and their functions is an emerging topic, with most papers published in the last few years (e.g., [26, 244, 182]). While the well-connected pore space within the streambed water may remain oxic, anoxic microzones form due to heterogeneity in sediment or organic carbon availability [44, 113, 244]. These previously uncharacterized zones along hyporheic flowpaths may help predictions of unexplained water-quality phenomena such as anaerobic denitrification and production of N_2 and N_2O within bulk oxic hyporheic zones. Briggs et al. [26]'s numerical models suggested that pore-network structure controls residence times and thus microzone formation, where zones of lower hydraulic conductivity will have longer porewater residence times and higher respiration rates. However, they did not explore varying nutrient conditions or the role of bioclogging on how these zones may evolve through time. Roy Chowdhury [44] developed one of the first numerical models that combined hydraulics and microbial conditions to explore microzone formation. These models indicated microzone existence and distributions are not controlled by residence time alone, but by interactions between hydraulic flux, nutrient concentrations and biomass, and are thus dynamic in space and time. Under all conditions explored that considered biomass growth, anoxic microzones would perish after only a few days as bioclogging would largely occur in the downwelling parts of the hyporheic zone, shifting the system from advection- to diffusion-dominated transport and removing all oxic regions in the hyporheic zone. Similarly, Hampton et al. [113]'s experiments demonstrate that hydrologic fluxes systematically shift the location of the bulk oxic-anoxic interface and that there are transport and reaction timescales that favor incomplete denitrification such that N_2O production rates may be higher at intermediate residence times rather than the shortest or longest ones. The role of the hyporheic zone and the presence of anoxic microzones on processing metals in the environment also remains an important area of research [90, 91, 130].

On top of these issues, changes in climate are altering groundwater levels and fluxes, which can have concurrent and cascading effects on groundwater-surface water interactions and may magnify or mitigate issues critical to ecosystem health and stability. For example, agricultural intensification has increased application of both irrigation and fertilizer use across the world. Despite worldwide efforts to reduce watershed inputs of nitrate and phosphorus and quantify legacy nutrient stores, agricultural lands are predicted to expand in the future to try to meet the food demands of an increasing population. An implication of the legacy nutrients and the expansion of agriculture is the continued storage and release of nutrients to surface-water systems. Concurrently, irrigation practices—including shifts between irrigation methods from flood to sprinkler, etc. —and changing climatic conditions affect both groundwater and surface-water systems, which will affect hyporheic flows (e.g., [260, 309]). These impacts on flow regimes and hyporheic flow are relatively understudied and the importance of their effects are generally unknown. Numerical models can be used to understand how water resources will respond to changes in water use, climate, and land use, amongst other disturbances, and help predict and mitigate potential consequences of these changes. However, numerical models represent our basic knowledge of how a system functions based on the data we physically collect can help to constrain them. Thus, advances in understanding how groundwater-surface water interactions will change in the Anthropocene and the influence these changes may have on hyporheic zone dynamics depend on being able to address some of the outstanding knowledge gaps, such as:

- How does the hyporheic zone control the fate and transport of metals?
- How will changes in land cover, land use, irrigation practices, and climate impact baseflow?
- How can we readily capture the spatial and temporal heterogeneous shifts in groundwater-surface water interactions?

5 Groundwater in the Deep CZ

Groundwater extends to kilometers depth in the Earth's crust [135], comprising the largest store of water on the continents [76]. Here we focus on the meteoric component of groundwater in the CZ, defined as waters that come from precipitation, in contrast to magmatic, metamorphic, or marine-derived fluids. Groundwater-flow systems within unweathered parent materials (e.g., fractured bedrock) are often distinct (i.e., chemistry, fluid circulation patterns, transit times) from water in overlying regolith [55]. There is evidence that meteoric recharge circulates over 1–2 km depth in sedimentary and crystalline bedrock based on geophysical and geochemical observations from boreholes and thermal springs [36, 196, 223, 238, 292] (Fig. 4), implying that the CZ may extend much deeper in some settings than commonly assumed. Groundwater circulation down to these depths is primarily driven by topography, with variable influence from hydraulic anisotropy associated with three-dimensional geology (e.g., [48, 238]); solute transport may be dominated by advection or diffusion depending on the permeability structure (e.g., [159]).

Deep CZ groundwater may represent the regional water table or can be part of stacked aquifer systems (e.g., layered volcanic or sedimentary rocks), in which case there are multiple geochemically distinct stores of groundwater [255, 304]. For example, in volcanic settings, layered lava flows with highly anisotropic hydraulic permeabilities likely dictate groundwater flow pathways [184], and can also lead to long-lived perched aquifers. Layered permeability associated with complex lithology

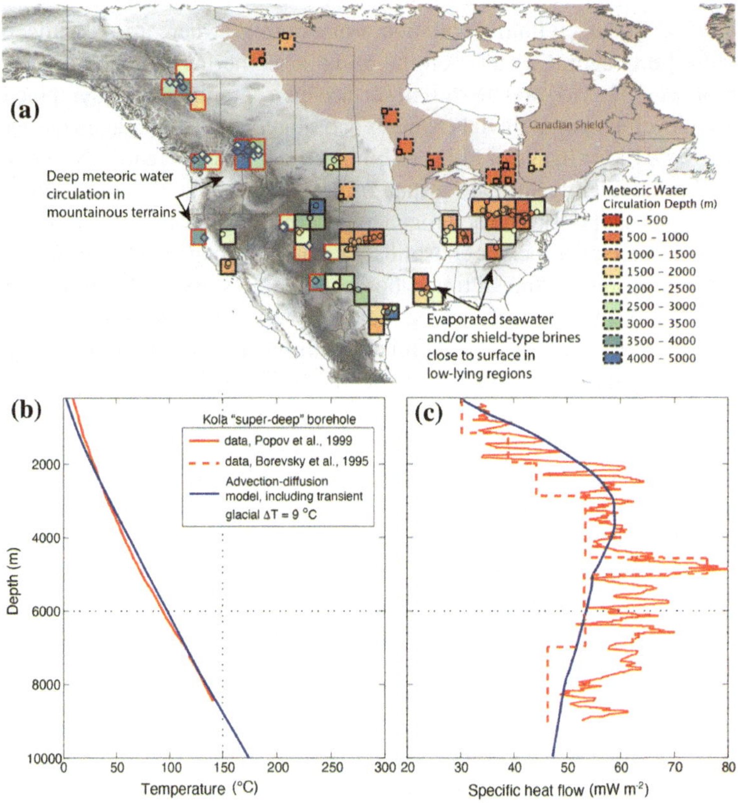

Fig. 4 **a** Depth of meteoric water circulation across North America based on water stable-isotope signatures (modified from [196]). **b**, **c** Data and representative model fits (modified from Vogt et al. [292]) from the "super-deep" borehole SG-3 on the Kola peninsula at the northern rim of the Fennoscandian Shield near the Finnish-Russian border. **b** Measured temperature versus depth (red curve), fit with a transient heat advection–diffusion model over the last 80 ka that includes a boxcar-shaped 9 °C surface temperature decrease associated with the Last Glacial Maximum. **c** Inferred vertical specific heat flow from two studies of the Kola temperature data (red curves) and model fit. Through a Bayesian inverse approach, Vogt et al. [292] infer notable vertical variations in permeability and fluid flow to depths of 7 km in this continental setting, and non-negligible coupling between paleoclimate and fluid advection in near-surface temperatures

has been inferred in deep boreholes within cratons as well (e.g., Vogt et al. [292]) and could play a role in deep CZ circulation generally. Multiscale fault and fracture networks, whether they arise from primary bedrock structure or faulting/tectonic deformation, also play a primary role in aqueous fluid transport even at kilometer depths [68]. The permeability of fault or fracture networks can evolve through geologic time as groundwater reacts with the rocks (increasing permeability, as in karst) or secondary minerals are precipitated (reducing permeability). Faults and fractures are not static and may open or close in response to gradual deformation of the solid Earth, or to rapid perturbation such as earthquakes [136, 268, 299].

Permeability and deep CZ processes likely coevolve with the topography at the surface that drives deep fluid circulation. For example, where a > 1 km deep CZ is inferred in the Oregon Cascades on the basis of isothermal borehole temperature-depth profiles [137, 275], a state shift in surface hydrographs and topographic form is also seen. Development of fluvial channels and ridge/valley topography initiates only on lava flows ~1 Myr in age or older, as high vertical transport capacity exhibited by younger flows evidently shifts to a state of dominant horizontal permeability (lava flow tops/bottoms) [144]. The heterogeneous permeability structure of the deep CZ, within unweathered bedrock or consolidated materials, can lead to variable groundwater residence times or transit times. For example, in some cases groundwater age distributions indicate relatively young waters (<60 years, based on the presence of modern age tracers, such as tritium) mixed with older groundwater (<50,000 years old, based on radiocarbon) up to ~1 km depth [19, 87, 141, 304].

Weathering within saturated fractured bedrock or porous media is controlled by the transit times and heterogeneous flowpaths of groundwater and its chemistry. Weathering reactions are commonly mediated by microbial communities that may leverage surficial inputs of dissolved gases (e.g., O_2, CO_2) and photosynthetically derived carbon or, at some (unknown) depth, be entirely cutoff from near-surface inputs [2, 16, 163]. These deeper microbial communities rely on lithogenic sources of energy to fuel chemolithoautotrophic growth and directly or indirectly enhance weathering reactions [2, 165, 208, 256]. Recharge events, such as deep infiltration of spring snowmelt, can enhance microbial activity and mineral weathering by creating hydrologic connectivity to the near-surface and delivering freshwater and carbon sources to the deep CZ [163, 213] that alter chemical potential and equilibrium. Thus, while it is generally thought that at increasing depths in the CZ, mineral weathering often becomes transport limited by the relatively long transit times of deeper flow systems [181] there are exceptions. This can lead to the development of weathering fronts from the "bottom up" towards the surface in locations where more permeable layers are present at depth.

Actively circulating groundwater that re-emerges at the surface on timescales of years to millennia can transport solutes derived from weathering in the deep CZ to streams [34, 160, 228, 254]. The depth of actively circulating groundwater varies as a function of topographic gradients, fluid density, and permeability distributions (e.g., fracture orientation, presence of confining units), with the greatest circulation depths (~1–4 km) in mountainous terrains [86, 196]. The contribution of groundwater, in terms of water volume, solutes and/or dissolved gases to streams varies as a function of the local climate, catchment characteristics (e.g., relief, geometry and subsurface architecture), and location within the catchment. In general, groundwater contributions to surface waters increase with decreasing elevation and increasing spatial scales [52, 88]. Groundwater has also been shown to contribute notably to streamflow even during high discharge events [283]. Deep CZ groundwater may be disconnected from streamflow seasonally or more permanently in areas where the regional water table is below the stream channel. Infiltration events or wetter periods that increase the water table elevation can lead to greater groundwater-surface water connectivity [195].

To date, most of our observations of groundwater in the deep CZ have come from a few, spatially disparate measurements (e.g., boreholes, wells, springs). More recent geophysical studies have enabled broader spatial and temporal resolution, and in some cases depth resolution, of the physical architecture (e.g., fracture distribution) of the deeper CZ and groundwater transport (St. Clair et al. 2015; Riebe et al. 2017; Holbrook et al. 2014; Parsekian et al. 2014), but often still only to some 10 s of meters. One tool that covers large spatial areas and extend to significant depths is airborne electromagnetic (AEM) surveys, which can unravel features of the subsurface architecture up to ~500 m depth in high resistivity rocks such as basalts (e.g., [80]) and up to ~300 m low resistivity materials such as sedimentary units (e.g., [204]).

Recent scientific drilling campaigns to collect core materials and install monitoring wells, guided by geophysical results, have helped to illuminate the CZ down to approximately 40 to 150 m (e.g., [33, 55, 131, 205]). Direct observations have been made of fluid and rock chemical composition, microbial activity, and rock hydraulic and electric properties, informing CZ structure and function. The use of downhole sensors to measure real-time groundwater levels and chemistry (e.g., pH, temperature, CO_2, O_2) have provided vital information on deep CZ dynamics (e.g., [185, 242, 304]). In addition, more thorough analysis of groundwater-age distributions, applications of isotopic tracers, and development of reactive solute-transport models, have greatly improved our understanding of groundwater flowpaths, transit times, and weathering reactions [186, 265]. Similar scientific or commercial drilling and fluid and rock characterization has been done previously in a few terrestrial locations to several km. These locations may provide opportunities for the CZ community to probe processes deeper and possibly more isolated from the surface both at present and in the geologic past (e.g., [64, 65]).

Scientists are just beginning to illuminate the deep CZ–from the water table to the ‘bottom’ of the CZ–to understand its processes, dynamics, and hydrologic connections with near-surface environments, with many outstanding questions remaining, including:

- How do near-surface inputs of water, gases, carbon, nutrients, solutes and energy influence porosity development and groundwater storage in the deep CZ?
- What is the depth of actively circulating groundwater that delivers lithogenic solutes from the deep CZ to surface waters and how does it vary depending on climate, landscape characteristics and subsurface architecture?

- How do we measure and incorporate subsurface heterogeneity into our models of deep CZ processes, particularly in fracture-dominated systems? And, how do we account for 'hot spots' of weathering that may occur at relatively small spatial scales (e.g., within fractures), but account for a large portion of solute fluxes at the hillslope to catchment scales?
- What are the feedbacks between long-term landscape evolution, surface hydrology, and deep CZ processes?

6 Interbasin Flows: Their Importance for Understanding Large-Scale Linkages in the CZ

An important and relatively under-studied aspect of groundwater flow in the CZ is "interbasin groundwater flow" (IGF), groundwater flow beneath surface topographic divides from recharge in one hydrologic basin (watershed or catchment) to discharge in another. IGF is one consequence of regional groundwater flow at the scale of tens to hundreds of kilometers.

Occurrence of IGF relies on the existence of "losing" and "gaining" catchments, the places where IGF flowlines start and end, respectively. As such, IGF represents an important connection between hydrogeology and catchment science. IGF complicates the often assumed or hoped-for condition that a small watershed is a self-contained unit of study for water resources, with inputs and outputs only across its upper surface and a stream export. The connection between IGF and catchments is often seen as an unwanted complication in catchment science, but is usually discovered and studied through catchment research. For example, IGF has been studied using measurements and models of watershed hydrology and water budgets [96, 206, 217, 251, 311], and chemical or isotopic signals in catchments where IGF discharges to streams and springs, including major ions and isotopes of oxygen, helium, carbon, chlorine, and strontium [88, 92, 95, 98, 99, 267].

Hydrogeologists have long been aware of rapid long-distance and hard-to-predict subsurface interbasin flow associated with obvious geological heterogeneities in karst aquifers. Toth [288] opened up the field by showing that IGF can occur even in homogeneous isotropic media, if the system is thick enough and if there are water table variations (sinusoidal in his paper) imposed on a regional trend in hydraulic head. This motivated decades of diverse work in regional groundwater flow, groundwater-surface water interaction, and catchment science aimed at better understanding IGF and its implications for water quality, design and interpretation of hydrologic studies, ecosystem C and N budgets, and more. Exploring and synthesizing work on IGF, Fan [71] argued that we as a community should conceptualize catchments as semi-closed hydrologic units that are nested on top of the larger regional hydrogeologic system.

One conundrum that IGF creates for biogeochemists is how to interpret water chemistry data, both in places where IGF is recognized, and where it occurs but goes unrecognized. For example, chemical weathering rates are sometimes estimated from the rate of stream export of major ions, expressed as solute mass per watershed area per year (e.g., [104]). Such weathering estimates could be biased high in gaining watersheds and low in losing watersheds unless the IGF is known and accounted for. Solute fluxes by IGF may be large: in a lowland rainforest in Costa Rica, IGF accounted for 90–99% of the major ion inputs to the Arboleda watershed, and atmospheric deposition only 1–10% [96]. Oviedo-Vargas [215] found high rates of CO_2 emission from the Arboleda stream to air. This result might be wrongly interpreted in terms of ecosystem respiration or another biogeochemical process if it were not known that the cause is IGF; the stream receives old groundwater with elevated CO_2 from outgassing of the underlying volcanic system. More broadly, in terms of the overall C budget of this lowland rainforest (Fig. 5) [97, 215], the Arboleda watershed has a net ecosystem exchange (NEE) of about 250 g C per m^2 of watershed per year (a net input of carbon from the atmosphere). With stream degassing of 300 gC/m^2yr (thought not to be captured by the NEE estimate) and stream export of 700 gC/m^2yr, the watershed would appear as a net C source if IGF is not considered (250–700–300 = −750 gC/m^2yr). Including the C input from IGF (860 gC/m^2yr) changes the interpretation of the watershed from a large source to a small sink (about 110 gC/m^2yr), a fundamentally different view of the ecosystem C budget and its connection to climate.

The degree to which IGF is sensitive to changing climatic conditions is relatively unknown. Schaller and Fan [251] used a hydrologic budget approach to assess whether 1555 watersheds in the 48 states of the contiguous U.S. were losing or gaining water by IGF. IGF was more prominent in drier climates; the largest deviations from a "closed" water budget (i.e., a budget with no IGF) occurred in watersheds with lower annual rainfall, especially below 500 mm [251]. Groundwater simulation results in a regional-scale vertical cross section showed that a lower water-table position under low recharge led to groundwater discharge that was less evenly distributed regionally among local elevation minima (small valleys) and more focused as IGF farther downgradient at lower elevation in the region. In other words, lower recharge led to drying of higher-elevation groundwater-discharge areas, and a greater proportion of

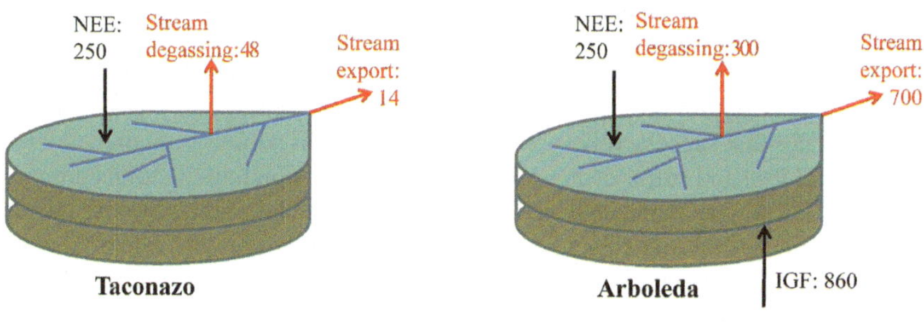

Fig. 5 Carbon flux schematic for adjacent lowland rainforest watersheds at La Selva biological station in Costa Rica, updated from [97]. The arboleda receives inputs by IGF, the Taconazo does not. Black arrows are inputs, red are outputs. Units are g C per m^2 of watershed area per year. Stream export includes both DIC and DOC. NEE represents an average for 1998–2000 data from Loescher et al. [172], stream degassing values are from Oviedo Vargas et al. [215]

discharge as IGF. These findings may be particularly relevant to changing groundwater conditions under drier climates, for example in the western U.S. And impacts may involve feedbacks from groundwater to climate; for example, persistent groundwater discharge (often associated with IGF discharge areas) may slightly increase annual rainfall and notably reduce the amplitude of year-to-year variation in rainfall, through precipitation recycling feedbacks between wetter soils and evaporation [21].

Given the strong controls IGF has on spatial redistribution of water and nutrients, which impacts CZ functioning, several key knowledge gaps should be addressed:

- How can and should IGF influence the interpretation of water chemical data?
- What is the connection between IGF and climate?
- How sensitive is CZ functioning to shifts in IGF?

7 Coastal Groundwater Process in the CZ

Coastlines are a key interface between land and sea where hydrodynamics and strong geochemical and ecosystem gradients intensify critical zone processes. Groundwater plays an important role in these processes [245], as aquifers host zones of mixing between water of terrestrial and marine origin with a wide range of residence times and geochemical characteristics. Groundwater is also a major flux to coastal surface waters, rivaling or exceeding rivers in volumetric discharge [162] and contributing substantial solute loads [277] that affect marine ecosystems [166].

Coastal systems are complex, with ecological and biogeochemical processes tightly coupled to hydrological forcings that act on timescales from seconds (i.e., waves) to millennia (i.e., glacial-cycle sea-level change). An example of this coupling is apparent in intertidal beach aquifers. There, storms, tides, and waves cause runup of seawater that infiltrates the beach face, mixes with through-flowing freshwater, and circulates seaward (Fig. 6, processes 1 and 2). Beneath the freshwater is another mixing zone where saltwater convection is driven by density gradients along a deeper freshwater-saltwater interface (Fig. 6, process 3). Infiltrating seawater contributes oxygen and organic matter, whereas through-flowing freshwater is often anoxic, low in carbon, and high in iron, nutrients, and other land-derived solutes. Mixing of these waters within the beach groundwater system creates redox gradients and reaction zones that successively support reactions such as aerobic oxidation, denitrification, and iron and sulfur cycling [40, 150] and corresponding microbial communities [192]. The geochemical system is strongly connected to the physical system, as hydrologic setting and subsurface characteristics determine the rate and extent of geochemical transformations [100, 122]. Hydrologic transience also enhances mixing and reactivity. For example, movement of the intertidal circulation cell (Fig. 6) due to tidal and seasonal shifts causes contact between mobile and immobile reactants [152].

Groundwater also plays an important role in the evolution of ecosystems at the land-sea margin. Long-timescale dynamics like sea-level rise and land-use change as well as short-timescale oscillations and events such as tides, seasons, and storms,

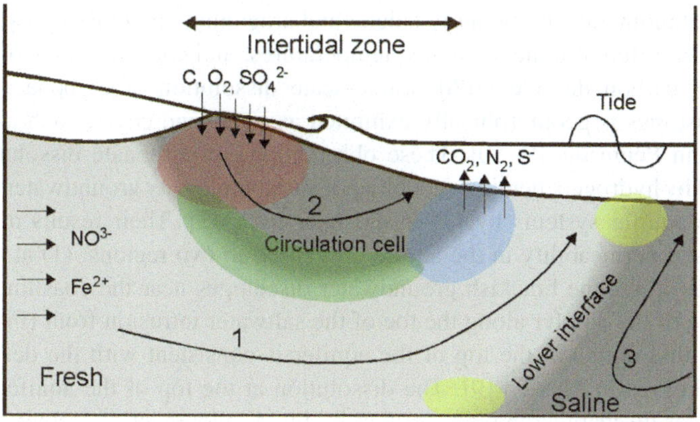

Fig. 6 Processes governed by the interaction of fresh and saline groundwater in an intertidal beach aquifer. Darker shading indicates saline water. Flow paths are shown as numbered black arrows: (1) freshwater driven by the upland hydraulic gradient; (2) intertidal saltwater circulation driven by tides and waves, and (3) offshore circulation driven by density gradients. Key geochemical zones influenced by freshwater-saltwater mixing and circulation are shaded in color. Red: swash zone where seawater carrying reactants such as carbon, oxygen, and sulfate infiltrate into the beach leading to aerobic respiration; Green: zone where denitrification and other mixing-driven reactions occur. Blue: discharge zone where reduced species carried in anoxic groundwater (e.g., Fe(II) react with oxygenated saltwater forming precipitates (e.g., Fe(oxy)hydroxides); Yellow: zones where carbonate dissolution alters aquifer porosity and permeability [123]. Adapted from and Heiss et al. [245]

are actively changing coastal environments [94, 286]. Tree mortality, reduction in crop yields, and other ecosystem changes occur as a result of flooding and salinization [4, 282, 307]. Rising sea level raises the hydrologic base level, increasing water table elevations, shrinking vadose zones and drowning plant roots. Increases in groundwater and soil salinity are driven by sea-level rise, inland hydrologic shifts such as pumping, and episodic extreme high tides and storms. The episodic events superposed on longer-timescale changes can reduce ecosystem resilience to a tipping point of large-scale change [149, 286]. While ghost forests [153] and crop damage are stark features of the landscape, concurrent inland migration of salt marshes brings ecosystem services [56] such as flood mitigation [93] and carbon sequestration [43], and counteracts marsh loss due to drowning and erosion [154].

An example of the strong two-way coupling between subsurface hydrology and geochemistry occurs in coastal carbonate aquifers. Due to their typically high hydraulic conductivity, coastal carbonates are particularly susceptible to saltwater intrusion (i.e., inland movement of the lower freshwater-saltwater interface; Fig. 6). Coastal CZ processes in carbonates involve mineral dissolution, precipitation and diagenetic reactions that can lead to a redistribution of porosity and hydraulic conductivity as well as a release of phosphorus which can stimulate ecosystem productivity. Carbonate mineral dissolution in regions where groundwater has varying salinity have been well documented to result from undersaturated conditions with respect to carbonate minerals, particularly calcite and aragonite, upon mixing two groundwaters with different ionic strengths, temperatures, and concentrations of calcium, carbonate, carbon dioxide [306]. Large-scale dissolution of carbonate minerals along coastlines is geomorphically exhibited as caves and coves, such as those in the Yucatan Peninsula [9, 10]. These observations of carbonate dissolution were supported by hydrogeochemical modeling of variable-density groundwater in coastal carbonate aquifer systems by [241] and later by [233]. Their results determined porosity and permeability in the aquifer increased in two regions: (1) at the top of the aquifer where the brackish groundwater discharges near the coastline, and (2) at the base of the aquifer along the toe of the saltwater intrusion front (Fig. 6). The enhanced dissolution at the top of the aquifer is consistent with the development of coastal caves in Mexico [9]. The dissolution at the top of the aquifer near the groundwater discharge zone is amplified by the development of a convection cell, which results in a positive mineral-dissolution feedback loop where the increase in aquifer permeability leads to increased seawater intrusion into the aquifer and further dissolution.

Other reactions observed in coastal carbonate CZ regions exposed to seawater intrusion include calcite precipitation and aragonite neomorphism (conversion of aragonite to calcite), processes that can lead to a decrease in the porosity and hydraulic conductivity and reduce carbonate weathering in the CZ [9]. From example, supersaturated conditions with respect to calcite were observed in brackish groundwaters collected in Florida [212] and Spain [311]. A variety of hypotheses have been put forth to describe the differences in saturation states observed between the various carbonate CZs affected by saltwater intrusion such as groundwater-table fluctuations, water-to-rock ratio or initial porosity/permeability distribution, changes in the partial pressures of carbon dioxide, and the amount of organic carbon present [212, 233, 311].

Saltwater intrusion into coastal carbonate CZs has been found to release phosphorus from the carbonate bedrock/soils stimulating primary production. The release of phosphorus from the bedrock occurs by two processes, desorption upon exposure to very low concentrations of salt, and the dissolution of calcium carbonate minerals with different mixtures of fresh and saltwater [82, 83, 224]. Saltwater intrusion and the associated release of phosphorus has been found to stimulate primary production along the carbonate CZ of the Florida coastal Everglades [156, 225]. In freshwater conditions, the limestone bedrock has a strong affinity to adsorb phosphorus, most likely contributing to the naturally oligotrophic conditions in the Everglades [84]. However, as sea-level continues to rise, greater portions of the freshwater aquifer will be exposed to saltwater thereby increasing the availability of phosphorus in this CZ. Overall, this will push the system further away from oligotrophy and alter biotic processes critical to the functions of the Everglades.

These examples highlight the complex role of groundwater in the coastal CZ. Improved understanding of the feedbacks between hydrology, ecology, geochemistry, and geomorphology of these systems through observational networks and development of coupled process models [300] will allow us to tackle questions such as:

- How do physical forcings, geochemical gradients, and geomorphological settings alter landscapes and elemental fluxes along the land-sea margin?
- How will future changes in climate and land use alter coastal systems, and what are the dominant timescales of change?
- How are ecosystem processes co-evolving with porosity and hydraulic conductivity of coastal carbonate CZ systems exposed to seawater intrusion?
- How does the release of phosphorus from coastal carbonate aquifers contribute to ecosystem processes in CZ areas?

8 CZ Vision for Addressing Society's Water and Ecological Problems

Illustrated above is the fact that groundwater is a vital natural resource for humans and ecosystems around the world, and that a CZ approach is fundamental to tackling current and future groundwater threats and feedbacks to other environmental systems. Ensuring access to potable groundwater is also critical to global economic health. For example, groundwater is a major contributor to streamflow generation, which plays an important economic role in regional fisheries. Streamflow also strongly interacts with silviculture and agriculture practices; thus the management of one resource economy influences the other. Knowledge of where plants access water in the subsurface, the amount of water stored in diverse lithologies, and the sensitivities of these fluxes to more variable weather patterns creates the need to co-develop land-use practices that are evaluated and updated to account for emerging technologies, social and economic needs, and our understanding of CZ science. Hidden within this hydrologic problem are the impacts on terrestrial carbon dynamics. Layered throughout the world is the legacy of land ownership and Indigenous rights. Intertwined within our natural systems is constructed infrastructure, altered flows, and chemical controls. Developing groundwater solutions for such complex systems requires bringing diverse minds to the table and thinking about the entire system from the top of the canopy down to the depths of circulating groundwater.

Groundwater and vadose-zone moisture play a fundamental role in sustaining life on Earth. The contents and questions outlined above can be simplified into three areas highlighted by the U.S. National Academies *Earth and Time Report* (2021): (1) How does the CZ influence climate? (2) How is the water cycle changing and what does this mean for CZ function? And (3) What role do hydrologic changes play in altering how biogeochemical cycles evolve? Addressing these questions requires coordination among national and international agencies, institutions of higher education, and profit and non-profit organizations. For example, uniting existing geophysical and drill-core data repositories from mineral extraction companies and national geological agencies with satellite (e.g., GRACE) and airborne (e.g., AEM) geophysical data and targeted field observations could allow us to construct transferable knowledge on Earth's near surface geologic architecture and what controls it. Offering managers, stakeholders, and policy makers the foundation for developing sustainable groundwater extraction plans.

One important component to deriving solutions are observatories—places where groundwater is measured along with other CZ variables (e.g., geologic structure, rooting distributions, net primary productivity) to develop a process-based understanding. Groundwater is among the most difficult CZ variables to measure, characterize, and predict. Thus, observatories play an important role in both the collection of physical measurements and a place where scientists from diverse disciplinary and methodological backgrounds can get together to co-develop and synthesize these perspectives, and generate new hypotheses and models for improved application outside of the narrow confines of the observatories themselves. Additionally, observatories are locations where we can understand the impacts of climate or weather (e.g., drought, fire, flooding) and land-cover disturbance on groundwater resources, and untangle the effects of compound or co-occurring disturbances. For example, how will groundwater quantity and quality respond to drought and insect infestation followed by fire in a forested terrain? Or how do we manage groundwater resources in areas where streamflow is declining, mean annual precipitation is increasing, warmer temperatures are prompting higher evapotranspiration rates, and the biota are shifting from grass to woody dominance? Given that the response of biota and climate-carbon cycle feedbacks appears to be accelerating at the global scale [170, 266] the need for observatories that quantify the impact of these changes on groundwater is necessary for projecting water security into the future.

Acknowledgements We acknowledge funding support from several U.S. National Science Foundation programs: Signals in the Soils Hydrology Directorate (PLS: EAR-2024388), Frontiers Research in Earth Science (FRES; PLS and KS: EAR-2121694; JM EAR-2120733), Research Coordination Networks (PLS and KS: EAR-1904527), Critical Zone Collaborative Network (PLS: EAR-1904527, HRB: EAR-2012669, KS and LL: EAR-2012730; HM: EAR-2012484), Long-Term Ecological Research Program [RMP: DEB-9910514, DBI-0620409, DEB-1237517, DEB-1832229, DEB-2025954] and Hydrologic Sciences Program (DPG: EAR-0049047, −0439732, and −1744714, HRB: EAR-1446161).

References

1. Acharya BS, Halihan T, Zou CB et al (2017) Vegetation controls on the Spatio-temporal heterogeneity of deep moisture in the unsaturated zone: a hydrogeophysical evaluation. Sci Rep 7:1499. https://doi.org/10.1038/s41598-017-01662-y
2. Akob DM, Küsel K (2011) Where microorganisms meet rocks in the Earth's critical zone. Biogeosciences 8(12):3531–3543
3. Allen-Diaz BH (1991) Water table and plant species relationships in Sierra Nevada Meadows. Am Midl Nat 126(1):30–43. https://doi.org/10.2307/2426147
4. Anderson F, Al-Thani N (2016) Effect of sea level rise and groundwater withdrawal on seawater intrusion in the gulf coast aquifer: implications for agriculture. J Geosci Environ Prot 4(04):116
5. Anderson SP, Bales RC, Duffy CJ (2008) Critical zone observatories: building a network to advance interdisciplinary study of Earth surface processes. Mineral Mag 72(1):7–10

6. Anderson SP, von Blanckenburg F, White AF (2007) Physical and chemical controls on the critical zone. Elements 3(5):315–319
7. Anyah RO, Weaver CP, Miguez-Macho G, Fan Y, Robock A (2008) Incorporating water table dynamics in climate modeling: 3 simulated groundwater influence on coupled land-atmosphere variability. J Geophys Res: Atmos 113(D7)
8. Argus DF, Landerer FW, Wiese DN, Martens HR et al (2017) Sustained water loss in California's mountain ranges during severe drought from 2012 to 2015 inferred from GPS. J Geophys Res: Solid Earth 122(12):10–559
9. Back W, Hanshaw BB, Herman JS, Van Driel JN (1986) Differential dissolution of a Pleistocene reef in the ground-water mixing zone of coastal Yucatan. Mexico Geol 14(2):137–140
10. Back W, Hanshaw BB, Pyle TE, Plummer LN, Weidie AE (1979) Geochemical significance of groundwater discharge and carbonate solution to the formation of Caleta Xel Ha, Quintana Roo. Mex Water Resour Res 15(6):1521 – 1535
11. Bales RC, Hopmans JW, O'Geen AT, Meadows M, Hartsough PC, Kirchner P, Hunsaker CT, Beaudette D (2011) Soil moisture response to snowmelt and rainfall in a Sierra Nevada Mixed-Conifer Forest. Vadose Zone J 10(3):786–799
12. Barbeta A, Peñuelas J (2017) Relative contribution of groundwater to plant transpiration estimated with stable isotopes. Sci Rep 7(1):1–10. https://doi.org/10.1038/s41598-017-09643-x
13. Barnard HR, Graham CB, Van Verseveld WJ, Brooks JR, Bond BJ, McDonnell JJ (2010) Mechanistic assessment of hillslope transpiration controls of diel subsurface flow: a steady-state irrigation approach. Ecohydrol: Ecosyst, Land Water Process Interact, Ecohydrogeomorphology 3(2):133–142
14. Bazilevskaya E, Rother G, Mildner DF, Pavich M, Cole D, Bhatt MP, Brantley SL (2015) How oxidation and dissolution in diabase and granite control porosity during weathering. Soil Sci Soc Am J 79(1):55–73
15. Befus KM, Barnard PL, Hoover DJ, Hart JF, Voss CI (2020) Increasing threat of coastal groundwater hazards from sea-level rise in California. Nat Clim Chang 10(10):946–952
16. Benk SA, Yan L, Lehmann R, Roth V, Schwab VF, Totsche KU, Küsel K, Gleixner G (2019) Fueling diversity in the subsurface: composition and age of dissolved organic matter in the critical zone. Front Earth Sci 296
17. Bernhardt ES, Blaszczak JR, Ficken CD, Fork ML, Kaiser KE, Seybold EC (2017) Control points in ecosystems: moving beyond the hot spot hot moment concept. Ecosystems 20:665–682
18. Bestland E, Milgate S, Chittleborough D, VanLeeuwen J, Pichler M, Soloninka L (2009) The significance and lag-time of deep through flow: an example from a small, ephemeral catchment with contrasting soil types in the Adelaide Hills, South Australia. Hydrol Earth Syst Sci 13:1201–1214
19. Bethke CM, Johnson TM (2008) Groundwater age and groundwater age dating. Annu Rev Earth Planet Sci 36(1):121–152. https://doi.org/10.1146/annurev.earth.36.031207.124210
20. Betts AK, Ball JH, Beljaars ACM, Miller MJ, Viterbo PA (1996) The land surface-atmosphere interaction: a review based on observational and global modeling perspectives. J Geophys Res 101(D3):7209–7225. https://doi.org/10.1029/95JD02135
21. Bierkens MFP, van den Hurk JJM (2007) Groundwater convergence as a possible mechanism for multi-year persistence in rainfall. Geophys Res Lett 34(L02402). https://doi.org/10.1029/2006GL028396
22. Billings SA, Sullivan L, Hirmas D, Nippert L, Richter DD (2022) The CZ as an ecological problem: how the interplay of biotic and abiotic actors determines the functioning of Earth's living skin in critical zone and ecosystems dynamics Eds Springer volume
23. Blöschl G, Bierkens MFP, Chambel A, Cudennec C, Destouni G, Fiori A et al (2019) Twenty-three unsolved problems in hydrology (UPH)–a community perspective. Hydrol Sci J 764(10):1141–1158

24. Brantley SL, Goldhaber MB, Ragnarsdottir KV (2007) Crossing disciplines and scales to understand the critical zone. Elements 3(5):307–314
25. Briggs MA, Day-Lewis FD, Mahmood Poor Dehkordy F, Hampton T et al (2018) Direct observations of hydrologic exchange occurring with less-mobile porosity and the development of anoxic microzones in sandy lakebed sediments. Water Resour Res. https://doi.org/10.1029/2018WR022823
26. Briggs MA, Day-Lewis FD, Zarnetske JP, Harvey JW (2015) A physical explanation for the development of redox microzones in hyporheic flow. Geophys Res Lett 42(11):4402–4410
27. Brookfield AE, Macpherson GL, Covington MD (2017) Effects of changing meteoric precipitation patterns on groundwater temperature in Karst environments. Groundwater 55(2):227–236
28. Brookfield AE, Hansen AT, Sullivan PL, Czuba JA, Kirk MF, Li L, Newcomer ME, Wilkinson G (2021) Predicting algal blooms: are we overlooking groundwater? Sci Total Environ 769 (144442). https://doi.org/10.1016/j.scitotenv.2020.144442
29. Brookfield AE, Hill MC, Rodell M, Loomis BD, Stotler RL, Porter ME, Bohling GC (2018) In situ and GRACE-based groundwater observations, similarities, discrepancies, and evaluation in the high plains aquifer in Kansas. Water Resour Res 54:8034–8044
30. Brooks PD, Chorover J, Fan Y, Godsey SE, Maxwell RM, McNamara JP, Tague C (2015) Hydrological partitioning in the critical zone: recent advances and opportunities for developing transferable understanding of water cycle dynamics. Water Resour Res 51(9):6973–6987. https://doi.org/10.1002/2015WR017039
31. Burgess SS, Adams MA, Turner NC, Ong CK (1998) The redistribution of soil water by tree root systems. Oecologia 115(3):306–311
32. Burri NM, Weatherl R, Moeck C, Schirmer M (2019) A review of threats to groundwater quality in the anthropocene. Sci Total Environ 684:136–154
33. Buss HL, Brantley SL, Scatena FN, Bazilievskaya EA, Blum A, Schulz M, Cole D (2013) Probing the deep critical zone beneath the Luquillo experimental Forest, Puerto Rico. Earth Surf Proc Land 38(10):1170–1186
34. Buss HLC, Lara M, Moore OW, Kurtz AC, Schulz MS, White AF (2017) Lithological Influences on contemporary and long-term regolith weathering at the Luquillo critical zone observatory. Geochim Cosmochim Acta 196:224–251
35. Butler JJ, Kluitenberg K, Whittemore DO, Loheide SP, Jin W, Billinger MA, Zhan X (2007) A field investigation of phreatophyte-induced fluctuations in the water table. Water Resour Res 43. https://doi.org/10.1029/2005WR004627
36. Büttner G, Huenges E (2003) The heat transfer in the region of the Mauna Kea (Hawaii)—constraints from borehole temperature measurements and coupled thermo-hydraulic modeling. Tectonophysics 371(1–4), 23–40
37. Caldwell MM, Dawson TE, Richards JH (1998) Hydraulic lift: consequences of water efflux from the roots of plants. Oecologia 113:151–161
38. Callesen I, Harrison R, Stupak I, Hatten J, Raulund-Rasmussen K, Boyle J, Zabowski D (2016) Carbon storage and nutrient mobilization from soil minerals by deep roots and rhizospheres. For Ecol Manage 359:322–331
39. Champ DR, Gulens J, Jackson RE (1979) Oxidation–reduction sequences in ground water flow systems. Can J Earth Sci 16(1):12–23
40. Charette MA, Sholkovitz ER (2002) Oxidative precipitation of groundwater-derived ferrous iron in the subterranean estuary of a coastal bay. Geophys Res Lett 29(10):85–91
41. Chen F, Dudhia J (2001) Coupling an advanced land surface-hydrology model with the Penn State–NCAR MM5 modeling system part I: model implementation and sensitivity. Mon Weather Rev 129:569–585. https://doi.org/10.1175/1520-0493(2001)129%3c0569:CAALSH%3e2.0s.CO;2
42. Chimner R, Resh S (2014) Are riparian bur oak phreatophytic? A stable water isotope study in homestead national monument, Nebraska. Nat Areas J 34:56–64. https://doi.org/10.3375/043.034.0107

43. Chmura GL, Anisfeld SC, Cahoon DR, Lynch JC (2003) Global carbon sequestration in tidal, saline wetland soils. Global Biogeochem Cycles 17(4):12
44. Chowdhury SR, Zarnetske JP, Phanikumar MS, Briggs MA, Day-Lewis FD, Singha K (2020) Formation criteria for hyporheic anoxic microzones: assessing interactions of hydraulics, nutrients and biofilms. Water Resour Res. https://doi.org/10.1029/2019WR025971
45. Chu C, Jones NE, Mandrak NE, Piggott AR, Minns CK (2008) The influence of air temperature, groundwater discharge, and climate change on the thermal diversity of stream fishes in southern Ontario watersheds. Can J Fish Aquat Sci 65(2):297–308
46. Chui TFM, Low SY, Liong S (2011) An ecohydrological model for studying groundwater-vegetation interactions in wetlands. J Hydrol 409(1–2):291–304
47. Ciruzzi DM, Loheide SP II (2021) Groundwater subsidizes tree growth and groundwater use in sandy humid forests. Ecohydrol 14:e2294. https://doi.org/10.1016/j.scitotenv.2023.162577
48. Clauser C, Giese P, Huenges E, Kohl T, Lehmann H, Rybach L, Zoth G (1997) The thermal regime of the crystalline continental crust: implications from the KTB. J Geophys Res: Solid Earth 102(B8):18417–18441
49. Cleverly J, Dahm C, Thibault J, Mcdonnell D, Coonrod J (2006) Riparian ecohydrology: regulation of water flux from the ground to the atmosphere in the Middle Rio Grande, New Mexico. Hydrol Process 20:3207–3225. https://doi.org/10.1002/hyp.6328
50. Coble AA, Barnard H, Du E, Johnson S, Jones J, Keppeler E, Wagenbrenner J (2020) Long-term hydrological response to forest harvest during seasonal low flow: potential implications for current forest practices. Sci Total Environ 730(138926)
51. Condon LE, Atchley AL, Maxwell RM (2020a) Evapotranspiration depletes groundwater under warming over the contiguous United States. Nat Commun 11(1):873
52. Condon LE, Maxwell RM (2015) Evaluating the relationship between topography and groundwater using outputs from a continental-scale integrated hydrology model. Water Resour Res 51(8):6602–6621
53. Condon LE, Maxwell RM (2019) Simulating the sensitivity of evapotranspiration and streamflow to large-scale groundwater depletion. Sci Adv 5(6):eaav4574. https://doi.org/10.1126/sciadv.aav4574
54. Condon LE, Markovich KH, Kelleher CA, McDonnell JJ, Ferguson G, McIntosh JC (2020) Where is the bottom of a watershed? Water Resour Res 56(3):e2019WR026010
55. Coralie R, François C, Daniel V, Thierry L, Yann L, Julien A, Luc A (2021) Characterization of groundwater circulations in a headwater catchment from an analysis of chemical concentrations, Sr-Nd-U isotope ratios, and CFC, SF6 gas tracers (Strengbach CZO, France). Appl Geochem 105030
56. Costanza R, d'Arge R, De Groot R, Farber S, Grasso M, Hannon B, Van Den Belt M (1997) The value of the world's ecosystem services and natural capital. Nature 387(6630):253–260
57. Cuthbert MO, Gleeson T, Moosdorf N, Befus KM, Schneider A, Hartmann J, Lehner B (2019) Global patterns and dynamics of climate–groundwater interactions. Nat Clim Chang 9(2):137–141
58. Dawson TE (1993) Hydraulic lift and water use by plants: implications for water balance, performance and plant-plant interactions. Oecologia 95(4):565–574
59. de Graaf IE Gleeson T, Van Beek LPH, Sutanudjaja EH, Bierkens MF (2019) Environmental flow limits to global groundwater pumping. Nature 574(7776):90–94
60. Domec JC, King JS, Noormets A, Treasure E, Gavazzi MJ, Sun G, McNulty SG (2010) Hydraulic redistribution of soil water by roots affects whole-stand evapotranspiration and net ecosystem carbon exchange. New Phytol 187:171–183. https://doi.org/10.1111/j.1469-8137.2010.03245.x
61. Domec JC, Warren JM, Meinzer FC, Brooks JR, Coulombe R (2004) Native root xylem embolism and stomatal closure in stands of douglas-fir and ponderosa pine: mitigation by hydraulic redistribution. Oecologia 141:7–16. https://doi.org/10.1007/s00442-004-1621-4
62. Dong X, Cohen ML, Martin JB, McLaughlin DL, Murray AB, Ward ND, Flint MK, Heffernan JB (2019) Ecohydrologic processes and soil thickness feedbacks control limestone-weathering rates in a karst landscape. Chem Geol 527:118774. https://doi.org/10.1016/j.chemgeo.2018.05.021

63. Drake H, Ivarsson M, Bengtson S, Heim C, Siljeström S, Whitehouse MJ, Åström ME (2017) Anaerobic consortia of fungi and sulfate reducing bacteria in deep granite fractures. Nat Commun 8(1):1–9
64. Drake H, Tullborg E-L, MacKenzie AB (2009) Detecting the near-surface redox front in crystalline bedrock using fracture mineral distribution, geochemistry and U-series disequilibrium. Appl Geochem 24:1023–1039
65. Druhan JL, Fernandez N, Wang J, Dietrich WE, Rempe D (2017) Seasonal shifts in the solute ion ratios of vadose zone rock moisture from the Eel river critical zone observatory. Acta Geochimica 36:385–388
66. Duff JH, Triska FJ (1990) Denitrification in sediments from the hyporheic zone adjacent to a small forested stream. Can J Fish Aquat Sci 47(6):1140–1147
67. Elberling B, Larsen F, Christensen S, Postma D (1998) Gas transport in a confined unsaturated zone during atmospheric pressure cycles. Water Resour Res 34(11):2855−2862
68. Emmermann R, Lauterjung J (1997) The German continental deep drilling program KTB: overview and major results. J Geophys Res: Solid Earth 102(B8):18179–18201
69. Evaristo J, McDonnell JJ (2017) Prevalence and magnitude of groundwater use by vegetation: a global stable isotope meta-analysis. Sci Rep 7(1):1–12
70. Famiglietti JS, Ferguson G (2021) The hidden crisis beneath our feet. Science 372(6540):344–345
71. Fan Y (2019) Are catchments leaky? WIREs Water 6:e1386. https://doi.org/10.1002/wat2.1386
72. Fan Y, Li H, Miguez-Macho G (2013) Global patterns of groundwater table depth. Science 339(6122):940–943
73. Faunt CC, Sneed M, Traum J, Brandt JT (2016) Water availability and land subsidence in the Central Valley, California, USA. Hydrogeol J 24(3):675–684
74. Fellows AW, Goulden ML (2017) Mapping and understanding dry season soil water drawdown by California montane vegetation. Ecohydrology 10(1):e1772
75. Ferguson G, Cuthbert MO, Befus K, Gleeson T, McIntosh JC (2020) Rethinking groundwater age. Nat Geosci 13(9):592–594
76. Ferguson G, McIntosh JC, Warr O, Sherwood Lollar B, Ballentine CJ, Famiglietti JS et al (2021) Crustal groundwater volumes greater than previously thought. Geophys Res Lett 48:e2021GL093549. https://doi.org/10.1029/2021GL093549
77. Ferguson IM, Maxwell RM (2010) The role of groundwater in watershed response and land surface feedbacks under climate change Water. Resour Res 46. https://doi.org/10.1029/2009WR008616
78. Field JP, Breshears DD, Law DJ, Villegas JC, López-Hoffman L et al (2015) Critical zone services: expanding context, constraints, and currency beyond ecosystem services. Vadose Zone J 14(1):1–7
79. Findlay S (1995) Importance of surface-subsurface exchange in stream ecosystems: the hyporheic zone. Limnol Oceanogr 40(1):159–164
80. Finn CA, Sisson TW, Deszcz-Pan M (2001) Aerogeophysical measurements reveal collapse prone hydrothermally altered zones at mount rainier Volcano, Washington. In: 14th EEGS symposium on the application of geophysics to engineering and environmental problems. European Association of Geoscientists and Engineers, pp cp−192
81. Flinchum BA, Holbrook WS, Parsekian AD, Carr BJ (2019) Characterizing the critical zone using borehole and surface nuclear magnetic resonance. Vadose Zone J 18(1):1–18
82. Flower H, Rains M, Lewis D, Zhang JZ, Price R (2016) Control of phosphorus concentration through adsorption and desorption in shallow groundwater of subtropical carbonate estuary. Estuar Coast Shelf Sci 169:238–247
83. Flower H, Rains M, Lewis D, Zhang JZ, Price R (2017) Saltwater intrusion as potential driver of phosphorus release from limestone bedrock in a coastal aquifer. Estuar Coast Shelf Sci 184:166–176
84. Flower H, Rains M, Taşcı Y, Zhang JZ, Trout K, Lewis D, Das A, Dalton R (2022) Why is calcite a strong phosphorus sink in freshwater? Investigating the adsorption mechanism using batch experiments and surface complexation modeling. Chemosphere 286:131596

85. Foley JA, Ramankutty N, Brauman KA, Cassidy ES, Gerber JS, Johnston M, Zaks DP (2011) Solutions for a cultivated planet. Nature 478(7369):337–342
86. Frisbee MD, Tolley DG, Wilson JL (2017) Field estimates of groundwater circulation depths in two mountainous watersheds in the western US and the effect of deep circulation on solute concentrations in streamflow. Water Resour Res 53(4):2693–2715
87. Frisbee MD, Wilson JL, Gomez-Velez JD, Phillips FM, Campbell AR (2013) Are we missing the tail (and the tale) of residence time distributions in watersheds? Geophys Res Lett 40(17):4633–4637
88. Frisbee MD, Phillips FM, Campbell AR, Liu F, Sanchez SA (2011) Streamflow generation in a large, alpine watershed in the southern Rocky Mountains of Colorado: is streamflow generation simply the aggregation of hillslope runoff responses? Water Resour Res 47(6)
89. Fu C, Wang G, Bible K, Goulden ML, Saleska SR, Scott RL, Cardon ZG (2018) Hydraulic redistribution affects modeled carbon cycling via soil microbial activity and suppressed fire. Glob Change Biol 24(8):3472–3485
90. Fuller CC, Harvey JW (2000) Reactive uptake of trace metals in the hyporheic zone of a mining-contaminated stream, Pinal Creek, Arizona. Environ Sci Technol 34(7):1150–1155
91. Gandy CJ, Smith JWN, Jarvis AP (2007) Attenuation of mining-derived pollutants in the hyporheic zone: a review. Sci Total Environ 373(2–3):435–446
92. Ganong CN, Small GE, Ardón M, McDowell WH, Genereux DP, Duff JH, Pringle CM (2015) Interbasin flow of geothermally-modified groundwater stabilizes stream exports of biologically important solutes against variation in precipitation. Freshw Sci 34(1):276–286. https://doi.org/10.1086/679739
93. Gedan KB, Kirwan ML, Wolanski E, Barbier EB, Silliman BR (2011) The present and future role of coastal wetland vegetation in protecting shorelines: answering recent challenges to the paradigm. Clim Change 106(1):7–29
94. Gedan KB, Silliman BR, Bertness MD (2009) Centuries of human-driven change in salt marsh ecosystems. Ann Rev Mar Sci 1:117–141
95. Genereux DP (2004) Comparison of naturally-occurring chloride and oxygen-18 as tracers of interbasin groundwater transfer in lowland rainforest, Costa Rica. J Hydrol 295:17–27. https://doi.org/10.1016/j.jhydrol.2004.02.020
96. Genereux DP, Jordan MT, Carbonell D (2005) A paired-watershed budget study to quantify interbasin groundwater flow in a lowland rainforest, Costa Rica. Water Resour Res 41:W04011. https://doi.org/10.1029/2004WR003635
97. Genereux DP, Nagy L, Osburn CL, Oberbauer SF (2013) A connection to deep groundwater alters ecosystem carbon fluxes and budget: example from a Costa Rican rainforest. Geophys Res Lett 40(10):2066–2070. https://doi.org/10.1002/grl.50423
98. Genereux DP, Webb M, Solomon DK (2009) The chemical and isotopic signature of old groundwater and magmatic solutes in a Costa Rican rainforest: evidence from carbon, helium, and chlorine. Water Resour Res 45:W08413. https://doi.org/10.1029/2008WR007630
99. Genereux DP, Wood SJ, Pringle CM (2002) Chemical tracing of interbasin groundwater transfer in the lowland rainforest of Costa Rica. J Hydrol 258:163–178. https://doi.org/10.1016/S0022-1694(01)00568-6
100. Geng X, Heiss JW, Michael HA, Li H, Raubenheimer B, Boufadel MC (2021) Geochemical fluxes in sandy beach aquifers: modulation due to major physical stressors, geologic heterogeneity, and nearshore morphology. Earth Sci Rev 221:103800
101. van Genuchten MT (1980) A closed-form equation for predicting the hydraulic conductivity of unsaturated soils. Soil Sci Soc Am J 44(5):892–898
102. Gleeson T, Wada Y, Bierkens MF, Van Beek LP (2012) Water balance of global aquifers revealed by groundwater footprint. Nature 488(7410):197–200
103. Gleeson T, Villholth K, Taylor R, Perrone D, Hyndman D (2019) Groundwater: a call to action. Nature 76(7786):213−213
104. Goldsmith ST, Carey AE, Johnson BM, Welch SA, Lyons WB, McDowell WH, Pigott JS (2010) Stream geochemistry, chemical weathering and CO_2 consumption potential of andesitic terrains, Dominica, Lesser Antilles. Geochim Cosmochim Acta 74(1):85–103. https://doi.org/10.1016/j.gca.2009.10.009

105. Gou S, Miller G (2014) A groundwater-soil-plant-atmosphere continuum approach for modelling water stress, uptake, and hydraulic redistribution in phreatophytic vegetation. Ecohydrol 7:1029–1041. https://doi.org/10.1002/eco.1427
106. Grant GE, Dietrich WE (2017) The frontier beneath our feet. Water Resour Res 53(4):2605–2609. https://doi.org/10.1002/2017WR020835
107. Gribovszki Z, Kalicz P, Szilágyi J, Kucsara M (2008) Riparian zone evapotranspiration estimation from diurnal groundwater level fluctuations. J Hydrol 349(1–2), 6–17
108. Gu C, Hornberger GM, Herman JS, Mills AL (2008) Effect of freshets on the flux of groundwater nitrate through streambed sediments. Water Resour Res 44(5)
109. Gurdak JJ, Walvoord MA, McMahon PB (2008) Susceptibility to enhanced chemical migration from depression-focused preferential flow, High Plains Aquifer. Vadose Zone J 7(4):1218–1230
110. Haghighi E, Damm A, Jimenez-Martinez J (2021) Root hydraulic redistribution underlies the insensitivity of soil respiration to combined heat and drought. Appl Soil Ecol 167:104155
111. Hahm WJ, Rempe DM, Dralle DN, Dawson TE, Dietrich WE (2019) Oak transpiration drawn from the weathered bedrock vadose zone in the summer dry season. Water Resour Res 56(11):e2020WR027419. https://doi.org/10.1029/2020WR027419
112. Hammersmark CT, Dobrowski SZ, Rains MC, Mount JF (2010) Simulated effects of stream restoration on the distribution of wet-meadow vegetation. Restor Ecol 18(6):882–893
113. Hampton TB, Zarnetske JP, Briggs MA et al (2019) Residence time controls on the fate of nitrogen in flow-through lakebed sediments. J Geophys Res-Biogeosciences. https://doi.org/10.1029/2018JG004741
114. Harmon R, Barnard H, Day-Lewis FD, Mao D, Singha K (2021) Exploring environmental factors that drive diel variations in tree water storage using wavelet analysis. Front Water. https://doi.org/10.3389/frwa.2021.682285
115. Harmon R, Barnard HR, Singha K (2020) Water table depth and bedrock permeability control magnitude and timing of transpiration-induced diel fluctuations in groundwater. Water Resour Res 56(5):e2019WR025967
116. Hartmann A, Jasechko S, Gleeson T et al (2021) Risk of groundwater contamination widely underestimated because of fast flow into aquifers. Proc Natl Acad Sci 118(20)
117. Hartsough P, Tyler SW, Sterling J, Walvoord M (2001) A 146 kyr record of nitrogen flux from desert soil profiles as inferred from vadose zone pore waters. Geophys Res Lett 28(15):2955–2958
118. Harvey JW, Böhlke JK, Voytek MA, Scott D, Tobias CR (2013) Hyporheic zone denitrification: controls on effective reaction depth and contribution to whole-stream mass balance. Water Resour Res 49(10):6298–6316
119. Harvey J, Gooseff M (2015) River corridor science: Hydrologic exchange and ecological consequences from bedforms to basins. Water Resour Res 51(9):6893–6922
120. Hauser E, Richter DD, Markewitz D, Brecheisen Z, Billings SA (2020) Persistent anthropogenic legacies structure depth dependence of regenerating rooting systems and their functions. Biogeochem 147(3):259–275
121. Hayashi M, Rosenberry DO (2002) Effects of ground water exchange on the hydrology and ecology of surface water. Ground Water 40(3):309–316
122. Heiss JW, Michael HA, Koneshloo M (2020) Denitrification hotspots in intertidal mixing zones linked to geologic heterogeneity. Environ Res Lett 15(8):084015
123. Heiss JW, Post VE, Laattoe T, Russoniello CJ, Michael HA (2017) Physical controls on biogeochemical processes in intertidal zones of beach aquifers. Water Resour Res 53(11):9225−9244
124. Hendrickx JM, Flury M (2001) Uniform and preferential flow mechanisms in the vadose zone. In: Conceptual models of flow and transport in the fractured vadose zone. National Academy Press, Washington, D.C., pp 149–187
125. Henszey RJ, Pfeiffer K, Keough JR (2004) Linking surface- and ground-water levels to riparian grassland species along the Platte river in Central Nebraska, USA. Wetlands 24(3):665–687

126. Hinckley ELS, Ebel BA, Barnes RT, Anderson RS, Williams MW, Anderson SP (2014) Aspect control of water movement on hillslopes near the rain–snow transition of the Colorado Front Range. Hydrol Process 28(1):74–85
127. Hinkle ME (1994) Environmental conditions affecting concentrations of He, CO_2, O_2, and N_2 in soil gases. Appl Geochem 9:53–63
128. Hinnell AC, Ferré TA (2008) Considerations for measuring pressure head and water content in inclined boreholes. Water Resour Res 44:W12444. https://doi.org/10.1029/2008WR007122
129. Hinsinger P, Plassard C, Jaillard B (2006) Rhizosphere: a new frontier for soil biogeochemistry. J Geochem Explor 88(1–3):210–213
130. Hoagland B, Navarre-Sitchler A, Cowie R, Singha K (2020) Groundwater-stream connectivity mediates metal(loid) geochemistry in the hyporheic zone of streams impacted by historic mining and acid rock drainage. Front Water. https://doi.org/10.3389/frwa.2020.600409
131. Holbrook WS, Marcon V, Bacon AR, Brantley SL, Carr BJ, Flinchum BA, Riebe CS (2019) Links between physical and chemical weathering inferred from a 65-m-deep borehole through Earth's critical zone. Sci Rep 9(1):1–11
132. Holmes RM, Jones JB, Fisher SG, Grimm NB (1996) Denitrification in a nitrogen-limited stream ecosystem. Biogeochem 33(2):125–146
133. Houghton RA, Nassikas AA (2018) Negative emissions from stopping deforestation and forest degradation, globally. Glob Change Biol 24(1):350–359
134. Hu J, Moore DJ, Burns SP, Monson RK (2010) Longer growing seasons lead to less carbon sequestration by a subalpine forest. Glob Change Biol 16(2):771–783
135. Ingebritsen SE, Manning CE (1999) Geological implications of a permeability-depth curve for the continental crust. Geol 27(12):1107–1110
136. Ingebritsen SE, Manning CE (2010) Permeability of the continental crust: dynamic variations inferred from seismicity and metamorphism. Geofluids 10(1–2):193–205
137. Ingebritsen SE, Sherrod DR, Mariner RH (1989) Heat flow and hydrothermal circulation in the cascade range, north-central Oregon. Science 243(4897):1458–1462
138. Intergovernmental Panel on Climate Change (IPCC) (2022) Summary for Policymakers. In: Global warming of 1.5°C: IPCC special report on impacts of global warming of 1.5°C above pre-industrial levels in context of strengthening response to climate change, sustainable development, and efforts to eradicate poverty. Cambridge University Press, Cambridge, pp 1–24. https://doi.org/10.1017/9781009157940.001
139. Jackson RB, Sperry JS, Dawson TE (2000) Root water uptake and transport: using physiological processes in global predictions. Trends Plant Sci 5:482–488. https://doi.org/10.1016/S1360-1385(00)01766-0
140. Jasechko S, Perrone D (2021) Global groundwater wells at risk of running dry. Science 372(6540):418–421
141. Jasechko S, Perrone D, Befus KM, Bayani Cardenas M et al (2017) Global aquifers dominated by fossil groundwaters but wells vulnerable to modern contamination. Nat Geosci 10(6):425–429. https://doi.org/10.1038/ngeo2943
142. Jasechko S, Perrone D, Seybold H, Fan Y, Kirchner JW (2020) Groundwater level observations in 250,000 coastal US wells reveal scope of potential seawater intrusion. Nat Commun 11(1):1–9
143. Jasechko S, Seybold H, Perrone D, Fan Y, Kirchner JW (2021) Widespread potential loss of streamflow into underlying aquifers across the USA. Nature 591(7850):391–395
144. Jefferson A, Grant GE, Lewis SL, Lancaster ST (2010) Coevolution of hydrology and topography on a basalt landscape in the Oregon Cascade Range, USA. Earth Surf Process Land: J Br Geomorphol Res Group 35(7):803–816
145. Johnson B, Malama B, Barrash W, Flores AN (2013) Recognizing and modeling variable drawdown due to evapotranspiration in a semiarid riparian zone considering local differences in vegetation and distance from a river source. Water Resour Res 49(2):1030–1039
146. Jordan T, Ashley GM, Barton MD, Burges SJ et al (2001) Basic research opportunities in Earth science. National Academy Press, Washington, D.C.

147. Kampf S, Markus J, Heath J, Moore C (2015) Snowmelt runoff and soil moisture dynamics on steep subalpine hillslopes. Hydrol Process 29(5):712–723
148. Katz BS, Stotler RL, Hirmas D, Ludvigson G, Smith JJ, Whittemore DO (2016) Geochemical recharge estimation and the effects of a declining water table. Vadose Zone J 15(10). https://doi.org/10.2136/vzj2016.04.0031
149. Kearney WS, Fernandes A, Fagherazzi S (2019) Sea-level rise and storm surges structure coastal forests into persistence and regeneration niches. PLoS ONE. https://doi.org/10.1371/journal.pone.0215977
150. Kim KH, Heiss JW, Michael HA, Cai WJ, Laattoe T, Post VE, Ullman WJ (2017) Spatial patterns of groundwater biogeochemical reactivity in an intertidal beach aquifer. J Geophys Res Biogeosci 122(10):2548–2562
151. Kim KH, Michael HA, Field EK, Ullman WJ (2019) Hydrologic shifts create complex transient distributions of particulate organic carbon and biogeochemical responses in beach aquifers. J Geophys Res Biogeosci 124(10):3024–3038
152. Kim KH, Heiss JW, Geng X, Michael HA (2020) Modeling hydrologic controls on particulate organic carbon contributions to beach aquifer biogeochemical reactivity. Water Resour Res 56(10):e2020WR027306
153. Kirwan ML, Gedan KB (2019) Sea-level driven land conversion and the formation of ghost forests. Nat Clim Chang 9(6):450–457
154. Kirwan ML, Walters DC, Reay WG, Carr JA (2016) Sea level driven marsh expansion in a coupled model of marsh erosion and migration. Geophys Res Lett 43(9):4366–4373
155. Kløve B, Ala-Aho P, Bertrand G, Gurdak JJ, Kupfersberger H, Kværner J, Pulido-Velazquez M (2014) Climate change impacts on groundwater and dependent ecosystems. J Hydrol 518:250–266
156. Koch GR, Childers DL, Staehr PA, Price RM, Davis SE, Gaiser EE (2012) Hydrological conditions control P loading and aquatic metabolism in an oligotrophic, subtropical estuary. Estuaries Coasts 35(1):292–307
157. Kollet SJ, Maxwell RM (2008) Capturing the influence of groundwater dynamics on land surface processes using an integrated, distributed watershed model. Water Resour Res 44(W02402). https://doi.org/10.1029/2007WR006004
158. Kuang X, Jiao JJ, Li H (2013) Review on airflow in unsaturated zones induced by natural forcings. Water Resour Res 49:6137–6165
159. Kuntz B, Rubin S, Berkowitz B, Singha K (2011) Quantifying solute transport behavior at the Shale hills critical zone observatory. Vadose Zone J 10 https://doi.org/10.2136/vzj2010.0130
160. Kurtz AC, Lugolobi F Salvucci G (2011) Germanium-silicon as a flow path tracer: application to the Rio Icacos Watershed. Water Resour Res 47(6)
161. Kurz-Besson C, Otieno D, Lobo Do Vale R, Siegwolf R et al (2006) Hydraulic lift in cork oak trees in a savannah-type mediterranean ecosystem and its contribution to the local water balance. Plant Soil 282:361–378. https://doi.org/10.1007/s11104-006-0005-4
162. Kwon EY, Kim G, Primeau F, Moore WS, Cho HM, DeVries T, Cho YK (2014) Global estimate of submarine groundwater discharge based on an observationally constrained radium isotope model. Geophys Res Lett 41(23):8438–8444
163. Küsel K, Totsche KU, Trumbore SE, Lehmann R, Steinhäuser C, Herrmann M (2016) How deep can surface signals be traced in the critical zone? Merging biodiversity with biogeochemistry research in a central German Muschelkalk landscape. Front Earth Sci 4
164. Law BE, Hudiburg TW, Berner LT, Kent JJ, Buotte PC, Harmon ME (2018) Land use strategies to mitigate climate change in carbon dense temperate forests. Proc Natl Acad Sci
165. Lazar CS, Lehmann R, Stoll W, Rosenberger J, Totsche KU, Küsel K (2019) The endolithic bacterial diversity of shallow bedrock ecosystems. Sci Total Environ 679:35–44. https://doi.org/10.1016/j.scitotenv.2019.04.281
166. Lecher AL, Mackey KR (2018) Synthesizing the effects of submarine groundwater discharge on marine biota. Hydrol 5(4)

167. Lee E, Kumar P, Knowles JF, Minor RL, Tran N, Barron-Gafford GA, Scott RL (2021) Convergent hydraulic redistribution and groundwater access supported facilitative dependency between trees and grasses in a semi-arid environment. Water Resour Res 57:1–20. https://doi.org/10.1029/2020WR028103
168. Levitt DG, Newell DL, Stone WJ, Wykoff DS (2005) Surface water-groundwater connection at the Los Alamos Canyon Weir site: part 1 monitoring site installation and tracer tests. Vadose Zone J 4(3):708–717
169. Lewis WC (1991) Atmospheric pressure changes and cave airflow: a review. Natl Speleol Soc Bull 53:1–12
170. Lewis SL, Maslin MA (2015) Defining the anthropocene. Nature 519(7542):171–180
171. Li J, Zhan H, Huang G, You K (2011) Tide-induced airflow in a two-layered coastal land with atmospheric pressure fluctuations. Adv Water Resour 34:649–658
172. Loescher HW, Oberbauer SF, Gholz HL, Clark DB (2003) Environmental controls on net ecosystem level carbon exchange and productivity in a Central American tropical wet forest. Glob Change Biol 9:396–412. https://doi.org/10.1046/j.1365-2486.2003.00599.x
173. Loheide SP II, Booth EG (2011) Effects of changing channel morphology on vegetation, groundwater, and soil moisture regimes in groundwater dependent ecosystems. Geomorphol 126(3–4):364–376. https://doi.org/10.1016/j.geomorph.2010.04.016
174. Loheide SP II, Gorelick SM (2007) Riparian hydroecology: a coupled model of the observed interactions between groundwater flow and meadow vegetation patterning. Water Resour Res 43:W07414. https://doi.org/10.1029/2006WR005233
175. Loheide SP, Butler Jr JJ, Gorelick SM (2005) Estimation of groundwater consumption by phreatophytes using diurnal water table fluctuations: a saturated-unsaturated flow assessment. Water Resour Res 41(7)
176. Loheide II SP, Gorelick SM (2007) Riparian hydroecology: a coupled model of the observed interactions between groundwater flow and meadow vegetation patterning. Water Resour Res 43(7). https://doi.org/10.1029/2006WR005233
177. Loheide II SP (2008) A method for estimating subdaily evapotranspiration of shallow groundwater using diurnal water table fluctuations. Ecohydrology 1(1):59–66. https://doi.org/10.1002/eco.7
178. Lowry CS, Loheide SP II (2010) Groundwater dependent vegetation: quantifying the groundwater subsidy. Water Resour Res 46:W06202. https://doi.org/10.1029/2009WR008874
179. Lowry CS, Loheide II SP, Moore CE, Lundquist JD (2011) Groundwater controls on vegetation composition and patterning in mountain meadows. Water Resour Res 47, W00J11. https://doi.org/10.1029/2010WR010086
180. Lowry CS, Loheide II SP (2010) Groundwater-dependent vegetation: quantifying the groundwater subsidy. Water Resour Res 46(6). https://doi.org/10.1029/2009WR008874
181. Maher K (2010) The dependence of chemical weathering rates on fluid residence time. Earth Planet Sci Lett 294(1–2):101–110
182. Mahmood Poor Dehkordy F, Briggs MA, Day-Lewis FD, Bagtzoglou AC (2018) Simulation of less-mobile porosity dynamics in contrasting sediment water interface porous media. Hydrol Process 32(13):2030–2043
183. MahmoodPoor Dehkordy F, Briggs MA, Day-Lewis FD, Singha K et al (2019) Multi-scale preferential flow processes in an urban streambed under variable hydraulic conditions. J Hydrol. https://doi.org/10.1016/j.jhydrol.2019.03.022
184. Manga M (1997) A model for discharge in spring-dominated streams and implications for the transmissivity and recharge of Quaternary volcanics in the Oregon Cascades. Water Resour Res 33(8):1813–1822
185. Manning AH, Caine JS (2007) Groundwater noble gas, age, and temperature signatures in an Alpine watershed: valuable tools in conceptual model development. Water Resour Res 43:W04404. https://doi.org/10.1029/2006WR005349
186. Manning AH, Clark JF, Diaz SH, Rademacher LK, Earman S, Niel Plummer L (2012) Evolution of groundwater age in a mountain watershed over a period of thirteen years. J Hydrol 460–461, 13–28. https://doi.org/10.1016/jjhydrol201206030

187. Mares R, Barnard HR, Mao D, Revil A, Singha K (2016) Examining diel patterns of soil and xylem moisture using electrical resistivity imaging. J Hydrol 12. https://doi.org/10.1016/j.jhydrol.2012.06.030
188. Massman WJ, Frank JM (2006) Advective transport of CO_2 in permeable media induced by atmospheric pressure fluctuations: 2 Observational evidence under snowpacks. J Geophys Res Biogeosci 111(G3):G03005. https://doi.org/10.1029/2006JG000164
189. Massmann J, Farrier DF (1992) Effects of atmospheric pressures on gas transport in the vadose zone. Water Resour Res 28(3):777–791
190. Maxwell RM, Condon LE (2016) Connections between groundwater flow and transpiration partitioning. Science 353(6297):377–380
191. Maxwell RM, Kollet SJ (2008) Interdependence of groundwater dynamics and land energy feedbacks under climate change. Nat Geosci 1(10):665–669
192. McAllister SM, Barnett JM, Heiss JW et al (2015) Dynamic hydrologic and biogeochemical processes drive microbially enhanced iron and sulfur cycling within the intertidal mixing zone of a beach aquifer. Limnol Oceanogr 60(1):329–345
193. McCormick EL, Dralle DN, Hahm WJ, Tune AK, Schmidt LM, Chadwick KD, Rempe DM (2021) Widespread woody plant use of water stored in bedrock. Nature 597(7875):225–229
194. McDonnell JJ (2003) Where does water go when it rains? Moving beyond the variable source area concept of rainfall-runoff response. Hydrol Process 17:1869–1875
195. McIntosh JC, Schaumberg C, Perdrial J, Harpold A et al (2017) Geochemical evolution of the critical zone across variable time scales informs concentration-discharge relationships: Jemez river basin critical zone observatory. Water Resour Res 53:4169–4196. https://doi.org/10.1002/2016WR019712
196. McIntosh JC, Ferguson G (2021) Deep meteoric water circulation in Earth's crust. Geophys Res Lett 48:e2020GL090461 https://doi.org/10.1029/2020GL090461
197. McMahon MJ, Christy AD (2000) Root growth, calcite precipitation, and gas and water movement in fractures and macropores: a review with field observations. Ohio J Sci 100:88–93
198. McMahon PB, Dennehy KF, Bruce BW, Böhlke JK, Michel RL, Gurdak JJ, Hurlbut DB (2006) Storage and transit time of chemicals in thick unsaturated zones under rangeland and irrigated cropland, high plains, United States. Water Resour Res 42:W03413. https://doi.org/10.1029/2005WR004417
199. Meinzer OE (1927) Plants as indicators of ground water. US Geol Surv Water-Supply Pap 577:95. https://doi.org/10.3133/wsp577
200. Meyboom P (1967) Groundwater studies in the Assiniboine river drainage basin—part II: hydrologic characteristics of phreatophytic vegetation in south-central Saskatchewan. Geol Surv Can 139
201. Miguez-Macho G, Fan Y (2021) Spatiotemporal origin of soil water taken up by vegetation. Nature 1–5
202. Miller GR, Chen X, Rubin Y, Ma S, Baldocchi DD (2010) Groundwater uptake by woody vegetation in a semiarid oak savanna. Water Resour Res 46(10)
203. Mimura N, Nurse L, McLean RF, Agard J et al (2007) Small Islands. In: Climate change 2007: impacts, adaptation and vulnerability, contribution of working group II to the fourth assessment report of the intergovernmental panel on climate change. Cambridge University Press, Cambridge, UK, pp 687–716
204. Minsley BJ, Rigby JR, James SR, Burton BL, Knierim KJ, Pace MD, Kress WH (2021) Airborne geophysical surveys of the lower Mississippi Valley demonstrate system-scale mapping of subsurface architecture. Commun Earth Environ 2(1):1–14
205. Moravec BG, White AM, Root RA, Sanchez A, Olshansky Yaras BK, Chorover J (2020) Resolving deep critical zone architecture in complex volcanic terrain. J Geophys Res: Earth Surf 125(1):e2019JF005189
206. Munoz E, Arumi JL, Wagener T, Oyarzún R, Parra V (2016) Unraveling complex hydrogeological processes in Andean basins in south-central Chile: an integrated assessment to understand hydrological dissimilarity. Hydrol Process 30:4934–4943. https://doi.org/10.1002/hyp.11032

207. NRC-NAS (2012) Challenges and opportunities in hydrologic sciences. The National Academies Press, Washington D.C.
208. Napieralski SA, Buss HL, Brantley SL, Lee S, Xu H, Roden EE (2019) Microbial chemolithotrophy mediates oxidative weathering of granitic bedrock. Proc Natl Acad Sci 116(52):26394–26401
209. National Academies of Sciences, Engineering, and Medicine (2001) Conceptual models of flow and transport in the fractured vadose zone. The National Academies Press, Washington, D.C. https://doi.org/10.17226/10102
210. Neeper DA (2002) Investigation of the vadose zone using barometric pressure cycles. J Contam Hydrol 54(1–2):59–80
211. Neumann RB, Gardon ZG (2012) The magnitude of hydraulic redistribution by plant roots: a review and synthesis of empirical and modeling studies. New Phytol 194:337–352. https://doi.org/10.1111/j.1469-8137.2012.04088.x
212. Ogurcak DE, Price RM (2019) Groundwater geochemistry fluctuations along a fresh-saltwater gradient on the carbonate islands of the lower Florida Keys. Chem Geol 527:118925
213. Olshansky Y, Knowles JF, Barron-Gafford GA, Rasmussen C, Abramson N, Chorover J (2019) Soil fluid biogeochemical response to climatic events. J Geophys Res Biogeosci 124(9):2866–2882
214. Orellana F, Verma P, Loheide II SP, Daly E (2012) Monitoring and modeling water-vegetation interactions in groundwater-dependent ecosystems. Rev Geophys 50:RG3003. https://doi.org/10.1029/2011RG000383
215. Oviedo-Vargas D, Genereux DP, Dierick D, Oberbauer SF (2015) The effect of regional groundwater on carbon dioxide and methane emissions from a lowland rainforest stream in Costa Rica. J Geophys Res Biogeosci 120:2579–2596. https://doi.org/10.1002/2015JG003009
216. O'Geen A, Safeeq M, Wagenbrenner J, Stacy E et al (2018) Southern Sierra critical zone observatory and Kings river experimental watersheds: a synthesis of measurements, new insights, and future directions. Vadose Zone J 17:180081. https://doi.org/10.2136/vzj2018.04.0081
217. Pacheco FAL (2015) Regional groundwater flow in hard rocks. Sci Total Environ 506–507:182–195. https://doi.org/10.1016/j.scitotenv.2014.11.008
218. Parsekian AD, Singha K, Minsley BJ, Holbrook WS, Slater L (2015) Multiscale geophysical imaging of the critical zone. Rev Geophys 53(1):1–26
219. Pelletier JD, Barron-Gafford GA, Breshears DD, Brooks PD et al (2013) Coevolution of vegetation, soils, and topography in the sky islands of southern Arizona. J Geophys Res Earth Surf 118(2):741–758
220. Penna D, Borga M, Norbiato D, Dalla Fontana G (2009) Hillslope scale soil moisture variability in a steep alpine terrain. J Hydrol 364(3):311–327
221. Phillips FM (1994) Environmental tracers for water movement in desert soils of the arid American southwest. Soil Sci Soc Am J 58:15–24
222. Phillips FM, Walvoord MA, Small EE (2004) Effects of environmental change on groundwater recharge in the desert southwest. In: Hogan JF, Phillips FM, Scanlon BR (eds) Groundwater recharge in a desert environment: the southwestern United States
223. Popov YA, Pevzner SL, Pimenov VP, Romushkevich RA (1999) New geothermal data from the Kola superdeep well SG-3. Tectonophysics 306(3–4):345–366
224. Price RM, Savabi MR, Jolicoeur JL, Roy S (2010) Adsorption and desorption of phosphate on limestone in experiments simulating seawater intrusion. Appl Geochem 25(7):1085–1091
225. Price RM, Swart PK, Fourqurean JW (2006) Coastal groundwater discharge–an additional source of phosphorus for the oligotrophic wetlands of the Everglades. Hydrobiologia 569(1):23–36
226. Price RM, Herman JS (1991) Geochemical investigation of salt-water intrusion into a coastal carbonate aquifer: Mallorca, Spain. Geol Soc Am Bull 103(10):1270–1279. https://doi.org/10.1029/009WSA15

227. Price RM, Herman JS (1991) Geochemical investigation of salt water intrusion into a coastal carbonate aquifer: Mallorca, Spain. Geol Soc Am Bull 103(10):1270–1279. https://doi.org/10.1130/00167606(1991)103%3C1270:GIOSWI%3E2.3.CO;2
228. Rad SD, Allègre CJ Louvat P (2007) Hidden Erosion on volcanic Islands. Earth Planet Sci Lett 262(1):109–124
229. Rains MC, Mount JF, Larsen EW (2004) Simulated changes in shallow groundwater and vegetation distributions under different reservoir operations scenarios. Ecol Appl 14(1):192–207. https://doi.org/10.1890/02-5307
230. Rempe DM, Dietrich WE (2014) A bottom-up control on fresh-bedrock topography under landscapes. PNAS 111(18):6576–6581
231. Rempe DM, Dietrich WE (2018) Direct observations of rock moisture, a hidden component of the hydrologic cycle. PNAS 115(11):2664–2669
232. Rey DM, Hinckley ES, Walvoord M, Singha K (2021) Integrating observations and models to determine the effect of seasonally frozen ground on hydrologic partitioning in alpine hillslopes in the Colorado Rocky Mountains. USA Hydrologic Processes. https://doi.org/10.1002/hyp.14374
233. Rezaei M, Sanz E, Raeisi E, Ayora C, Vázquez-Suñé E, Carrera J (2005) Reactive transport modeling of calcite dissolution in the fresh-salt water mixing zone. J Hydrol 311(1–4):282–298
234. Richards LA, Gardner WR, Ogata G (1956) Physical process determining water loss from soil. Soil Sci Soc Am Proc 20:310–314
235. Rimon Y, Dahan O, Nativ R (2011) Vadose zone water pressure variation during infiltration events. Vadose Zone J 10:1105–1112
236. Rimon Y, Dahan O, Nativ R, Geyer S (2007) Water percolation through the deep vadose zone and groundwater recharge: preliminary results based on a new vadose zone monitoring system. Water Resour Res 43(5):W05402. https://doi.org/10.1029/2006WR004855
237. Robinson TW (1958) Phreatophytes. U.S. Geol Surv Water Supply Pap 1423(84). https://doi.org/10.3133/ofr55152
238. Saar MO (2011) Geothermal heat as a tracer of large-scale groundwater flow and as a means to determine permeability fields. Hydrogeol J 19(1):31
239. Salve R, Rempe DM, Dietrich WE (2012) Rain, rock moisture dynamics, and the rapid response of perched groundwater in weathered, fractured argillite underlying a steep hillslope. Water Resour Res 48(11):W11528. https://doi.org/10.1029/2012WR012583
240. Sandvig RM, Phillips FM (2006) Ecohydrological controls on soil moisture fluxes in arid to semiarid vadose zone. Water Resour Res 42:W08422. https://doi.org/10.1029/2005WR004644
241. Sanford WE, Konikow LF (1989) Simulation of calcite dissolution and porosity changes in saltwater mixing zones in coastal aquifers. Water Resour Res 25(4):655–667
242. Saraceno J, Kulongoski JT, Mathany TM (2018) A novel high-frequency groundwater quality monitoring system. Environ Monit Assess 190(8):1–14
243. Sardans J, Peñuelas J (2014) Hydraulic redistribution by plants and nutrient stoichiometry: shifts under global change. Ecohydrol 7(1):1–20
244. Sawyer AH (2015) Enhanced removal of groundwater-borne nitrate in heterogeneous aquatic sediments. Geophys Res Lett 42(2):403–410
245. Sawyer AH, Michael HA, Schroth AW (2016) From soil to sea: the role of groundwater in coastal critical zone processes. Wiley Interdiscip Rev Water 3(5):706–726
246. Scanlon BR, Healy RW, Cook PG (2002) Choosing appropriate techniques for quantifying groundwater recharge. Hydrogeol J 10:18–39
247. Scanlon BR, Keese K, Reedy RC, Simunek J, Andraski BJ (2003) Variations in flow and transport in thick desert vadose zones in response to paleoclimatic forcing (0–90 kyr): field measurements, modeling, and uncertainties. Water Resour Res 39(7):1179. https://doi.org/10.1029/2002WR001604
248. Scanlon BR, Reedy RC, Stonestrom DA, Prudic DE, Dennehy KF (2005) Impact of land use and land cover change on groundwater recharge and quality in the southwestern US. Glob Change Biol 11:1577–1593

249. Scanlon BR, Reedy RC, Tachovsky JA (2007) Semiarid unsaturated zone chloride profiles: Archives of past land use change impacts on water resources in the southern High Plains, United States. Water Resour Res 43(6)
250. Scanlon BR, Goldsmith RS (1997) Field study of spatial variability in unsaturated flow beneath and adjacent to playas. Water Resour Res 33(10):2239–2252
251. Schaller MF, Fan Y (2009) River basins as groundwater exporters and importers: implications for water cycle and climate modeling. J Geophys Res-Atmos 114:D04103. https://doi.org/10.1029/2008JD010636
252. Schmid W, Hanson RT, Leake SA, Hughes JD, Niswonger RG (2014) Feedback of land subsidence on the movement and conjunctive use of water resources. Environ Model Softw 62:253–270
253. Schmidt L, Rempe D (2020) Quantifying dynamic water storage in unsaturated bedrock with borehole nuclear magnetic resonance. Geophys Res Lett 47(22):e2020GL089600
254. Schopka HH, Derry LA (2012) Chemical weathering fluxes from Volcanic Islands and the importance of groundwater: the Hawaiian example. Earth Planet Sci Lett 339–340:67–78. https://doi.org/10.1016/j.epsl.2012.05.028
255. Seyfried M, Lohse K, Marks D, Flerchinger G, Pierson F, Holbrook WS (2018) Reynolds creek experimental watershed and critical zone observatory. Vadose Zone J 17(1):1–20. https://doi.org/10.2136/vzj2018.07.0129
256. Shelobolina E, Xu H, Konishi H, Kukkadapu R, Wu T, Blöthe M, Roden E (2012) Microbial lithotrophic oxidation of structural Fe(II) in biotite. Appl Environ Microbiol 78(16):5746–5752. https://doi.org/10.1128/AEM.01034-12
257. Siebert S, Burke J, Faures JM, Frenken K, Hoogeveen J, Döll P, Portmann FT (2010) Groundwater use for irrigation–a global inventory. Hydrol Earth Syst Sci 14(10):1863–1880
258. Silvertown J, Dodd M, Gowing D (2001) Phylogeny and the niche structure of meadow plant communities. J Ecol 89(3):428–435. https://doi.org/10.1046/j.1365-2745.2001.00553.x
259. Singer MB, Stella JC, Dufour S, Piegay H, Wilson RJS, Johnstone L (2012) Contrasting water-uptake and growth responses to drought in co-occurring riparian tree species. Ecohydrol 6(3):402–412. https://doi.org/10.1002/eco.1283
260. Smilovic M, Gleeson T, Adamowski J, Langhorn C (2019) More food with less water–optimizing agricultural water use. Adv Water Resour. https://doi.org/10.1016/j.advwatres.2018.09.016
261. Smith JWN, Bonell M, Gibert J, McDowell WH, Sudicky EA, Turner JV, Harris RC (2008) Groundwater–surface water interactions, nutrient fluxes and ecological response in river corridors: translating science into effective environmental management. Hydrol Process: Int J 22(1):151–157
262. Soylu ME, Loheide II SP, Kucharik CJ (2017) Effects of root distribution and root water compensation on simulated water use in maize influenced by shallow groundwater. Vadose Zone J 16(10). https://doi.org/10.2136/vzj2017.06.0118
263. Soylu ME, Loheide II SP, Kucharik CJ (2017) Effects of root distribution and root water compensation on simulated water use in maize influenced by shallow groundwater. Vadose Zone J 16(10):vzj2017.06.0118. https://doi.org/10.2136/vzj2017.06.0118
264. Spane FA (2002) Considering barometric pressure in groundwater flow investigations. Water Resour Res 38(6):1078. https://doi.org/10.1029/2001WR000701
265. Sprenger M, Stumpp C, Weiler M, Aeschbach W et al (2019) The demographics of water: a review of water ages in the critical zone. Rev Geophys 57(3):800–834. https://doi.org/10.1029/2018RG000633
266. Steffen W, Broadgate W, Deutsch L, Gaffney O, Ludwig C (2015) The trajectory of the anthropocene: the great acceleration. Anthropocene Rev 2(1):81–98
267. Stewart-Maddox NS, Frisbee MD, Andronicos CL, Genereux DP, Meyers ZP (2018) Identifying the regional extent and geochemical evolution of interbasin groundwater flow using geochemical inverse modeling and $^{87}Sr/^{86}Sr$ ratios in a complex conglomeratic aquifer. Chem Geol 500:20–29. https://doi.org/10.1016/j.chemgeo.2018.07.026

268. Stober I, Bucher K (2015) Hydraulic conductivity of fractured upper crust: insights from hydraulic tests in boreholes and fluid-rock interaction in crystalline basement rocks. Geofluids 15(1–2):161–178
269. Stonestrom DA, Prudic DE, Walvoord, MA et al (2007) Focused ground-water recharge in the Amargosa desert basin. USGS professional paper 1703. In: Stonestrom DA, Constantz J, Ferré TPA, Leake SA (eds) Groundwater recharge in the arid and semiarid southwestern United States. Chapter E:107–136. https://doi.org/10.3133/pp1703E
270. Sullivan PL, Engel V, Ross MS, Price RM (2014a) The influence of vegetation on the hydrodynamics and geomorphology of a tree Island in everglades national park (Florida, United States). Ecohydrol 7(2):727–744
271. Sullivan PL, Macpherson GL, Martin JB, Price RM (2019) Evolution of carbonate and karst critical zones. Chem Geol 527:119223. https://doi.org/10.1016/j.chemgeo.2019.06.023
272. Sullivan PL, Price RM, Miralles-Wilhelm F et al (2014b) The role of recharge and evapotranspiration as hydraulic drivers of ion concentrations in shallow groundwater on Everglades tree islands, Florida (USA). Hydrol Process 28(2):293–304
273. Sullivan PL, Price RM, Ross MS, Stoffella SL, Sah JP, Scinto LJ, Sklar FH (2016) Trees: a powerful geomorphic agent governing the landscape evolution of a subtropical wetland. Biogeochem 128(3):369–384
274. Sullivan PL, Billings S, Hirmas D, Li L, Zhang X, Ziegler S, Wen H (2022) Embracing the dynamic nature of soil structure: a paradigm illuminating the role of life in critical zones of the Anthropocene. Earth-Sci Rev 103873
275. Swanberg CA, Walkey WC, Combs J (1988) Core hole drilling and the "rain curtain" phenomenon at Newberry Volcano, Oregon. J Geophys Research. Solid Earth 93(B9):10163–10173
276. Swetnam T, Brooks PD, Barnard HR, Harpold AA, Gallo E (2017) Revisiting environmental gradient ecology: how topography trumps climate in determining forest carbon reservoir size. Ecosphere 8(4)
277. Taniguchi M, Dulai H, Burnett KM, Santos IR, Sugimoto R, Stieglitz T, Burnett WC (2019) Submarine groundwater discharge: updates on its measurement techniques, geophysical drivers, magnitudes, and effects. Front Environ Sci 7:141
278. Taylor RG, Scanlon B, Döll P, Rodell M, Van Beek R, Wada Y, Treidel H (2013) Ground water and climate change. Nat Clim Chang 3(4):322–329
279. Theis CV (1941) The effect of a well on the flow of a nearby stream. Eos, Trans Am Geophys Union 22(3):734–738
280. Thompson SE, Harman CJ, Troch PA, Brooks PD, Sivapalan M (2011) Spatial scale dependence of ecohydrologically mediated water balance partitioning: a synthesis framework for catchment ecohydrology. Water Resour Res 47
281. Thorley RM, Taylor LL, Banwart SA, Leake JR, Beerling DJ (2015) The role of forest trees and their mycorrhizal fungi in carbonate rock weathering and its significance for global carbon cycling. Plant, Cell Environ 38(9):1947–1961
282. Tosi L, Da Lio C, Bergamasco A, Cosma M, Cavallina C, Fasson A, Donnici S (2022) Sensitivity, hazard, and vulnerability of farmlands to saltwater intrusion in low-lying coastal areas of Venice, Italy. Water 14(1):64
283. Tran H, Zhang J, Cohard JM, Condon LE, Maxwell RM (2020) Simulating groundwater-streamflow connections in the upper Colorado river basin. Groundwater 58(3):392–405
284. Triska FJ, Duff JH, Avanzino RJ (1993) Patterns of hydrological exchange and nutrient transformation in the hyporheic zone of a gravel-bottom stream: examining terrestrial—aquatic linkages. Freshw Biol 29(2):259–274
285. Tromp-van Meerveld H, McDonnell J (2006) On the interrelations between topography, soil depth, soil moisture, transpiration rates and species distribution at the hillslope scale. Adv Water Resour 29(2):293–310. https://doi.org/10.1016/j.advwatres.2005.02.016
286. Tully K, Gedan K, Epanchin-Niell R, Strong A, Bernhardt ES, BenDor T, Weston NB (2019) The invisible flood: The chemistry, ecology, and social implications of coastal saltwater intrusion. Bioscience 69(5):368–378

287. Tyler SW, Chapman JB, Conrad S, Hammermeister DP, Blout D, Miller J, Ginnani JM (1996) Soil water flux on the Nevada test site: spatial and temporal variations over the last 120,000 years. Water Resour Res 32:1481–1499
288. Tóth J (1963) A theoretical analysis of groundwater flow in small drainage basins. J Geophys Res 68(16):4795–4812. https://doi.org/10.1029/JZ068i016p04795
289. Tóth J (1999) Groundwater as a geologic agent: an overview of the causes, Processes, and manifestations. Hydrogeol 7(1):1–14
290. UK Environment Agency (2004) Hydroecology: integration for modern regulation. Bristol, UK
291. Visser A, Thaw M, Deinhart A, Bibby R, Safeeq M, Conklin M, Esser B, Van der Velde Y (2019) Cosmogenic isotopes unravel the hydrochronology and water storage dynamics of the Southern Sierra Critical Zone. Water Resour Res 55(2):1429–1450
292. Vogt C, Mottaghy D, Rath V, Marquart G, Dijkshoorn L, Wolf A, Clauser C (2014) Vertical variation in heat flow on the Kola Peninsula: palaeoclimate or fluid flow?. Geophys J Int 199(2):829–843
293. Voytek EB, Barnard H, Jougnot D, Singha K (2019) Transpiration- and precipitation-induced subsurface water flow observed using the self-potential method. Hydrol Process. https://doi.org/10.1002/hyp.13453
294. Wald JA, Graham RC, Schoeneberger PJ (2012) Distribution and properties of soft weathered bedrock at $\leq$ 1 m depth in the contiguous United States. Earth Surf Proc Land 38(6):614–626
295. Walvoord MA, Phillips FM, Stonestrom DA, Evans RD, Hartsough PC, Newman BD, Striegel RG (2003) A reservoir of nitrate beneath desert soils. Science 302(7):1021–1024
296. Walvoord MA, Phillips FM, Tyler SW, Hartsough PC (2002b) Deep arid system hydrodynamics 2 application to paleohydrologic reconstruction using vadose zone profiles from the northern Mojave Desert. Water Resour Res 38(12):1291. https://doi.org/10.1029/2001WR000825
297. Walvoord MA, Plummer MA, Phillips FM, Wolfsberg AV (2002a) Deep arid system hydrodynamics 1 equilibrium states and response times in thick desert vadose zones. Water Resour Res 38(12):1308. https://doi.org/10.1029/2001WR000824
298. Walvoord MA, Scanlon BR (2004) Hydrologic processes in deep Vadose zones in interdrainage arid environments. In: Hogan JF, Phillips FM, Scanlon BR (eds) Groundwater recharge in a desert environment: the southwestern United States. https://doi.org/10.1029/009WSA02
299. Wang CY, Manga M (2015) New streams and springs after the 2014 Mw6.0 South Napa earthquake. Nat Commun 6
300. Ward ND, Megonigal JP, Bond-Lamberty B et al (2020) Representing the function and sensitivity of coastal interfaces in Earth system models. Nat Commun 11(1):1–14
301. Weeks EP (1979) Barometric fluctuations in wells tapping deep unconfined aquifers. Water Resour Res 15(5):1167–1176
302. Weeks EP (2005) The Lisse effect revisited. Groundwater 40:652–656
303. Western AW, Blöschl G (1999) On the spatial scaling of soil moisture. J Hydrol 217(3):203–224
304. White A, Moravec B, McIntosh J, Olshansky Y, Paras B, Sanchez RA, Chorover J (2019) Distinct stores and the routing of water in the deep critical zone of a snow-dominated volcanic catchment. Hydrol Earth Syst Sci 23(11):4661–4683
305. White WN (1932) A method of estimating ground-water supplies based on discharge by plants and evaporation from soil: results of investigations in Escalante Valley, Utah. Water Supply Pap 659-A. https://doi.org/10.3133/wsp659A
306. Wigley TML, Plummer LN (1976) Mixing of carbonate waters. Geochim Cosmochim Acta 40(9):989–995
307. Williams K, Ewel KC, Stumpf RP, Putz FE, Workman TW (1999) Sea-level rise and coastal forest retreat on the west coast of Florida, USA. Ecol 80(6):2045–2063
308. Williamson TN, Newman BD, Graham RC, Shouse PJ (2004) Regolith water in zero-order chaparral and perennial grass watersheds four decades after vegetation conversion. Vadose Zone J 3:1007–1016. https://doi.org/10.2136/vzj2004.1007

309. Winter TC (1999) Relation of streams, lakes, and wetlands to groundwater flow systems. Hydrogeol J 7(1):28–45
310. Winter TC, Harvey JW, Franke OL, Alley WM (1999) Ground water and surface water: a single resource. USGS Circular 1139. https://doi.org/10.3133/cir1139
311. Zanon C, Genereux DP, Oberbauer SF (2014) Use of a watershed hydrologic model to estimate interbasin groundwater flow in a Costa Rican rainforest. Hydrol Process 28:3670–3680. https://doi.org/10.1002/hyp.9917
312. Zarnetske JP, Haggerty R, Wondzell SM, Baker MA (2011) Dynamics of nitrate production and removal as a function of residence time in the hyporheic zone. J Geophys Res: Biogeosci 116(G1)
313. Zarnetske JP, Haggerty R, Wondzell SM (2015) Coupling multiscale observations to evaluate hyporheic nitrate removal at the reach scale. Freshw Sci 34(1):172–186
314. Zipper SC, Hammond JC, Shanafield M, Zimmer M, Datry T, Jones CN, Allen DC et al (2021) Pervasive changes in stream intermittency across the United States. Environ Res Lett 16(8):084033
315. Zipper SC, Soylu ME, Booth EG, Loheide SP (2015) Untangling the effects of shallow groundwater and soil texture as drivers of subfield-scale yield variability. Water Resourc Res 51(8):6338–6358

Bedrock, Regolith, and Soil Continuum: Pedogenic Perspectives and Factors Controlling Spatial Distribution of Soils in the Critical Zone

Ashlee Dere and Talitha Santini

1 What is Soil?

Soil is a basic and fundamental resource for sustaining life on Earth as we know it. However, unlike other basic and fundamental resources like water and air, the definition of what constitutes soil is less clear cut. In part, this is due to the dynamic nature of soils, underpinned by their position at the interface of lithosphere, atmosphere, hydrosphere, and biosphere in the Critical Zone. Human activity in the Anthropocene has further expanded the definition of soil by widening the range of available parent materials at Earth's surface and how Critical Zone processes interact with these materials. Defining what soil is (and is not) is critical to understanding, management, and conservation. Our objective here is not to stoke the fires of debate about what soil 'should' be, but rather to present the most widely used definitions (Table 1), and briefly explore their common elements and utility.

Soil definitions can be framed based on **composition** ("minerals and organic matter"), **genesis** ("formed as a result of additions, losses, transfers, and transformations"), **spatial arrangement and position** ("horizons", "layers", "at the Earth's surface"), and **functions** ("ability to support rooted plants"), and some definitions combine elements of all four frames. Favored definitions may be aligned with disciplinary background and expertise: mineralogists and geochemists may favor compositionally driven definitions; pedologists and physicists may favor genetic and spatially driven definitions; engineers may favor composition and functionally driven definitions; and biologists may favor functionally driven definitions.

A. Dere (✉)
College of Arts and Sciences, University of NE–Omaha, Omaha, NE, USA
e-mail: adere@unomaha.edu

T. Santini
School of Agriculture and Environment, The University of Western Australia, Crawley, WA, Australia
e-mail: talitha.santini@uwa.edu.au

T. White and A. Provenzale (eds.), *Critical Zone and Ecosystem Dynamics*, Advances in Critical Zone Science, https://doi.org/10.1007/978-3-031-69076-1_6

Table 1 Collated soil definitions highlighting common themes in defining soil

Definition	Source
"The layer(s) of generally loose mineral and/or organic material that are affected by physical, chemical, and/or biological processes at or near the planetary surface and usually hold liquids, gases, and biota and support plants."	van Es [21]
"…a natural, three-dimensional body that has formed at the Earth's surface, through the interactions of at least five soil-forming factors (climate, biota, relief, parent materials, and time)." "…composed of clastic particles (mineral matter), organic materials in various stages of decay, living organisms, water (or ice), and gases within pores of various sizes."	Schaetzl [13]
"…natural bodies resulting from biogeochemical processes on the land surfaces of Earth." "Soils have both inorganic and organic components and thus are influenced by the composition of the geologic material in which they form and the organic ecosystems that form in and over them. Both the inorganic and organic components respond to inputs of energy and water as dictated by the climatic conditions at the location where they form." "The upper boundary of soil is the interface with the air. The lower boundary that separates soil from "not-true-soil" (C and R horizons) is not easily defined."	Buol [3]
"…any material within 2 m of the Earth's surface that is in contact with the atmosphere, excluding living organisms, areas with continuous ice not covered by any other material, and water bodies deeper than 2 m."	IUSS Working Group World Reference Base 9
"For engineering purposes, soil is considered to be any loose sedimentary deposit, such as gravel, sand, silt, clay or a mixture of these materials. It should not be confused with the geological definition of soil, which is the weathered organic material on the surface, or topsoil. Topsoil is generally removed before any engineering projects are carried out."	Smith [16]
"Soils consist of solid particles, enclosing voids or pores. The voids may be filled with air or water or both."	Bodó and Jones [1]

Regardless of the favored definition, three key elements of 'soil' recognized in most definitions are: (1) some degree of physical and chemical weathering compared to the parent material from which the soil has formed; (2) some incorporation of organic matter and/or organisms compared to the parent material; and (3) the position of soil at the interface between the atmosphere, lithosphere, hydrosphere, and biosphere (Table 1). These key elements capture the basics of soil formation and its role within the Critical Zone. Soil forms from the action of weathering, driven by Earth surface and biological processes, on a parent material. Parent material can be consolidated bedrock or reworked material such as glacial till, aeolian sands, or even industrial wastes. Some weathering processes inherently add organic matter; other weathering processes facilitate the subsequent ingress/proliferation of living organisms and accumulation of organic matter. Understanding what soil is, how it

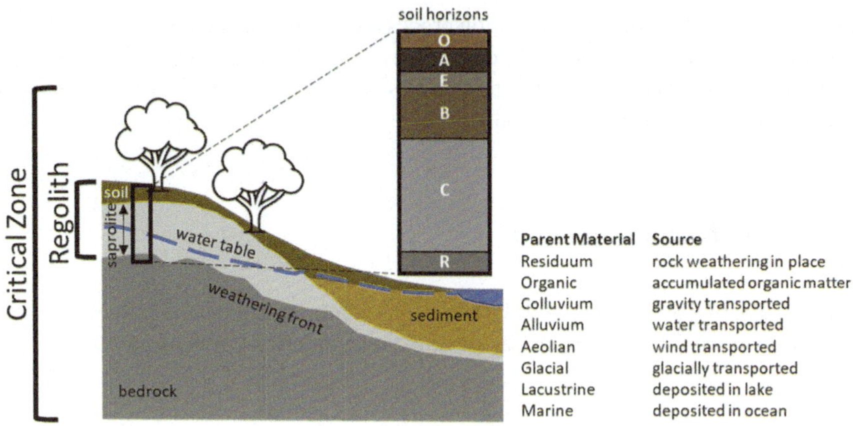

Fig. 1 Conceptual diagram of regolith and soil horizonation within the critical zone and possible parent materials and sources from which soils can form

is formed and transformed, and how to measure, describe, and quantify it enables a full appreciation of its responses and contributions to Critical Zone processes.

Soil can be further described, classified, and positioned within the Critical Zone based on depth and space-related properties. As **parent material** (the material from which the soil is forming) is weathered, it may remain *in situ*, or be transported spatially by Earth surface processes such as wind, water, and gravity movement. Parent material remaining *in situ* is called **residuum** whereas parent material moved elsewhere is named according to the process by which it is deposited (Fig. 1). Weathering processes act upon the parent material to create **horizons** in soil, distinct layers that differ in chemical, mineralogical, physical, biological, and morphological properties with depth from the soil surface to bedrock (Fig. 1). The presence of horizons is a key marker of soil development and differentiation from parent material, and forms a central part of soil description as explored in Sect. 4.

Having defined and positioned the soil body within the Critical Zone, we now need to consider the Critical Zone processes acting upon parent materials that form and transform soils.

2 Soil Forming Models of Pedogenesis

Weathering of parent material by Earth surface and Critical Zone processes forms soil. Without these processes, we would have only rock. Weathering processes can be split into **chemical weathering** (changing the mineralogical and chemical composition of parent material) and **physical weathering** (changing the physical properties of parent material, such as fragmentation, comminution, and particle size). Physical weathering assists chemical weathering by increasing exposed surface area and porosity, in turn accelerating mineral dissolution processes and increasing fluid exchange with water and atmospheric gases. Although the types and intensities of various weathering processes influence rates and trajectories of soil formation, several other factors also contribute to soil formation outcomes.

2.1 State Factor Equation

Historically, one of the most useful frameworks summarizing factors influencing soil formation in the CZ is the State Factor Equation formalized by Jenny [10]:

$$S = f(Cl, O, R, P, T, \ldots) \tag{1}$$

In this equation, the type of soil (S) at any given location is the result of five main environmental factors, including the **climate** (Cl: largely temperature and precipitation), the **organisms** (O) living in the soil, the topographic position or **relief** (R), the **parent material** (P) from which the soil is forming, and the amount of **time** (T) over which the other factors have been influencing the soil (Fig. 2). The growing influence of human activity in the Anthropocene on all five original soil forming factors have led to the proposal of human activity (H) as the sixth soil forming factor [6, 22]:

$$S = f(Cl, O, R, P, T, H) \tag{2}$$

As written, the equation cannot be solved quantitatively, but it nonetheless provides a powerful framework for sampling soils across environmental gradients to understand the environmental factors influencing soil formation and function. For example, if all variables except one are held constant, the influence of the variable of interest can be more directly explored. A study where only climate varies is a **climosequence**; only organisms a **biosequence**; only relief a **toposequence**; only parent material a **lithosequence**; and only time a **chronosequence**. It is extremely difficult to constrain all variables (i.e., organisms necessarily vary with climate, time is often variable and poorly constrained, etc.), but the framework helps to conceptualize the major environmental gradients, and select sampling sites that are most

Fig. 2 Illustration of the variables in the state factor equation that determine the type of soil formed in any given critical zone

useful for understanding which factors drive soil evolution and which soil properties they influence. Careful consideration of these variables and earnest effort to select sampling sites that vary only by the variable of interest can facilitate interpretation of CZ processes. Equally, describing and classifying soils based on their properties can give insights into their genesis. Predictable relationships between the five soil forming factors and major soil properties have been defined, such that dominant soil forming processes can be inferred from soil properties and functions observed today [12]. Careful soil description and identification of even subtle differences among spatially separated sites thought to be similar in their exposure to the five soil forming factors can reveal insights into changes in the intensity of these factors and CZ ecosystem dynamics.

2.2 Simonson's Model

A useful framework for considering soil processes within the CZ is the Simonson model [15]:

$$S = f(a, r, t_1, t_2) \tag{3}$$

This model considers that soil (S) horizon differentiation arises because of four main processes at work, including **additions** (a) to the soil, **removals** (r) from the soil, **translocations** (t_1) from one part of the soil to another, and **transformations** (t_2) within the soil (Fig. 3). Additions include energy, carbon, water, and dust/eroded soil; removals include soil erosion and elemental leaching; transformations are chemical reactions or mineral weathering that creates new clay minerals; translocations move carbon, clay, or elements in the soil, often from the surface to deeper horizons but can be moved in three dimensions. Human activity influences all processes in this model. While perhaps less instructive for experimental and sampling design, this model considers the major processes responsible for the eventual soil and observed horizonation and can therefore be very useful when considering drivers of soil formation. For example, inputs to a soil at a ridgetop topographic position would include energy, water, and carbon, but likely not eroded soil, whereas soil on a backslope would also be expected to receive soil input from upslope in addition to these inputs. Elements released by mineral weathering can either be lost to leaching or plant uptake or transformed by incorporation into new minerals or as organo-mineral complexes. Material from one part of the soil can be translocated by water or organisms. Human activities can accelerate removal of soil through erosion following land clearing, add nutrients via fertilizers, enhance translocations through tillage, and alter transformations by changing waterlogging regimes.

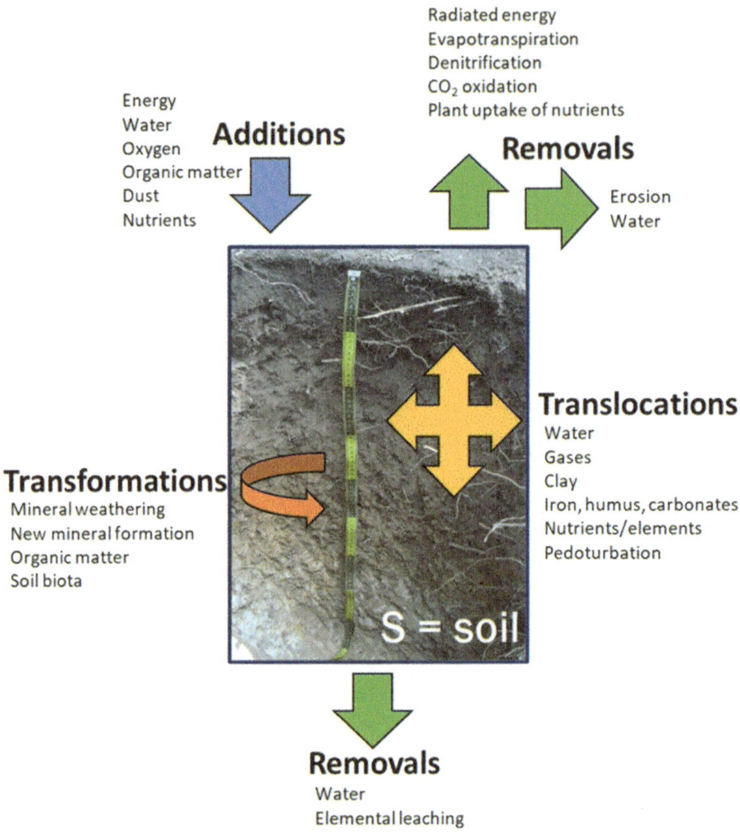

Fig. 3 The Simonson model considers additions, removals, transformations, and translocations as major drivers of soil horizonation and development

2.3 Sampling Soils in the CZ

Within the CZ, soils can be considered across a range of scales: molecular < soil particle < soil aggregates < soil horizon < pedon (soil column) < toposequence (catena, or soils across a hillslope) < regional physiography < pedosphere (all soils). Since soils cannot realistically be sampled at every point on the landscape, considering the scope and representativeness of the soil is key. For many CZ questions, soils are considered at the **pedon** or **hillslope** scale, where soils are sampled in landscape positions considered representative of key landscape features (Fig. 4).

An important consideration when sampling soils is water **flowpaths** as water will have high importance for chemical and biological reactions in soils and plays an important role in solid material transport downslope (Fig. 4). Soils on ridgetop topographic positions can largely be considered one-dimensional with respect to inputs: water, for example, would enter the soil surface and proceed vertically until

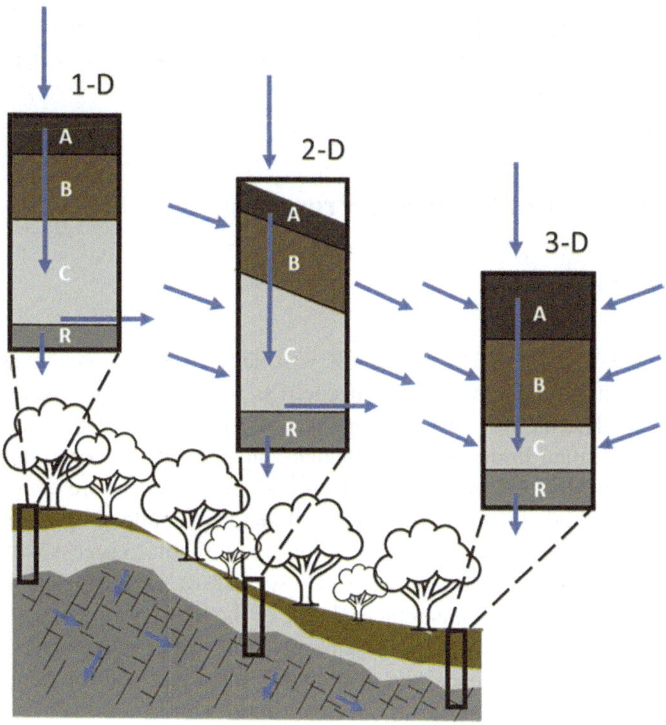

Fig. 4 Conceptual diagram of one-dimensional (1D), two-dimensional (2D), and three-dimensional (3-D) water flowpaths (blue arrows) through soils and fractured bedrock across a hillslope. Each black box is considered one pedon across the hillslope

it intercepts bedrock or groundwater, at which point it would flow downslope. Soils on slopes experience two-dimensional inputs, both from the soil surface and laterally as water and soil are transported downslope. Finally, convergent or footslope topographic positions experience three-dimensional flowpaths where water and soil material are transported vertically and laterally.

When selecting a representative soil sampling site, considering *a priori* data can help constrain soil forming factors. Previously published information can help identify sampling sites that are not only well represented across the landscape but also vary as little as possible except for the soil forming factor of interest. Resources such as soil maps (i.e., Web Soil Survey in the U.S.) provide information such as dominant soil type or other geochemical characteristics; geologic bedrock or surface maps can help constrain parent material; topographic maps help determine elevation, slope, and aspect; and vegetation maps can delineate dominant vegetation. Modern LiDAR (Light Detection And Ranging) data provides high resolution digital elevation models (typically ~1 m resolution) that are not only useful for mapping topography, but also for understanding landscape evolution and identifying potential chronosequences, understanding water movement across and through the soil profile, and refining pedological maps.

3 Soil Sample Collection and Processing

3.1 Field Procedures

Methods for extracting soil from the CZ will depend on the objectives of study (i.e., depth of interest, sample volume needed) as well as logistics (i.e., site access, time, budget). The number of sampling sites and replicates within each site will also vary but is often limited by time, budget, and site access constraints. Studies with single replicate soil sampling within sites along sequences are common where the sampling method is manual, terrain and climate are challenging, soils are deep, clayey, hard, or gravel-rich, sampling locations are difficult to access with vehicles, or for other practical and logistical reasons. In such cases, extra care should be taken to select a sampling point that is representative of the likely variation across the site by considering geomorphology (landscape position, water movement, inputs and outputs from erosion and deposition, water table and water bodies), vegetation and other organisms (roots, canopy cover, species, bioturbation), surface cover (leaf litter, gravel, pavement etc.), and site history (i.e., disturbance, land use).

3.2 Soil Cores

One option for soil sampling is to extract a soil core, where soil is removed vertically from the soil surface either manually using a hand auger that can remove soil in 5–20 cm increments (Fig. 5a and b) or by pushing a hollow tube into the ground using either mechanical force (e.g., with a geoprobe-type drilling rig) or using manual force with a weighted driver or hammer. Hand augering advantages include accessibility for most sites, minimal footprint, low equipment cost, and the potential to extract samples many tens of meters into the subsurface. Drawbacks include lack of visual context for soil continuity, homogenized soil samples over the extracted interval, possible contamination of shallow soil into deeper soil as the auger is extracted and replaced into the borehole, sample volume limited to auger bucket size, longer sampling time and potentially difficult manual labor, and limited success in rocky or hard soils where the auger cannot penetrate. Extracting a core mechanically can provide undisturbed soil for the length of the hollow tubing (often ~3 m sections), deeper soil sampling and more rapid soil extraction. Drawbacks include high cost for specialized equipment, difficulty accessing sites without road access, sample volume limited to core diameter, and compaction of soil in the tube.

Fig. 5 **a** Extracting a soil core using a hand auger in 10 cm increments (marked by colored tape on auger pole); **b** bulk soil extracted in 10 cm increments; **c** excavated soil pit with measuring tape and undisturbed section marked with orange tape

3.3 Soil Pits

Soil pits are the preferred, but often less practical, method of collecting soil samples in the CZ and can be excavated manually with shovels and picks or mechanically with a backhoe. Sampling can be completed at an interval of choice (commonly every 5 to 10 cm) or by genetic horizon (Fig. 5c). One major advantage of soil pits is that they provide important context for the spatial variability and vertical relationships of soil features. Furthermore, they can be dug manually at almost any site, and sample volume is not as limited as with soil coring in which multiple cores may need to be extracted to generate sufficient sample volume at each target depth. Limitations include time and effort to excavate the pit when backhoe access is not an option, large disturbance footprint, and sampling depth is often (although not always) limited to the upper 2 m given the instability of many soil pit walls. In some cases, this depth limitation can be mitigated by digging a soil pit to extract near surface information and then augering from the bottom of the soil pit to extract soil samples from deeper depths.

4 Preparing Soils for Analysis

4.1 *Drying, Splitting, Grinding*

Planning downstream analyses and then working backwards to determine the field sampling and storage conditions required preserves the integrity of the samples and ensures minimal soil wastage. Soils are commonly collected and stored in plastic zip-top, calico, or paper bags at ambient temperature for transport and future geochemical and physical analysis. A small subset of chemical and physical analyses may need refrigeration or preservation after sample collection. If soils are to be analyzed for biological components, such as microbial DNA, care should be taken to sterilize all sampling equipment and transport containers before sample collection, and then store samples either refrigerated at −4 °C or on dry ice at < −80 °C depending on downstream analyses. For most analyses, air drying samples (~25 °C) is appropriate for preparation, which can take several days to weeks depending on the soil moisture content. Once air dried, samples can be split for subsequent analysis by homogenizing the sample and using a riffle splitter that separates the sample into two, repeating splitting until the desired subsample volume is achieved. For some analyses, moisture should be removed by drying in an oven at 110 °C until soil weight is stable. For many soil analyses, soil samples are passed through a 2 mm sieve to separate rock fragments (>2 mm diameter) from fine soil particles (<2 mm). Fine soil particles are then analyzed further. Rock fragments are typically not analyzed further than determining their contribution to the soil matrix on a gravimetric (weight) or volumetric (volume) basis. This is because the large particle size of rock fragments is considered to render them relatively inert, having a much smaller contribution to properties such as total soil surface area, porosity, and water retention and release characteristics, compared with fine soil particles. For geochemical or mineralogical analyses, samples are often ground to pass through a 150-micron sieve or finer to ensure full quantification (e.g. total element concentrations). Others avoid grinding to capture the behavior of the soil in its natural (aggregated and/or particulate) state (e.g., pH, cation exchange capacity).

5 Soil Profile Descriptions

5.1 *Soil Morphology*

Soil morphology refers to the way that physical, chemical, and biological features are arranged within a soil profile. Detailed morphological observations can provide qualitative or semi-quantitative information about the soil forming factors and genesis. In fact, soil morphological observations can be extremely useful for interpreting other soil data (e.g., geochemistry) and determining optimal sampling and analysis intervals. Here we briefly describe key morphological features to observe in soils. For more detailed information see [14].

5.1.1 Horizons

Horizons refer to layers within the soil profile with distinct geochemical, mineralogical, physical, morphological, and biological properties. These can usually be visually observed during sampling (most completely in a soil pit), with further detailed classification then performed in the field or laboratory after initial description of the horizons. Horizons classically form roughly parallel to the soil surface driven by physical, chemical, or biological processes operating in the soil but can be angled relative to the contemporary soil surface due to, for example, erosion of the current soil surface or steep relief (Figs. 1 and 4). Sometimes horizons are formed by deposition of different parent material, as in a floodplain that receives episodic sediment. Generally, soils that are younger exhibit fewer distinguishable horizons; as soil development proceeds, horizons become more differentiated, then horizons tend to become poorly distinguished once again in older, highly weathered soils (e.g., tropical soils).

The first horizon that tends to form is the **A horizon**, distinguished by the addition of decomposed organic matter into the mineral soil material. In some CZ environments, such as those with high biomass production and slow organic matter decomposition rates, an **O horizon** can form on top of rock or an A horizon and may be the precursor to A horizon formation. This material is dominated by fresh, partially, and highly decomposed organic material, but has little mineral material. Material that has been only minimally chemically, physically, or biologically altered is considered the **C horizon**, and includes saprolite which retains the fabric of the bedrock but not the physical integrity, as well as sediments deposited immediately over bedrock. If the parent material is coherent rock, it is described as an **R horizon**. As parent material in either C or R horizons is weathered and transformed, a **B horizon** is formed. In environments with high precipitation and low soil water pH, an **E horizon** can form below an O or A horizon where organic matter and iron and aluminum oxides are stripped from this horizon and deposited in the horizons below. Transitional horizons contain properties of two major horizons, such as an AB or CB horizon, with the first listed indicating that those horizon characteristics are more dominant. Multiple horizons of the same type are common: there are often multiple distinguishable B horizons that differ in the degree of chemical and physical transformation; multiple unaltered C horizons which can result from a sequence of distinct parent materials; or horizons can be buried as new material is added to the CZ surface (e.g., floodplains, landslides).

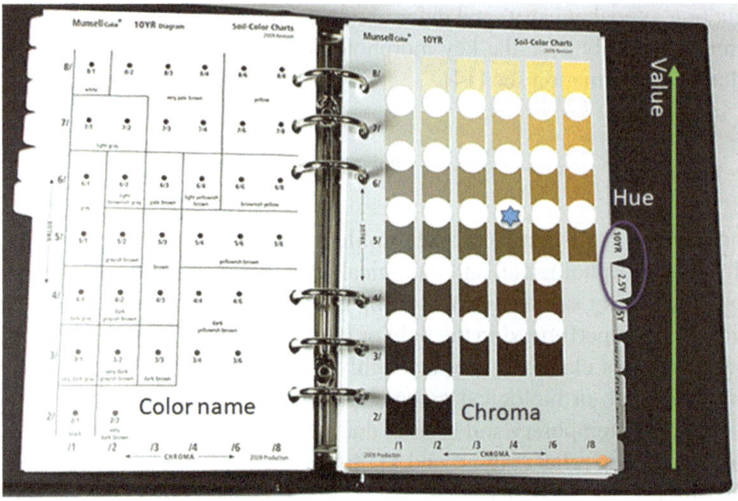

Fig. 6 A Munsell® color book provides a way to semi-quantitatively determine soil color. Hues are indicated on each page tab, value increases from bottom to top, and chroma increases from left to right. The blue star in this image indicates a Munsell color 10YR 5/4, a yellowish-brown color

5.1.2 Color

Soil color can be one of the most striking morphological observations and is especially useful for distinguishing horizons and providing clues about chemical transformations, organic matter additions, and parent material. **Dark brown or black** colors, usually observed in A horizons, arise from organic matter additions to the soil. **Reddish or yellow brown** colors indicate weathering of iron and aluminum oxide minerals, respectively. **Dull grey or green** colors result from iron reduction. **Bright white** colors can indicate calcium carbonate accumulation. Color can also be inherited from minerals in the parent material. The most common way to describe soil color is with a Munsell® color book that provides a semi-quantitative color value (Fig. 6).

5.1.3 Structure

Soil structure is the way soil naturally aggregates, or forms clumps of individual particles, and is especially important for gas and water flux through soils by providing porosity and conduits (Fig. 7). Soil that does not naturally aggregate in any way is considered **massive,** or **single-grained** if the soil texture is dominantly sand grains. Soil aggregated in small, rounded shapes is called **granular** structure, a common structure in A horizons with high root density and biological activity. B horizons commonly have **sub-angular blocky** (rounded edges) or **angular blocky** (sharp edges) structure. Structure with a longer y-axis than x-axis is considered **prismatic**,

Fig. 7 Examples of common types of soil structure, redoximorphic features (concentrations, depletions, and gleyed horizons), and evidence of clay movement in soils (clay films, slickensides)

and prisms with rounded tops are **columnar**, a structure most commonly observed in saline soils. **Platy** structure has a longer x-axis than y-axis and can be observed in compacted surface soils, leached E horizons, or inherited from a fissile parent material such as shale.

5.1.4 Moist Consistence

Moist consistence refers to the resistance of a soil aggregate, also called a ped, to crushing when moist (wet but not dripping water). Soils that are **very friable** or **friable** can be easily crushed, indicating they are not strongly aggregated or "glued" together. Soil aggregates with a **firm** moist consistence require moderate pressure to crush the aggregate between your thumb and forefinger, while aggregates with **very firm** or **extremely firm** moist consistence are very difficult to crush with your fingers and can require the use of a hammer to break apart. Soils that do not form aggregates at all, such as very sandy soils, have **loose** moist consistence.

5.1.5 Texture

Soil texture refers to the proportion of **sand**, **silt,** and **clay** sized particles in the fraction of the soil that passes through a 2 mm diameter sieve. Although size definitions can vary, the U.S. Natural Resource Conservation Service (NRCS) defines sand particles in the range of 2 mm to 53 microns, silt particles from 53 to 2 microns, and clay particles smaller than 2 microns [14]. In the field, soil texture can be estimated based on wetting and massaging the soil, feeling for grittiness from sand, silkiness from silt, and stickiness/firmness from clay. Texture is especially important for determining surface area for chemical reactions and water holding capacity; sandy soils with larger grains have less surface area to hold water or other nutrients, whereas small clay-sized particles have high surface area and are therefore more chemically reactive and can retain more water that prevents soils from draining. Coarse fragments (>2 mm) can be divided into gravel (2–76 mm), cobbles (76–250 mm), stones (250–600 mm) and boulders (>600 mm) under the NRCS system. Coarse fragments were traditionally assumed to be effectively unreactive due to their low contribution to the total surface area of the soil. Increasingly, coarse fragments are being recognized as functional, reactive parts of the soil and interrogated for their roles particularly in water and nutrient storage and release.

5.1.6 Redoximorphic Features

Redoximorphic features, sometimes referred to as "mottles", are zones of red or orange iron **concentrations** formed when iron oxidizes, and zones of **depletions** when iron reduces resulting in a dull grey color (or absence of color) (Fig. 7). Accumulation of reduced iron in waterlogged, oxygen-poor horizons can generate green and blue colors. Redoximorphic features can resemble colored spots or stripes; depletions can encompass entire soil horizons (called a **gleyed** soil horizon). These features indicate periods when the soil is saturated with water and oxygen is depleted from the soil water, causing iron to reduce, mobilize, and then oxidize once the water is more oxygenated. Concentrations and depletions are often observed adjacent to each other and are commonly found along roots or along the edges of soil structure where oxygen concentrations can fluctuate, in turn influencing the local or microsite iron chemistry.

5.1.7 Roots

Root density and distribution in soils is driven largely by the type of vegetation but observations about roots can provide information about soil conditions. Most roots are concentrated in the upper soil horizons, and vegetation with fine roots can significantly influence the amount of carbon added to the soil when the roots die and decay. Roots help create soil structure by breaking up the soil as they grow and gluing particles together with root exudates. Roots may be absent or growing laterally when soil horizons are dense and impenetrable; soil horizons that are often waterlogged will also lack roots that rely on oxygen.

5.1.8 Clay Films

Clay films, also called cutans, are coatings of clay particles on soil aggregate surfaces caused by the movement of clay particles by water in the soil (Fig. 7). Water moving through the soil can carry these small particles which build up on the surfaces of aggregates and along root channels. Clay films are usually an indication of advanced soil development, where the soil has had sufficient time and moisture to accumulate clay sized particles (either from weathering of larger particles or inherited from the parent material) and flush them deeper into the soil. Soils with a high clay content composed of expandable minerals can shrink and swell upon wetting and drying, creating **slickensides**, or a slip face along which soil slides.

5.1.9 Bulk Density

Bulk density refers to the mass of dry soil per volume and is an indicator of the amount of soil pore space available to hold water and gases or for biota (roots or organisms) to occupy. This measurement requires a known volume of soil be extracted from the soil, ideally at least from each horizon, and weighed after removing moisture (heated to 110 °C for ~24 h). Soil volume can be estimated using several methods: inserting a metal ring of known volume; measuring the dimensions of a carved block of soil; excavating a cavity, lining with plastic, and measuring the volume of water held by the cavity; and dipping a soil aggregate in paraffin wax to seal pores and measuring displacement in water [18]. Bulk density is an important measurement for determining quantities of nutrients or carbon in soils but can be difficult to estimate well. For example, pounding a ring or core can compact soil, artificially increasing bulk density; paraffin wax can be difficult to prepare and transport to the field site; rocks can make sampling difficult and must be accounted for in bulk density estimates; collecting a known volume of soil from deeper soils is challenging when soils are disaggregated by bucket augers or compacted by push cores.

5.2 Soil Physical, Geochemical, and Biological Analysis

Once sampled, soils can be analyzed for numerous physical, chemical, and biological properties that can provide information about soil function and evolution within the CZ. Table 2 lists some common assessments. See Soil Survey Staff [17, 18] for laboratory methods to determine some of these properties.

Table 2 Common physical, chemical, and biological properties measured in soils

Physical properties	Chemical properties	Biological properties
Particle size	Major element geochemistry	Microbial community composition
Bulk density	Trace element geochemistry	
Hydraulic conductivity	Organic/inorganic C/N	Microbial function (potential and expressed)
Aggregate stability	Total macro/micronutrients	
Dispersion and slaking	pH and salinity	Root density, length, and branching
Porosity	Cation exchange capacity	
Water retention and release	Cosmogenic nuclides	Enzyme activities
	Radiogenic nuclides	Root exudates
	Stable isotopes	Microbial respiration rate
	Water extractable elements	Microbial biomass
	Available macro/micronutrients	Microbial biomass C, N, P, K
	Bulk mineralogy	
	Clay mineralogy	

6 Interpreting Soil Biogeochemical Profiles

Soil biogeochemical profiles provide a powerful tool to interpret the role of soil in CZ ecosystem dynamics, as the release of elements from parent material, and subsequent loss by leaching or cycling by biota, can be inferred from geochemical profiles. For example, elements weathered from parent material and solubilized are accessible to vegetation or flushed to groundwater and ultimately streams, providing elements to aquatic ecosystems. Elements added to soils, such as carbon, provide energy sources for microorganisms, and contaminants from air pollution can accumulate in soils. Vegetation can concentrate plant-essential nutrients in near-surface soil horizons for reuse.

6.1 *Elemental Profiles*

One approach to evaluating soil geochemical transformations is by calculating the dimensionless coefficient tau, which compares elemental concentrations in the soil to initial concentrations in the parent material, normalized to an immobile element (Ti, Zr, Nb) that is not expected to be changed by loss or addition to soils (i.e., expected to remain constant) [2]. A negative tau value indicates that a soil has been depleted of an element relative to the initial parent material concentrations, whereas a positive tau value indicates that an element has been added or accumulated from a source other than the parent material (Fig. 8). Biota can deplete elements at depth and concentrate them at the surface, resulting in a biogenic tau profile. Alternatively, elements can be weathered from some minerals and transported deeper in the soil where they are incorporated into new minerals resulting in a depletion-enrichment tau profile. Finally, some elements are relatively inert and show neither depletion nor enrichment tau profiles. Advantages of a normalized approach such as tau for interpreting soil geochemistry include the ability to compare weathering extent across sites with varying elemental concentrations, and to calculate weathering rates if the age of soil exposure can be estimated. Limitations of this approach include accurately identifying the initial parent composition, especially in heterogeneous parent material.

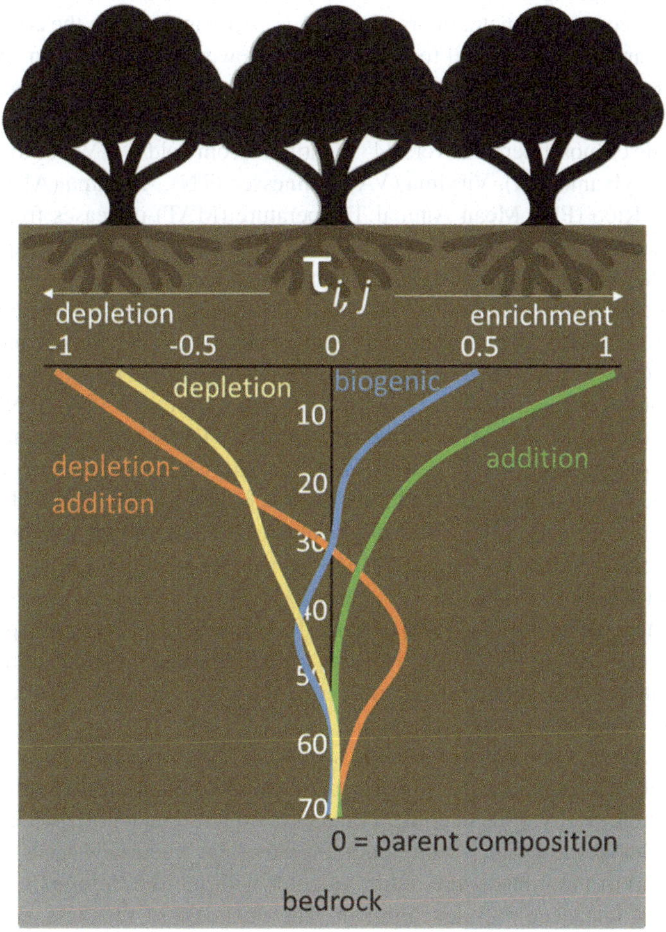

Fig. 8 Example tau profiles to interpret elemental depletion in soil geochemical profiles

7 Examples of Soil Studies in the Critical Zone

7.1 *Shale Climosequence*

Shale is a dominant rock at the Earth's surface and therefore a common parent material from which soils are forming. The wide geographic extent of shale bedrock allows for selecting study sites across a latitudinal climate gradient, or climosequence, to study the effect of climate on rock weathering and soil formation from shale parent material. In designing this study, an effort was made to hold all soil forming variables consistent except for climate; in practice, this is challenging given the complexity of landscapes and co-varying soil forming factors. However, carefully considering soil forming factors in selecting study site locations allows for qualitative and quantitative interpretations of the influence of climate on soil formation and weathering.

The shale climosequence covers 34° of latitude from cold/wet Wales to New York (NY), Pennsylvania (PA), Virginia (VA), Tennessee (TN), Alabama (AL), to warm/wet Puerto Rico (PR). Mean Annual Temperature (MAT) increases from north to south and mean annual precipitation (MAP) increases slightly across the Appalachian sites (PA, VA, TN, AL) and is very wet at each end member (Fig. 9a). To constrain parent material, or the material weathering to form soil, study sites were located in the basal 10 m of the Silurian Rose Hill Shale Formation, an iron-rich and organic-poor shale (Fig. 9a). Although all sites are forested, vegetation inevitably varies with climate; conifer tree species are more prevalent in northern sites (Wales, NY), mixed deciduous vegetation dominates in the mid-latitudes (PA, VA, TN, AL) and tropical vegetation occupies the southernmost site (PR). Study sites are located on ridgetop topographic positions to consider the simplest one-dimensional inputs of water and energy on soil formation (Fig. 4). Finally, the duration of soil formation inevitably varies in northern latitudes as past glaciations reset or modified Earth's surface: in Wales, past glaciations stripped soils to bedrock; in NY, glaciers deposited ground-up shale on top of bedrock; in PA, periglacial conditions effectively stripped soils to bedrock. Constraining the duration of soil formation, or soil residence time, using past glaciation events and cosmogenic isotopes can help untangle the influence of time on soil weathering and development.

Shale derived soils across the climosequence are deeper and show increasing evidence of soil formation as temperatures increase from north to south (Fig. 9b; [4]). Morphologically, soils in colder climates exhibit minimal horizon development and color changes associated with mineral weathering, whereas warmer sites on the southern end of the transect are redder in color with more developed horizonation indicative of enhanced soil development. The thickness of the soils, estimated as the depth to which a hand auger could be driven into the soil, also increases from north to south across the climosequence, indicating deeper bedrock weathering and transformation to soil on these southern ridgetop topographic positions (Fig. 9c). For example, clay content increases across the climosequence, from average clay content of about 20% in Wales and PA, to ~25% in VA and over 40–50% at the southern sites, which influences soil hydrology and the development of redoximorphic features as

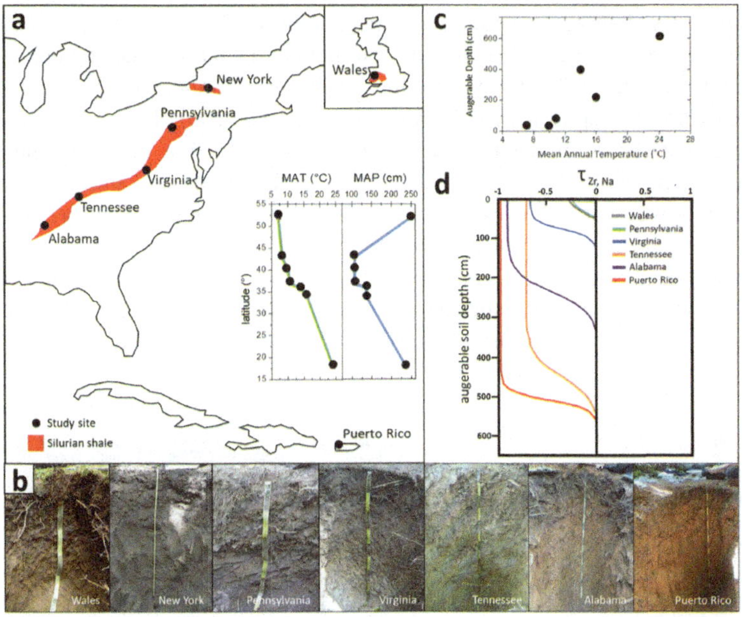

Fig. 9 **a** Map of the shale transect study across a northern hemisphere climosequence. **b** Photos of soils from each site show enhanced soil morphological development, **c** deeper soils, and **d** enhanced Na depletion, an indicator of feldspar weathering, at the surface and with depth, at warmer sites across the climosequence

clay content inhibits water movement (Fig. 9b, [5]). Soils are increasingly weathered from north to south at the soil surface and with depth. For example, Na, an element released during feldspar weathering, is about 20% depleted at the soil surface in Wales and PA, increasing to 70–95% in VA, TN, and AL, respectively, and 100% depletion in PR down to 5 m depth (Fig. 9d, [4]). Calculating such elemental depletion and measuring bulk quantitative mineralogy, we can interpret that feldspar weathering may be the first weathering reaction that initiates soil development, but the weathering of more abundant chlorite and illite minerals to vermiculite and hydroxy-interlayered vermiculite is more likely controlling regolith thickness across the climosequence [5]. A similar analysis of trace metal and rare earth element (REE) geochemistry documents minimal elemental depletion at northern sites while southern sites show depletion-enrichment profiles indicative of REE mobility, but not loss, in the soil profiles in warmer climates [11].

7.2 K'Gari-Cooloola Coastal Dune Chronosequence

The sandy coastal dune podzols (Spodosols) of K'Gari (formerly known as Fraser Island) and the Cooloola Coast in eastern Australia have formed through repeated deposition, weathering, erosion, and redeposition of beach sands over millennia (Fig. 10a and b). Weathering of the sand parent material has produced thick, brightly colored B horizons visible from kilometers away, where dunes have been eroded and exposed along the aptly named Rainbow Beach (Fig. 10b). The impressive depth and dramatic colors of the B horizons are the culmination of deep weathering, forming giant podzols over 20 m deep. A state factors-based sampling approach can be employed here to investigate how soil horizonation develops over time and ask how long it takes for these horizons to form.

The K'gari-Cooloola dunes developed from a consistent and homogenous parent material (beach sand) and under a (relatively) consistent climate. The three other factors in the State Factor equation (time, relief, and organisms) can be easily

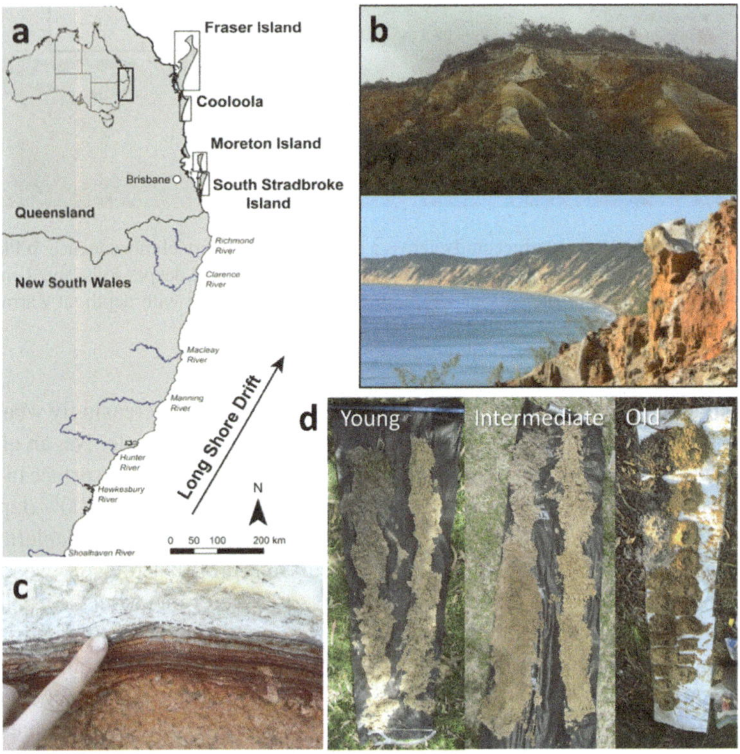

Fig. 10 **a** Map of the K'Gari-Cooloola sand mass study region, showing the Hawkesbury River to the south as the source of parent material sands from which the podzols have developed; **b** buried podzols in dunes are exposed during coastal erosion, forming cliffs that reveal colored layers of E (white), B (orange, red, and brown), and C horizons (pale yellow); **c** an example of one of the many types of colorful B horizons observed in the K'Gari-Cooloola region, showing white silica cementation over red–orange iron cemented layers; **d** example soil profiles retrieved by augering to depth, with progressive thickening of the E horizon and intensification of color in the developing B horizons with age. Augered samples are laid out with surface samples at top left and deepest sample at the bottom right of each photo

constrained. This dune system therefore provides numerous excellent opportunities to study chronosequences (time), toposequences (relief), and biosequences (organisms). To answer our questions about horizonation over time, we would select a chronosequence consisting of parent materials of different ages.

The first step is to identify sampling sites that represent a spread of soil ages from youngest to oldest. At this site, relative dune ages were inferred from geomorphological context, with dune systems generally increasing in age toward the west of K'Gari as fresh sand continues to accumulate on the east and especially the south-east coast (Fig. 10a). Detailed techniques such as optically stimulated luminescence (OSL) dating can then quantify absolute age, and this can then be related back to podzol profile properties. This approach has been used by [7] to show that multiple periods of dune building and erosion have occurred. The stronger consistence of B horizons, due to their inherited iron and aluminum oxides and organic matter, has often resisted erosion during episodic dune deflation (Fig. 10c). These recalcitrant B horizons were then buried under fresh dune sand deposits in which new podzols formed. Some B horizons are palimpsests, enriched with younger illuviated material from fresh dune deposits that have infused into the existing, older B horizon. The climatic conditions during podzolization as well as position relative to the water table, and relative enrichment of iron, aluminum, and organic matter, all influence the color, structure, and consistence of B horizons [20]. Characteristic podzol A, E, and B horizons thicken over time as the weathered profile deepens (Fig. 10d); although the A horizons rarely exceed 1 m deep [8], E horizons up to 14 m deep and B horizons up to 20 m deep were recorded [8]. OSL dating indicated that sand in these horizons was emplaced 100,000–800,000 years ago–thus, deep weathering is required to form deep horizons [8].

References

1. Bodó B, Jones C (2013) Introduction to soil mechanics. John Wiley and Sons, West Sussex, UK
2. Brantley SL, Goldhaber MB, Ragnarsdottir KV (2007) Crossing disciplines and scales to understand the critical zone. Elements 3(5):307–314
3. Buol SW, Southard RJ, Graham RC, McDaniel PA (2011) Soil genesis and classification. John Wiley and Sons, West Sussex, UK
4. Dere AL, White TS, April RH, Reynolds B, Miller TE, Knapp EP, McKay LD, Brantley SL (2013) Climate dependence of feldspar weathering in shale soils along a latitudinal gradient. Geochim Cosmochim Acta 122:101–126
5. Dere AL, White TS, April RH, Brantley SL (2016) Mineralogical transformations and soil development in shale across a latitudinal climosequence. Soil Sci Soc Am J 80(3)
6. Dror I, Yaron B, Berkowitz B (2022) The human impact on all soil-forming factors during the anthropocene. ACS Environmental Au 2:11–19
7. Ellerton D, Rittenour T, Miot da Silva G, Gontz A, Shulmeister J, Hesp P, Welsh KJ (2018) Late-Holocene cliff-top blowout activation and evolution in the Cooloola sand mass, south-east Queensland, Australia. Holocene 28(11):1697–1711

8. Ellerton D, Rittenour T, Shulmeister J, Gontz A, Welsh KJ, Patton N (2020) An 800 kyr record of dune emplacement in relationship to high sea level forcing, Cooloola Sand Mass, Queensland, Australia. Geomorphol 354:106999
9. IUSS Working Group WRB (2015) World reference base for soil resources 2014, Update 2015. In: International soil classification system for naming soils and creating legends for soil maps. World Soil Resources Reports No. 106, Rome: FAO
10. Jenny H (1941) Factors of soil formation: a system of quantitative pedology. Republished in 1994. New York: Dover Publications
11. Jin L, Ma L, Dere A, White T, Mathur R, Brantley SL (2017) REE mobility and fractionation during shale weathering along a climate gradient. Chem Geol 466:352–379
12. Lin H (2011) Three principles of soil change and pedogenesis in time and space. Soil Sci Soc Am J 75(6):2049–2070
13. Schaetzl RJ, Thompson ML (2015) Soils. Cambridge University Press, Cambridge
14. Schoeneberger PJ, Wysocki DA, and Benham EC (eds) (2012) Field book for describing and sampling soils. Government Printing Office
15. Simonson RW (1959) Outline of a generalized theory of soil genesis. Soil Sci Soc Am J 23(2):152–156
16. Smith MJ (1981) Soil Mechanics, 4th edn. Routledge Taylor and Francis Group, New York
17. Soil Survey Staff (2022) Kellogg soil survey laboratory methods manual. Soil survey investigations report no. 42, version 6.0. U.S. department of agriculture, natural resources conservation service
18. Soil Survey Staff (2014) Soil survey field and laboratory methods manual. Soil survey investigations report no. 51, version 2.0. R. Burt and soil survey staff (ed.). U.S.
19. Staff, Soil Survey (1999) Soil taxonomy: a basic system of soil classification for making and interpreting soil surveys. Agriculture handbook 436
20. Thompson CH (1992) Genesis of podzols on coastal dunes in southern Queensland I field relationships and profile morphology. Soil Research 30(5):593–613
21. van Es H (2017) A new definition of soil CSA. News 62:20–21
22. Yaalon DH, Yaron B (1966) Framework for man-made soil changes–an outline of metapedogenesis. Soil Sci 102:272–277

Soil Organic Matter Dynamics in the Critical Zone

Kimber Moreland, Nicholas C. Dove, Qina Yan, Tingyu Hou, Morgan E. Barnes, Stephen C. Hart, Timothy Filley, Praveen Kumar, and Asmeret Asefaw Berhe

1 Introduction

> "Soil speaks to us through the colors and sculptures of its profile, thereby revealing its personality. Observing soils, studying them, and reflecting on them induces respect, if not wonder. All of us relate to soil unconsciously in our daily nourishments that make us participants in the continuous flow of nutrient atoms that originate in the soil."
>
> My friend, the soil – Interview with Hans Jenny, 1984.

Soil is the thin layer of Earth's surface that is the heart of the critical zone (CZ) and lays the foundation for life. Soil is made up of particulate matter and voids, where the particulate matter is composed of mineral and organic components, and the voids are where gasses and water reside. Soil organic matter (OM) is a mixture

K. Moreland (✉)
Lawrence Livermore National Laboratory, Livermore, CA, USA
e-mail: kmoreland@ucmerced.edu

N. C. Dove
Biosciences Division, Oak Ridge National Laboratory, Oak Ridge, TN, USA

Q. Yan
Lawrence Berkeley National Laboratory, Berkeley, CA, USA

T. Hou · T. Filley
Department of Geography and Environmental Sustainability, University of Oklahoma, Norman, OK, USA

M. E. Barnes
Pacific Northwest National Lab, Richland, WA, USA

K. Moreland · S. C. Hart · A. Asefaw Berhe
Department of Life and Environmental Sciences and Sierra Nevada Research Institute, University of California, Merced, CA, USA

P. Kumar
Department of Civil & Environmental Engineering, University of Illinois, Urbana, IL, USA

T. White and A. Provenzale (eds.), *Critical Zone and Ecosystem Dynamics*, Advances in Critical Zone Science, https://doi.org/10.1007/978-3-031-69076-1_7

of the decaying remains of once-living plants, microorganisms, insects, and animals, and typically accounts for only 1–6% of soil weight [187]. Even though soil OM makes up only a small portion of the soil, it has significant importance on its function and that of the CZ. As noted above by Hans Jenny soil OM is the main energy and nutrient source for microorganisms and is key for soil's water holding capacity, tilth, and physical stability. Hence, our understanding of the CZ would not be complete without an appreciation of soil OM and the linked ecosystem dynamics.

Soil is the most complex biomaterial in the earth system [197], where most of the chemical and structural complexity is attributed to the structure of soil OM [122]. Soil OM is derived from the altered remains of plant, microbial tissues, macrofauna, and mesofauna structures. The actual chemistry and environmental reactivity of the remains of formerly living organisms that make up soil OM are dictated by several factors, including their chemical composition, physical structure, and associations with cations and minerals. These interacting factors lead to a complex and dynamic continuum of decomposition products available for reuse by organisms or temporarily stabilized in soil. Chemically, soil OM is a complex amalgam of molecules of various sizes and stages of microbial, thermal, and geological alteration. These include chemically recognizable carbohydrates, lipids, waxes, lignin, amino acids, amino sugars, and proteins, among other biologically based compounds, as well as compounds that are so altered by microbes, disturbance, and geology that they cannot be traced to their original chemical or biotic source.

These compounds are often quantified using the individual and relative elemental composition of hydrogen (H), carbon (C), oxygen (O), nitrogen (N), sulfur (S), and phosphorus (P), among others. About half of soil OM is composed of carbon (C) and, as a result, soils store more C than the atmosphere and vegetation combined [111]. Globally, the soil system stores ~3000 Pg organic C that plays essential roles for maintenance of ecosystem productivity, regulation of the Earth's climate, and more [10, 89].

Our ability to predict the response of this expansive C pool to natural and human perturbations (e.g., due to climate change and land management) is challenged by the complexity of soil OM, including the processes that drive its formation, persistence, and loss. The concept of soil OM persistence has become central to contemporary discussions of soil OM dynamics and implies long-term preservation resulting from ecological, biological, and physiochemical conditions and interactions that soil OM encounters in the environment [160]. Currently, human activity releases 9.2–11.5 Pg of C into the atmosphere every year. This release has raised the atmospheric CO_2 concentration from 280 ppm prior to the industrial revolution to its current 419 ppm, causing overall warming in Earth's climate [62]. To achieve a global goal of limiting atmospheric warming to 1.5 °C compared to pre-industrial levels requires a reduction of global CO_2 emissions by 50% before 2030 [84]. Managing soils to maximize sequestration of atmospheric CO_2 in soil has emerged as one of the most important components of the solutions for climate change mitigation [212]. At the same time, concerns remain on the potential for soil to exacerbate climate change through the release of large amounts of C as CO_2 and CH_4 [108].

Soil OC storage results when the rate of input of plant, microbial, and animal derived C is faster than the rate of loss due to decomposition, leaching, methanogenesis, and erosion. Stability, and potential persistence, may be indicated by higher mean residence times (MRT, average time a C atom at steady state spends in the soil pool). Soil organic C stabilization is a relative term and is defined as a decrease in the potential for loss of OM by microbial respiration, erosion, or leaching [169]. Soil OC is often grouped into three different pools that vary in their susceptibility to microbial decomposition, which emerge as different MRTs. These include an active (MRT < 1 year), slow (10^1–10^2 years), and passive pool (>10^2 years; [135]). However, these operational pools do not explicitly define the mechanisms of soil OM stabilization.

Our understanding of the mechanisms that control the persistence of soil OM has evolved. The current paradigm maintains that persistence of soil OM is primarily a function of ecosystem conditions where OM can be protected from loss via physical isolation (i.e., within aggregation) from decomposers, chemical attachment (i.e., sorption or intercalation) to soil minerals, transformation to highly condensed aromatic C by fire and geologic processes, and freeze/thaw cycles [53, 110, 115, 160]. Soil OM can be preserved and stabilized within soil aggregates through a variety of biophysical mechanisms [17], including physical isolation by aggregation [164], soil mineral association [102], and cation complexation [102, 104, 107]. Hence, stabilization mechanisms and soil OM persistence are controlled by interaction of multiple biological and physicochemical conditions and processes operating in the environment [11, 54, 160].

Over varying spatial and temporal scales, the formation, stabilization, and persistence of soil OM can be related to the state factors of soil formation (climate, topography, biota, parent material, and time, see [87]). The primary controls on soil OC inputs include rate of primary production, deposition, rate of OM transfer to the soil pool [8, 18], while rate C output from the soil pool is controlled by the rate of microbial decomposition, erosion, and leaching. Climate (i.e., temperature and precipitation) is a strong control on both primary production (i.e., plant controls) and decomposition of soil OC (i.e., microbial controls), which influences soil OM inputs and outputs, respectively. In addition, soil biota (composition, activity, physical location) [160], and below- and above-ground inputs [23, 50, 77, 162] affect decomposition, nutrient cycling, and plant productivity. Another control of OM dynamics is the type of parent material that controls soil pH, hydrologic regimes in soil, nutrient availability, soil depth, and more, all of which influence C inputs and outputs [143]. Topography is also an important control as it modifies microclimate (moisture, temperature, irradiance), erosion, vegetation type, and soil type at the landscape scale [19, 208]. The last of the five soil forming factors is time, which determines the relative importance of the aforementioned factors' effects on soil OM production and decomposition.

The CZ approach allows us to consider the role of Jenny's state factors across the whole ecosystem (biological and physiochemical components), from the bottom of the circulating groundwater to the top of the vegetation, and their interactions [87]. In this chapter, we focus on key controls on soil OM dynamics using a CZ framework in ecosystems by considering a myriad of interacting factors and multiple scales of time and space [147]. We highlight important advancements on the understanding of soil OM dynamics using three thematic areas. In the following sections, we present syntheses of the latest findings that demonstrate that, (1) deep soil OC constitutes large soil C storage that may be vulnerable to decomposition if disturbed; (2) the decomposability of pyrolyzed OM (PyOM) from wildfire is dependent on the complex interactions between the chemical composition of the PyOM and the physical, chemical, and biological properties of the surrounding environment; and (3) increasing rate of soil erosion due to human activities and its interaction with surface microtopographic features imposes important controls on the turnover rate of important biogeochemical cycles.

1.1 Deep Soil Organic Matter Dynamics in the Critical Zone

One of the major difficulties in understanding deep soil OC dynamics is that the standard definition of what should constitute "deep" soil OC is unspecified because the variables that control C stock (i.e., its vertical distribution, and persistence) vary along the soil profile with no set boundaries. The definition of deep varies; some studies have adopted a boundary between topsoil and subsoil based on location of the A horizon, or assigned 30 cm as a cutoff [89]. Operationally, topsoil and subsoil refers to the depth of soil rather than the formation of pedogenic horizons. For instance, the 0–20 cm soil layer is often called topsoil and below 30 cm is subsoil [61, 153]. For the purposes of this chapter, deep soil C will be considered below 30 cm or the A horizon (note some topsoils are only 1–10 cm thick).

In the Southern Sierra Critical Zone Observatory deep soil layers can account for up to 70% of the OC stored in soils making deep soil OC an important aspect of the CZ [129]. Key ecosystem services that deep soils provide are drought tolerance, long term OC sequestration, nutrient cycling, erosion protection, and supporting deeply rooted plants [73]. Deep soil is often overlooked in biogeochemical studies, but it could be the key to understanding major controls on aboveground biomass and how the entire ecosystem may respond to changes in future climate [148].

Deep soil C dynamics have been the focus of a very limited number of studies for several reasons. Deep soils have lower concentrations of C, N, and other plant essential nutrients than topsoils do, they have a lower microbial diversity and activity, they are more difficult to sample, and topsoils are considered more important for agricultural purposes [47, 48]. Focus on the topsoil also reflects the widely held assumption that the topsoil was more important in terms of soil OC flux and the soils below 20 or 30 cm were presumed to change at such a slow rate that it was ignored [147]. In the past, soil studies focused only as deep as a plow could cultivate; however, our current understanding shows that many soils are tens of meters deep [146]. [161] estimates that, globally, the average soil depth is 2 m, with some soils extending to an astonishing 50 m. The average soil sampling depth of manuscripts published from 2000 to 2020 in the top four soil science journals is 24 cm, illustrating that there is a knowledge gap concerning the role that deep soil plays in CZ processes [196]. In this section, we will discuss how OC reaches the deep soil, why deep soil OC can have a longer MRT than surface soil at the same location, what controls deep soil OC, and the importance of deep soil OC in an ecosystem and global change context. Finally, we assert that understanding soil OC throughout the whole soil profile is inherent to a CZ perspective of OM dynamics, which will ultimately improve future predictions of soil OC storage and persistence.

1.1.1 How Does Organic Carbon Get to Deep Soil Layers?

Both modern and ancient C can be mobilized deeper into soil by dissolved OM, root translocation, bioturbation, physical translocation (i.e., erosion), and can already by present due to parent material that contains OC. The relative importance of these sources and processes change with climate, land-use, soil type etc., [153]. The nature of soil OM is transformed over time through the process of microbial and chemical mineralization of the plant, microbial, and animal remains [155]. The process of mineralization renders some soil OM and litter mobile [105], leading to both dissolved and particulate soil OM distribution down the soil profile through diffusive, preferential, and gravitational flow paths [29, 94]. [63], showed that dissolved OC (DOC) had a greater microbial signature with depth, suggesting that deep soil OC was a mixture of plant matter and microbial remains in various stages of advanced decomposition. It has been suggested that translocation of DOC is one of the main mechanisms in which OM moves into deeper soils [93].

Deep soil OC amount can be influenced by deep roots and their associated mycorrhizae. For example, root allocation patterns from multiple vegetation types have been found to control vertical soil OC distribution [86, 89]. This is because deep roots can modify their environment by enhancing bedrock weathering to create additional deep soil, can contribute to the formation of hydrological flow paths that increase vertical transport of DOC, and can provide a source of soil OM through their own turnover or by C-containing root exudates [175]. If these roots form relationships with mycorrhizae, they have greater access to nutrients from the weathered bedrock and mineral-associated OC, further increasing the depth at which they influence soil OM [6, 27]. This downward migration of soil OM from the topsoil and input of C from roots in deep soil layers can contribute to a large C stock in deep soil layers.

Bioturbation by animals such as earthworms, termites, and arthropods, and plant roots physically move soils and mix them with fresh and old OM. Bioturbation alters fundamental properties of soil such as porosity, OC amount, and particle-size distribution [191]. Bioturbation is often cited as a major process influencing the vertical distribution of soil OM [9, 177]. During feeding, burrowing, and defecation, these animals directly move OM and physically mix the soil, increasing soil OC amount. These activities are complex, and the abundance of organisms does not generally follow the abundance of their potential food resources (soil OM) [55]. Biologically-mediated soil burial rates range between 1–2 mm/y and this burial of soil OM can create more persistent soil OC [153]. Overall, bioturbation significantly impacts the vertical distribution of soil OC, but also alters the chemical character of soil through selective feeding, enteric enzymatic processes, and species interactions [38, 59, 123] through mineral interactions and changes in aggregate dynamics [124].

Erosion, volcanic history, and fire activity can alter the physical transport of soil OM into deeper soil layers. Colluvial (soil transported from upslope via gravity) soils can be enriched in soil OC in comparison to the soils of upslope areas due to the deposition and progressive burial of soil OC [185]. Buried soil OC (especially ancient buried soils known as Paleosols) was found to be more persistent (slow decomposition due to chemical chemical composition) compared to soil OC sampled at the same depth at a stable location [120]. Volcanic soils (e.g., many Andisols) are rich in colloidal iron/aluminum oxide-OM complexes that move into the subsoil, especially in pasturelands, increasing the persistence of OC [133]. Plant material that has been burnt under low oxygen conditions (i.e., pyrolysis) creates pyrogenic OM (PyOM) with very high MRT [66]. PyOM in the dissolved phase moves down the soil profile into deeper soil layers with one study noting that up to 46% of the initial deposited PyOM was subject to vertical transport [151]. Buried, volcanic, or PyOM moving into the subsoil has the potential to be more persistent due to its inherent or acquired physico-chemical properties, attachment to minerals, and lower microbial activity compared to surface soil, making the subsoil contain more persistent OC [15] (See Wildfire and Erosion sections for more information on deep soil, PyOM, and physically transported OM).

Another mechanism for the creation of deep soil OC is related to emplacement of geologic C from depositional processes (e.g., glacially eroded and deposited shale C) or through primary weathering and soil formation on parent rock containing metamorphic C (e.g., graphite) [95, 198]. This OC is also referred to as geogenic OC, and is defined as OC that originates from deposition during sedimentation and rock formation [95]. The contribution of geogenic OC to total soil OC may vary based on the mineralogy and geologic history of the parent material. Approximately 65% of the continental Earth's surface is covered with sediments and sedimentary rocks, suggesting that a large portion of soil OC is geogenic [7]. One study has found that the total global amount of soil geogenic C makes up around 80% of soil OC in a glacier-derived soil, implying that the contribution of geogenic OC to global soil C stocks is large in certain locations [181]. Moreover, [95] found that the greatest possible contribution of geogenic OC was 15.4% for loess (94 cm depth) and 21.5% for Red Sandstone (74 cm depth). It is estimated that the upper 1 m of Earth's surface contains ~106 Mt OC from petrogenic sources that are often millions of years old and could lead to overestimation of soil C MRT [36, 206]. These few studies illustrate that common and abundant terrestrial parent materials with low amounts of sedimentary OC constitute a considerable proportion of soil OC stocks, and should be considered in understanding OC in deep soils.

1.1.2 Persistence of Deep Soil Organic Matter

Generally, with increased depth, soil OC persistence increases [56, 136, 158129]. In the Southern Sierra Critical Zone Observatory one study showed that the MRT of C in subsoils can reach 20,000 y [129], highlighting the importance of understanding factors that regulate subsoil C persistence and therefore the long-term storage of OM in the CZ [13, 21, 89, 129, 136, 137, 153]. Soil OM can become less accessible to microorganisms through mineral association which is both physical occlusion within soil microaggregates (<50-63 μm) and chemical attachment to reactive mineral surfaces [80, 112, 170]. Aggregates physically protect soil OM by forming physical barriers between microbes and enzymes consequently influencing microbial turnover [164]. The mineral associated chemical bonds are more difficult for decomposers and their enzymes to break down compared to non-bound OM, allowing C to persist for longer periods of time in soil, especially deeper in soil where microbial activity is likely to be reduced, compared to near-surface soil layers [101, 106]. Recent evidence indicates that the majority of mineral-bound soil OM is microbially transformed from plant detrital C, consisting of microbial necromass fragments [96, 109, 184]. Deep soils are an ideal location for the accumulation of microbially processed soil OM in the form of DOC to be translocated and potentially stored for longer periods of time compared to the topsoil. In deep soil layers, microbial residue can be further protected by sorption to mineral surfaces or occlusion within aggregates [106, 144, 165]. Since deep soil environments typically have less microbial density and greater nutrient limitation on microbial metabolism they may be key in long-term persistence of soil OM [33, 154, 168]. However, research suggests that occlusion in micro-aggregates or sorption to mineral surfaces alone cannot alone explain long-term persistence, as they are reversible [61, 97, 207]. Recent studies have advocated for soil OC persistence to be understood as a more dynamic system by which there is a continual movement of organic molecules within the physical soil matrix that are transformed by microorganisms [54, 114].

Despite longer MRTs of deep soil mineral-associated and occluded OM, these forms of OM are not inherently stable or permanent [116], but rather decompose slower than OM in topsoil [30, 103]. For example, roots can be multiple meters or more deep and root exudates can exchange with or liberate previously mineral-associated OM making the OM accessible to decomposition [97]. In addition, input of fresh OM to deep soil layers can prime the loss of old, previously stabilized deep soil OM [11]. Geomorphology can also control the abundance and persistence of soil OM, especially in erosional landscapes. Hunter et al. [208] found gentle and broad hill slopes with slower erosion rates have organic matter that persist longer than steep and sharpe hill slopes with fast erosion rates. Furthermore, physical disturbance (e.g., freeze–thaw and wet-dry cycles, bioturbation, and deep tillage) can disrupt deep mineral-associated OM [57, 112]. There is also evidence that ingestion of mineral-bound, or micro-aggregated soil OM by geophagus animals (e.g., earthworms) can cause replacement/exchange of previously stabilized OM [38] and these animals are found below the A horizon. Consequently, deep soil OM can be understood as a dynamic pool of persistent, and microbially altered OM, that is continually altered through its interactions with both soil microbes, the soil matrix, and the surrounding environment [54, 192].

1.1.3 Why is Deep Soil Organic Matter Important from a Climate Change Perspective?

Because the soil stores more C than the atmosphere and vegetation combined, soil has become a promising piece to be considered in climate change mitigation practices ([62, 211, 212]. However, it is currently unknown how decomposition of deep soil OM will respond to rising temperatures over large scales (both spatial and temporal), leading to large uncertainties in model projections. Soils are modeled to warm 4.5 ± 1.1 °C down to 1 m by the end of the twenty-first century under IPCC Representative Concentration Pathway 8.5 (2018). Indeed, soil temperature down to 3.2 m has been found to increase 31% more than air temperature over the past 50 years [199]. Taken together, subsoil is anticipated to continue to warm, which will have important implications for the fate of deep soil OM.

Deep soil OM has been thought to be more stable. However, recent studies have shown that deep soil OM can be vulnerable to microbial decomposition [57, 61, 120, 178]. For instance, while some oligotrophic members of the microbial community are enriched at depth [28], laboratory rates of C and N mineralization of added substrates were as fast in the subsoils as in surface soils in an old-growth forest [92]. Furthermore, across large ecological gradients, over half of extracellular enzyme activity in the upper meter occurs below 20 cm [47]. [81] and [171] show that with 4 °C warming, deep soil layers could account for 3% total ecosystem respiration, which represents 30% of current anthropogenic emissions. Furthermore, in this same experiment, microbial acclimation in the subsoil was less than that of the surface, suggesting that enhanced subsoil respiration due to warming could continue into the future [52]. The effect of global warming on deep soil OC storage suggests that, in regions with a MAT above 5 °C, even a 1 °C temperature increase could lead to a loss of over 10% of soil OC [100]. This suggests that when favorable conditions for microbial decomposition present themselves (i.e., warmer temperatures, more water), deep soil C may be susceptible to mineralization. A few studies do suggest that there will be a lag or muted warming response from deep soil microbes due to decreased microbial composition and activity with depth [210]. Therefore, some deep soils will be vulnerable to soil OC loss and could contribute to an increase in CO_2 efflux in the coming decades depending on the extent of deep warming and the microbial response. It should be noted that in some systems elevation and total thickness of the soil and saprock could play an important role in deep soils response to warming [209]. Understanding that deep C may be vulnerable to warming and could exacerbate a positive climate feedback loop makes minimizing current emissions of greenhouse gasses and protection of deep soils critical.

1.2 Wildfire-Organic Matter Interactions in the Critical Zone

Wildfire is a globally widespread phenomenon that impacts OM dynamics in a variety of ways, affecting C storage and chemistry across the landscape. Each year, wildfire burns over 400 Mha of Earth's surface [67], releasing 2.5 Pg of C into Earth's atmosphere, or 4% of yearly global net primary production [182]. However, wildfire also results in changes in vegetation, soil structure, and nutrients that continue to affect OM dynamics decades after disturbance [32, 49, 75]. For example, one of the most conspicuous effects of wildfire is its impact on vegetation, where fire can result in reduced tree cover and OM inputs, affecting the physical conditions for decomposition and CO_2 efflux [131]. Similarly, changes in vegetation may also affect the chemistry of root exudates [75] and soil stability [3], further impacting ecosystem C stocks. Thus, understanding the impacts of wildfire holistically, through a CZ framework, is imperative in constraining long-term predictions of global C storage, especially within a global change framework.

In many ecosystems, wildfire regimes, or the frequency, size, and severity of how fires burn, are changing, which limits our ability to predict how wildfire impacts soil C stocks. For instance, hotter, drier conditions in the western United States (US) due to climate change has increased the wildfire season and wildfire activity [188]. Globally, this results in a 19% increase in fire season length and a doubling of global burnable area [91]. Furthermore, decades of wildfire suppression in the western US has led to dense forests that burn at higher severities than their historic counterparts [127]. These increases in wildfire size, severity, and frequency have ushered in a new era of "mega-fires" that not only pose increased risk to humans, but also ecosystems and C stocks [5]. Conversely, the suppression of natural grassland wildfires, such as in many rangeland systems, can result in rapid ecosystem shifts as trees invade creating woodlands and forests [39, 150]. This globally significant land use-driven change in the wildfire regime of grasslands causes reduced inputs of long-lived PyOM into soil, a shift in the plant and microbial chemistry input to the soil, and a reorganization of soil OM aggregate apportionment that could alter C stocks of both shallow and deep C [58, 116]. These changes imply that C budgets derived from historical trends may not be accurate in the coming decades. Hence, by understanding the mechanistic connections between the different impacts of wildfire on OM and other ecosystem properties, we can better predict OM dynamics under fire regimes that are different from their historical counterparts.

In this section, we review the interactions between wildfire and OM dynamics from a CZ perspective that includes OM chemistry, coupled elemental cycles, and the terrestrial-aquatic interfaces. We show that by incorporating the full impact of fire (besides the direct impacts—e.g., immediate combustion and pyrolysis of OM), we can better understand OM dynamics and persistence in the face of changing wildfire regimes.

1.2.1 Persistence of Pyrogenic Carbon

One of the ways in which wildfire directly impacts the C cycle is through the combustion and pyrolysis of OM at or above the soil surface. PyOM describes thermally altered OM derived from incomplete combustion during burning. Generally, PyOM and specifically pyrogenic C (PyC) is thought to cycle slowly due to its condensed polyaromatic chemistry and the hydrophobic nature of these moieties [24, 159]. However, the chemical components and reactivity of PyOM and PyC vary along a continuum, which is influenced by pyrolysis temperature, O_2 concentrations, and source biomass, resulting in MRT that vary in orders of magnitude (e.g., decades to millennia) [65, 78, 79, 98].

The chemical composition of PyOM is primarily controlled by biomass source and pyrolysis temperature [78, 98]. At moderate temperatures (<500 °C), biomass source plays a relatively larger role in determining in PyOM composition,however, at higher temperatures, differences among biomass sources becomes difficult to discern [78]. Higher temperatures also lead to greater aromaticity, lower molar ratios of hydrogen-to-C (H:C) and oxygen-to-C (O:C), greater surface area, and reduced solubility in water [24]. Therefore, as pyrolysis temperatures increase, the decomposability of PyOM declines, which has been shown in numerous laboratory incubations [3, 157].

Because of the low-density characteristic of PyOM, PyOM is prone to movement across the landscape and vertically into the soil profile [4]. The resulting distribution has been posited to determine the persistence of this C pool [121]. As mentioned earlier in this chapter, if OM travels to significant depths, decomposition may be slowed due to increased mineral protection, reduced microbial biomass, and, in hydric soils, reduced O_2 concentrations. Furthermore, movement of PyOM into depositional landscapes, where it can be buried, or into aquatic systems, where it moves into the ocean or is incorporated into stream sediments (where oxygen is limited), can increase PyC persistence [2]. Alternatively, surficial PyOM may be moderately degradable if microbial communities, which possess the capability to produce oxidative enzymes, have the ability to recolonize and recover after fire and the nutritional conditions are favorable [200].

The recovery of microbial communities that can degrade PyOM at the surface, however, may take decades to occur, especially under changing fire regimes [51]. The ability to produce oxidative enzymes necessary for PyOM degradation is constrained to specific members of the microbial community. Generally, these community members are considered white-rot fungi within the class Agaricomycetes [60, 64],however, certain bacteria also possess the capability to produce these enzymes [42]. Whether these microbes are present immediately after fire, before movement of PyOM, could be critical in determining the degree of PyC persistence. However, fungi are generally more sensitive to heat than bacteria and may take longer to recover [140]. Nevertheless, the recovery and presence of microbes with the capabilities to produce oxidative enzymes is likely a key determinant of PyC persistence.

1.2.2 Coupled Elemental Cycles

It is well known that the C cycle is intimately linked to the cycling of other elements, particularly N and P, due to stoichiometric constraints on microbial and plant biomass [28, 35, 76]. For instance, inadequate N or P may limit microbial growth and activity [31], thereby reducing respiration leading to increased C storage, or reducing the accumulation of microbial products on soil surfaces lessening C persistence [104].

However, the effect of fire on N and P is nuanced with bi-directional feedbacks to the C cycle. Therefore, a more complete understanding of the impact of wildfire on OM dynamics must include the impacts of fire on other elements besides C.

Across ecosystems, wildfire greatly disrupts most soil nutrient pools [32]. For example, wildfire generally leads to an overall loss in the ecosystem N pool through combustion of OM, which releases N oxides (NO_X) into the atmosphere, and through volatilization of aqueous N compounds (e.g., ammonia)[117]. While wildfire may reduce *overall* soil N pools, which are largely dominated by organic N, wildfire generally results in increased *inorganic* N due to pyrolysis and combustion of C-bound nutrients and ash deposition [34, 71]. Similarly, P may also be lost to the atmosphere—albeit at higher temperatures (700 °C compared to 200 °C) [69, 141], but inorganic P is generally enhanced due to similar mechanisms [176]. Furthermore, burning also increases the availability of other cations necessary for plant and microbial growth, such as calcium, magnesium, and potassium [68], which also increase pH and microbial activity. These nutrient "flushes" are generally short-lived (i.e., return to pre-fire levels within a decade) as plant uptake and nutrient leaching lead to depletion of these inorganic pools [75]. In the longer term, post-fire vegetation with different C:N:P stoichiometries and the impact of wildfire on the soil microclimate may influence post-fire inorganic nutrient pools [75]. For instance, canopy openings after crown-fires allow for the colonization of grasses and N-fixing shrubs [90], both of which have lower C:N:P ratios than trees leading to litter that is more nutrient-rich [26]. However, the availability of nutrients derived from fresh litter is largely constrained by post-fire microbial decomposition.

Post-fire microbial activity both controls and is regulated by C and nutrient availability, and feedback among elemental cycles likely influences soil OC recovery after fire. In the early stages of post-fire recovery when nutrient availability is relatively high, microbes are unlikely to be nutrient limited, and activity is more likely C limited. Under C limitation, ammonium immobilization is reduced, leading to enhanced nitrification [50]. Indeed, the post-fire nitrate flush generally occurs in the years following the ammonium flush as nitrification increases [166]. This, in part, likely contributes to multi-decadal reductions in soil N concentrations [49, 138, 186]. Because N is a limiting element for plant growth in many temperate forests [113], reductions in soil N have the potential to encumber post-fire revegetation and accumulation of soil OM. At the same time, N limitation may also reduce microbial activity and respiration leading to C persistence [31]. Opposing feedbacks, thus, make it difficult to predict how soil OC pools will recover, but further research on coupled elemental cycles in post-fire environments may help to constrain predictions and biogeochemical models.

An additional factor in soil recovery trajectory relates to the influence of PyOM additions to soil on the decay rate of background soil OM or a "priming effect"[65, 189, 190]. Many time series assessing the interactions of soil properties, nutrient status, and PyOM source and temperature on the stability of background soil OM show an early enhanced loss of native C (positive priming) and a subsequent slowing of microbial decay (i.e., negative and positive priming). However, there is an inconsistent effect of PyOM on background soil C across the literature [157, 202, 203, 204, 205] as background edaphic properties may play a considerable role [201].

1.2.3 Terrestrial-Aquatic Interfaces

A critical zone perspective of post-fire soil OM dynamics recognizes that PyOM may become transported, in particulate or dissolved forms, from burned landscapes to aquatic ecosystems. Similarly, when severe fire clears vegetation, denuded landscapes may be more susceptible to erosion, leading to transport of unburned OM. Combined, these impacts are considered some of the largest drivers of aquatic impairment by directly impacting the burned area and indirectly affecting water quality in downstream ecosystems [12]. Alternatively, fire-driven inputs of nutrients from erosion to streams may sustain productivity of oligotrophic mountain streams for decades [126, 139]. Post wildfire, PyOM is transported to surface waters via erosion and surface runoff [2, 4, 37, 45, 152], wet and dry deposition [24], and subsurface flow [72]. The amount of PyC entering aquatic ecosystems is significant in that globally, on average, rivers transport 44–108 Tg PyC y^{-1} [24].

When PyOM enters the aquatic environment, it undergoes different processes than on land, affecting its persistence. In aquatic environments, oxidative enzymes necessary for the degradation of condensed, aromatic ring structures common in PyOM are less common than on land, potentially due to lower fungal abundances in aquatic environments [70]. However, certain aquatic bacteria are able to produce these enzymes [118] and degrade PyOM [130]. PyOM degradation in aquatic ecosystems appears to be dependent on the size of the PyOM, with dissolved PyOM particles (DPyOM) having significantly shorter MRT compared to the particulate PyOM fraction (PPyOM). Multiple studies have found that DPyC is relatively degradable with MRTs less than 1,500 y [149], while the PPyC has MRTs averaging 3,700 ± 400 y [37]. In fact, soil incubation studies of solid PyOM have found that the mass of water extractable PyC (i.e., DPyC), compares with the amount of CO_2-C liberated in the fast cycling soil C pools [64, 65]. When compared to the large range of residence times of terrestrial PyC (300–1000 y); [65, 115], it is clear that, if PyOM enters the aquatic environment, PyOM will likely persist in the environment for a longer time. Nevertheless, direct comparisons of PyC residence times among terrestrial and aquatic environments, including different size fractions, are lacking, which limits our ability to create accurate C budgets after wildfire and predict PyC persistence on a global scale (Fig. 1).

1.3 Organic Matter Erosion in the Critical Zone

Soil erosion plays a key role in controlling the dynamics of soil OM as Earth's surface is dominated by dynamic landscapes, with over 64% of land area having

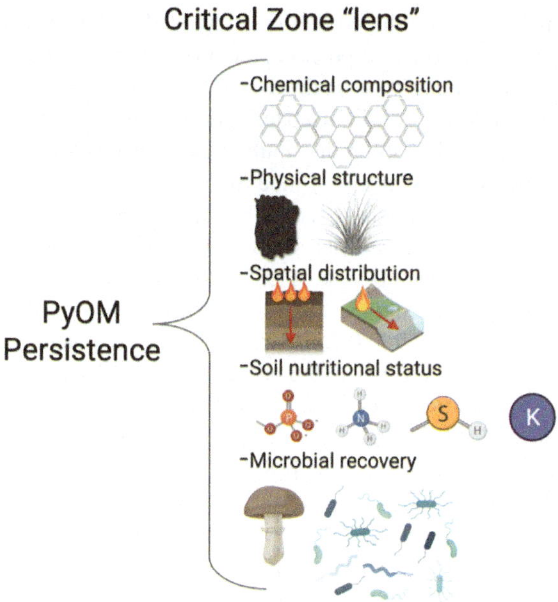

Fig. 1 View of pyrogenic carbon (PyC) persistence through a Critical Zone "lens." Under a Critical Zone lens, the persistence of PyC, in addition to the chemical composition of the PyC, is holistically determined by multiple CZ properties, including the physical structure of the PyOM, the spatial distribution of PyC, the soil nutritional status, and the post-fire microbial recovery.

slopes steeper than 8% [174]. Since the start of agriculture, and especially in the past two centuries (i.e., the industrial era), the rate of soil erosion has accelerated by tens to thousands of times compared to pre-industrial rates due to anthropogenic activities [128]. The rate of soil erosion is expected to increase with the climate change-induced intensification of the hydrologic cycle [14, 193]. The main causes and agents of soil erosion occur are due to many reasons, including rain splash, bioturbation (e.g., tree-throw, burrowing mammals such as gophers, and soil engineers such as earthworms), landslides, intensive agriculture, deforestation, and wind and water-driven transport [156, 173].

Soil erosion is spatially and temporally ubiquitous. Soil erosion is a physical process that laterally transports topsoil and associated soil OM due to both natural and human accelerated processes. Globally, soil erosion redistributes on the order of 75 Gt of soil and 5 Gt of C, which is slightly higher than the net land uptake of atmospheric C (~3.4 Gt of C) [18, 156, 172].

Soil erosion can induce terrestrial sequestration of atmospheric CO_2 (i.e., a net soil C sink) [18, 20, 172]. In any given hillslope and catchment, soil erosion results in the removal of topsoil and exposure of subsoil at erosional landform positions and, after multiple erosional events, leads to burial of the transported soil at depositional landform positions [17, 18]. Typically, when soil C is eroded from source catchments, 70–90% of the eroded topsoil and associated soil C is deposited locally within the same or adjacent topographic sequences while the remaining proportion is translocated to aquatic ecosystems [18, 172]. Specifically, soil erosion in the industrial era has led to increased storage of 4.4 Gt C per year of the eroded soil OC in depositional landform positions [172]. Soil erosion can lead to a net sink of atmospheric CO_2, as long as: (1) the eroded C is replaced, at least partially, by new photosynthate at erosional landform positions, and (2) at least some of the eroded C is stabilized in slower cycling pools in the depositional landform positions [18, 74]. Estimates of the size of the erosion-induced terrestrial C sink range from 0.12 to 1.5 Gt C per year [46]. Soil C erosion and redistribution in intensively managed agricultural areas, like the upper Midwest of the United States, is an evolving, dynamic process that responds to changes in land use technology and best management practices [134, 195].

Erosion-induced land surface OM redistribution could contribute to a large portion of the terrestrial C sink. A 2022-2023 global assessment showed that the annual net atmospheric C increase 5.2 ± 0.2 Gt C/y) is a result of input (source) from burning fossil fuel (~9.9 ± 0.5 Gt C/y) and land-use change and forestry (1.3 ± 0.7 Gt C/y) and output (sink) from the ocean (2.9 ± 0.4 Gt C/y) and land (3.3 ± 0.8 Gt C/y) [62]. However, to balance such an equation, a hidden C sink totalling must be present. Several studies argue that the erosion-induced soil C sink is an important part of the hidden or missing C sink [18, 167, 172]. The land uptake of atmospheric C driven by the soil OC redistribution is estimated to be up to half of the redistributed C amount in erosional sites [195]. An improved understanding of the dynamic fates of eroded soil C during erosion, transportation, and deposition is key to understanding whether eroded C represents a source or sink and at what magnitude.

1.3.1 Soil Erosion and Deposition Controls on Soil Organic Carbon Physical Redistribution in Hillslopes

Soil transport (erosion and deposition) accounts for soil OC's physical relocation from eroding to depositional landform positions. Field measurements show significantly higher OM stocks and older soil OC in depositional sites, compared to eroding ones [17, 195, 208]. The rate of soil OC redistribution is not necessarily linearly proportional to the rate of soil transport [194]. When soil materials are detached from the erosional sites and are deposited at local convergent sites, low-lying areas, or in nearby streams, the detached soil particles may contain organic fragments, aggregates, or isolated mineral grains. The soil transport process breaks up soil aggregate structure and exposes OM, and preferentially transfers light and fine particulates from eroding slopes downwards [83, 134]. Therefore, soil OC is unevenly redistributed and buried at low-lying depositional sites along with mineral soil deposition [1, 17]. For example, in a first-order watershed in an Intensively Managed Landscape Critical Zone Observatory (IML CZO) in Clear Creek, Iowa, USA, the areas with the highest exchange rates of soil OC redistribution groups are at sites where soil transport rates are much slower. This is because even though soil transport rates are slow, meaning the influx and efflux of soil material at a specific location are similar, the influx and efflux of soil OM can be equally high, which results in elevated soil OC redistribution [194]. Further, the nonlinearity between soil transport rate and soil OC redistribution rate tends to be stronger in high erosional lands (e.g., agricultural fields) [194]. In order to fully understand the fate of erosional or deposited soil OM, we need a

CZ perspective that integrates inherited microtopography from both geological and modern land-use processes [119].

Rapid soil erosion and deposition can also lead to spatial heterogeneity of soil OC vertically across a landscape [22, 195]. Studies at IML-CZO in Clear Creek, Iowa, [195] and Calhoun Critical Zone Observatory in South Carolina, USA [22], both show that the fast burial of soil OC from eroded sites results in a higher OC content from 10–60 cm in depositional sites [22, 195]. In general, a soil OC profile at depositional areas tends to have a bumped or a 'nose' shape where the nose represents the fast burial due to anthropogenic activities and associated environmental change. On the other hand, a vertical soil OC profile at an erosional area corresponds to a shallower soil depth as the surface (0–5 cm) soil OC is eroded and C from deep soil layers is exposed. Overall, the erosion-induced spatial heterogeneity of soil OC vertical profiles could play a significant role in determining the fate of C and it could be an important factor to understand in order to capture the fate of C.

Accelerated soil erosion enhances the breakdown of soil aggregates, which can reduce the stability of soil OC protected within them [82, 83]. Soil erosion can make aggregate-protected OM vulnerable through tillage, runoff, and rain in intensively managed landscapes [82, 83, 183]. One case study conducted at IML-CZO compares hillslope (< 5° slope) erosion under different orientation directions of tillage (i.e., tillage rows oriented with or perpendicular to the hill slope). The human-induced orientation caused a differential soil erosion rate, vertical C loss (e.g., 30 times lower for the perpendicular case), and a sorting effect in that light particulate and fine mineral-bound OM was found more at toe slope areas compared to upper lands (Fig.2b) [83]. Such physical–chemical sorting impacts were not evident in flat (< 1°) land surfaces in the floodplain. Further, a raindrop-induced disruption of aggregates can result in local depletion of soil OM as organic fragments are detached from the larger protective aggregate structures, redistributed across the landscape, and exposed to the surface with more active microbial activities. These rain drop-induced impacts also scaled with the intensity of tillage management [82, 83]. Human-induced breakdown of soil aggregates has a high potential for C loss from soils.

1.3.2 Soil Erosion and Deposition Controls on the Biogeochemistry of Redistributed Soil Organic Carbon in Hillslopes

In hillslope erosional areas (usually the shoulder and backslope areas), the exposure of previous subsurface soils can lead to a local C sink [44, 195]. This is mainly because exposed subsoils are subject to dynamic replacement by C input from plant litter inputs due to photosynthesis [18, 74], and these typically C-poor subsoils generally have higher C saturation capacity than surface soils [163]. Further, sites with faster soil erosion rates correspond to greater soil OC replacement, indicating a stronger C sink. Radiocarbon measurements have shown a greater soil OC replacement in intensively managed agricultural ecosystems (which have faster soil erosion rates) compared to grasslands and forests [16, 125, 179, 180]. Overall, the biogeochemical

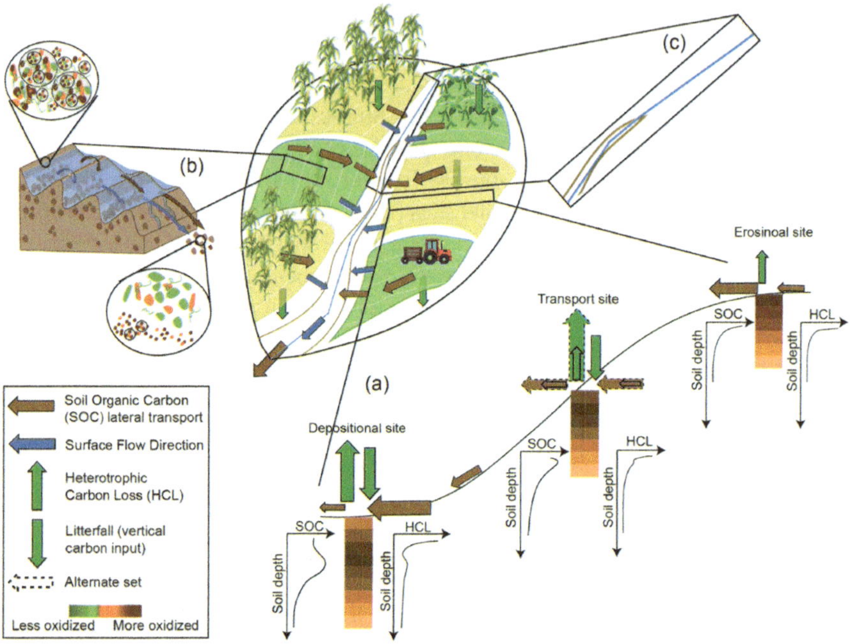

Fig. 2 Schematic of conceptualized processes of the impacts of high soil erosion on soil OC dynamics in an intensively managed landscape (adapted from [83] and [194]). **a** The profiles of soil OC and soil OC decomposition, and the C exchange between soils and the atmosphere at different locations at a hillslope. "Alternative set" here means that the arrows that represent the C fluxes both laterally and vertically have two scenarios at sites where the lateral influx and efflux of soil material are similar. Specifically, if the lateral influx and efflux of soil material are both small, the corresponding Heterotrophic Carbon Loss (HCL) is also small; if the lateral influx and efflux of the soil material are both quite large, the corresponding HCL tends to be high. Note that the width of arrows represents the relative magnitude of water, sediment, and OC fluxes. **b** "Preferential entrainment" of soil aggregates (meaning a selective transport of lighter/finer particulate organic material and soil particles, and deposition of denser soil particles within furrows) due to human activities-induced soil roughness orientation. The surface orientation will influence soil geochemistry, spatial heterogeneity, and stream input chemistry. **c** Stream channels straightening for drainage and farming purposes.

reactions of soil OM in erosional areas are influenced by the exposure to subsurface soils.

In depositional areas (usually foot slope and toe slope areas), buried soil OM experiences fundamentally different biogeochemical processes compared to its source location at eroding sites. Soil OM detached from eroding sites is quickly (from days to decades) decomposed during transport [85, 173], with a high spatial variation of the release rates of CO_2 (due to microbial decomposition) along a 2-D hillslope [194]. When soil OM reaches a depositional site, burial of the eroded soil protects OM. This burial protection slows down the decomposition rate due to the increased occurrence of water-saturated conditions, thus lowering oxygen availability, limiting microbial activity for aerobic oxidation [46, 167, 172]. A study conducted in a zero-order grassland watershed exhibited greater soil OM storage at the depositional area, where this soil OM contained older C compared to the eroding slopes [17, 18]. Additional enrichment of reactive minerals is favorable for the accumulation of old C in depositional settings. For example, amorphous sesquioxides, crystalline, and complexed aluminum and iron oxyhydroxides stabilize soil OM in depositional areas both chemically and physically [19]. However, even though the soil OM is protected and may have a slower decomposition coefficient, the rate of decomposition amount (which is the product of the decomposition coefficient and soil OC storage) could still be higher at depositional sites compared to erosional sites because of the high soil OC storage [194]. Overall, the biogeochemical reactions of soil OM at depositional areas are strongly influenced by soil burial and the local environment.

1.3.3 Transport of OM in Terrestrial-Aquatic Interfaces

Rainstorms are responsible for the majority of fluxes of soil OC particulate organic C (POC) and DOC transported from terrestrial source areas into streams. This transport process happens over a relatively short period of a year. A study covering 31 forested watersheds in the eastern US showed that over 50% of the annual DOC export to streams occurs during large storm events that make up only 5% of a water year [145]. In an agricultural watershed with ditches and subsurface tile drainages implemented in the Midwest, US (Fig. 2c), ~80% of annual DOC exports occurred 20% of the time during a water year [40, 41]. Further, the transport of soil OC to streams is strongly influenced by rainstorm intensity (mm/time). Specifically, disproportionately greater export in both DOC and POC occurs during high-flow periods, compared to baseflow in both agricultural [41, 132, 142] and minimally managed ecosystems [43, 88, 145]. Of course, generalizations related to the export of terrestrial C are difficult as the concentration and source of POC and DOC at any point in a stream is a complex function of: storm intensity, duration, and location, the connectivity of the drainage network, amplifying or mitigating functions of critical interfaces such as hillslopes and flood plains, and the spatial heterogeneity of land cover/land use [25, 99].

2 Conclusion

In the past, distinct groups of scientists studied aboveground plant dynamics and belowground processes [147]. CZ science aims to bridge the above- and belowground divide, bringing together teams of scientists to study the whole system and explore the key processes that connect the upper and lower parts of the CZ. The goal of this chapter was to show the importance of synthesizing the soil system by using three thematic areas exploring the nature and dynamics of OM in deep soils, and impacts of physical perturbation of the CZ due to wildfire and erosion.

Including soil OM dynamics in the CZ perspective allows better integration of the roles of above- and below-ground processes in determining OM stock, nature, persistence and fluxes through the whole soil profile, from the soil-atmosphere to the soil–bedrock interfaces [147]. Current local and global estimates of soil C stocks and stability are biased towards topsoil, resulting in inaccurate accounting of terrestrial C sequestration potential, sources of CO_2 efflux into the atmosphere, and our understanding of the overall vulnerability/stability of soil OC to climate change. Similarly, it is important to recognize that landscapes and their ability to store OM are dynamic and are influenced by disturbances such as wildfire and erosion. The impacts of such disturbances are greatly influenced by multiple CZ factors. Thus, the inclusion of deep soil OM and its interaction with soil constituents, as well as disturbances to the topsoil and subsoil, into global C modeling efforts will help address some of the current uncertainty in the fate of C in a changing climate.

Acknowledgements Kimber Moreland was supported by the U.S. Department of Energy, Livermore National Laboratory under Contract DE-AC52-07NA27344. Morgan Barnes was supported by the U.S. Department of Energy, Office of Biological and Environmental Research, Environmental System Science program through the River Corridor Scientific Focus Area project at the Pacific Northwest National Laboratory. Praveen Kumar acknowledges support from NSF Grant EAR 2012850.

List of Acronyms

C carbon
OC organic carbon
OM organic matter
OC organic carbon
PyOM pyrogenic organic matter
PyC pyrogenic carbon
POC particulate organic carbon
DOC Dissolved organic carbon
HCL heterotrophic carbon loss

References

1. Abban B, (Thanos) Papanicolaou AN, Cowles MK et al (2016) An enhanced Bayesian fingerprinting framework for studying sediment source dynamics in intensively managed landscapes. Water Resour Res 52:4646–4673.https://doi.org/10.1002/2015WR018030
2. Abney RB, Berhe AA (2018) Pyrogenic carbon erosion: implications for stock and persistence of pyrogenic carbon in soil. Front Earth Sci 6.https://doi.org/10.3389/feart.2018.00026
3. Abney RB, Jin L, Berhe AA (2019) Soil properties and combustion temperature: Controls on the decomposition rate of pyrogenic organic matter. CATENA 182:104–127. https://doi.org/10.1016/j.catena.2019.104127

4. Abney RB, Sanderman J, Johnson D et al (2017) Post-wildfire erosion in mountainous terrain leads to rapid and major redistribution of soil organic carbon. Front Earth Sci 5.https://doi.org/10.3389/feart.2017.00099
5. Adams MA (2013) Mega-fires, tipping points and ecosystem services: Managing forests and woodlands in an uncertain future. For Ecol Manage 294:250–261. https://doi.org/10.1016/j.foreco.2012.11.039
6. Allen MF (2007) Mycorrhizal fungi: highways for water and nutrients in arid soils. Vadose zone journal 6:291–297. https://doi.org/10.2136/vzj2006.0068
7. Amiotte Suchet P, Probst J-L, Ludwig W (2003) Worldwide distribution of continental rock lithology: implications for the atmospheric/soil CO_2 uptake by continental weathering and alkalinity river transport to the oceans. Global Biogeochem Cycles 17.https://doi.org/10.1029/2002GB001891
8. Amundson R, Berhe AA, Hopmans JW et al (2015) Soil science. Soil and human security in the 21st century. Science 348:1261071. https://doi.org/10.1126/science.1261071
9. Amundson R, Richter DD, Humphreys GS et al (2007) Coupling between biota and earth materials in the critical zone. Elements 3:327–332. https://doi.org/10.2113/gselements.3.5.327
10. Amundson R (2001) The carbon budget in soils. Annu Rev Earth Planet Sci 29:535–562. https://doi.org/10.1146/annurev.earth.29.1.535
11. Bailey V, Pries C, Lajtha K (2019) What do we know about soil carbon destabilization? Environ Res Lett. https://doi.org/10.1088/1748-9326/ab2c11
12. Ball G, Regier P, González-Pinzón R et al (2021) Wildfires increasingly impact western US fluvial networks. Nat Commun 12:2484. https://doi.org/10.1038/s41467-021-22747-3
13. Batjes NH (1996) Total carbon and nitrogen in the soils of the world. Eur J Soil Sci 47:151–163. https://doi.org/10.1111/j.1365-2389.1996.tb01386.x
14. Berc J, Lawford R, Bruce J et al (2003) Conservation implications of climate change: soil erosion and runoff from cropland. Soil Water Conserv Soc
15. Berhe AA, Barnes RT, Six J, Marín-Spiotta E (2018) Role of soil erosion in biogeochemical cycling of essential elements: carbon, nitrogen, and phosphorus. Annu Rev Earth Planet Sci 46:521–548. https://doi.org/10.1146/annurev-earth-082517-010018
16. Berhe AA, Harden JW, Torn MS, Harte J (2008) Linking soil organic matter dynamics and erosion-induced terrestrial carbon sequestration at different landform positions. J Geophys Res 113.https://doi.org/10.1029/2008JG000751
17. Berhe AA, Harden JW, Torn MS et al (2012) Persistence of soil organic matter in eroding versus depositional landform positions. J Geophys Res 117.https://doi.org/10.1029/2011JG001790
18. Berhe AA, Harte J, Harden JW, Torn MS (2007) The significance of the erosion-induced terrestrial carbon sink. Bioscience 57:337–346. https://doi.org/10.1641/B570408
19. Berhe AA, Kleber M (2013) Erosion, deposition, and the persistence of soil organic matter: mechanistic considerations and problems with terminology. Earth Surf Process Landforms 38:908–912. https://doi.org/10.1002/esp.3408
20. Billings SA, Buddemeier RW, deB. Richter D et al (2010) A simple method for estimating the influence of eroding soil profiles on atmospheric CO_2. Global Biogeochem Cycles 24. https://doi.org/10.1029/2009GB003560
21. Billings SA, de Souza LFT (2020) Earth's soil harbors ancient carbon. Nat Geosci. https://doi.org/10.1038/s41561-020-0614-1
22. Billings SA, Richter D de B, Ziegler SE, et al (2019) Distinct contributions of eroding and depositional profiles to land-atmosphere CO_2 exchange in two contrasting forests.https://doi.org/10.3389/feart.2019.00036
23. Bird JA, Torn MS (2006) Fine roots vs. needles: a comparison of 13C and 15N dynamics in a ponderosa pine forest soil. Biogeochemistry 79:361–382. https://doi.org/10.1007/s10533-005-5632-y
24. Bird MI, Wynn JG, Saiz G et al (2015) The pyrogenic carbon cycle. Annu Rev Earth Planet Sci 43:273–298. https://doi.org/10.1146/annurev-earth-060614-105038

25. Blair NE, Bettis EA, Filley TR et al (2021) The spatiotemporal evolution of storm pulse particulate organic carbon in a low gradient, agriculturally dominated watershed. Frontier Water 3.https://doi.org/10.3389/frwa.2021.600649
26. Boyle SI, Hart SC, Kaye JP, Waldrop MP (2005) Restoration and canopy type influence soil microflora in a ponderosa pine forest. Soil Sci Soc Am J 69:1627–1638. https://doi.org/10.2136/sssaj2005.0029
27. Brantley SL, Eissenstat DM, Marshall JA et al (2017) Reviews and syntheses: On the roles trees play in building and plumbing the Critical Zone. Biogeosciences Discussions:1–41. https://doi.org/10.5194/bg-2017-61
28. Brewer TE, Aronson EL, Arogyaswamy K et al (2019) Ecological and genomic attributes of novel bacterial taxa that thrive in subsurface soil horizons. mBio 10. https://doi.org/10.1128/mBio.01318-19
29. Bundt M, Widmer F, Pesaro M et al (2001) Preferential flow paths: biological 'hot spots' in soils. Soil Biol Biochem 33:729–738. https://doi.org/10.1016/S0038-0717(00)00218-2
30. Cagnarini C, Blyth E, Emmett BA et al (2019) Zones of influence for soil organic matter dynamics: a conceptual framework for data and models. Glob Chang Biol. https://doi.org/10.1111/gcb.14787
31. Camenzind T, Hättenschwiler S, Treseder KK et al (2017) Nutrient limitation of soil microbial processes in tropical forests. Ecol Monogr 88:4–21. https://doi.org/10.1002/ecm.1279
32. Certini G (2005) Effects of fire on properties of forest soils: a review. Oecologia 143:1–10. https://doi.org/10.1007/s00442-004-1788-8
33. Chabbi A, Kögel-Knabner I, Rumpel C (2009) Stabilized carbon in subsoil horizons is located in spatially distinct parts of the soil profile. Soil Biol Biochem 41:256–261. https://doi.org/10.1016/j.soilbio.2008.10.033
34. Christensen NL (1973) Fire and the nitrogen cycle in california chaparral. Science 181:66–68. https://doi.org/10.1126/science.181.4094.66
35. Cleveland CC, Liptzin D (2007) C:N: P stoichiometry in soil: is there a "Redfield ratio" for the microbial biomass? Biogeochemistry 85:235–252. https://doi.org/10.1007/s10533-007-9132-0
36. Copard Y, Amiotte-Suchet P, Di-Giovanni C (2007) Storage and release of fossil organic carbon related to weathering of sedimentary rocks. Earth Planet Sci Lett 258(1–2):345–357
37. Coppola AI, Wiedemeier DB, Galy V et al (2018) Global-scale evidence for the refractory nature of riverine black carbon. Nature Geosci 11:584–588. https://doi.org/10.1038/s41561-018-0159-8
38. Crow SE, Filley TR, McCormick M et al (2009) Earthworms, stand age, and species composition interact to influence particulate organic matter chemistry during forest succession. Biogeochemistry 92:61–82. https://doi.org/10.1007/s10533-008-9260-1
39. Dai X, Boutton TW, Glaser B et al (2005) Black carbon in a temperate mixed-grass savanna. Soil Biol Biochem 37:1879–1881. https://doi.org/10.1016/j.soilbio.2005.02.021
40. Dalzell BJ, Filley TR, Harbor JM (2005) Flood pulse influences on terrestrial organic matter export from an agricultural watershed. J Geophys Res 110.https://doi.org/10.1029/2005JG000043
41. Dalzell BJ, Filley TR, Harbor JM (2007) The role of hydrology in annual organic carbon loads and terrestrial organic matter export from a midwestern agricultural watershed. Geochim Cosmochim Acta 71:1448–1462. https://doi.org/10.1016/j.gca.2006.12.009
42. de Gonzalo G, Colpa DI, Habib MHM, Fraaije MW (2016) Bacterial enzymes involved in lignin degradation. J Biotechnol 236:110–119. https://doi.org/10.1016/j.jbiotec.2016.08.011
43. Dhillon GS, Inamdar S (2014) Storm event patterns of particulate organic carbon (POC) for large storms and differences with dissolved organic carbon (DOC). Biogeochemistry 118:61–81. https://doi.org/10.1007/s10533-013-9905-6
44. Dialynas YG, Bras RL, deB. Richter D (2017) Hydro-geomorphic perturbations on the soil-atmosphere CO_2 exchange: how (un)certain are our balances?https://doi.org/10.1002/2016WR019411

45. Dittmar T, de Rezende CE, Manecki M et al (2012) Continuous flux of dissolved black carbon from a vanished tropical forest biome. Nat Geosci 5:618–622. https://doi.org/10.1038/ngeo1541
46. Doetterl S, Berhe AA, Nadeu E et al (2016) Erosion, deposition and soil carbon: A review of process-level controls, experimental tools and models to address C cycling in dynamic landscapes. Earth Sci Rev 154:102–122. https://doi.org/10.1016/j.earscirev.2015.12.005
47. Dove NC, Arogyaswamy K, Billings SA et al (2020a) Continental-scale patterns of extracellular enzyme activity in the subsoil: an overlooked reservoir of microbial activity. Environ Res Lett 15:1040a1. https://doi.org/10.1088/1748-9326/abb0b3
48. Dove NC, Barnes ME, Moreland K et al (2021) Depth dependence of climatic controls on soil microbial community activity and composition. ISME Communications 1:78. https://doi.org/10.1038/s43705-021-00081-5
49. Dove NC, Safford HD, Bohlman GN et al (2020) High-severity wildfire leads to multi-decadal impacts on soil biogeochemistry in mixed-conifer forests. Ecol Appl 30:e02072. https://doi.org/10.1002/eap.2072
50. Dove NC, Stark JM, Newman GS, Hart SC (2019). Carbon control on terrestrial ecosystem function across contrasting site productivities: the carbon connection revisited. Ecology, e02695.
51. Dove NC, Taş N, Hart SC (2022). Ecological and genomic responses of soil microbiomes to high-severity wildfire: linking community assembly to functional potential. ISME J.
52. Dove NC, Torn MS, Hart SC, Taş N (2021) Metabolic capabilities mute positive response to direct and indirect impacts of warming throughout the soil profile. Nat Commun 12:2089. https://doi.org/10.1038/s41467-021-22408-5
53. Dungait JAJ, Hopkins DW, Gregory AS, Whitmore AP (2012) Soil organic matter turnover is governed by accessibility not recalcitrance. Glob Change Biol 18:1781–1796. https://doi.org/10.1111/j.1365-2486.2012.02665
54. Dynarski KA, Bossio DA, Scow KM (2020) Dynamic stability of soil carbon: reassessing the "permanence" of soil carbon sequestration. Front Environ Sci 8.https://doi.org/10.3389/fenvs.2020.514701
55. Ekschmitt K, Kandeler E, Poll C et al (2008) Soil-carbon preservation through habitat constraints and biological limitations on decomposer activity. Z Pflanzenernähr Bodenk 171:27–35. https://doi.org/10.1002/jpln.200700051
56. Eusterhues K, Rumpel C, Kleber M, Kögel-Knabner I (2003) Stabilization of soil organic matter by interactions with minerals as revealed by mineral dissolution and oxidative degradation. Org Geochem 34:1591–1600. https://doi.org/10.1016/j.orggeochem.2003.08.007
57. Fierer N, Schimel JP (2002) Effects of drying–rewetting frequency on soil carbon and nitrogen transformations. Soil Biol Biochem 34:777–787. https://doi.org/10.1016/S0038-0717(02)00007-X
58. Filley TR, Boutton TW, Liao JD et al (2008) Chemical changes to nonaggregated particulate soil organic matter following grassland-to-woodland transition in a subtropical savanna. J Geophys Res 113.https://doi.org/10.1029/2007JG000564
59. Filley TR, McCormick MK, Crow SE et al (2008b) Comparison of the chemical alteration trajectory of *Liriodendron tulipifera L.* leaf litter among forests with different earthworm abundance. J Geophys Res 113. https://doi.org/10.1029/2007JG000542
60. Floudas D, Binder M, Riley R et al (2012) The Paleozoic origin of enzymatic lignin decomposition reconstructed from 31 fungal genomes. Science 336:1715–1719. https://doi.org/10.1126/science.1221748
61. Fontaine S, Barot S, Barré P et al (2007) Stability of organic carbon in deep soil layers controlled by fresh carbon supply. Nature 450:277–280. https://doi.org/10.1038/nature06275
62. Friedlingstein P, O'Sullivan M, Jones MW, Andrew RM, Bakker DCE, Hauck J, Landschützer P et al (2023) Global carbon budget 2023. Earth Syst Sci Data 15(12):5301–5369
63. Gabor RS, Eilers K, McKnight DM et al (2014) From the litter layer to the saprolite: Chemical changes in water-soluble soil organic matter and their correlation to microbial community composition. Soil Biol Biochem 68:166–176. https://doi.org/10.1016/j.soilbio.2013.09.029

64. Gibson C, Berry TD, Wang R et al (2016) Weathering of pyrogenic organic matter induces fungal oxidative enzyme response in single culture inoculation experiments. Org Geochem 92:32–41. https://doi.org/10.1016/j.orggeochem.2015.12.003
65. Gibson C, Hatton P-J, Bird JA et al (2018) Tree taxa and pyrolysis temperature interact to control pyrogenic organic matter induced native soil organic carbon priming. Soil Biol Biochem 119:174–183. https://doi.org/10.1016/j.soilbio.2018.01.022
66. Gibson C, Hatton P-J, Bird J et al (2018) Interacting Controls of Pyrolysis Temperature and Plant Taxa on the Degradability of PyOM in Fire-Prone Northern Temperate Forest Soil. Soil Syst 2:48. https://doi.org/10.3390/soilsystems2030048
67. Giglio L, Randerson JT, van der Werf GR (2013) Analysis of daily, monthly, and annual burned area using the fourth-generation global fire emissions database (GFED4). J Geophys Res Biogeosci 118:317–328. https://doi.org/10.1002/jgrg.20042
68. Gray DM, Dighton J (2006) Mineralization of forest litter nutrients by heat and combustion. Soil Biol Biochem 38:1469–1477. https://doi.org/10.1016/j.soilbio.2005.11.003
69. Grier CC (1975) Wildfire effects on nutrient distribution and leaching in a coniferous ecosystem. Can J For Res 5:599–607. https://doi.org/10.1139/x75-087
70. Grossart H-P, Rojas-Jimenez K (2016) Aquatic fungi: targeting the forgotten in microbial ecology. Curr Opin Microbiol 31:140–145. https://doi.org/10.1016/j.mib.2016.03.016
71. Grove TS, O'Connell AM, Dimmock GM (1986) Nutrient changes in surface soils after an intense fire in jarrah (*Eucalyptus marginata* Donn ex Sm.) forest. Austral Ecol 11:303–317. https://doi.org/10.1111/j.1442-9993.1986.tb01400.x
72. Güereña DT, Lehmann J, Walter T et al (2015) Terrestrial pyrogenic carbon export to fluvial ecosystems: Lessons learned from the White Nile watershed of East Africa. Global Biogeochem Cycles 29:1911–1928. https://doi.org/10.1002/2015GB005095
73. Hahm WJ, Riebe CS, Lukens CE, Araki S (2014) Bedrock composition regulates mountain ecosystems and landscape evolution. Proc Natl Acad Sci USA 111:3338–3343. https://doi.org/10.1073/pnas.1315667111
74. Harden JW, Sharpe JM, Parton WJ et al (1999) Dynamic replacement and loss of soil carbon on eroding cropland. Global Biogeochem Cycles 13:885–901. https://doi.org/10.1029/1999GB900061
75. Hart SC, DeLuca TH, Newman GS et al (2005) Post-fire vegetative dynamics as drivers of microbial community structure and function in forest soils. For Ecol Manage 220:166–184. https://doi.org/10.1016/j.foreco.2005.08.012
76. Hart SC, Nason GE, Myrold DD, Perry DA (1994) Dynamics of Gross Nitrogen Transformations in an Old-Growth Forest: The Carbon Connection. Ecology 75:880–891. https://doi.org/10.2307/1939413
77. Hart SC (1990) Control of Decomposition Processes and Nutrient Flow in a California Forest and Grassland. University of California
78. Hatton P-J, Chatterjee S, Filley TR et al (2016) Tree taxa and pyrolysis temperature interact to control the efficacy of pyrogenic organic matter formation. Biogeochemistry 130:103–116. https://doi.org/10.1007/s10533-016-0245-1
79. Hedges JI, Eglinton G, Hatcher PG et al (2000) The molecularly-uncharacterized component of nonliving organic matter in natural environments. Org Geochem 31:945–958. https://doi.org/10.1016/S0146-6380(00)00096-6
80. Hemingway JD, Rothman DH, Grant KE et al (2019) Mineral protection regulates long-term global preservation of natural organic carbon. Nature 570:228–231. https://doi.org/10.1038/s41586-019-1280-6
81. Hicks Pries CE, Castanha C, Porras RC, Torn MS (2017) The whole-soil carbon flux in response to warming. Science 355:1420–1423. https://doi.org/10.1126/science.aal1319
82. Hou T, Berry TD, Singh S et al (2018) Control of tillage disturbance on the chemistry and proportion of raindrop-liberated particles from soil aggregates. Geoderma 330:19–29. https://doi.org/10.1016/j.geoderma.2018.05.013
83. Hou T, Filley TR, Tong Y et al (2021) Tillage-induced surface soil roughness controls the chemistry and physics of eroded particles at early erosion stage. Soil and Tillage Research 207:104807. https://doi.org/10.1016/j.still.2020.104807

84. IPCC (2018) Global Warming of 1.5 °C. Intergovernmental Panel on Climate Change, Geneva, Switzerland
85. Jacinthe PA, Lal R, Owens LB, Hothem DL (2004) Transport of labile carbon in runoff as affected by land use and rainfall characteristics. Soil and Tillage Research 77:111–123. https://doi.org/10.1016/j.still.2003.11.004
86. Jackson RB, Canadell J, Ehleringer JR et al (1996) A global analysis of root distributions for terrestrial biomes. Oecologia 108:389–411. https://doi.org/10.1007/BF00333714
87. Jenny H (1994) Factors Of Soil Formation: A System Of Quantitative Pedology (Dover Earth Science). Dover Publications, New York
88. Jeong J-J, Bartsch S, Fleckenstein JH et al (2012) Differential storm responses of dissolved and particulate organic carbon in a mountainous headwater stream, investigated by high-frequency, in situ optical measurements. J Geophys Res 117.https://doi.org/10.1029/2012JG001999
89. Jobbágy EG, Jackson RB (2000) The Vertical Distribution of Soil Organic Carbon and its Relation to Climate and Vegetation. Ecological Applications
90. Johnson DW, Curtis PS (2001) Effects of forest management on soil C and N storage: meta analysis. For Ecol Manage 140:227–238. https://doi.org/10.1016/S0378-1127(00)00282-6
91. Jolly WM, Cochrane MA, Freeborn PH et al (2015) Climate-induced variations in global wildfire danger from 1979 to 2013. Nat Commun 6:7537. https://doi.org/10.1038/ncomms8537
92. Jones DL, Magthab EA, Gleeson DB et al (2018) Microbial competition for nitrogen and carbon is as intense in the subsoil as in the topsoil. Soil Biol Biochem 117:72–82. https://doi.org/10.1016/j.soilbio.2017.10.024
93. Kaiser K, Guggenberger G (2000) The role of DOM sorption to mineral surfaces in the preservation of organic matter in soils. Org Geochem 31:711–725. https://doi.org/10.1016/S0146-6380(00)00046-2
94. Kaiser K, Kalbitz K (2012) Cycling downwards – dissolved organic matter in soils. Soil Biol Biochem 52:29–32. https://doi.org/10.1016/j.soilbio.2012.04.002
95. Kalks F, Noren G, Mueller CW et al (2021) Geogenic organic carbon in terrestrial sediments and its contribution to total soil carbon. SOIL 7:347–362. https://doi.org/10.5194/soil-7-347-2021
96. Kallenbach CM, Frey SD, Grandy AS (2016) Direct evidence for microbial-derived soil organic matter formation and its ecophysiological controls. Nat Commun 7:13630. https://doi.org/10.1038/ncomms13630
97. Keiluweit M, Bougoure JJ, Nico PS et al (2015) Mineral protection of soil carbon counteracted by root exudates. Nat Clim Chang 5:588–595. https://doi.org/10.1038/nclimate2580
98. Keiluweit M, Nico PS, Johnson MG, Kleber M (2010) Dynamic molecular structure of plant biomass-derived black carbon (biochar). Environ Sci Technol 44:1247–1253. https://doi.org/10.1021/es9031419
99. Kim J, Blair NE, Ward AS, Goff K (2020) Storm-induced dynamics of particulate organic carbon in clear creek, Iowa: an intensively managed landscape critical zone observatory story. Front Water 2.https://doi.org/10.3389/frwa.2020.578261
100. Kirschbaum MU (2000) Will changes in soil organic carbon act as a positive or negative feedback on global warming? Biogeochemistry 48:21–50
101. Kleber M, Eusterhues K, Keiluweit M, et al (2015) Mineral–organic associations: formation, properties, and relevance in soil environments. Elsevier:1–140
102. Kleber M, Johnson MG (2010) Advances in understanding the molecular structure of soil organic matter. In: Advances in Agronomy v106. Elsevier:77–142
103. Kleber M, Nico PS, Plante A et al (2011) Old and stable soil organic matter is not necessarily chemically recalcitrant: implications for modeling concepts and temperature sensitivity. Glob Chang Biol 17:1097–1107. https://doi.org/10.1111/j.1365-2486.2010.02278.x
104. Kleber M, Sollins P, Sutton R (2007) A conceptual model of organo-mineral interactions in soils: self-assembly of organic molecular fragments into zonal structures on mineral surfaces. Biogeochemistry 85:9–24. https://doi.org/10.1007/s10533-007-9103-5

105. Klotzbücher T, Kaiser K, Filley TR, Kalbitz K (2013) Processes controlling the production of aromatic water-soluble organic matter during litter decomposition. Soil Biol Biochem 67:133–139. https://doi.org/10.1016/j.soilbio.2013.08.003
106. Kögel-Knabner I, Guggenberger G, Kleber M et al (2008) Organo-mineral associations in temperate soils: Integrating biology, mineralogy, and organic matter chemistry. Z Pflanzenernähr Bodenk 171:61–82. https://doi.org/10.1002/jpln.200700048
107. Kögel-Knabner I (2002) The macromolecular organic composition of plant and microbial residues as inputs to soil organic matter. Soil Biol Biochem 34:139–162. https://doi.org/10.1016/S0038-0717(01)00158-4
108. Kopittke PM, Berhe AA, Carrillo Y et al (2022) Ensuring planetary survival: the centrality of organic carbon in balancing the multifunctional nature of soils. Crit Rev Environ Sci Technol:1–17.https://doi.org/10.1080/10643389.2021.2024484
109. Kopittke PM, Dalal RC, Hoeschen C et al (2020) Soil organic matter is stabilized by organo-mineral associations through two key processes: The role of the carbon to nitrogen ratio. Geoderma 357:113974. https://doi.org/10.1016/j.geoderma.2019.113974
110. Kravchenko AN, Guber AK, Razavi BS et al (2019) Microbial spatial footprint as a driver of soil carbon stabilization. Nat Commun 10:3121. https://doi.org/10.1038/s41467-019-11057-4
111. Lal R (2004) Soil carbon sequestration to mitigate climate change. Geoderma 123:1–22. https://doi.org/10.1016/j.geoderma.2004.01.032
112. Lavallee JM, Soong JL, Cotrufo MF (2020) Conceptualizing soil organic matter into particulate and mineral-associated forms to address global change in the 21st century. Glob Chang Biol 26:261–273. https://doi.org/10.1111/gcb.14859
113. LeBauer DS, Treseder KK (2008) Nitrogen limitation of net primary productivity in terrestrial ecosystems is globally distributed. Ecology 89:371–379. https://doi.org/10.1890/06-2057.1
114. Lehmann J, Hansel CM, Kaiser C et al (2020) Persistence of soil organic carbon caused by functional complexity. Nat Geosci. https://doi.org/10.1038/s41561-020-0612-3
115. Lehmann J, Kleber M (2015) The contentious nature of soil organic matter. Nature 528:60–68. https://doi.org/10.1038/nature16069
116. Liao JD, Boutton TW, Jastrow JD (2006) Organic matter turnover in soil physical fractions following woody plant invasion of grassland: Evidence from natural 13C and 15N. Soil Biol Biochem 38:3197–3210. https://doi.org/10.1016/j.soilbio.2006.04.004
117. Lindaas J, Pollack IB, Garofalo LA et al (2021) Emissions of reactive nitrogen from western U.S. wildfires during summer 2018. J Geophys Res 126. https://doi.org/10.1029/2020JD032657
118. Li J, Wang P, Salam N et al (2020) Unraveling bacteria-mediated degradation of lignin-derived aromatic compounds in a freshwater environment. Sci Total Environ 749:141236. https://doi.org/10.1016/j.scitotenv.2020.141236
119. Li M, Foster EJ, Le PVV et al (2020) A new dynamic wetness index (DWI) predicts soil moisture persistence and correlates with key indicators of surface soil geochemistry. Geoderma 368:114239. https://doi.org/10.1016/j.geoderma.2020.114239
120. Marin-Spiotta E, Chaopricha NT, Plante AF et al (2014) Long-term stabilization of deep soil carbon by fire and burial during early Holocene climate change. Nature Geosci 7:428–432. https://doi.org/10.1038/ngeo2169
121. Masiello CA, Berhe AA (2020) First interactions with the hydrologic cycle determine pyrogenic carbon's fate in the Earth system. Earth Surf Process Landforms 45:2394–2398. https://doi.org/10.1002/esp.4925
122. Masoom H, Courtier-Murias D, Farooq H et al (2016) Soil organic matter in its native state: unraveling the most complex biomaterial on earth. Environ Sci Technol 50:1670–1680. https://doi.org/10.1021/acs.est.5b03410
123. Ma Y, Filley TR, Johnston CT et al (2013) The combined controls of land use legacy and earthworm activity on soil organic matter chemistry and particle association during afforestation. Org Geochem 58:56–68. https://doi.org/10.1016/j.orggeochem.2013.02.010

124. Ma Y, McCormick MK, Szlavecz K, Filley TR (2019) Controls on soil organic carbon stability and temperature sensitivity with increased aboveground litter input in deciduous forests of different forest ages. Soil Biol Biochem 134:90–99. https://doi.org/10.1016/j.soilbio.2019.03.020
125. McCorkle EP, Berhe AA, Hunsaker CT et al (2016) Tracing the source of soil organic matter eroded from temperate forest catchments using carbon and nitrogen isotopes. Chem Geol 445:172–184. https://doi.org/10.1016/j.chemgeo.2016.04.025
126. Meyer GA, Pierce JL (2003) Climatic controls on fire-induced sediment pulses in Yellowstone National Park and central Idaho: a long-term perspective. For Ecol Manage 178:89–104. https://doi.org/10.1016/S0378-1127(03)00055-0
127. Miller JD, Safford HD, Crimmins M, Thode AE (2009) Quantitative evidence for increasing forest fire severity in the sierra nevada and southern cascade mountains, california and nevada, USA. Ecosystems 12:16–32. https://doi.org/10.1007/s10021-008-9201-9
128. Montgomery DR (2007) Soil erosion and agricultural sustainability. Proc Natl Acad Sci USA 104:13268–13272. https://doi.org/10.1073/pnas.0611508104
129. Moreland KC, Tian Z, Berhe AA et al (2021) Deep in the Sierra Nevada critical zone: Saprock represents a large terrestrial organic carbon stock. Environ Res Lett. https://doi.org/10.1088/1748-9326/ac3bfe
130. Myers-Pigg AN, Louchouarn P, Amon RMW et al (2015) Labile pyrogenic dissolved organic carbon in major Siberian Arctic rivers: Implications for wildfire-stream metabolic linkages. Geophys Res Lett 42:377–385. https://doi.org/10.1002/2014GL062762
131. Neary DG, Klopatek CC, DeBano LF, Ffolliott PF (1999) Fire effects on belowground sustainability: a review and synthesis. For Ecol Manage 122:51–71. https://doi.org/10.1016/S0378-1127(99)00032-8
132. Oeurng C, Sauvage S, Coynel A et al (2011) Fluvial transport of suspended sediment and organic carbon during flood events in a large agricultural catchment in southwest France. Hydrol Process 25:2365–2378. https://doi.org/10.1002/hyp.7999
133. Osher LJ, Matson PA, Amundson R (2003) Effect of land use change on soil carbon in Hawaii. Springer Science and Business Media LLC 65:213–232. https://doi.org/10.1023/a:1026048612540
134. Papanicolaou ANT, Wacha KM, Abban BK et al (2015) From soilscapes to landscapes: A landscape-oriented approach to simulate soil organic carbon dynamics in intensively managed landscapes. J Geophys Res Biogeosci 120:2375–2401. https://doi.org/10.1002/2015JG003078
135. Parton WJ, Schimel DS, Cole CV, Ojima DS (1987) Analysis of Factors Controlling Soil Organic Matter Levels in Great Plains Grasslands1. Soil Sci Soc Am J 51:1173. https://doi.org/10.2136/sssaj1987.03615995005100050015x
136. Paul EA, Follett RF, Leavitt SW et al (1997) Radiocarbon dating for determination of soil organic matter pool sizes and dynamics. Soil Sci Soc Am J 61:1058. https://doi.org/10.2136/sssaj1997.03615995006100040011x
137. Paul EA (2016) The nature and dynamics of soil organic matter: Plant inputs, microbial transformations, and organic matter stabilization. Soil Biol Biochem 98:109–126. https://doi.org/10.1016/j.soilbio.2016.04.001
138. Pellegrini AFA, Ahlström A, Hobbie SE et al (2018) Fire frequency drives decadal changes in soil carbon and nitrogen and ecosystem productivity. Nature 553:194–198. https://doi.org/10.1038/nature24668
139. Pierce J, Meyer G (2008) Long-term fire history from alluvial fan sediments: the role of drought and climate variability, and implications for management of Rocky Mountain forests. Int J Wildland Fire 17:84. https://doi.org/10.1071/WF07027
140. Pingree MRA, Kobziar LN (2019) The myth of the biological threshold: A review of biological responses to soil heating associated with wildland fire. For Ecol Manage 432:1022–1029. https://doi.org/10.1016/j.foreco.2018.10.032
141. Raison RJ, Khanna PK, Woods PV (1985) Mechanisms of element transfer to the atmosphere during vegetation fires. Can J For Res 15:132–140. https://doi.org/10.1139/x85-022

142. Ramos TB, Rodrigues S, Branco MA et al (2015) Temporal variability of soil organic carbon transport in the Enxoé agricultural watershed. Environ Earth Sci 73:6663–6676. https://doi.org/10.1007/s12665-014-3888-z
143. Rasmussen C, Southard RJ, Horwath WR (2006) Mineral control of organic carbon mineralization in a range of temperate conifer forest soils. Global Change Biol 12:834–847. https://doi.org/10.1111/j.1365-2486.2006.01132.x
144. Rasmussen C, Torn MS, Southard RJ (2005) Mineral assemblage and aggregates control carbon dynamics in a california conifer forest. Soil Sci Soc Am J 69:1711. https://doi.org/10.2136/sssaj2005.0040
145. Raymond PA, Saiers JE (2010) Event controlled DOC export from forested watersheds. Biogeochemistry 100:197–209. https://doi.org/10.1007/s10533-010-9416-7
146. Richter DD, Markewitz D (1995) How deep is soil? Bioscience 45:600–609. https://doi.org/10.2307/1312764
147. Richter D, deB, Billings SA, (2015) "One physical system": Tansley's ecosystem as Earth's critical zone. New Phytol 206:900–912. https://doi.org/10.1111/nph.13338
148. Riebe CS, Hahm WJ, Brantley SL (2017) Controls on deep critical zone architecture: a historical review and four testable hypotheses. Earth Surf Process Landforms 42:128–156. https://doi.org/10.1002/esp.4052
149. Riedel T, Zark M, Vähätalo AV et al (2016) Molecular signatures of biogeochemical transformations in dissolved organic matter from ten world rivers. Front Earth Sci 4. https://doi.org/10.3389/feart.2016.00085
150. Rodionov A, Amelung W, Peinemann N et al (2010) Black carbon in grassland ecosystems of the world. Global Biogeochem Cycles 24.https://doi.org/10.1029/2009GB003669
151. Rumpel C, Ba A, Darboux F et al (2009) Erosion budget and process selectivity of black carbon at meter scale. Geoderma 154:131–137. https://doi.org/10.1016/j.geoderma.2009.10.006
152. Rumpel C, Chaplot V, Planchon O et al (2006) Preferential erosion of black carbon on steep slopes with slash and burn agriculture. CATENA 65:30–40. https://doi.org/10.1016/j.catena.2005.09.005
153. Rumpel C, Kögel-Knabner I (2011) Deep soil organic matter—a key but poorly understood component of terrestrial C cycle. Plant Soil 338:143–158. https://doi.org/10.1007/s11104-010-0391-5
154. Salomã C, Nunan N, Pouteau V et al (2010) Carbon dynamics in topsoil and in subsoil may be controlled by different regulatory mechanisms. Glob Chang Biol 16:416–426. https://doi.org/10.1111/j.1365-2486.2009.01884.x
155. Sanderman J, Amundson R (2003) Biogeochemistry of decomposition and detrital processing. In: Treatise on Geochemistry. Elsevier:249–316
156. Sanderman J, Berhe AA (2017) Biogeochemistry: The soil carbon erosion paradox. Nat Clim Chang 7:317–319. https://doi.org/10.1038/nclimate3281
157. Santos F, Torn MS, Bird JA (2012) Biological degradation of pyrogenic organic matter in temperate forest soils. Soil Biol Biochem 51:115–124. https://doi.org/10.1016/j.soilbio.2012.04.005
158. Scharpenseel HW, Becker-Heidmann P, Neue HU, Tsutsuki K (1989) Bomb-carbon, 14C-dating and 13C — Measurements as tracers of organic matter dynamics as well as of morphogenetic and turbation processes. Semantic Scholar
159. Schmidt MWI, Noack AG (2000) Black carbon in soils and sediments: Analysis, distribution, implications, and current challenges. Global Biogeochem Cycles 14:777–793. https://doi.org/10.1029/1999GB001208
160. Schmidt MWI, Torn MS, Abiven S et al (2011) Persistence of soil organic matter as an ecosystem property. Nature 478:49–56. https://doi.org/10.1038/nature10386
161. Shangguan W, Hengl T, Mendes de Jesus J et al (2017) Mapping the global depth to bedrock for land surface modeling. J Adv Model Earth Syst 9:65–88. https://doi.org/10.1002/2016MS000686
162. Silver WL, Miya RK (2001) Global patterns in root decomposition: comparisons of climate and litter quality effects. Oecologia 129:407–419. https://doi.org/10.1007/s004420100740

163. Six J, Conant RT, Paul EA, Paustian K (2002) Stabilization Mechanisms of Soil Organic Matter: Implications for C-Saturation of Soils. Springer Science and Business Media LLC. https://doi.org/10.1023/a:1016125726789
164. Six J, Elliott ET, Paustian K (2000) Soil macroaggregate turnover and microaggregate formation: a mechanism for C sequestration under no-tillage agriculture. Soil Biol Biochem 32:2099–2103. https://doi.org/10.1016/S0038-0717(00)00179-6
165. Six J, Paustian K (2014) Aggregate-associated soil organic matter as an ecosystem property and a measurement tool. Soil Biol Biochem 68:A4–A9. https://doi.org/10.1016/j.soilbio.2013.06.014
166. Smithwick EAH, Turner MG, Mack MC, Chapin FS (2005) Postfire Soil N Cycling in Northern Conifer Forests Affected by Severe, Stand-Replacing Wildfires. Ecosystems 8:163–181. https://doi.org/10.1007/s10021-004-0097-8
167. Smith SV, Renwick WH, Buddemeier RW, Crossland CJ (2001) Budgets of soil erosion and deposition for sediments and sedimentary organic carbon across the conterminous United States. Global Biogeochem Cycles 15:697–707. https://doi.org/10.1029/2000GB001341
168. Sokol NW, Bradford MA (2019) Microbial formation of stable soil carbon is more efficient from belowground than aboveground input. Nature Geosci 12:46–53. https://doi.org/10.1038/s41561-018-0258-6
169. Sollins P, Homann P, Caldwell BA (1996) Stabilization and destabilization of soil organic matter: mechanisms and controls. Geoderma 74:65–105. https://doi.org/10.1016/S0016-7061(96)00036-5
170. Sollins P, Kramer MG, Swanston C et al (2009) Sequential density fractionation across soils of contrasting mineralogy: evidence for both microbial- and mineral-controlled soil organic matter stabilization. Biogeochemistry 96:209–231. https://doi.org/10.1007/s10533-009-9359-z
171. Soong JL, Castanha C, Hicks Pries CE et al (2021) Five years of whole-soil warming led to loss of subsoil carbon stocks and increased CO_2 efflux. Sci Adv 7.https://doi.org/10.1126/sciadv.abd1343
172. Stallard RF (1998) Terrestrial sedimentation and the carbon cycle: Coupling weathering and erosion to carbon burial. Global Biogeochem Cycles 12:231–257. https://doi.org/10.1029/98GB00741
173. Starr GC, Lal R, Malone R et al (2000) Modeling soil carbon transported by water erosion processes. Land Degrad Dev 11:83–91. https://doi.org/10.1002/(SICI)1099-145X(200001/02)11:1%3c83::AID-LDR370%3e3.0.CO;2-W
174. Staub B, Rosensweig C (1992) Global Zobler soil type, soil texture, surface slope, and other properties: Digital raster data on a 1-degree geographic (lat/long) 180x360 grid, in Global ecosystems database version 2.0. ORNL Distributed Active Archive Center. https://doi.org/10.3334/ornldaac/418
175. Stone EL, Kalisz PJ (1991) On the maximum extent of tree roots. For Ecol Manage 46:59–102. https://doi.org/10.1016/0378-1127(91)90245-Q
176. St John TV, Rundel PW (1976) The role of fire as a mineralizing agent in a Sierran coniferous forest. Oecologia 25:35–45. https://doi.org/10.1007/BF00345032
177. Tonneijck FH, Jongmans AG (2008) The influence of bioturbation on the vertical distribution of soil organic matter in volcanic ash soils: a case study in northern Ecuador. Eur J Soil Sci 59:1063–1075. https://doi.org/10.1111/j.1365-2389.2008.01061.x
178. Torres-Sallan G, Schulte RPO, Lanigan GJ et al (2017) Clay illuviation provides a long-term sink for C sequestration in subsoils. Sci Rep 7:45635. https://doi.org/10.1038/srep45635
179. VandenBygaart AJ, Gregorich EG, Helgason BL (2015) Cropland C erosion and burial: Is buried soil organic matter biodegradable? Geoderma 239–240:240–249. https://doi.org/10.1016/j.geoderma.2014.10.011
180. VandenBygaart AJ, Kroetsch D, Gregorich EG, Lobb D (2012) Soil C erosion and burial in cropland. Glob Change Biol 18:1441–1452. https://doi.org/10.1111/j.1365-2486.2011.02604.x

181. van der Voort TS, Mannu U, Hagedorn F et al (2019) Dynamics of deep soil carbon – insights from ^{14}C time series across a climatic gradient. Biogeosciences 16:3233–3246. https://doi.org/10.5194/bg-16-3233-2019
182. van der Werf GR, Randerson JT, Giglio L et al (2006) Interannual variability in global biomass burning emissions from 1997 to 2004. Atmos Chem Phys 6:3423–3441. https://doi.org/10.5194/acp-6-3423-2006
183. Wacha K, Papanicolaou A, Giannopoulos C et al (2018) The role of hydraulic connectivity and management on soil aggregate size and stability in the clear creek watershed, iowa. Geosciences (Basel) 8:470. https://doi.org/10.3390/geosciences8120470
184. Wang B, An S, Liang C et al (2021) Microbial necromass as the source of soil organic carbon in global ecosystems. Soil Biol Biochem 162:108422. https://doi.org/10.1016/j.soilbio.2021.108422
185. Wang Q, Wang Y, Wang S et al (2014) Fresh carbon and nitrogen inputs alter organic carbon mineralization and microbial community in forest deep soil layers. Soil Biol Biochem 72:145–151. https://doi.org/10.1016/j.soilbio.2014.01.020
186. Wan S, Hui D, Luo Y (2001) Fire effects on Nitrogen pools and dynamics in terrestrail ecosystems: a meta-analysis. Ecol Appl 11:1349–1365. https://doi.org/10.1890/1051-0761(2001)011[1349:FEONPA]2.0.CO;2
187. Weil R, Brady N (2016) The nature and properties of soils (15th Edition), 15th edn. Pearson, Columbus
188. Westerling AL, Hidalgo HG, Cayan DR, Swetnam TW (2006) Warming and earlier spring increase western U.S. forest wildfire activity. Science 313:940–943. https://doi.org/10.1126/science.1128834
189. Whitman T, Singh BP, Zimmerman AR et al (2015) Priming effects in biochar-amended soils: implications of biochar-soil organic matter interactions for carbon storage. In: Lehmann J, Joseph S (eds) Biochar for environmental management: science, technology and implementation. Routledge, pp 455–488
190. Whitman T, Zhu Z, Lehmann J (2014) Carbon mineralizability determines interactive effects on mineralization of pyrogenic organic matter and soil organic carbon. Environ Sci Technol 48:13727–13734. https://doi.org/10.1021/es503331y
191. Wilkinson MT, Richards PJ, Humphreys GS (2009) Breaking ground: Pedological, geological, and ecological implications of soil bioturbation. Earth Sci Rev 97:257–272. https://doi.org/10.1016/j.earscirev.2009.09.005
192. Woolf D, Lehmann J (2019) Microbial models with minimal mineral protection can explain long-term soil organic carbon persistence. Sci Rep 9:6522. https://doi.org/10.1038/s41598-019-43026-8
193. Yang D, Kanae S, Oki T et al (2003) Global potential soil erosion with reference to land use and climate changes. Hydrol Process 17:2913–2928. https://doi.org/10.1002/hyp.1441
194. Yan Q, Kumar P (2021) Impacts of landscape evolution on heterotrophic carbon loss in intensively managed landscapes. Front Water 3.https://doi.org/10.3389/frwa.2021.666278
195. Yan Q, Le PVV, Woo DK et al (2019) Three-dimensional modeling of the coevolution of landscape and soil organic carbon. Water Resour Res 55:1218–1241. https://doi.org/10.1029/2018WR023634
196. Yost JL, Hartemink AE (2020) How deep is the soil studied – an analysis of four soil science journals. Plant Soil. https://doi.org/10.1007/s11104-020-04550-z
197. Young IM, Crawford JW (2004) Interactions and self-organization in the soil-microbe complex. Science 304:1634–1637. https://doi.org/10.1126/science.1097394
198. Zethof JHT, Leue M, Vogel C et al (2019) Identifying and quantifying geogenic organic carbon in soils – the case of graphite. https://doi.org/10.5194/soil-2019-30
199. Zhang H, Wang E, Zhou D et al (2016) Rising soil temperature in China and its potential ecological impact. Sci Rep 6:35530. https://doi.org/10.1038/srep35530
200. Zimmermann M, Bird MI, Wurster C et al (2012) Rapid degradation of pyrogenic carbon. Glob Change Biol 18:3306–3316. https://doi.org/10.1111/j.1365-2486.2012.02796.x

201. Zimmerman AR, Gao B, Ahn M-Y (2011) Positive and negative carbon mineralization priming effects among a variety of biochar-amended soils. Soil Biol Biochem 43:1169–1179. https://doi.org/10.1016/j.soilbio.2011.02.005
202. Stewart CE, Zheng J, Botte J, Cotrufo MF (2013) Co-generated fast pyrolysis biochar mitigates green-house gas emissions and increases carbon sequestration in temperate soils. GCB Bioener 5(2):53–164
203. Singh N, Abiven S, Maestrini B, Bird JA, Torn MS, Schmidt MWI (2014) Transformation and stabilization of pyrogenic organic matter in a temperate forest field experiment. Glob Change Biol 20(5):1629–1642
204. Maestrini B, Herrmann AM, Nannipieri P, Schmidt MWI, Abiven S (2014) Ryegrass-derived pyrogenic organic matter changes organic carbon and nitrogen mineralization in a temperate forest soil. Soil Biol Biochem 69:291–301
205. Kuzyakov Y, Bogomolova I, Glaser B (2014) Biochar stability in soil: Decomposition during eight years and transformation as assessed by compound-specific 14C analysis. Soil Biol Biochem 70:229–236
206. Grant KE, Hilton RG, Galy VV (2023) Global patterns of radiocarbon depletion in subsoil linked to rock-derived organic carbon. Geochem Perspect Let 25:36–40
207. Santos F, Abney R, Barnes M, Bogie N, Ghezzehei TA, Jin L, Moreland K et al (2019) The role of the physical properties of soil in determining biogeochemical responses to soil warming. Ecosystem consequences of soil warming (pp. 209–244). Elsevier
208. Hunter BD, Roering JJ, Silva LCR, Moreland KC (2024) Geomorphic controls on the abundance and persistence of soil organic carbon pools in erosional landscapes. Nature Geosci 17(2):151–157
209. Yang Y, Berhe AA, Barnes ME, Moreland KC, Tian Z, Kelly AE, Bales RC, et al. (2022) Climate warming alters nutrient storage in seasonally dry forests: insights from a 2300 m Elevation Gradient. Global biogeochemical cycles
210. Dove NC, Barnes ME, Moreland K, Graham RC, Berhe AA, Hart SC (2021) Depth dependence of climatic controls on soil microbial community activity and composition. ISME Commun 1(1):78
211. Hicks Pries C, Ryals R, Zhu B, Min K, Cooper A, Goldsmith S, Pett-Ridge J et al (2023) The deep soil organic carbon response to global change. Ann Rev Ecol Evolut Syst 54(1)
212. Longbottom T, Wahab L, Min K, Jurusik A, Moreland K, Dolui M, Thao T et al (2022) What's Soil Got to Do with Climate Change? GSA Today 32(5):4–10

The Role of Soil Aggregation in the Critical Zone

Maria A. Lilli, Fotini E. Stamati, and Nikolaos P. Nikolaidis

Keywords Soil degradation · Soil aggregation · Soil restoration · Soil fertility · Nature based solutions

1 Soil Degradation—A Global Threat

Humanity is facing significant challenges in responding to increasing demands in food, water and energy worldwide. Currently, the global cereal balance has a marginal surplus with the world cereal equivalent food demand to be around 14,886 million tons, while its production to be 15,970 million tons. The required increases to support the increasing population and lifestyle changes by 2050 is of the order of 30% [25]. The rapid increase in world population, expected to reach 9,8 billion in 2050, is exerting a significant strain in increasing cereal production if we want to avoid famine at a large scale. At the same time, global water demand is projected also to reach 5,425 km^3, due to increases in the demand for agriculture, industry, energy production and domestic use [11]. In addition, climate change will adversely affect water availability through increasing temperature, changes in precipitation, melting of snow, floods and droughts, reducing ecosystem services [35]. In other words, grand challenges such as climate change, food and water security and population growth are coupled and together are threatening the security and services of water, ecosystem, food and energy nexus at an accelerating rate.

At the core of the Water, Ecosystem, Food, Energy (WEFE) Nexus are the soils which will be impacted significantly by the effects of climate changes and the associated global changes and at the same time will be part of the solution to mitigate

M. A. Lilli (✉) · F. E. Stamati · N. P. Nikolaidis
School of Chemical and Environmental Engineering, Technical University of Crete, 73100 Chania, Greece
e-mail: mlilli@isc.tuc.gr

T. White and A. Provenzale (eds.), *Critical Zone and Ecosystem Dynamics*, Advances in Critical Zone Science, https://doi.org/10.1007/978-3-031-69076-1_8

these impacts and adapt. Soils have many different essential functions that provide ecosystem services and life-sustaining benefits, related for example to the fact that 95% of food is produced from soils [35]. Soils sequester carbon from the atmosphere, filter, buffer and transform water, nutrients and contaminants; provide habitat and gene pool and produce fuel and fibre [7]. Soil degradation is the results of land use conversion and agricultural intensification and it can be the precursor of desertification with soil organic carbon (SOC) and nutrient decline, loss of biodiversity, loss of soil fertility [6] and loss of soil structural stability [27]. These soil threats impact global food security. The major drivers of soil degradation are climate change, land-use change and above ground biodiversity change [25] which impact significantly soil functions and the ecosystem services soils provide.

The onset of soil degradation is primarily the conversion of forested or grass lands to agriculture as well as the continuing agricultural practices. Conversion of native vegetated lands to croplands induces SOM losses due to gradual changes in land management like plowing and less litter input. On a global basis biological oxidation is considered to be the dominant pathway of carbon loss [28, 29], while leaching of dissolved organic carbon cannot contribute more than 10% of the SOC loss [30].

On the other hand, erosion can be viewed as a transport and deposition process that influences the areas in which oxidation takes place [16]. Under plowing, macro-aggregates are partly rapidly destroyed, while physically protected bio-available SOM becomes bio-accessible and can be microbially oxidized [4]. Particulate organic carbon (POC) is the fraction mostly affected by plowing in the short term [37]. In addition, incoming OM is re-exposed faster than it is occluded due to the higher macro-aggregate turnover. Therefore more OM is mineralized and less stabilized in micro-aggregates and silt–clay sized soil fractions [33] in the medium- to long-term time scale [9].

SOC is directly related to soil fertility and biomass production. Loss of OM from the soil is a sign of soil degradation and reduction of soil functions and the resulting soil services. The average rate of SOC stock decline (t C/ha y) after the conversion of a native land to cropland has been found to decrease in time. Variations depend on the initial SOC content and apportionment to its pools which is highly depended on previous land use type, the size and litter quality and the decomposition regime which is primarily determined by the climatic conditions and the texture among other factors [22]. Figure 1 shows the reduction of carbon and nitrogen content after 69 years subsequent to conversion to agricultural lands. These chrono sequence data illustrate that soil fertility is lost within a few years after the conversion. The type of the previous land use, forest or grassland, the initial SOC content, and the clay content were identified. In all cases, the first year rate of decline for SOC stock (2–7 t SOC/ha year) was found to be one order of magnitude greater than the average rate observed after 10 (Grassland to Cropland poor in clay), 20–30 years (Grassland to Cropland rich in clay), or more (Forest to Cropland) subsequent to the conversion (<0.5 t SOC/ha year), depending on the size of the initial OM stock, and decrease in time following a logarithmic pattern (Fig. 1). The available data regarding the rate of TN stock decline indicate similar patterns of decline. The rate of depletion of the SOC pool varies significantly between the tropics and the temperate ecoregions [14].

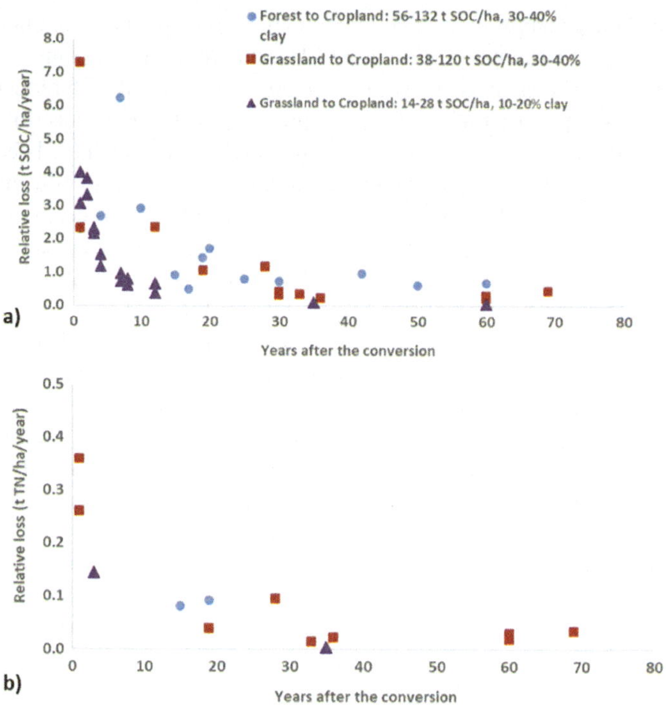

Fig. 1 Relative decline of soil **a** OC and **b** TN, after the conversion of native land to cropland (normalized to 20 cm soil depth)

Decline of SOM stocks is related to poor soil fertility and structural stability (particle aggregation) in several agricultural areas of the world, making soils more susceptible to erosion and desertification with global consequences for food security, climate change, biodiversity, water quality, and agricultural economy [15, 21]. Addressing the adverse effects of soil degradation and actively restoring soil functions and services is at the heart of the response to mitigate and adapt to the impacts of climate change as well as optimize the WEFE Nexus for a sustainable living.

2 Soil Restoration—Improving Soil Functions

The United Nations has recognized that 33% of the soils on earth are already degraded, and over 90% could be degraded by 2050, emphasizing the importance of altering this trend by restoring soil structure [35]. The EU published the EU Biodiversity Strategy for 2030 where the protection of soil fertility, reduction of soil erosion, increasing of soil organic matter and soil restoration are between the main priorities. Crop rotation, organic carbon addition with associated nutrient elements, reduced tillage, use of cover crops, controlled grazing are among the most commonly nature based practices for sustainable agriculture [23]. Incorporation of plant residues and organic amendments, such as compost and manure, combined with reduced tilling, enhance carbon sequestration [17] and plant production [19] as well as soil hydraulic properties [36]. During the last three decades, these practices have been increasingly described as agroecological practices, aiming to produce significant amounts of food, and contribute to improving the sustainability of agroecosystems while being based on various ecological processes and ecosystem services such as nutrient cycling, biological N fixation, natural regulation of pests, soil and water conservation, biodiversity conservation, and carbon sequestration [38].

Soil structure is one of the main factors controlling the ability of soil to support plant growth, the movement of water and the cycling of carbon and nutrients. Since the early 1900s, soil scientists have recognized that soil particles combine to form larger aggregates. Soil aggregation, largely responsible for soil structure, is fundamental for soil functioning and agricultural productivity. Soil aggregation is a dynamic and complex process that includes formation, stabilization and destabilization, as well as the disruption of soil macro and micro-aggregates (Fig. 2) [18].

Soil water stable aggregates (WSA) can physically stabilize SOM and protect it from decomposition [32]. An "agronomically valuable" soil is the soil where greater than 60% of the aggregated particle mass is in the range between 0.25 and 10 mm [5]. An appropriate management of organic matter additions like composts to soils would increase aggregate stability and thus reduce crusting and erosion problems [1].

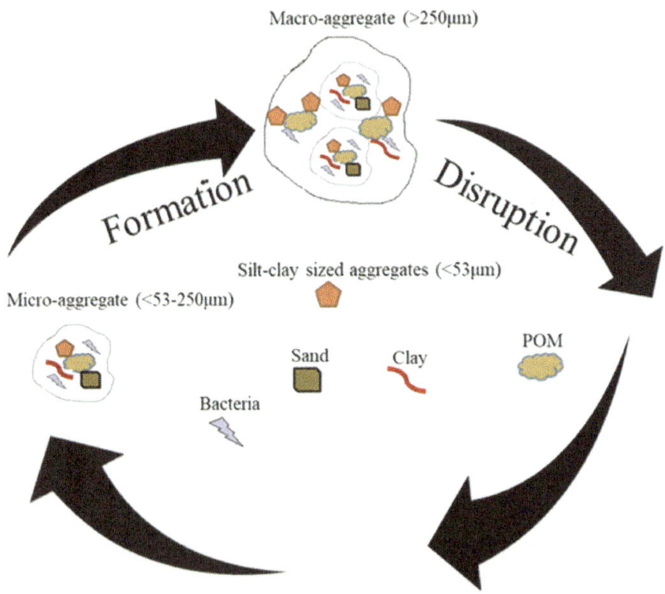

Fig. 2 Macro-aggregate life cycle. Modified from [34]

Soil aggregates characterized as the smallest ecological units, are formed by the inclusion of micro-aggregates within the macro-aggregates. Macro-aggregates (particle diameter >250 μm) are composed of clay- and silt-sized aggregates, microbial by products, and bacterial decay. Microbial decomposers of plant residues provide the binding for soil particles and smaller aggregates to form macro-aggregates around particulate organic matter (POM). Macro-aggregated POM is further decomposed and results in the formation of micro-aggregates within the macro-aggregates by encapsulating the finely fragmented POM in silt–clay sized aggregates. The microbial activity decreases gradually as the incorporated organic matter inside macro-aggregates is biodegraded and this procedure results in less stable macro-aggregates. The unprotected macro-aggregates are prone to slaking, which leads to their disruption and the simultaneous release of micro-aggregates and silt–clay sized aggregates. The cycle of aggregation and disruption can be continued, when new fresh plant residues enter the soil and as a result new macro-aggregates are formed (Fig. 2) [3, 26, 34].

Figure 3 illustrates the aggregation process and the impact of SOM addition on soil structure and carbon and nitrogen content in the soils. These are data collected by [34] from two nearby agricultural fields in Crete, Greece, one set aside for 30 years and the other cultivated crop land. Figure 3a presents the soil water stable aggregate distribution (g sand-free aggregate/100 g soil) for both fields. The fraction of WSA in the set aside field increased from 45.8 to 65.5% during the 30 years of carbon addition while at the same time the carbon (Fig. 3b) and the nitrogen (Fig. 3c) contents in the WSA particles increased from 16.7 to 37.4 g C/Kg for carbon and from 1.22 to 2.3 g N/Kg for nitrogen. Values of WSA greater than 60% have been regarded as indication of "agricultural valuable soils" [7, 23] relating in this way improvements in soil structure to soil fertility.

The question though remains on how fast can SOM additions improve soil structure and fertility and thus restore soil degradation? In order to elucidate this matter, we used data derived from a 4-year field-scale horticulture experiment where tomato plants were grown using different treatments of commercial mineral fertilizers (at a rate of 200 kg N ha^{-1}), compost (at a rate of 50 t dry weight ha^{-1}), sheep and goat manure (at the rate of 50 t dry weight ha^{-1}), and a 30% manure–70% compost mixture amendment at the same rate [12]. The study was conducted over the years 2011 to 2014 within the study area of the Koiliaris Critical Zone Observatory (CZO) at an experimental field located 14 km south-east from Chania, Greece (35° 260 1400 N, 24° 080 3400 E). The experimental field was a former orange tree plantation which had been abandoned for approximately 25 years. One month before the start of the experiment the vegetation was totally removed and the field was ploughed. Tomato seedlings were transplanted in early August and harvested in November in the first year of the experiment, and in subsequent years the growing period was from early May until October. Tillage was applied before the transplanting of the tomato seedlings (months 8, 17, 29, 41) to incorporate the organic amendments to the soil and prepare it for planting. The most productive year was the third for all treatments, whereas the highest biomass production among them was observed in the manure and in the compost–manure mixture treatments. Plants treated with compost

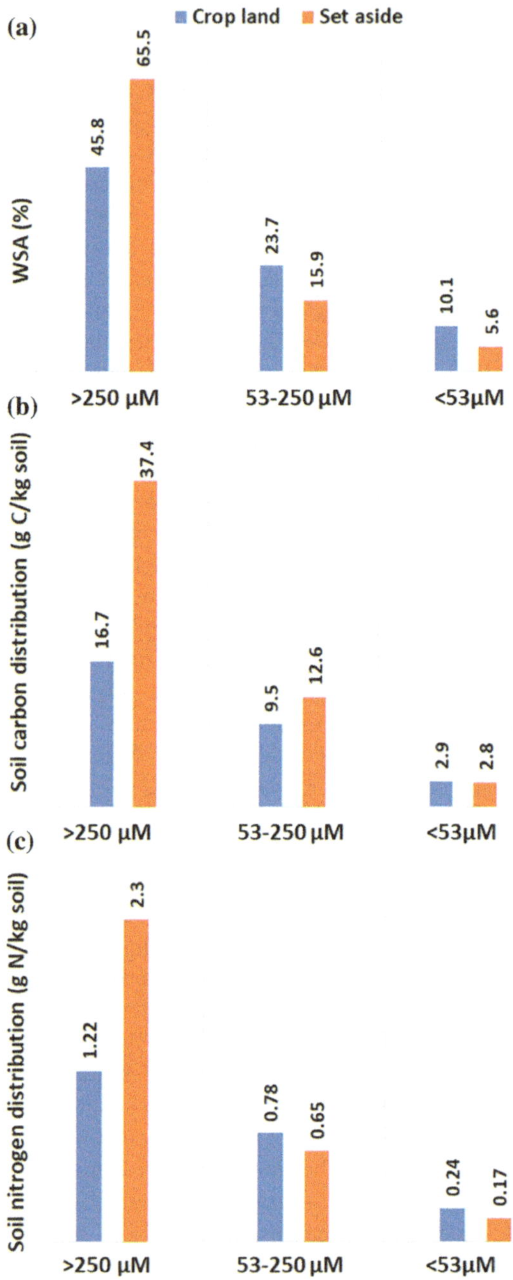

Fig. 3 **a** Soil water stable aggregate distribution (g sand-free aggregate/100 g soil), as well **b** carbon and **c** nitrogen distribution among aggregates fractions (g/kg). Data adapted from [34]

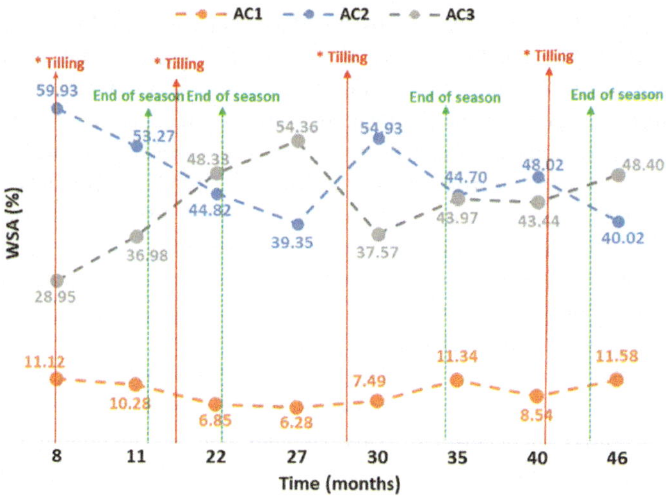

Fig. 4 Soil water stable aggregate evolution. Data adapted from [12]

showed lower C accumulated in the aboveground biomass to that produced by plants treated with manure or with the compost–manure mixture, but similar quantities to the inorganic fertilization treatment, thus suggesting that organic amendments can have the potential of providing plants with sufficient quantities of nutrients to sustain plant production.

Figure 4 presents the changes in soil structure in the compost–manure mixture treatment. At the end of the 4-year experiment, AC3 content had increased significantly. In general, the highest increase was observed in treatments that received the carbon amendments and specifically in the treatment presented in Fig. 4, where the increase reached 72.2% relative to the initial AC3 mass fraction. Overall, the increase of water-stable macroaggregates in the carbon amended plots was almost two times higher than in treatment of inorganic fertilization, indicating the beneficial effect of organic amendments on soil structure development. The WSA distribution during the experiment was affected by tillage. These results suggest that soil aggregation process is fast and soil structure can be achieved in the time scale of weeks to months.

3 Tools for Soil Restoration

The extend of soil degradation worldwide necessitates the need for tools that can model carbon sequestration and guide the actions of land restoration. One such model is the Carbon, Aggregation, and Structure Turnover (CAST) model [34]. The CAST model simulates the mechanisms of aggregation assuming three size classes of aggregates (silt–clay-sized aggregates (AC1, <53 μm), microaggregates (AC2; 53–250 μm), and macroaggregates (AC3; >250 μm)) as well as the five carbon pools used by the RothC model (Decomposable Plant Material—DPM, Resistant Plant Material—RPM, Microbial Biomass—BIO, Humified Organic Matter—HUM and Inert Organic Matter—IOM).. Soil carbon pools are coupled with the aggregate and structure turnover modules and the rates of aggregation, disaggregation and carbon reactions are calibrated using field data from sites. The model can predict the changes in particle size aggregation, porosity and bulk density. The CAST model can identify the factors affecting carbon sequestration and turnover in soils and ascertain the response of soil to land use, climate change and agro-ecological practices [3, 19, 26].

To illustrate the use of the CAST model and how it can be used to model soil restoration practices, data from a case study site (Milia, a seventeenth century mountain settlement in Western Crete, Greece) were used.

Milia represents a strongly eroded Eutric Lithosol soil in which restoration practices including terraces construction and spreading of organic amendments have been applied [26]. The residents produce compost plant material and animal manure, mix it with an empirical ratio close to 2.5/1 and re-compost it to produce the soil organic amendment to restore the fertility of the terraces. The applied compost contained 14.5% °C. Two terraces were selected for modelling with different compost application practices. The frequency of compost application was twice per year, through tilling, corresponding to a total application of 8 t C/ha. As far as the 1st terrace (Fig. 5a), the compost was applied every year for 10 years, and as far as the 2nd terrace (Fig. 5b), it was applied for 8 years followed by 2-years fallow. Figure 6 presents the results of the CAST model simulation of the evolution of the SOC of the soil, the mass of aggregate type AC3, the aggregate type AC3 plus the coarse POM of the non-aggregated pools (cPOM), the porosity and bulk density for the two terraces.

The CAST model has been used to successfully simulate the evolution of carbon in the soil as well as the changes in soil structure using data from Greece, Iowa, USA [34], Dama glacier, Switzerland [2], China [19] and other European sites [26] and it can be used as a tool to assess soil restoration practices.

4 Soil Degradation at the Basin Level: The Example of Koiliaris CZO

The degree of soil degradation worldwide suggests that soil restoration assessments should be considered at the landscape level. Such assessments should identify the areas that are prone to soil degradation and consider Nature Based Solutions (NBS) that would address the problem in a sustainable way and provide resilience to the ecosystem. NBS are defined as actions inspired by, supported by, or copied from nature [8] and have been used extensively to improve the ecosystem services provided by landscapes and to address global societal challenges [10]. At the same time, one should identify NBS that address all the impacts to the WEFE Nexus. The fact that one NBS cannot solve a soil degradation issue alone, it is suggested to use NBS

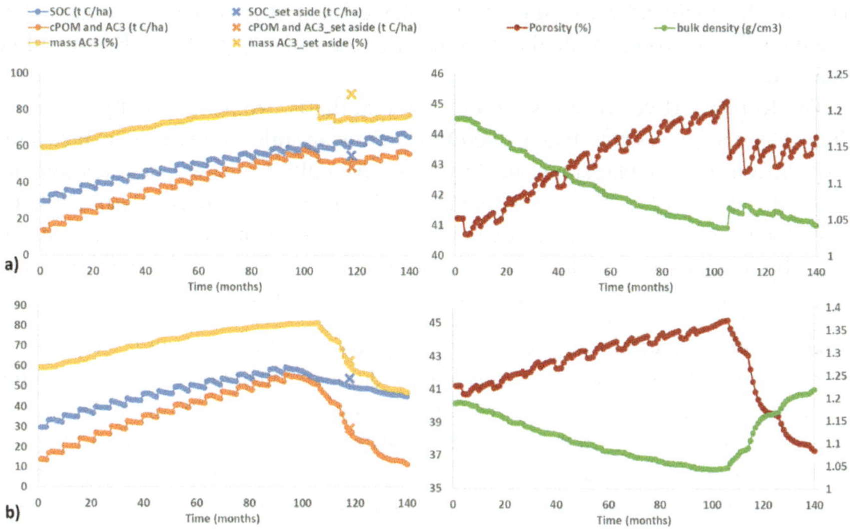

Fig. 5 Evolution of the organic carbon content of the soil, the mass of aggregate type AC3, the aggregate type AC3 plus the coarse POM of the nonaggregated pools, the porosity and bulk density for the **a** 1st and **b** 2nd terrace in Milia, Greece. Points depicted with X indicate the measurements from the terraces and the initial conditions were measurements from un-distrurbed areas. The data used are unpublished data

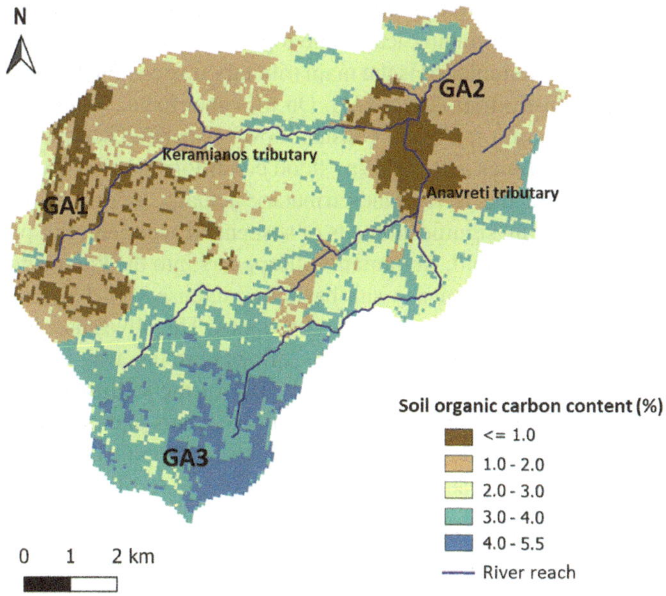

Fig. 6 Soil organic carbon content in Koiliaris CZO, Chania, Crete, Greece. Soil depth of the samples 0–30 cm

bundles which will collective support the enhancement of ecosystem services and its future resilience. To illustrate these issues, the case of Koiliaris River Basin will be analyzed.

The Koiliaris River Basin is a CZO (www.koiliaris-czo.tuc.gr) on the island of Crete that represents severely degraded Mediterranean soils due to different pressures derived from land management namely, intensive cultivation, soil degradation and livestock grazing [20]. The main challenges that need to be addressed in Koiliaris CZO, focus on three geographic areas (GA) within the basin (GA1: the western part of the basin, GA2: the northeastern part of the basin and GA3: the southern part of the basin) (Fig. 6) and relate to the aforementioned pressures.

Intense soil degradation A temporary tributary (Keramianos) and a temporary spring (Anavreti), along with two episodic tributaries, contribute to the Stylos spring discharge to form the Koiliaris River. The two episodic tributaries drain the eastern part of the watershed, while the temporary tributary (Keramianos) flows over the western part. The Keramianos tributary drains a small sub-catchment that generates surface runoff due to the schist geologic formation of the area and the respective shallow soils. Keramianos stream flows along a karstic gorge and then over an alluvial plain before joining the Koiliaris River. Schist formations are quite friable and in combination with the steep slopes and the adaptation of intensive agricultural practices that are common in the area of Keramianos sub-basin, the topsoil becomes extremely brittle and easily erodible. More specifically, due to the abandonment of traditional agricultural practices over the years, tractors now enter and plow the terraces, leading occasionally to their collapse, and exacerbating the erodibility of the soils. In addition, overgrazing leaves the topsoil unprotected and vulnerable to surface runoff. Thus, Keramianos is the main tributary responsible for the bulk of the sediment transport in Koiliaris River [31]. On the other hand, waters derived from the karstic springs, have a relatively constant–low–suspended sediment concentration. Only two to four times a year, during flood conditions when daily precipitation exceeds 120 mm [13], the Keramianos tributary actually merges with the Stylos Springs flow, transferring significant loads of suspended sediment. For precipitation rates lower than this threshold, Keramianos disappears into the karstic gorge [24].

Significant areas of Keramianos sub-basin has intense soil degradation that is related to the cultivation of olive trees in steep slopes. The olives have been planted in terraces without the associated stone walls, allowing in this way the soil to erode. Nerantzaki et al. [24] have estimated that 63–70% of the sediment transport (wet and dry year respectively) is generated during flood events and originates from the Keramianos sub-basin. The erosion rate ranges between 0.55 to 1.6 ton/ha/yr for the sub-basin, causing land degradation and reduction of soil fertility.

Agricultural practices and soil degradation During the first few years after conversion from forest or grassland to crop land, rapid loss of bio-available carbon and nitrogen and deterioration of soil stability has been observed, accompanied by a significant shift in soil fertility. The Cretan landscape has been transformed during the past 50–60 years from a low intensity agrarian landscape with low impact agricultural practices, and high biodiversity, to a mechanized, high intensity agriculture and properties with monocultures. Agricultural practices have been the primary cause of land degradation (soil tillage, no organic matter addition to soil, high pesticide and herbicide use). Figure 6 shows that the soil organic carbon content is low in Koiliaris river basin, and primarily in GA1 and GA2 (0–2%), where the land is cultivated. Addition of crop residue and organic matter to soils is the only way towards a sustainable agriculture. Soil fertility improves faster if in addition to bio-fertilization, farmers use traditional agricultural practices such as crop rotations and legume row plantings within olive trees and orchards [25].

Biodiversity degradation due to free grazing livestock The Koiliaris River basin offers a unique morphologic situation due to its karstic hydrogeology draining the upland grazing areas through karstic springs (GA3). Approximately 25,000 sheep and goats are grazing in the karstic highlands of Koiliaris River (Fig. 2), causing significant livestock impacts. Thus, sheep and goats' free grazing create environmental pressures and significant loss of biodiversity at the higher elevations of the basin.

Actions that use natural processes in a resource efficient manner are an ideal and promising choice in the case of Koiliaris CZO in order to address the abovementioned challenges and optimize the WEFE nexus. NBS bundles are proposed to implement to the three areas of interest in Koiliaris CZO in order to optimize the Nexus. The proposed NBS can contribute to (a) erosion control, (b) biodiversity protection and (c) soil regeneration strategies.

(a) NBS bundle for erosion control—Focus on olive groves

- Vegetated stonewalls for terrace restoration and soil protection.
- Soil revegetation below the olive trees- terraces.
- Fertilization of olive groves with organic matter for soil structure regeneration.
- Plant trees and hedges at lower part (bufferzone) of the slope before river.

(b) NBS for high elevation biodiversity protection

- Remove environmental pressures from sheep and goat free grazing
    - Move to lower elevation to animal house
    - Utilize manure
    - Let the system regenerate

(c) NBS bundles for soil regeneration

- Adopt agroecological practices.
- Use soil organic amendments, organic fertilization and conservation measures for soil improvement.
- Incorporate manure, compost, biosolids, or crop residues to enhance carbon storage.
- Implement the riparian forest restoration project.

The selected NBS will enhance the WEFE nexus of the watershed. In addition, soil organic matter incorporation in the soil will improve soil structure and increase soil water retention. The organized management of livestock will let the natural regeneration of biodiversity in high elevations while it will provide manure for the organic fertilization of the agricultural fields in the valley. Agroecological practices will also enhance the biodiversity in the valley. Finally, the regenerated soil will increase soil fertility and biomass production. These NBS will enhance the WEFE Nexus and improve the life and well-being of the people in the area.

5 Conclusions

The underlying solution to the grand challenges' humanity is facing is the reversing of soil degradation and the restoration of soil fertility and health. The basic premise of soil restoration is the addition of organic matter and the use of agroecological practices. Soil organic matter amendments improve soil structure, sequester more carbon and nutrients and enhance the ecosystem services provided by soils and landscapes. The changes brought to soil by SOM amendments take place in the time scales of weeks to months. Modeling tools such as the CAST model can be used to simulate the impacts of SOM additions to soil structure, carbon turnover and sequestration and assess their evolution in time during restoration. Nature based solutions should predominantly used to assess soil restoration and such studies should be design at the landscape level. The case of Koiliaris CZO clearly illustrates the path to sustainable land use management that would result in the improvement of well-being of the people in the area.

Acknowledgements This work is funded by the PRIMA programme supported by the European Union. GA n° [2041] [LENSES] [Call 2020 Section 1 Nexus IA].

References

1. Abiven S, Menasseri S, Chenu C (2009) The effects of organic input over time on soil aggregate stability – a literature analysis. Soil Biol Biochem 41:1–12
2. Andrianaki M, Bernasconi SM, Nikolaidis NP (2017) Quantifying the incipient development of soil structure and functions within a glacial forefield chronosequence (Chapter Eight). In: Banwart SA, Sparks DL (eds) Advances in agronomy, vol 142. Academic Press, pp 215–239. ISSN 0065-2113, ISBN 9780128122228. https://doi.org/10.1016/bs.agron.2016.10.013
3. Apostolakis A, Panakoulia S, Nikolaidis NP, Paranychianakis NV (2017) Shifts in soil structure and soil organic matter in a chronosequence of set-aside fields. Soil Tillage Res 174:113–119. ISSN 0167-1987. https://doi.org/10.1016/j.still.2017.07.004
4. Balesdent J, Besnard E, Arrouays D, Chenu C (1998) The dynamics of carbon in particle-size fractions of soil in a forest-cultivation sequence. Plant Soil 201:49–57
5. Banwart S, Bernasconi SM, Bloem J, Blum W et al (2011) Assessing soil processes and function in critical zone observatories: hypotheses and experimental design. Vadose Zone J 10:974–987

6. Banwart SA, Nikolaidis NP, Zhu YG, Peacock CL, Sparks DL (2019) Soil functions: connecting earth's critical zone. Annu Rev Earth Planet Sci 47:333–359
7. Banwart S, Noellemeyer E, Milne E (2014) Soil Carbon: science, management and policy for multiple benefits. In: Scope, vol 71. CABI, Wallingford. https://doi.org/10.1079/9781780645322.0000
8. Cohen-Shacham E, Walters G, Janzen C, Maginnis S (eds) (2016) Naturebased solutions to address global societal challenges. IUCN, Gland, Switzerland
9. Don A, Scholten T, Schulze ED (2009) Conversion of cropland into grassland: implications for soil organic-carbon stocks in two soils with different texture. J Plant Nutr Soil Sci 172:53–62
10. Faivre N, Fritz M, Freitas T, de Boissezon B, Vandewoestijne S (2017) Nature-based solutions in the EU: innovating with nature to address social economic and environmental challenges. Environ Res 159:509–518
11. Islam SMF, Karim Z (2019) World's demand for food and water: the consequences of climate change. In: Farahani MHDA, Vatanpour V, Taheri AH (eds) Desalination—challenges and opportunities. IntechOpe
12. Kotronakis M, Giannakis GV, Nikolaidis NP, Rowe EC, Valstar J, Paranychianakis NV, Banwart SA (2017) Chapter eleven - modeling the impact of carbon amendments on soil ecosystem functions using the 1D-ICZ model. In: Banwart SA, Sparks DL (eds) Advances in agronomy, vol 142. Academic Press, Cambridge, Massachusetts, pp 315–352. https://doi.org/10.1016/bs.agron.2016.10.010
13. Kourgialas NN, Karatzas GP, Nikolaidis NP (2010) An integrated framework for the hydrologic simulation of a complex geomorphological river basin. J Hydrol 381:308–321. https://doi.org/10.1016/j.jhydrol.2009.12.003
14. Lal R (2001) World cropland soils as a source or sink for atmospheric carbon. Adv Agron 71:145–191
15. Lal R (2004) Soil carbon sequestration impacts on global climate change and food security. Science 304:1623–1627
16. Lal R (2008) Carbon sequestration. Philos Trans R Soc 363:815–830
17. Lal R (2015) Restoring soil quality to mitigate soil degradation. Sustainability 7(5):5875–5895. https://doi.org/10.3390/su7055875
18. Lehmann A, Leifheit EF, Rillig MC (2017) Mycorrhizas and soil aggregation. In: Collins Johnson N, Gehring C, Jansa J (eds) Mycorrhizal mediation of soil: fertility structure and carbon storage storage. Elsevier, Cambridge, Massachusetts, pp 241–262
19. Li S, Li Y, Li X, Tian X, Zhao A, Wang S, Wang S, Shi J (2016) Effect of straw management on carbon sequestration and grain production in a maize-wheat cropping system in Anthrosol of the Guanzhong Plain. Soil Tillage Res 157:43–51. https://doi.org/10.1016/j.still.2015.11.002
20. Lilli MA, Efstathiou D, Moraetis D, Schuite J, Nerantzaki SD, Nikolaidis NP (2020) A multidisciplinary approach to understand hydrologic and geochemical processes at Koiliaris critical zone observatory. Water 12:2474
21. Mafongoya PL, Bationo A, Kihara J, Wasw BS (2006) Appropriate technologies to replenish soil fertility in southern Africa. Nutr Cycl Agroecosyst 76:137–151
22. Mann LK (1986) Changes in soil carbon storage after cultivation. Soil Sci 142:279–288
23. Milne E, Banwart SA, Noellemeyer E, Abson DJ, Ballabio C et al (2015) Soil carbon, multiple benefits. Environ Dev 13(33–38):646. https://doi.org/10.1016/j.envdev.2014.11.005
24. Nerantzaki SD, Giannakis GV, Efstathiou D, Nikolaidis NP, Sibetheros IA, Zacharias KGP (2015) Modeling suspended sediment transport and assessing the impacts of climate change in a karstic Mediterranean watershed. Sci Total Environ 538:288–297. https://doi.org/10.1016/j.scitotenv.2015.07.092
25. Nikolaidis NP (2011) Human impacts on soils: tipping points and knowledge gaps. Appl Geochem 26:S230–S233. ISSN 0883-2927. https://doi.org/10.1016/j.apgeochem.2011.03.111
26. Panakoulia SK, Nikolaidis NP, Paranychianakis NV, Menon M, Schiefer J, Lair GJ, Krám P, Banwart SA (2017) Factors controlling soil structure dynamics and carbon sequestration across different climatic and lithological conditions (Chapter Nine). In: Advances in agronomy, vol 142, pp 241–276, ISSN 0065-2113, ISBN 9780128122228. https://doi.org/10.1016/bs.agron.2016.10.008

27. Quijano L, Six J, Navas A, Van Oost K (2019) Effect of soil redistribution on soil aggregate stability and soil organic carbon in Mediterranean cultivated soils. Geophys Res Abstr V21, EGU2019-5950-1
28. Rasmussen PE, Goulding KWT, Brown JR, Grace PR, Janzen HH, Körschens M (1998) Long-term agroecosystem experiments: assessing agricultural sustainability and global change. 282:893–896
29. Rees RM, Bingham IJ, Baddeley JA, Watson CA (2005) The role of plants and land management in sequestering soil carbon in temperate arable and grassland. Geoderma 128:130–154
30. Schulze ED, Freibauer A (2005) Carbon unlocked from soils. Nature 437:205–206
31. Sibetheros IA, Nerantzaki S, Efstathiou D, Giannakis G, Nikolaidis NP (2013) Sediment transport in the Koiliaris River of Crete. Procedia Technol 8:315–323. https://doi.org/10.1016/j.protcy.2013.11.042
32. Six J, Conant RT, Paul EA, Paustian K (2002) Stabilization mechanisms of soil organic matter: implications for C-saturation of soils. Plant Soil 241(2):155–161
33. Six J, Elliott ET, Paustian K (2000) Soil macroaggregate turnover and microaggregate formation: a mechanism for carbon sequestration under no tillage agriculture. Short Comm Soil Biology and Biochemistry 32:2099–2103
34. Stamati FE, Nikolaidis NP, Banwart S, Winfried EH, Blum WEH (2013) A coupled carbon aggregation and structure turnover (CAST) model for topsoils. Geoderma 211–212:51–64. https://doi.org/10.1016/j.geoderma.2013.06.014
35. Sullivan PL, Billings SA, Hirmas D, Li L, Zhang X, Ziegler S et al (2022) Embracing the dynamic nature of soil structure: A paradigm illuminating the role of life in critical zones of the Anthropocene. Earth-Sci Rev 225:103873, ISSN 0012-8252. https://doi.org/10.1016/j.earscirev.2021.103873
36. Udom BE, Nuga BO, Adesodun JK (2016) Water-stable aggregates and aggregate-associated organic carbon and nitrogen after three annual applications of poultry manure and spent mushroom wastes. Appl Soil Ecol 101:5–10. https://doi.org/10.1016/j.apsoil.2016.01.007
37. Wagai R, Mayer LM, Kitayama K, Knicker H (2008) Climate and parent material controls on organic matter storage in surface soils: three-pool density-separation approach. Geoderma 147:23–33
38. Wezel A, Casagrande M, Celette F et al (2014) Agroecological practices for sustainable agriculture. A review. Agron Sustain Dev 34:1–20. https://doi.org/10.1007/s13593-013-0180-7

Microbial Patterns and Processes in the Critical Zone

Emma L. Aronson, H. Linton Freund, and Mia R. Maltz

1 Microbial Habitats

Microorganisms are found in all environments, including the whole range of the critical zone (CZ), from deep subsurface rock fissures to the surface of leaves of the tallest trees. Microorganisms play a range of vital roles in the CZ, including facilitating nutrient cycles and decomposition. A current "grand challenge" for critical zone research is to "resolve the interactions and feedbacks between biota and Earth materials that give rise to critical zone structure and its life-supporting functions, as well as how the structure and these functions change with time, climate, and rock type" [36]. To that end, in this chapter, we will explore the range of diversity, composition, and function of microorganisms in the critical zone. The main focus of this chapter will be the soil, as it is the most dominant, diverse, and highly studied habitat of the critical zone. Litter on the surface of the soil is the typical habitat for many microbial decomposers, which will be examined to a lesser extent here. Additionally, wind erosion of soil surfaces introduces microbes into the atmosphere, creating a distinct microbial community that we will briefly describe. Finally, we will touch on the subsurface environment at the interface of soil formation and hydrology, which comprises an understudied set of microbial habitats, and is only starting to be explored.

From a microbial perspective, even a continuous forest or large grassland contains a wide range of **microhabitats** within the critical zone; these microhabitats promote the development and survival of immense microbial diversity. The

E. L. Aronson (✉) · H. L. Freund · M. R. Maltz
Department of Microbiology and Plant Pathology, University of California, Riverside, CA, USA
e-mail: emmaa@ucr.edu

M. R. Maltz
Department of Plant Sciences and Landscape Architecture, University of Connecticut, Storrs, CT, USA

T. White and A. Provenzale (eds.), *Critical Zone and Ecosystem Dynamics*, Advances in Critical Zone Science, https://doi.org/10.1007/978-3-031-69076-1_9

leaves of trees host a range of **photosynthetic** and **chemolithotrophic** microorganisms. Photosynthetic microorganisms derive their energy from the Sun, whereas chemolithotrophic microorganisms derive energy by facilitating chemical reactions of inorganic molecules, like minerals. The bark of trees and the rocks exposed on the ground host mosses, liverworts [120], and other primitive plants, as well as **lichens**, which are symbiotic associations between a fungus and a photosynthetic partner (e.g., algae or cyanobacteria), living among fungal filaments in mutualistic assemblages [99]. The litter that accumulates at the soil surface is host to a range of **decomposers**, a diverse group of bacteria and fungi that breakdown C and other nutrients in the system [100]. Climatic factors and litter chemistry interact to influence decomposition rates [5], as decomposers mineralize essential elements and form soil organic matter. These processes maintain primary productivity while replenishing pools of nutrients available for plant uptake [135].

1.1 Biogeography Drives Microbial Assembly

Biogeography is the study of the distribution of biodiversity over space and time [22, 23, 40], with the combined goals of understanding what organisms are found where, at what abundance, and the mechanisms and history associated with observed trends [41, 92]. Biogeography offers an important perspective on the mechanisms that impact diversity, including speciation, extinction, dispersal, and species interactions [60, 97]. While scientists have been studying the biogeography of larger organisms, including plants and animals, for hundreds of years, the study of microbial biogeography is in its infancy [49, 56, 58]. It is only through the recent biogeography is in its infancy [49, 56, 58]. It is only through the recent development of relatively fast and low-cost methods for amplicon and **metagenomic sequencing** of microbial DNA within the past few decades that we have been able to begin to unravel the patterns of microbial distributions in space and time. Metagenomic sequencing differs from amplicon sequencing in that it involves breaking DNA into small pieces, sequencing all of those pieces, and then using **bioinformatics** to attempt to assemble the sequenced DNA into operons, genes, or full **genomes**.

All studies of microbial biogeography address one or more of the four crucial processes that determine community assembly. These processes, credited as Vellend's Four Processes for Community Ecology, are diversification—generation of new genetic variation, dispersal—movement across space, selection—fitness-based changes in community structure, and drift—random or stochastic variation in relative abundances of populations in the community over time [56, 149],). For microorganisms, there are a few key differences that complicate applying these processes in the same way as we do with larger organisms [110]. In particular, mechanisms like passive dispersal are more important for microorganisms compared to larger organisms [60]. Many microbes have evolved to take advantage of wind, water, or host-assisted dispersal, while others are **cosmopolitan** or global in their distributions [80]. In addition, many microorganisms can remain dormant for days, months, and even at decadal to millennial time scales, then, along with subsequent hospitable conditions, these microorganisms return to a vegetative state [105], which is rare or impossible for larger organisms. In addition, microbes can readily incorporate DNA from the environment, viruses, and other microorganisms through HGT. HGT allows for microbes to adapt to their environments, and thus plays an important role in how selective pressures structure microbial communities [29, 110]. Furthermore, we know that environmental microbial communities include dormant microorganisms that may have drifted in from far away, as well as deceased or decayed organisms that have left behind copious **relic DNA**. Considering that microbial DNA needs to be sequenced to study a given microbial community, the introduction of foreign living or dead organisms can complicate our interpretation of the community patterns we observe in the sequenced samples. These confounding factors, coupled with key differences between microorganisms and larger organisms, complicate our understanding of microbial communities across the critical zone.

2 Soil is the Main Microbial Habitat in the CZ

The soil itself has many microhabitats, starting at the top with the organic horizon that exists in many different soil profiles. The organic horizon is home to **humus**, which is the partially decomposed, slightly acidic, brown or black layer at the top of the soil column. This layer includes many diverse, small, often unidentifiable, decomposed pieces of the plants, animals, and microbes in the ecosystem, as well as a range of chemical components. The chemical and physical variability of humus allows it to host a highly diverse variety of decomposers and chemolithotrophs who use the various chemical components [140].

Across the soil column, plant roots, microbial biomass, and fungal necromass [142] contribute to soil organic carbon (C) pools [37] and provide substrates, residues, or amino acids that prime decomposition and enzymatic processes. The diverse composition of mineral and organic sources for microbial energy and C sources throughout the soil column drives the assembly of microbial communities across different depths and in different ecosystems. Microbial composition and processes near the soil surface may differ from those in deep subsoils. An array of abiotic factors throughout the soil column select for the survival of specific microbial taxa. Some microbial taxa may be more abundant in deeper subsoils and more rare in surface horizons. In a steppe ecosystem within inner Mongolia, Zhao et al. [161] found that the abundances of certain bacterial and fungal taxa (i.e., Actinobacteria and Ascomycetes positively) correlate with increasing depth or soil organic carbon. Likewise, in our cross-Critical Zone Observatory research, a subset of soil microbial taxa consistently increased in relative abundance with depth in the soil profile (i.e., Chloroflexi, Nitrospirae, Euryarchaeota, and candidate phyla GAL15 and Dormibacteraeota); ostensibly, because of life history strategies or genomic features that optimized their ability to persist in deep, low nutrient environments [28] (Fig. 1). Although some taxa with beneficial traits or novel adaptations to oligotrophic conditions may increase in abundance with soil depth, a majority of bacterial, archaeal, and fungal taxa tend to decrease, in both diversity and abundance, with depth (e.g., taxa from Acidobacteria, Ascomycetes, and Basidiomycetes; [161]). At deeper soil depths, fewer taxa in our data set had matches to publicly available genomes, indicating that the bacterial and archaeal taxa found in deeper soil horizons are less represented in genomic databases than those found in surface soils. Patterns related to how microbial taxa change with depth may be ecologically significant or elucidate strategies for co-existence or partitioning **niche** space. In a microbial context, a niche is the combination of chemical and biological factors that determine where microorganisms can live.

At the bottom of the soil profile, the interface with the bedrock and subsurface hydrology can be very variable and complex. Microbial processes at depth, including

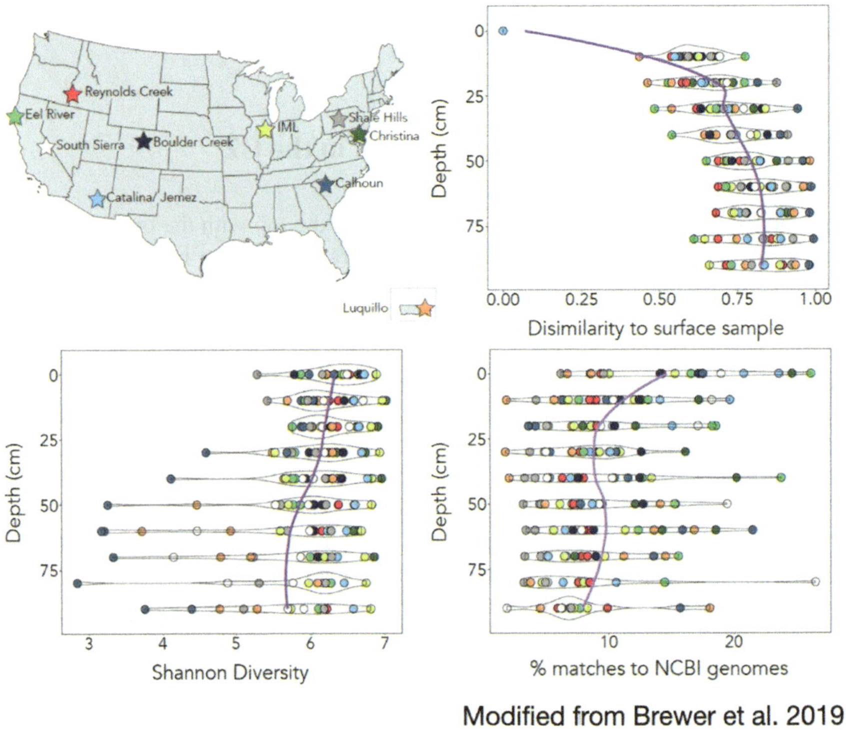

Fig. 1 Reproduced with permission from [28]. **a** Site map of sampling locations. We analyzed bacterial and archaeal communities from two soil profiles located at each of 10 different CZOs across the United States, at 10-cm intervals down to 1 m in depth (where possible). **b** Bray–Curtis dissimilarity to surface samples increases with depth. **c** Bacterial and archaeal diversity generally decreases with depth. Colors of points match the colors of the CZO sites indicated in panel A with two profiles sampled per site (n_20). **d** At deeper soil depths, we found that fewer taxa in our data set had matches to publicly available genomes. The purple trend lines represent smoothed conditional means, generated using the loess modeling method

microbial production of organic acids (OAs), may facilitate geobiological weathering, liberating minerals and nutrients from parent material. For instance, plant-supporting mycorrhizal fungal production of OAs may facilitate mineral dissolution and resource acquisition for plants, and therefore significantly influence plant productivity, soil microbial nutrient cycling, and C storage in forested ecosystems [125, 146]. Bacterial metabolic properties and biofilm production are essential in weathering, mineral formation, and in generating components of sedimentary rocks, including carbonates, bedded silicates, and phosphates [126]. Biomolecules from heterotrophic microbes, such as bacterially derived amino acids, also contribute substantially to biogeochemical cycling [64] in the litter layer and throughout the soil column. Interactions among biotic organisms and parent material (e.g., geologic materials) vary across ecosystems, often as a function of climate, topography, and time [12, 127, 157]; these state factors influence carbon density [116], litter decomposition, and soil formation.

Soil depth, degree of weathering, and diagnostic features used to classify soil taxonomic orders are correlated with both soil microbial community composition [13] and extracellular enzyme activity [44]. This enzymatic reservoir is related to microbial biomass and deep soil carbon, as illustrated in our work across Critical Zone Observatories. Microbial abundance and biomass allocation patterns may inform our ability to understand soil C storage and whole-profile enzyme dynamics. Despite low microbial abundance in deep soils, these microhabitats can be diverse in composition, function, and distribution.

3 Microbial Niches Are Shaped by Environmental Gradients

Microorganisms inhabit a range of **niches** within each ecosystem. Examples of environmental gradients that shape microbial niches across habitats are pH, temperatures (both averages and extremes), salinity, nutrient availability and access, and complex biological interactions between organisms that make up **trophic webs** [67, 72, 77, 155, 158]. While microbial communities are found in a range of systems beyond the soil environment, e.g., aquatic systems, these systems are beyond the scope of this chapter. However, many of the same niches, and the gradients that intersect to form these niches, structure microbial communities and functions across these diverse systems.

Across soils at a continental to global scale, it has been shown that the primary driver of phylogenetic patterns in bacteria is soil pH [82] (Fig. 2). The pH of a soil, or any environment, is a measure of how acidic or basic the medium is. pH correlates with the **Oxidation–Reduction Potential (ORP), or redox potential**, which refers to the overall composition of different oxidizing and reducing substances in the medium. The concentrations of elements contributing to the redox potential of an environment is difficult to quantify, thus it is often given as an electric potential

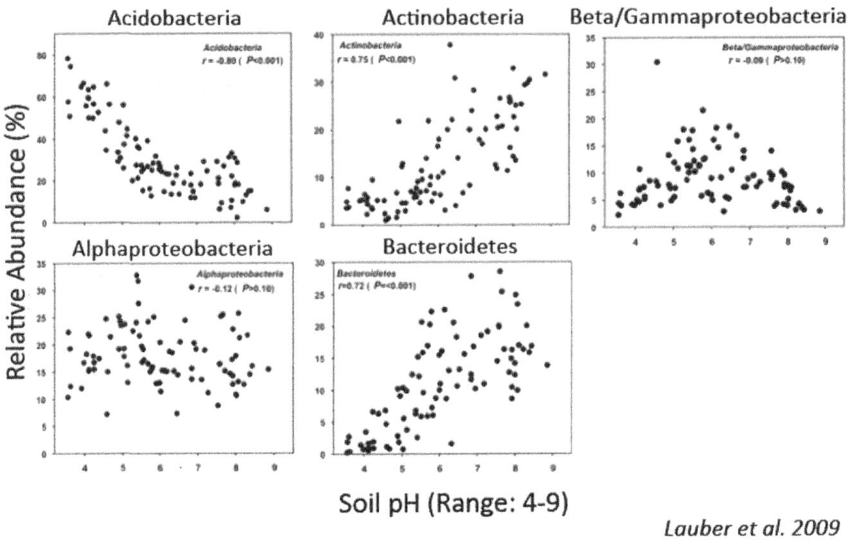

Fig. 2 Reproduced with permission from [82]. Relative abundances of dominant soil bacterial phyla and subphyla across 88 sequenced soils from North and South America

in millivolts, accounting for the oxygen, iron, and other oxidizing and reducing substances [63]. Microorganisms experience these redox conditions by interacting with these oxidizing and reducing substances, in much more complex ways than scientists can evaluate across a wide range of sites. However, we can use pH as one method to approximate this distribution of microbial niches across these geochemical gradients, particularly in soil. To a large extent, the suitability of a particular niche for a specific microorganism is defined by its pH and redox potential.

3.1 The Climate Influences Environmental Gradients

Although oxidation–reduction potential and pH of the local environment are very important in determining where a microorganism can survive, there are a plethora of natural gradients that define the niches of microorganisms [34]. Some of these gradients are regulated by the climate, such as temperature and moisture. Temperature is experienced by microorganisms on a microscopic scale, and is therefore determined by ambient average temperature and extremes, as well as temperature gradients in the soil column. Moisture is governed by rainfall amounts and patterns, as well as soil composition, higher clay particle distribution in soil can both block moisture from moving through the column due to their small size, and hold on to moisture for longer, as a result of their ionic composition [139]. Many of the features of variable soil environments result in gradients of substrate availability, and these components interact. For example, temperature essentially limits substrate availability because lower temperatures slow down chemical reactions, while low moisture is a limiting factor in dry conditions [145].

The response of microbial functional groups to temperature and moisture gradients may influence their abundance and dispersal [141]. Soil microbial activity slows or ceases at temperatures approaching biological zero (5 °C), with every incremental 10 °C increase, microbial processes and respiration rates may double until temperature optima is reached for that particular microbial group (e.g., 35 °C; [26]). Microbial abundance varies by temperature, with few taxa surviving at high temperatures; these taxa, including Ammonia-Oxidizing Archaea (AOA), may exert functional differences, such as increased AOA activity at higher temperatures [86]. Some plant growth promoting rhizobacteria (PGRPs), such as *Burkholderia*, can protect host plants through cold acclimation and temperature tolerance [50] or by altering **abscisic acid (ABA)** concentrations and osmotic potential under stressful environmental conditions, such as drought [87]. Microbes may produce compatible osmolytes, such as trehalose, which can function like antifreeze or engine coolant in a motor vehicle, it can modulate bacterial stress response, aiding microbes faced with either heat shock or osmotic stress and protecting cells from freezing, desiccation, thermal injury, or oxidative damage [28, 55].

Mycorrhizal fungi ameliorate stress responses, in turn, differently for host plants in **xeric**, or dry, systems than in more **mesic**, humid, environments. Because of their physical and biochemical properties, arbuscular mycorrhizal (AM) fungi are known for their abilities to access and transfer phosphorus (P) to their host. As P is required for photosynthesis, increases in P uptake may indirectly increase host plants' drought tolerance [6]. AM fungi may also directly ameliorate drought stress on host plants by altering root morphologies or stomatal conductance. Plants respire water through their stomata with guard cells open or close to allow water and gas exchange through stomata; stomatal closing is initiated by high ABA. At lower soil moisture levels, plants that are mycorrhizal may rely more heavily on mycorrhizal symbionts that forage for water in dry soil environments; these mycorrhizal plants also maintain higher ABA concentrations than may be found in nonmycorrhizal plants, ostensibly to control stomata and conserve plant water stores under dry conditions [45].

Habitat conditions, such as aerobic gradients, soil porosity, and horizons (e.g., organic vs. mineral soil layers), climatic factors (e.g., mean annual temperature), or topographic features and elevation gradients also play important roles in the distribution of root-associated or free-living microorganisms [76, 88]. Different types of mushrooms may fruit in the tropics, as compared to fungal **sporocarps** detected from either boreal forests or arid deserts. Along elevational and latitudinal gradients, climatic factors often vary clinally, which contributes to structuring both soil bacterial and fungal communities at large geographic scales [141]. With inundation and high soil moisture levels, AM and ectomycorrhizal fungal colonization of plant roots and mycorrhizal diversity tends to decrease [134, 147]. Conditions in wet forests may raise mycorrhizal root tips up towards, or even above, soil surfaces [17], seeking out optimal aerobic conditions in avoidance of anoxic zones [88]. These ectomycorrhizal mushroom-forming fungi play important roles in forests that receive heavy rain [138], fungal growth and basidiospore dispersal is not only stimulated by rain, but these fungi, in turn, also release large quantities of spores that influence water condensation in clouds and likewise enhance precipitation [62].

Water in the soil plays a vital role in connecting microorganisms to the chemicals they need for the oxidation and reduction reactions, as well as C and other nutrient sources that sustain them. While there are **macronutrients** that all life requires, including C, N, and phosphorus (P), most microorganisms require some combination of micronutrients, many of which act as electron acceptors or donors for redox reactions. The composition of microbial communities is strongly driven by edaphic factors, host plant abundance, and nutrient availability, specifically the C:N of surrounding vegetation [18, 35, 138]. Both bacterial and fungal growth respond to varying C:N concentrations in surrounding substrates or growth media [124], since plant species and litter inputs differ in their chemistry and structure, as plant composition changes, litter quality affects nutrient availability and microbial processes.

Water concentrations in the soil column further contribute to another important soil gradient, that of oxygen availability, of which **anoxia**, or a lack of oxygen, is one extreme (see Sect. 1.3.3 for more context). In modern microorganisms, there is a division between those that are intolerant of any oxygen and require anoxic conditions, and those that can tolerate full atmospheric oxygen, with microoxic microorganisms falling in between. In most cases, saturated soils, like wetlands and deeper soil layers, are home to anoxic and microoxic microorganisms. In addition, some soil microaggregates can hold onto small amounts of water and resist drying out completely for weeks or months, providing microhabitats for anoxic or microoxic bacteria to reside in drier soils [139].

3.2 PH Explains Soil Bacterial Abundances at Global Scales

A major breakthrough in the study of microbial **biogeography** (see Sect. 1.5) was the finding by [82] that across diverse and distant soils, pH was the best predictor of relative abundance trends in major soil bacterial phyla and subphyla. These authors performed microbial community analysis via **amplicon sequencing** of 88 soils sampled from a wide range of sites across North and South America. Amplicon sequencing is a form of DNA analysis that takes advantage of conserved genes across large phylogenetic groups. Typically, as was done in [82], regions of the 16S rRNA gene in bacteria are amplified using **Polymerase Chain Reaction** (PCR) and sequenced. All known members of the bacteria and archaea have these genes as they encode part of their ribosomes, and there are regions of their sequences that change very little from species to species. However, there are also variable regions within these ribosomal genes that can be used to identify and distinguish organisms at the phylum or class level, and, in some cases, down to their genus or species [130]. To do this, the sequenced amplicons of this gene are compared to the 16S rRNA gene of known isolates and previously described organisms in enormous public databases used for this purpose.

In the case of [82], they found that the distribution of *Acidobacteria, Actinobacteria*, and *Bacteroidetes* depended on the pH of the soil sample. In particular, *Acidobacteria* were found to be relatively more abundant in acidic soils, *Actinobacteria* and *Bacteroidetes* in alkaline soils, and *Beta-* and *Gammaproteobacteria* in neutral soils. In contrast, the relative abundance of *Alphaproteobacteria* were not structured by soil pH. These trends are demonstrated in Fig. 2, reproduced here with permission from [82].

3.3 Life is Influenced by the Oxygenic Atmosphere

When life first evolved on the earth around 3.5–3.9 bya, the atmosphere and the ocean were completely anoxic. When microorganisms first began performing oxygenic photosynthesis around 3 bya, oxygen reacted with exposed minerals and metals, and did not accumulate in the atmosphere [66, 71]. This type of photosynthesis was far more efficient energetically than previously evolved methods and it proliferated. After some tens or hundreds of millions of years, all reactive elements were oxidized, allowing O_2 to accumulate in the atmosphere and dissolve into the oceans [66]. This excess O_2 killed off most of the existing life at the time because microorganisms had not yet evolved antioxidant defenses, nor the ability to use oxygen for respiration or other processes [102]. After most life died off, the oxygen abated, and **anaerobic** life proliferated again [96]. This happened over and over, leaving red striations on sedimentary rocks from that period called **red beds** or **banded iron formations** [66, 117]. Over time, a great diversification of microbial life occurred. Some microorganisms evolved adaptations to the oxygen-rich atmosphere, including antioxidant defenses and uses for O_2 such as respiration [102]. Those with such adaptations are the ancestors of the incredible diversity of **aerobic** life today, including most eukaryotes. Other organisms found habitats and mechanisms that protected them from oxygen either partially, in the case of **facultative anaerobes** or **microaerophile** organisms that tolerate small concentrations of O_2, or completely, as in anaerobic organisms [106].

3.4 Other Gradients Shape Microbial Communities

Other gradients, like salinity and pressure, are less important in soils than in fresh and saltwater habitats. Pressure and salinity require special adaptations for microbial survival. Deep subsurface environments such as mines and geothermal brines (Sect. 1. 6) can have high levels of pressure, which is commonly seen in deep lakes and oceans. *Pseudothermotoga elfi* strain DSM9442, a bacterium previously isolated from a deep oil well, has been shown to shift its cellular morphology under high pressure, forming chains that exhibit greater viability under increasing hydrostatic pressure compared to single cells [123]. Salinity can be high in coastal and playa soils or those fed by salty water sources. Most microbes surviving in salt conditions demonstrate **osmotoleranace** for varying durations, which is likely spread through selective pressure on their genomic architecture or by sharing genes among bacterial taxa through horizontal gene transfer (HGT; [7]). Though salinity and pressure are not as influential as drivers of microbial diversification in soils as they are in some aquatic systems, the soil-aquatic interface is a set of microhabitats that is highly diverse. Further, it is critical to understand the drivers of microbial communities across non-soil systems so we can better describe soil environments, which are thought to be less extreme.

Microorganisms can only survive within a specific range of the environmental gradients described above. Those that survive in harsh environments which may threaten other organisms are known as **extremophiles**. Extremophiles are categorized by a variety of conditions that may promote their optimal growth. Thermophiles and hyperthermophiles survive best at over 50 °C and 80 °C, respectively, while psychrophiles can grow in temperatures below 15 °C [16, 91]. Halophiles thrive in environments high in salt concentration, such as hypersaline lakes, saltpans, and underground brines [148, 151, 153]. Barophiles, also known as peizophiles, persist in environments with greater than 0.1 MPa and are often found in deep sea environments with high hydrostatic pressure [19].

Many extremophiles are considered polyextremophiles due to their ability to withstand harsh, co-occurring conditions. For example, members of the *Sulfolobus* genus have been isolated from acidic hot springs from Yellowstone National Park, in the United States, to Italy, and are thus deemed thermoacidophiles [84]. Fungi in geothermal ecosystems enhance the thermotolerance of plants [31, 162] as well as exhibit traits, like the production of compatible osmolytes [112], to promote their own survival in extreme environments, such as hydrothermal vents.). Psychropiezophiles, organisms that can withstand low temperatures and high hydrostatic pressure, have been found between 1,000 m and 11,000 m below sea level [75, 113].

The ability for extremophiles to maintain homeostasis while surviving at the boundaries of these gradients gives us insight into the requirements of life on other planets. Microbial diversity of Spotted Lake in British Columbia was explored due to its concentration in magnesium sulfate salts, which are widely distributed in the sediments on Mars [115]. The concentration of magnesium sulfate salts in Spotted Lake allows this hypersaline system to serve as an analog to the ancient saline lakes found on Mars [115]. Research exploring metabolic and resistance adaptations of extremophiles within the critical zone will continue to shed light on the strategies and conditions that support life on this planet and beyond.

4 Soil Aggregation Creates Micro-Scale Habitats

Biological processes dominate the formation of soil aggregates (Fig. 3), which is driven by plant roots and soil microorganisms. Primarily, growing roots, bacterial biofilms, and hyphae in soils, as well as the exudates they release, entrain soil particles as they extend. Decomposer fungi and bacteria grow on plant material as they break down plant structural components. Filamentous decomposer fungi, which grow thread-like structures called hyphae, can stabilize soil aggregates, increase water filtration, and reduce erosion [32]. More fungal hyphae leads to more physical stabilization of soil aggregates [103]. Densely growing fungi, such as some ascomycetes that may be inefficient at breaking apart soil aggregates (i.e., taxa with low leucine aminopeptidase activity) contribute substantially to soil aggregation [83]. **Arbuscular mycorrhizal** (AM) fungal hyphae and ectomycorrhizal (ECM) fungi, which form mutualisms with many plants, enmesh fine particles into macroaggregates as they seek nutrients and water throughout the soil column [9]. Some microorganisms, especially symbiotic root-associated mycorrhizal fungi, release sticky glycoproteins, such as glomalin. Glomalin and other glycoproteins are hydrophobic, insoluble in water, resistant to heat, and hold aggregates together. Glomalin in soils has been shown to form approximately 10% of the stable organic C that soils hold long-term [131, 143]. The physical and biochemical properties of AM fungi and glomalin, in turn, reduce soil erosion [95]. Further, it has been shown that increases in glomalin production within many ecosystems occur under elevated CO_2 conditions [119]. ECM grow more abundantly at elevated CO_2 conditions, which could correspond with increased C storage from ECM deposits, along with ericoid mycorrhizal residues or other microbially-derived C compounds [142]. This stored C may offset greenhouse gas emissions, as these residues, including polysaccharides, chitin, glucans, peptidoglycans, or other remnants of microbial cell walls, may form recalcitrant materials that are not easily respired into CO_2 and may persist long-term in the soil [114]. Finally, many soil fauna generate cohesive pellets and casts as they bioturbate the soil, which can form components of aggregates as well [38].

5 Rock Weathering Creates Regolith

For hundreds of years, researchers have been trying to quantify and better understand the role of life, in terms of microorganisms, plants, and fauna, in soil development at the rock **weathering front** [33, 137, 144, 150]. However, solid insights are elusive, and this is an active area of inquiry [21, 27, 51, 90, 107]. **Weathering** of rock is considered the combination of processes that convert parent material to **regolith**. Often, the contribution of the weathering of rock to ecosystem functioning cannot be measured directly, but inferred based on calculated inputs and stream flow outputs. Regolith is often defined based on field sampling activities; the lower limit of regolith is the point of refusal for regular coring or digging [27]. Similarly, once a soil pit face

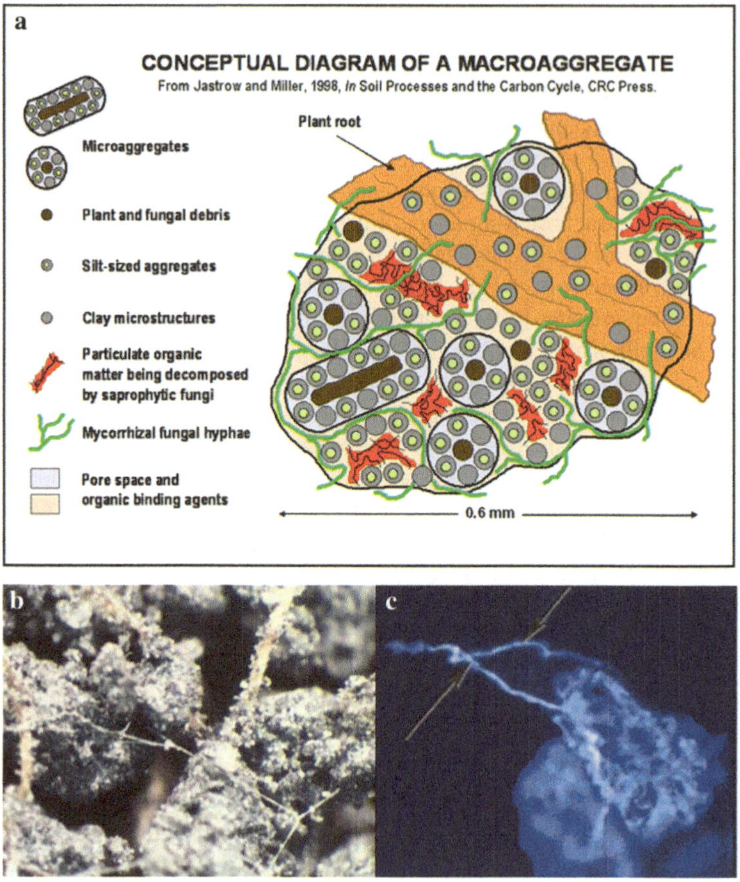

Fig. 3 Soil aggregates **a** Diagram reproduced from [73] illustrating the biological and abiotic components of soil aggregates; **b** Image by Michael Allen of soil aggregates; **c** Image by Michael Allen of stained mycorrhizal hyphae in a soil aggregate

is cleared, the upper limit of regolith is determined during the soil characterization. However, Brantley and Lebedeva [27] notes that there may be chemical differences between soil, regolith, and rock that are not as physically apparent. A wide variety of chemolithotrophic microorganisms assist in the conversion of rock to soil within the regolith. Mycorrhizal fungal hyphae may extend beyond the deep rhizosphere, tunneling into solid rock and exuding organic compounds that weather regolith and increase surface area [129]. In soils derived from rocks high in C, nitrogen (N), phosphorus (P), and other important macronutrients, it is likely that there are more diverse and abundant chemolithotrophs in the regolith and at the bedrock interface than where the bedrock is lower in these nutrients. Likewise, microorganisms may produce more organic acids in sites underlain by regolith containing limiting nutrients, resources, or trace minerals [90]. If regolith types with different macronutrient concentrations also house different microbial communities, then these factors may interact to alter rates of microbial geobiological weathering [46]. As these processes increase the bioavailability of limited nutrient resources, this may, in turn, mediate plant responses to global climate change. Overall, this frontier of research deserves further study [68].

6 Microorganisms Can Colonize Deep Subsurface Habitats

The location of **subsurface** zones is variable, depending on the habitat and climate. In Antarctic dry valleys, extremely cold temperatures prevent microorganisms from growing freely on the surface of rocks. However, there is a narrow subsurface zone of rock that is inhabited by unique lichens, which grow between the crystals of porous rocks, performing biogenic weathering of these rocks in-situ [53]. Microorganisms in the deep continental subsurface have been found in aquifers in basaltic and granitic rock. Many of these taxa function solely by H_2-driven chemolithoautotrophy [11], with H_2-consuming methanogenic archaea dominating these samples. Subterranean systems, such as caves, can house self-contained microbial trophic webs based entirely on chemolithotrophy. Caves are oligotrophic ecosystems with very low total C; yet, they host surprisingly diverse bacterial and (sometimes) archaeal communities [163]. Nitrification, denitrification, and sulfur transformations dominate these subsurface communities. Further, recent evidence points to cave evolution being driven by microbial weathering processes, in an analogous fashion to regolith and soil formation from rock at the bottom of the soil column [163].

While much of our discussion has centered on the soil as the primary habitat for microorganisms in the critical zone, deep subsurface microbial habitats are at least as interesting, if likely more sparsely populated by microorganisms and understudied at this time. In the subsurface, groundwater, geothermal, pressure, and geochemical gradients may be dominant in determining microbial composition and functioning. Much of the microbiology research that has been conducted on subsurface environments is connected to oil and gas extraction, as well as more recent work in relation to prospecting for geoengineering activities aimed at preventing climate change. Early work found thermophilic hydrogen-oxidizing bacteria, hyperthermophilic sulfidogenic archaea, and metal-reducing bacteria in hydrocarbon samples from subsurface sediments [133]. Further, analysis of extracted petroleum resources reveals that much of this substance is partially biodegraded, indicating strong microbial activities at depth [122]. Early culturing studies of extracted petroleum showed barophiles (Shewanella, Moritella spp.) dominated at ambient subsurface pressure conditions, while Pseudomonas-like sequences were prevalent at atmospheric pressure [156]. Microorganisms persisting in high pressure conditions have been shown to uptake larger concentrations of solutes (i.e., piezolytes) that help to stabilize proteins despite changes in external pressure [70]. Viable yeast cells have been detected deep in the continental crust [48], and the deep fossil record suggests that fungal metabolic activities may be facilitated by their symbiotic associations, as in the provision of organic compounds by chemoautotrophs in the deep oceanic crust [25].

While there has been limited research into the microbial communities of aquifers, the work that has been done shows a clear distinction between attached and suspended communities in these systems [52]. Interestingly, [52] showed that iron-reducing bacteria dominated in attached communities, while archaeal anaerobic methane oxidizers dominated in the suspended communities. They further showed that the chemical composition of groundwater greatly influenced the dominant microbial species. In sulfate-rich groundwater, sulfate-reducing *Desulfobacter* and *Desulfobulbus* were highly abundant, while in iron-rich waters, *Geobacter*, which are ferric-iron reducers, were most common. Finally, in low sulfate waters, the methanogenic archaea *Methanosarcina* and *Methanosaeta* were most common, and methane concentrations were high as a result [52].

7 The Atmosphere is a Poorly Understood Microbial Habitat

Environmental gradients also govern the microbial ecology in the atmosphere, specifically the aeolian (i.e., windblown) dust microbiome. Dust proliferation from the critical zone introduces microbes into the atmosphere, where they can persist attached to dust particulates or disperse to new ecosystems. Exogenous inputs into ecosystems from dust may serve as a source of microbial diversity and a resource subsidy to nutrient-poor systems. The dust microbiome is a unique community in that these microbes are exposed to multiple hostile and physical stressors, notably UV radiation, thermal and pressure changes, desiccation, wind shear stress, and nutrient scarcity. However, previous work has found that dust microbes can travel thousands of kilometers and remain viable in transit [93, 101]. Microbes can attach to dust particulates during transport, using them for physical protection against wind shear stress, as well as for nutrition [128, 136]. Dust microbes are equipped with a distinct set of traits that allow them to persist in such harsh conditions. For example, microbial extracellular structures such as lipopolysaccharides and exopolysaccharides adsorb and retain water on their cell surfaces, preventing desiccation and contributing to atmospheric processes such as cloud formation [43]. Metagenomes collected from aeolian dust over the Red Sea indicated that these microbes contain genes involved in repairing UV-damaged DNA, sporulation (i.e., the ability to form protective endospores), aerotaxis (i.e., movement due to increasing or decreasing oxygen concentration), and biofilm development [1].

Landscape topography and seasonal dynamics, as well as human activities, may influence the presence and abundance of microbes entrained into air currents and deposited into recipient ecosystems [94]. For instance, the variety of airborne microorganisms has been shown to correlate with dust chemistry, and the number of different microorganisms (i.e., **species richness**) from deposited dust varies with elevation. Collectively, the ongoing research into understanding dust microbes and their communities suggest that the geochemical gradients governing the atmosphere also select for the structural and functional diversity of the dust microbiome.

BOX 1—Global Dust Contributes Nutrients and Microorganisms to the California Sierra Nevada

Aeolian deposits laden with dust-derived microorganisms may have transformative effects on ecosystems far from where they originated, especially in highly weathered ecosystems constrained by local nutrient inputs. Within the Sierra Nevada of California, dry-season dust microbiomes differed by elevation, which corresponded to competing droughts in both Asian and San Joaquin Valley dust source areas [2, 94]. In fact, dust complexity and 16S rRNA gene OTU microbial taxa richness varied by site along the elevation gradient (Fig. 4). A collaborative team used isotope mixing models, calculated from neodymium (Nd) and strontium (Sr), and used Sr isotopic concentrations ($^{87}Sr/^{86}Sr$) to estimate dust heterogeneity, which was related to the amount of Asian dust in the sample. Dust heterogeneity peaked at the highest elevations within the Sierra Nevada. Although the dust deposited at higher elevations was more heterogeneous, it harbored fewer microbial taxa than less heterogenous dust collected at lower elevations in this montane ecosystem. Ostensibly, dust from local sources (i.e., California's San Joaquin Valley) harbored a greater variety of microbes than was found in deposited dust that traveled further distances to the Sierra Nevada of California and originated from the Gobi or Taklamakan Deserts in Asia [3]. Along the arduous transoceanic journey across the jet stream, this Asian dust may have jettisoned microorganisms through wet deposition, resulting in lower microbial taxa richness found in more heterogeneous dust. Although the drought events in dust source locations may have played an important role in long-distance microbial dispersal, interactions among temporal factors, weather patterns, microbial traits, and the landscape topography of deposition zones also determined which microorganisms persist in dust microbiomes following long-distance dispersal trajectories.

8 Microbial Metabolic Strategies and Biogeochemical Cycling

Microorganisms have evolved a range of metabolic strategies to gain energy and C from mediating chemical reactions. Overall, these can be categorized into a few main bins: **Heterotrophy**, a broad category, is when their energy is gained from consuming other organisms, either as a predator, parasite, or decomposer. Unlike heterotrophy, **autotrophy** is when microorganisms generate energy without any reliance on other organisms and utilize CO_2 as their C source. These categories can be further broken down into their respective energy sources. **Chemoorganotrophy** is a subset of heterotrophy where energy and C is specifically gained from organic molecule degradation, as in decomposition. Autotrophs can utilize

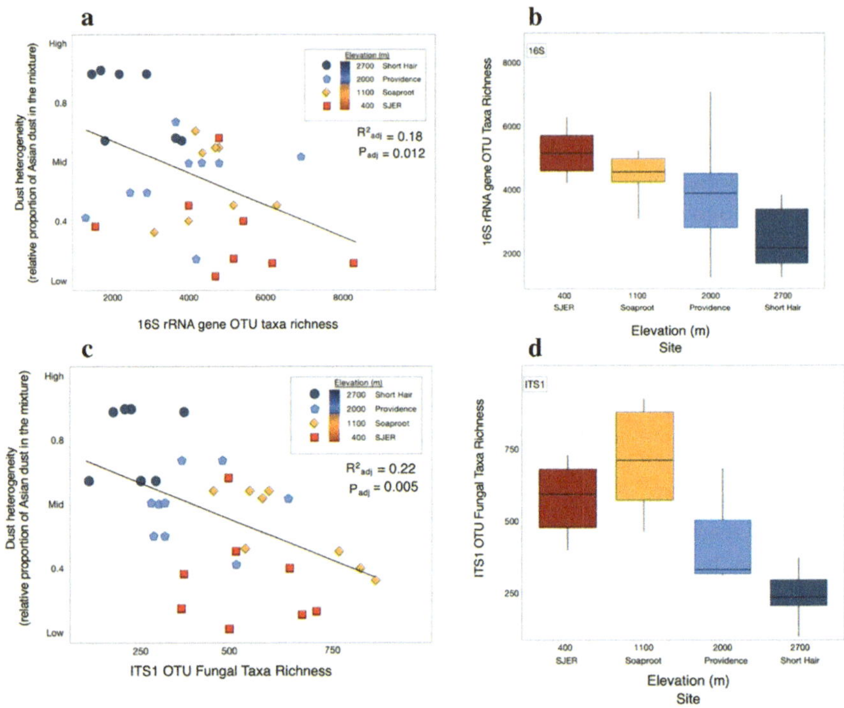

Fig. 4 Reproduced with permission from [94]. **a** Dust heterogeneity was correlated with 16S bacterial/archaeal OTU taxa richness; **b** 16S bacterial/archaeal OTU taxa richness by site along the elevation gradient; **c** Dust heterogeneity was correlated with ITS1 fungal OTU taxa richness; and **d** ITS1 fungal OTU taxa richness by site along the elevation gradient

phototrophy, where the energy is gained from the Sun, or **chemolithotrophy**, where the energy is gained by mediating reactions using inorganic molecules. Though most chemolithotrophs are autotrophs, there are a few chemolithoheterotrophs, which utilize inorganic compounds as energy sources and organic compounds as C sources. Additionally, some microorganisms can employ multiple metabolic strategies depending on their surroundings and are thus known as **mixotrophs** [69]. *Arcobacter anaerophilus* IR-1 is a known mixotrophic strain, containing the genetic machinery to perform chemoorganoheterotrophy, chemolithoheterotrophy, and chemolithoautotrophy [121]. Collectively, autotrophs and heterotrophs are responsible for C sequestration in soil ecosystems, yet respond very differently to abiotic changes in their surroundings [104].

Edaphic factors that vary with depth influence the assembly and metabolic strategies of the soil microbiome at depth. Diffusivity decreases with depth, preventing oxygen from accumulating below the soil surface and thus influencing where aerobic and anaerobic microbes reside in the soil column [61]. Phototrophy and chemoorganotrophy, along with aerobic respiration, are the dominant strategies towards the soil surface. Throughout most of the deeper soil and subsurface, chemolithotrophy is more dominant, and variable with in terms of the inorganic molecules used to gain energy, as well as by the different C sources available. Reduced soil diffusivity and porosity, coupled with decreased microbial abundance with increasing depth [28], allows for C to accumulate in deep soils [108]. C sequestration and the absence of oxygen deep in soil column selects for chemolithotrophs and anaerobes to drive nutrient cycling at these depths, using a variety of strategies that include methanogenesis, sulfate reduction, and iron reduction [108].

8.1 Microbial Energetics Drives Biogeochemistry

Most energy-yielding reactions that microorganisms mediate involve the transfer of electrons from a reduced **electron donor** to an oxidized **electron acceptor.** In aerobic respiration, the transfer of electrons occurs through a series of membrane-soluble carriers called an **electron transport system (ETS)**. There are 3 major classes of prokaryotic energy-acquiring processes using an ETS. Chemoorganotrophy involves organic electron donors, whereas chemolithotrophy involves inorganic electron donors. Phototrophy involves light captured by chlorophyll or similar molecules, usually coupled to splitting of H_2S or H_2O molecules. Oxidoreductases consist of multiple-protein complexes that include cytochromes as well as non-cytochrome proteins. **Cytochromes** are pigmented proteins whose absorbance spectrum shifts when there is a change in redox state.

Biogeochemical cycling encompasses the physical movement and chemical transformation of material by biochemical activities throughout the atmosphere, hydrosphere, and lithosphere. In particular, the microbially-relevant elements C, H, O, N, S, P, Fe, and Mo are often the target of studies on biogeochemical cycling in the critical zone [8, 78]. To better understand the microbial role in these cycles, it is important to recognize that the Earth is essentially a closed system, where these elements remain constant in amounts but change forms and states. Within this closed system, redox reactions are the source of all organic compounds, and the basis for all life [109]. Microbial contributions to biogeochemical cycling and the makeup of the Earth's atmosphere are great, due to their great numbers, ubiquity, diverse metabolic capabilities, and high rates of enzyme activity. In addition, the high surface to volume ratio of single-celled organisms leads to large gas exchange rates.

Fungal physical and biochemical properties may increase plant uptake of P, in particular. Mycorrhizal associations with plant roots not only lead to higher acid phosphatase activity, but the presence of fungal exudates and residues stimulate the expression of bacterial phosphatase genes [159]. These plant-microbial interactions and acquisition of limiting nutrients are inextricably linked with the activity and presence of these fungal symbionts, which play important roles in trace elemental, P, N, and C cycles.

Within the terrestrial biosphere, 75% of the fixed atmospheric CO_2 is deposited and stored within the soil and cycled through food webs [74]. In soil, CO_2 is released by microbial, animal, and tree root respiration, but various physico-chemical properties, along with the climate and weather of a particular system, can lead to differing rates of emissions, mediated by potential build-up of CO_2 belowground, as well as periodic release. Soil respiration, which is the rate of CO_2 release from soils, is one of the better understood biogeochemical fluxes.

BOX 2—Methane-Cycling Microorganisms Are Phylogenetically Diverse

As an example of an important set of microbially-mediated chemical reactions, we will explore the microbial methane (CH_4) cycle. Methane is one of the most potent human-released greenhouse gases and is also released geologically in many seeps and fissures worldwide. As a single carbon molecule, it is considered an organic molecule, even though its sources are often geological. In the atmosphere, each molecule of CH_4 may hold more than 25 times as much heat as each CO_2 molecule, but it is about 200 times less abundant in the atmosphere than CO_2 [85, 98, 152]. The main microbial groups that cycle this important gas are methanotrophs and methanogens. In microbiological nomenclature, "troph" often indicates consumption, while "gen" often indicates production.

Methanogens are a diverse, ancient group of methane-producing archaea, most of which are chemolithotrophic. Many of them are anaerobic and cannot live in the presence of oxygen, and are therefore common in saturated soils, wetlands, and aquatic systems. Methanogenesis is a dominant process in saturated soil. There are 4 main methanogenic reactions used by various methanogens:

$$
\begin{aligned}
&CO_2 \text{ and } H_2: \\
&CO_2 + 4H_2CH_4 + 2H_2O \\
&\text{Methanol and } H_2: \\
&CH_3OH + H_2CH_4 + H_2O \\
&\text{Methanol catabolism}: \\
&4CH_3OH_3CH_4 + HCO_3^- + H^+ + 2H_2O \\
&\text{Acetate oxidation}: \\
&CH_3CO_2^- + H^+CH_4 + 3H_2O
\end{aligned}
$$

Methanotrophs are chemoorganotrophs and have two main functional groups, those that are low capacity, but high-affinity, and those that are high capacity, low affinity [24]. The low capacity methanotrophs can perform methane oxidation at near-atmospheric concentrations, but at relatively low consumption rates. By contrast, the high capacity methanotrophs can consume methane at high rates, but only at higher concentrations. Most forms of methanotrophy require oxygen, and methanotrophs are common throughout well-drained soils, where oxygen, in the form of O_2, is abundant. Within wetland and aquatic systems, oxygen pockets and root oxygen leakage support methane oxidation. In addition, anaerobic oxidation of methane (AOM) by a consortium composed of anaerobic methanotrophic archaea (ANME) and sulfate-reducing bacteria (SRB), are able to use manganese ($Mn4^+$) and iron

($Fe3^{+}$) as electron acceptors. In addition, AOM is also performed by ANME through the reverse methanogenesis pathway [42].

In order to study methane production and consumption in soils, there are a range of methods available for in-depth analysis of microbially-mediated processes in the soil [15]. It is likely that most methanotrophs and methanogens are likely to be captured in data generated using 16S rRNA amplicon sequencing. In addition, amplicon sequencing can be used to explore the diversity of certain genes that are integral to the production and consumption of methane. For example, the *mcrA*, which encodes for enzyme Methyl Coenzyme M Reductase A, is a gene common to most forms of methanogens [54]. The *pmoA* gene, which encodes for the particulate methane oxidation A enzyme, is common to most methanotrophs [89]. These genes can be targeted both for diversity, using either amplicon sequencing and metagenomics, and abundance using quantitative PCR (qPCR). The data generated by qPCR shows the size of the methane-cycling population, and a technique called reverse-transcriptase qPCR can be used to quantify the activity of that population. These techniques can be used to target various genes to gain a deeper understanding of the microbial populations performing many different soil functions, and the activities of those populations.

Microbial contributions to biogeochemical cycling and the makeup of the Earth's atmosphere are great due to their large numbers, ubiquity, high surface to volume ratios, large gas exchange rates, diverse metabolic capabilities, and high rates of enzyme activity.

8.2 Decomposition is How Microorganisms Renew the World

Microorganisms play a range of vital roles in decomposition of leaf litter and all other dead organic material [20, 57]. Overall, a simple exponential decay curve equation can be used to describe the rates of decomposition of whole organic matter, which simplifies the contributions of multiple groups of microorganisms to the transformations that break down and release the nutrients and energy contained in the litter.

The primary variable that determines the rate of decomposition in an ecosystem is the **decomposition constant k**, in equations determining the amount of litter remaining after a certain amount of time, and affected primarily by climate and litter quality [160]. Litter quality refers to the amount of labile or available carbon and nutrients, compared to the amount of **recalcitrant** or difficult to break-down compounds it contains [160]. Recalcitrant compounds, such as lignin and cellulose can only be broken down by a limited set of microorganisms [111, 154]. Microbial metabolism of recalcitrant C compounds is determined by their capacity to produce specific litter decay enzymes [79]. Decomposition by heterotrophic microorganisms influences soil C-stocks and could have ramifications for atmospheric CO_2 concentrations and global change [10].

Lignin is a structural plant polymer, almost as abundant as **cellulose** in plants. Biosynthesis of lignin is not catalyzed by enzymes, but rather via spontaneous polymerization. The random nature of this sort of polymerization makes biodegradation extremely difficult. In plants, this polymerization occurs in tight association with cellulose and hemicellulose (i.e., holocelluloses). Holocelluloses are higher quality and more labile sources of organic C than lignin, which is considered a lower quality source of organic C in soil [65]. Unprocessed lignin is phytotoxic and is inedible to most herbivores and microbes [59]. Some filamentous microbes, such as white rot fungi and Actinomycete bacteria, decompose lignin within coarse woody debris, stabilize C in soil, and cycle nutrients throughout terrestrial ecosystems [30]. One important group of fungi, the brown rot fungi, incompletely processes lignin by depolymerizing lignocellulose via Fenton chemistry, completely decomposing holocellulose, and repolymerizing modified lignin into a stabilized form within the forest floor [14, 39]. Subsequently, brown rot fungi exude humic substances that condition soils and reduce lignin phytotoxicity [4, 118]. This brown rotted modified lignin is not only difficult for other microbes to break down but these brown rot humic residues contribute to long-term C-stabilization [47, 132]. Overall, lignin degradation happens slowly and is limited to aerobic environments. Most fungal lignin degraders (or depolymerizers) are in either the white- or brown rot groups within the Basidiomycetes. After white rot fungi degrade lignin, thready white cellulose fibers remain; likewise, brown rot fungi primarily degrade cellulose, leaving pithy lignin behind as cuboidal brown structures or fine brown powder. White and brown rot fungi work in concert, or sequentially, for more complete wood decomposition. Holocellulose decomposition is phylogenetically conserved within decomposer fungi, where closely related fungal taxa decompose cellulose at the same rate [81]. Global change is predicted to exacerbate environmental stress and increase CO_2 and other greenhouse gas emissions, therefore, research in the critical zone about organisms that influence woody debris decomposition, especially those that either serve as C sources or may sequester C below-ground may be particularly germane for determining the effects of these microbial groups on C budgets.

9 Conclusion

The critical zone is home to diverse and functionally vital microbial communities, which are structured by physicochemical properties, as well as nutrient availability and other gradients. In turn, microbial communities regulate biogeochemical cycling between pools in the critical zone and beyond, through multiple trophic strategies. Geochemical gradients across a variety of landscapes and niches within the critical zone select for a wide range of microbial metabolic processes within these landscapes, thus influencing the flux of nutrients throughout these ecosystems. In particular, the variation in available organic molecules and oxygen, as well as pH, structure microbial taxonomic and functional diversity within the critical zone. In conclusion, in order to understand interactions within the critical zone, we must consider the involvement of microbial communities and their mechanisms in these processes.

References

1. Aalismail NA, Ngugi DK, Díaz-Rúa R, Alam I, Cusack M, Duarte CM (2019) Functional metagenomic analysis of dust-associated microbiomes above the Red Sea. Sci Rep 9:1–12
2. Aarons SM, Arvin LJ, Aciego SM, Riebe CS, Johnson KR et al (2019) Competing droughts affect dust delivery to Sierra Nevada. Aeolian Res 41
3. Aciego SM, Riebe CS, Hart SC, Blakowski MA, Carey CJ et al (2017) Dust outpaces bedrock in nutrient supply to montane forest ecosystems. Nat Commun 8:14800
4. Adaskaveg JE, Blanchette RA, Gilbertson RL (1991) Decay of date palm wood by white-rot and brown-rot fungi. Can J Bot 69:615–629
5. Aerts R (1997) Climate, leaf litter chemistry and leaf litter decomposition in terrestrial ecosystems: a triangular relationship. Oikos 79:439–449
6. Ahmad Z, Waraich EA, ur Rehman MZ, Ayub MA, Usman M et al (2021) Foliar application of phosphorus enhances photosynthesis and biochemical characteristics of maize under drought stress. Phyton 90:503–514
7. Ahmed V, Verma MK, Gupta S, Mandhan V, Chauhan NS (2018) Metagenomic profiling of soil microbes to mine salt stress tolerance genes. Front Microbiol 9:1–11
8. Akob DM, Küsel K (2011) Where microorganisms meet rocks in the Earth's Critical Zone. Biogeosciences 8:3531–3543
9. Allen MF (2022) Mycorrhizal dynamics in ecological systems. Cambridge University Press
10. Allison SD (2006) Soil minerals and humic acids alter enzyme stability: implications for ecosystem processes. Biogeochemistry 81(3):361–373
11. Amend JP, Teske A (2005) Expanding frontiers in deep subsurface microbiology. Palaeogeogr Palaeoclimatol Palaeoecol 219(1):131–155
12. Amundson R, Jenny H (1997) On a state factor model of ecosystems. Bioscience 47(8):536–543
13. Arogyaswamy K (2020) The Influences of Geospatial Location and Depth on Soil Microbial Community Composition. Master's Thesis. University of California, Riverside
14. Arantes V, Goodell B (2010) Fungal attack on lignin and cellulose: Elucidation of brown- and white-rot mechanisms comparing biomimetic and in-vivo degradation patterns. IRG 41st Meeting. Biarritz, France
15. Aronson EL, Allison S, Helliker BR (2013) Environmental impacts on the diversity of methane-cycling microbes and their resultant function. Front Terr Microbiol. https://doi.org/10.3389/fmicb.2013.00225
16. Arora NK, Panosyan H (2019) Extremophiles: applications and roles in environmental sustainability. Environ Sustain 2:217–218
17. Baar J, Van Groenendael JM, Roelofs JGM (2000) Are ectomycorrhizal fungi associated with Alnus of importance for forest development in wet environments? Plant Biol 2:505–511
18. Bahram M, Hildebrand F, Forslund SK, Anderson JL, Soudzilovskaia NA et al (2018) Structure and function of the global topsoil microbiome. Nature 560(7717):233–237
19. Banerjee S, Pati S, Sengupta A, Shaheen S, Sarma J, Sulakshana PM, Tenguria S, Kumar A (2023) Adaptation strategies of piezophilic microbes. In: Kumar A, Tenguria S (eds) Bacterial survival in the hostile environment. Academic Press, pp 207–218. https://doi.org/10.1016/B978-0-323-91806-0.00015-1
20. Bani A, Pioli S, Ventura M, Panzacchi P, Borruso L, Tognetti R, Tonon G, Brusetti L (2018) The role of microbial community in the decomposition of leaf litter and deadwood. Appl Soil Ecol 126:75–84

21. Banwart SA, Nikolaidis NP, Zhu YG, Peacock CL, Sparks DL (2019) Soil functions: connecting earth's critical zone. Annu Rev Earth Planet Sci 47:333–359
22. Baas Becking LGM (1934) Geobiologie of inleiding Tot De milieukunde. W.P. Van Stockum and Zoon, The Hague
23. Beijerinck MW (1913) De infusies en de ontdekking der backteriën. Jaarboek van de Koninklijke Akademie voor Wetenschappen. Amsterdam, the Netherlands: Müller. (Reprinted in Verzamelde geschriften van M.W. Beijerinck, vijfde deel, pp 119–140
24. Bender M, Conrad R (1992) Kinetics of CH_4 oxidation in oxic soils exposed to ambient air or high CH_4 mixing ratios. FEMS Microbiol Ecol 10(4):261–269
25. Bengston S, Ivarsson M, Astolfo A, Belivanova V, Broman C, Marone F, Stampanoni M (2014) Deep-biosphere consortium of fungi and prokaryotes in Eocene subseafloor basalts. Geobiology 12(6):489–496
26. Brady NC, Weil RR (2004) Elements of the nature and properties of soils. Pearson Prentice Hall, New Jersey
27. Brantley SL, Lebedeva M (2011) Learning to read the chemistry of Regolith to understand the critical zone. Annu Rev Earth Planet Sci 39(1):387–416
28. Brewer TE, Aronson E, Arogyaswamy K, Billings SA, Botthoff JK, Campbell AN (2019) Ecological and genomic attributes of novel bacterial taxa that thrive in subsurface soil horizons. mBio 10:1–14
29. Brito IL (2021) Examining horizontal gene transfer in microbial communities. Nature Review Microbiology 19(7):442–453
30. Bugg TD, Ahmad M, Hardiman E, Rahmanpour R (2011) Pathways for degradation of lignin in bacteria and fungi. Nat Prod Rep 28(12):1883–1896
31. Bunn R, Lekberg Y, Zabinski C (2009) Arbuscular mycorrhizal fungi ameliorate temperature stress in thermophilic plants. Ecology 90(5):1378–1388
32. Caesar-ThonThat TC, Cochran V (2000) Soil aggregate stabilization by a saprophytic lignin-decomposing basidiomycete fungus. Microbiological aspects. Biol Fertil Soils 32:374–380
33. Chadwick OA, Derry LA, Vitousek PM, Huebert BJ, Hedin LO (1999) Changing sources of nutrients during four million years of ecosystem development. Nature 397:491–497
34. Chen J, Hanke A, Tegetmeyer HE, Kattelmann I, Sharma R, Hamann E, Hargesheimer T, Kraft B, Lenk S, Geelhoed JS, Hettich RL, Strous M (2017) Impacts of chemical gradients on microbial community structure. ISME J 11:920–931
35. Chen YL, Ding JZ, Peng YF, Li F, Yang GB, Liu L, Qin SQ, Fang K, Yang YH (2016) Patterns and drivers of soil microbial communities in Tibetan alpine and global terrestrial ecosystems. J Biogeogr 43:2027–2039
36. Chorover J, Aronson EL, McIntosh J, Roden E (2020) Life teems below the surface. Eos 101
37. Clemmensen KE, Bahr A, Ovaskainen O, Dahlberg A, Ekblad A, Wallander H, Stenlid J, Finlay RD, Wardle DA, Lindahl B (2013) Roots and associated fungi drive long-term carbon sequestration in boreal forest. Science 339(6127):1615–1618
38. Coleman DC, Wall DH (2015) Soil fauna: occurrence, biodiversity, and roles in ecosystem function. In: Paul EA (ed) Soil microbiology, ecology and biochemistry, 4th edn. Academic, Boston, pp 111–149
39. Cooke RC, Rayner AD (1984) Ecology of saprotrophic fungi. Addison-Wesley, Masschushetts
40. Cox CB, Moore PD, Ladle RJ (2016) Biogeography: an ecological and evolutionary approach. John Wiley and Sons, Ltd.
41. Crisp MD, Cook TSA, LG, (2011) Hypothesis testing in biogeography. Trends Ecol Evol 26(2):66–72
42. Cui M, Anzhou M, Hongyan Q, Xuliang Z, Zhuang G (2015) Anaerobic oxidation of methane: an "active" microbial process. Microbiology Open 4(1):1–11
43. Delort AM, Vaïtilingom M, Amato P, Sancelme M, Parazols M, Mailhot G, Laj P, Deguillaume L (2010) A short overview of the microbial population in clouds: potential roles in atmospheric chemistry and nucleation processes. Atmos Res 98:249–260
44. Dove NC, Arogyaswamy K, Billings SA, Botthoff JK, Carey CJ, Cisco C, DeForest J, Fairbanks D, Fierer N, Gallery RE, Kaye JP, Lohse KA, Maltz MR, Mayorga E, Pett-Ridge J, Yang

WH, Hart SC, Aronson EL (2020) Continental-scale patterns of extracellular enzyme activity in the subsoil: an overlooked reservoir of microbial activity. Environ Res Lett 15:1040a1
45. Duan X, Neuman DS, Reiber JM, Green CD, Saxton AM, Augé RM (1996) Mycorrhizal influence on hydraulic and hormonal factors implicated in the control of stomatal conductance during drought. J Exp Bot 47(10):1541–1550
46. Dynarski KA, Morford SL, Mitchell SA, Houlton BZ (2019) Bedrock nitrogen weathering stimulates biological nitrogen fixation. Ecology 100:e02741
47. Eastwood DC (2011) The Plant cell wall-decomposing machinery underlies functional diversity of forest fungi. Science 333(6043):762–765
48. Ekendahl S, O'neill AH, Thomsson E, Pedersen K (2003) Characterisation of yeasts isolated from deep igneous rock aquifers of the Fennoscandian Shield. Microb Ecol 46:416–428
49. Fenchel T (2003) Biogeography for Bacteria. Science 301:925–926
50. Fernandez O, Theocharis A, Bordiec S, Feil R, Jacquens L, Clément C, Barka EA (2012) *Burkholderia phytofirmans* PsJN acclimates grapevine to cold by modulating carbohydrate metabolism. Mol Plant Microbe Interact 25:496–504
51. Field JP, Breshears DD, Law DJ, Villegas JC, Lopez-Hoffman L, Brooks PD, Chorover J et al (2015) Critical zone services: expanding context, constraints, and currency beyond ecosystem services. Vadose Zone J 14(1)
52. Flynn TM, Sanford RA, Ryu H, Bethke CM, Levine AD, Ashbolt NJ, Santo Domingo JW (2013) Functional microbial diversity explains groundwater chemistry in a pristine aquifer. BMC Microbiol 13:146
53. Friedmann EI (1982) Endolithic Microorganisms in the Antarctic Cold Desert. Science 215(4536)
54. Friedrich MW (2005) Methyl-coenzyme M reductase genes: Unique functional markers for methanogenic and anaerobic methane-oxidizing Archaea. Academic Press Inc
55. Fung T, Kwong N, van der Zwan T, Wu M (2013) Residual glycogen metabolism in *Escherichia coli* is specific to the limiting macronutrient and varies during stationary phase. J Exp Microbiol Immunol 17:83–87
56. Gillespie RG, Baldwin BG, Waters JM, Fraser CI, Roderick NR, GK, (2012) Long-distance dispersal: a framework for hypothesis testing. Trends Ecol Evol 27(1):47–56
57. Glassman SI, Weihe C, Li J, Albright MBN, Looby CI, Martiny AC, Treseder KK, Allison SD, Martiny JBH (2018) Decomposition responses to climate depend on microbial community composition. Proc Natl Acad Sci USA 115:11994–11999
58. Green JL, Bohannan BJ, Whitaker RJ (2008) Microbial biogeography: from taxonomy to traits. Science 320(5879):1039–1043
59. Griffiths GS, Boddy L (1991) Fungal decomposition of attached angiosperm twigs, III: effect of water potential and temperature on fungal growth, survival, and decay of wood. New Phytol 117(2):259–269
60. Hanson CA, Fuhrman JA, Horner-Devine MC, Martiny JB (2012) Beyond biogeographic patterns: processes shaping the microbial landscape. Nat Rev Microbiol 10(7):497–506
61. Hao J, Chai YN, Lopes LD, Ordóñez RA, Wright EE, Archontoulis S, Schachtman DP (2021) The effects of soil depth on the structure of microbial communities in agricultural soils in Iowa (United States). Microb Ecol 87(4)
62. Hassett MO, Fischer MW, Money NP (2015) Mushrooms as rainmakers: How spores act as nuclei for raindrops. PLoS ONE 10:e0140407
63. Hem JD (1985) Study and Interpretation of the Chemical Characteristics of Natural Water, Third Edition
64. Hobara S, Osono T, Hirose D, Noro K, Hirota M, Benner R (2014) The roles of microorganisms in litter decomposition and soil formation. Biogeochemistry 118:471–486
65. Hofrichter M (2010) Fungal biodegradation of lignocelluloses: Brown-Rot Fungi. In The Mycota: industrial applications, vol 10, p 231. Springer
66. Holland HD (2006) The oxygenation of the atmosphere and oceans. Philos Trans R Soc Lond Biol Sci 361(1470)

67. Hollister EB, Engledow AS, Hammett AJM, Provin TL, Wilkinson HH, Gentry TJ (2010) Shifts in microbial community structure along an ecological gradient of hypersaline soils and sediments. ISME J 4:829–838
68. Houlton BZ, Morford SL (2015) A new synthesis for terrestrial nitrogen inputs. SOIL 1:381–397
69. Huang L, Xing X, Zhou P, Li Puma G (2021) Mixotrophic bacteria for environmental detoxification of contaminated waste and wastewater. Appl Microbiol Biotechnol 105:6627–6648
70. Ichiye T (2018) Enzymes from piezophiles. Semin Cell Dev Biol 84:138–146
71. Jabłońska J, Tawfik DS (2021) The evolution of oxygen-utilizing enzymes suggests early biosphere oxygenation. Nature Ecology and Evolution 5:442–448
72. Jiang X (2015) The impact of temperature, pH and environmental heterogeneity on prokaryotic diversity in Yellowstone National Park thermal springs. https://digitalrepository.unm.edu/biol_etds/57
73. Jastrow JD, Miller RM (1997) Soil aggregate stabilization and carbon sequestration: feedbacks through organomineral associations. Argonne National Laboratory
74. Jobbágy EG, Jackson RB (2000) The vertical distribution of soil organic carbon and its relation to climate and vegetation. Ecol Appl 10:423–436
75. Kaur J, Vishnu AL, Khipla N, Kaur J (2022) Microbial life in cold regions of the deep sea. survival strategies in cold-adapted microorganisms. Springer, Singapore, pp 63–86
76. Khokon AM, Schneider D, Daniel R, Polle A (2021) Soil layers matter: Vertical stratification of root-associated fungal assemblages in temperate forests reveals differences in habitat colonization. Microorganisms 9:2131
77. Kimbrel JA, Ballor N, Wu YW, David MM, Hazen TC, Simmons BA, Singer SW, Jansson JK (2018) Microbial community structure and functional potential along a hypersaline gradient. Front Microbiol 9:1492
78. King EK, Thompson A, Hodges C, Pett-Ridge JC (2014) Towards Understanding Temporal and Spatial Patterns of Molybdenum in the Critical Zone. Procedia Earth Planet Sci 10:56–62
79. Kirk TK, Farrell RL (1987) Enzymatic Combustion - The Microbial-degradation of lignin. Annu Rev Microbiol 41:465–505
80. Kivlin SN, Hawkes CV, Treseder KK (2011) Global diversity and distribution of arbuscular mycorrhizal fungi. Soil Biol Biochem 43(11):2294–2303
81. Kivlin SN, Treseder KK (2015) Initial phylogenetic relatedness of saprotrophic fungal communities affects subsequent litter decomposition rates. Microb Ecol 69:748–757
82. Lauber CL, Hamady M, Knight R, Fierer N (2009) Pyrosequencing-based assessment of soil pH as a predictor of soil bacterial community structure at the continental scale. Appl Environ Microbiol 75(15):5111–5120
83. Lehmann A, Zheng W, Ryo M, Soutschek MK, Roy J, Rongstock R, Rillig MC (2020) Fungal traits important for soil aggregation. Front Microbiol 10:2904
84. Lewis AM, Recalde A, Bräsen C, Counts JA, Nussbaum P, Bost J, Schocke L, Shen L, Willard DJ, Quax TEF, Peeters E, Siebers B, Albers SV, Kelly RM (2021) The biology of thermoacidophilic archaea from the order Sulfolobales. Oxford University Press
85. Lelieveld J, Crutzen PJ, Dentener FJ (1998) Changing concentration, lifetime and climate forcing of atmospheric methane. Tellus Ser B-Chem Phys Meteorol
86. Li H, Yang Q, Li J, Gao H, Li P, Zhou H (2015) The impact of temperature on microbial diversity and AOA activity in the Tengchong Geothermal Field, China. Sci Rep 5(17056)
87. Lievens L, Pollier J, Goossens A, Beyaert R, Staal J (2017) Abscisic acid as pathogen effector and immune regulator. Front Plant Sci 8(587)
88. van der Linde S, Suz LM, Orme CDL, Cox F, Andreae H, Asi E, Atkinson B et al (2018) Environment and host as large-scale controls of ectomycorrhizal fungi. Nature 558(7709):243–248
89. Luesken FA, Zhu B, van Alen TA, Butler MK, Diaz MR, Song B, Op den Camp HJM, Jetten MSM, Ettwig KF (2011) PmoA primers for detection of anaerobic methanotrophs. Appl Environ Microbiol 77:3877–3880

90. Lybrand RA, Austin JC, Fedenko J, Gallery RE, Rooney E, Schroeder PA, Zaharescu DG, Qafoku O (2019) A coupled microscopy approach to assess the nano-landscape of weathering. Sci Rep 9(1):5377
91. De Maayer P, Anderson D, Cary C, Cowan DA (2014) Some like it cold: Understanding the survival strategies of psychrophiles. Nature Publishing Group
92. MacArthur RH, Wilson EO (1967) The theory of island biogeography, vol. 1. Princeton university press
93. Maki T, Bin C, Kai K, Kawai K, Fujita K, Ohara K, Kobayashi F, Davaanyam E, Noda J, Minamoto Y, Shi G, Hasegawa H, Iwasaka Y (2019) Vertical distributions of airborne microorganisms over Asian dust source region of Taklimakan and Gobi Desert. Atmos Environ 214:116848
94. Maltz MR, Carey CJ, Freund HL, Botthoff JK, Hart SC, Stajich JE et al (2022) Landscape Topography and Regional Drought Alters Dust Microbiomes in the Sierra Nevada of California. Front Microbiol 1248
95. Mardhiah U, Caruso T, Gurnell A, Rillig MC (2016) Arbuscular mycorrhizal fungal hyphae reduce soil erosion by surface water flow in a greenhouse experiment. Appl Soil Ecol 99:137–140
96. Martin WF, Sousa FL (2016) Early microbial evolution: The age of anaerobes. Cold Spring Harb Perspect Biol 8(2):a018127
97. Martiny J, Bohannan B, Brown JH, Colwell RK, Fuhrman JA, Green JL, Horner-Devine MC, Kane M, Krumins JA, Kuske CR, Morin PJ, Naeem S, Ovreås L, Reysenbach AL, Smith VH, Staley JT (2006) Microbial biogeography: putting microorganisms on the map. Nat Rev Microbiol 4:102–112
98. IPCC, Masson-Delmotte V, Zhai P, Pirani A, Connors SL, Péan C, Berger S, Caud N, Chen Y, Goldfarb L, Gomis MI, Huang M, Leitzell K, Lonnoy E, Matthews JBR, Maycock TK, Waterfield T Yelekçi O, Yu R, and Zhou B (eds.)]. IPCC, 2021: Summary for Policymakers. In: Climate change 2021: the physical science basis. Contribution of working Group I to the sixth assessment report of the intergovernmental panel on climate change. Cambridge University Press, Cambridge and New York, pp 3−32. https://doi.org/10.1017/9781009157896.001
99. Matos P, Pinho P, Aragón G, Martínez I, Nunes A, Soares AMVM, Branquinho C (2015) Lichen traits responding to aridity. J Ecol 103:451–458
100. Matulich KL, Martiny JBH (2015) Microbial composition alters the response of litter decomposition to environmental change. Ecology 96:154–163
101. Mayol E, Arrieta JM, Jiménez MA, Martínez-Asensio A, Garcias-Bonet N, Dachs J, González-Gaya B, Royer SJ, Benítez-Barrios VM, Fraile-Nuez E, Duarte CM (2017) Long-range transport of airborne microbes over the global tropical and subtropical ocean. Nat Commun 8:1–8
102. Miller AF (2012) Superoxide dismutases: ancient enzymes and new insights. FEBS Lett 586(5):585–595. https://doi.org/10.1016/j.febslet.2011.10.048
103. Miller RM, Jastrow JD (1992) The role of mycorrhizal fungi in soil conservation. Mycorrhizae Sustain Agric 54:29–44
104. Moinet GYK, Cieraad E, Hunt JE, Fraser A, Turnbull MH, Whitehead D (2016) Soil heterotrophic respiration is insensitive to changes in soil water content but related to microbial access to organic matter. Geoderma 274:68–78
105. Morono Y, Ito M, Hoshino T, TeradaT, Hori T, Ikehara M, D'Hondt S, Inagaki F (2020) Aerobic microbial life persists in oxic marine sediment as old as 101.5 million years. Nat Commun 11
106. Morris RL, Schmidt TM (2013) Shallow breathing: Bacterial life at low O_2. Nat Rev Microbiol 11:205–212
107. Napieralski SA, Roden EE (2020) The weathering microbiome of an outcropping granodiorite. Front Microbiol 14(11):601907
108. Naylor D, McClure R, Jansson J (2022) Trends in microbial community composition and function by soil depth. Microorganisms 10(3):540

109. Nelson M, Allen J, Alling A, Dempster WF, Silverstone S (2003) Earth applications of closed ecological systems: relevance to the development of sustainability in our global biosphere. Adv Space Res 31:1649–1655
110. Nemergut DR, Schmidt SK, Fukami T, O'Neill SP, Bilinski TM et al (2013) Patterns and processes of microbial community assembly. Microbiol Mol Biol Rev 77:342–356
111. Nguyen STC, Freund HL, Kasanjian J, Berlemont R (2018) Function, distribution, and annotation of characterized cellulases, xylanases, and chitinases from CAZy. Appl Microbiol Biotechnol 102:1629–1637
112. Ni M, Yu JH (2007) A novel regulator couples sporogenesis and trehalose biogenesis in Aspergillus nidulans. PLoS ONE 2(10):e970
113. Nogi Y (2017) Microbial life in the deep sea: Psychropiezophiles. In: Psychrophiles: from biodiversity to biotechnology, 2nd edn. Springer International Publishing, pp 133–152
114. Paul EA (2007) 1 – Soil microbiology, ecology, and biochemistry in perspective. In: Soil microbiology, ecology, and biochemistry. Academic Press, pp 3–24
115. Pontefract A, Zhu TF, Walker VK, Hepburn H, Lui C, Zuber MT, Ruvkun G, Carr CE. (2017) Microbial diversity in a hypersaline sulfate lake: a terrestrial analog of ancient mars. Front Microbiol 8(SEP):1–12
116. Post WM, Emanuel WR, Zinke PJ, Stangenberger AG (1982) Soil carbon pools and world life zones. Nature 298:156–159
117. Posth NR, Konhauser KO, Kappler A (2013) Microbiological processes in banded iron formation deposition. Sedimentology 60:1733–1754
118. Rayner AD, Boddy L (1988) Fungal decomposition of wood: its biology and ecology. Wiley, Chichester
119. Rillig MC, Wright SF, Allen MF, Field CB (1999) Rise in carbon dioxide changes soil structure. Nature 400:628
120. Rimington WR, Duckett JG, Field KJ, Bidartondo MI, Pressel S (2020) The distribution and evolution of fungal symbioses in ancient lineages of land plants. Mycorrhiza 30:23–49
121. Roalkvam I, Drønen K, Stokke R, Daae FL, Dahle H, Steen IH (2015) Physiological and genomic characterization of Arcobacter anaerophilus IR-1 reveals new metabolic features in epsilonproteobacteria. Front Microbiol 6
122. Röling WFM, Head IM, Larter SR (2003) The microbiology of hydrocarbon degradation in subsurface: petroleum reservoirs: perspectives and prospects. Res Microbiol 154(5)
123. Roumagnac, M, Pradel N, Bartoli M, Garel M, Jones AA, Armougom F, Fenouil R, Tamburini C, Ollivier B, Summers ZM, Dolla A (2020) Responses to the hydrostatic pressure of surface and subsurface strains of pseudothermotoga elfii revealing the piezophilic nature of the strain originating from an oil-producing well. Front Microbiol 11
124. Rousk J, Bååth E (2007) Fungal and bacterial growth in soil with plant materials of different C/N ratios. FEMS Microbiol Ecol 62:258–267
125. Rousseau JD, Reid CP, English RJ (1992) Relationship between biomass of the mycorrhizal fungus Pisolithus tinctorius and phosphorus uptake in loblolly pine seedlings. Soil Biol Biochem 24:183–184
126. Rozanov AY, Astafieva MM (2009) The evolution of the early precambrian geobiological systems. Paleontol J 43:911–927
127. Rusakova E, Sukhacheva E, Hartemink AE (2022) Vasiliy Dokuchaev–A biographical sketch on the occasion of his 175th birthday. Geoderma 412:115718
128. Samake A, Uzu G, Martins JMF, Calas A, Vince E, Parat S, Jaffrezo JL (2017) The unexpected role of bioaerosols in the Oxidative Potential of PM. Sci Rep 7:1–10
129. Schmalenberger A, Duran AL, Bray AW, Bridge J, Bonneville S, Benning LG, Romero-Gonzalez ME, Leake JR, Banwart SA (2015) Oxalate secretion by ectomycorrhizal Paxillus involutus is mineral-specific and controls calcium weathering from minerals. Sci Rep 5:12187
130. Sevigny JL, Rothenheber D, Diaz KS, Zhang Y, Agustsson K, Bergeron RD, Thomas WK (2019) Marker genes as predictors of shared genomic functions. BMC Genomics 20:1–13
131. Singh AK, Zhu X, Chen C, Wu J, Yang B, Zakari S, Jiang XJ, Singh N, Liu W (2022) The role of glomalin in mitigation of multiple soil degradation problems. Taylor and Francis Ltd.

132. Song Z, Sadowsky M, Schilling JS (2012) Competition between two wood-degrading fungi with distinct influences on residues. Fed Eur Microbiol Soc: Microbiol Ecol 79(1):109–117
133. Stevens T (1997) Lithoautotrophy in the subsurface. FEMS Microbiol Rev 20:327–337
134. Sumorok B, Kiedrzynska E (2007) Mycorrhizal status of native willow species in the Pilica River floodplain along the moisture gradient. Wetlands: Monitoring, Modelling, and Management. Taylor and Francis Group, London, New York, pp 281–286
135. Swift MJ, Heal OW, Anderson JM (1979) Decomposition in terrestrial ecosystems. Blackwell, Oxford
136. Tang K, Huang Z, Huang J, Maki T, Zhang S, Ma X, Shi J, Bi J, Zhou T, Wang G, Zhang L (2018) Characterization of atmospheric bioaerosols along the transport pathway of Asian dust during the Dust-Bioaerosol 2016 Campaign. Atmos Chem Phys 18:7131–7148
137. Taylor LL, Banwart SA, Valdes PJ, Leake JR, Beerling DJ (2012) Evaluating the effects of terrestrial ecosystems, climate and carbon dioxide on weathering over geological time: a global-scale process-based approach. Philos Trans R Soc Lond Biol Sci 367(1588):565–582
138. Tedersoo L, Bahram M, Põlme S, Kõljalg U, Yorou NS, Wijesundera R et al (2014) Fungal biogeography. Global diversity and geography of soil fungi. Science 346:1256688
139. Totsche KU, Amelung W, Gerzabek MH, Guggenberger G, Klumpp E, Knief C et al (2018) Microaggregates in soils. Wiley-VCH Verlag
140. Trap J, Laval K, Akpa-Vinceslas M, Gangneux C, Bureau F, Decaëns T, Aubert M (2011) Humus macro-morphology and soil microbial community changes along a 130-yr-old Fagus sylvatica chronosequence. Soil Biol Biochem 43:1553–1562
141. Treseder KK, Maltz MR, Hawkins BA, Fierer N, Stajich JE, McGuire KL (2014) Evolutionary histories of soil fungi are reflected in their large-scale biogeography. Ecol Lett 17(9):1086–1093
142. Treseder KK, Holden SR (2013) Fungal carbon sequestration. American Association for the Advancement of Science
143. Treseder KK, Turner KM (2007) Glomalin in ecosystems. Soil Sci Soc Am J 71(4):1257–1266
144. Trudinger PA, Bubela B (1967) Microorganisms and the natural environment. Mineral. Deposita 2:147–157
145. Tucker CL, Reed SC (2016) Low soil moisture during hot periods drives apparent negative temperature sensitivity of soil respiration in a dryland ecosystem: a multi-model comparison. Biogeochemistry 128:155–169
146. Unestam T, Sun YP (1995) Extramatrical structures of hydrophobic and hydrophilic ectomycorrhizal fungi. Mycorrhiza 5:301–311
147. Vallino M, Fiorilli V, Bonfante P (2014) Rice flooding negatively impacts root branching and arbuscular mycorrhizal colonization, but not fungal viability. Plant, Cell Environ 37:557–572
148. Vavourakis CD, Ghai R, Rodriguez-Valera F, Sorokin DY, Tringe SG, Hugenholtz P, Muyzer G (2016) Metagenomic insights into the uncultured diversity and physiology of microbes in four hypersaline soda lake brines. Front Microbiol 7(FEB):1–18
149. Vellend M (2010) Conceptual synthesis in community ecology. Q Rev Biol 85:183–206
150. Vitousek P, Chadwick O, Matson P, Allison S, Derry L, Kettley L, Luers A, Mecking E, Monastra V Porder S (2003) Erosion and the rejuvenation of weathering-derived nutrient supply in an old tropical. Landsc Ecosyst 6 762 772
151. Voica DM, Bartha L, Banciu HL, Oren A (2016) Heavy metal resistance in halophilic Bacteria and Archaea. FEMS Microbiol Lett 363(14):1–9
152. Wahlen M (1993) The global methane cycle. Annu Rev Earth Planet Sci 21(1):407–426
153. Weingarten EA, Lawson LA, Jackson CR (2020) The saltpan microbiome is structured by sediment depth and minimally influenced by variable hydration. Microorganisms 8(4)
154. Wu X, Wu L, Liu Y, Zhang P, Li Q, Zhou J, Hess NJ Hazen TC, Yang W, Chakraborty R (2018) Microbial interactions with dissolved organic matter drive carbon dynamics and community succession. Front Microbiol 9
155. Wu Y, Zeng J, Zhu Q, Zhang Z, Lin X (2017) PH is the primary determinant of the bacterial community structure in agricultural soils impacted by polycyclic aromatic hydrocarbon pollution. Sci Rep 7

156. Yanagibayashi M, Nogi Y, Li L, Kato C (1999) Changes in the microbial community in Japan Trench sediment from a depth of 6292 m during cultivation without decompression. FEMS Microbiol Lett 170:271–279. https://doi.org/10.1111/j.1574-6968.1999.tb13384.x
157. Zavalishin AA (1958) Dokuchaev's doctrine of soil forming factors as a basis of the comparative geographical method of soil investigation. Pochvovedenie 9:39–47
158. Zhalnina K, Dias R, de Quadros PD, Davis-Richardson A et al (2014) Soil pH determines microbial diversity and composition in the park grass experiment. Microb Ecol 69:395–406
159. Zhang L, Feng G, Declerck S (2018) Signal beyond nutrient, fructose, exuded by an arbuscular mycorrhizal fungus triggers phytate mineralization by a phosphate solubilizing bacterium. ISME J 12:2339–2351
160. Zhang D, Hui D, Luo Y, Zhou G (2008) Rates of litter decomposition in terrestrial ecosystems: global patterns and controlling factors. J Plant Ecol 1:85–93
161. Zhao H, Zheng W, Zhang S, Gao W, Fan Y (2021) Soil microbial community variation with time and soil depth in Eurasian Steppe (Inner Mongolia, China). Ann Microbiol 71
162. Zhou WN, White JF, Soares MA, Torres MS, Zhou ZP, Li HY (2015) Diversity of fungi associated with plants growing in geothermal ecosystems and evaluation of their capacities to enhance thermotolerance of host plants. J Plant Interact 10(1):305–314
163. Zhu HZ, Jiang CY, Liu SJ (2022) Microbial roles in cave biogeochemical cycling. Front Microbiol 13:950005

Land Surface-Atmosphere CO_2 Fluxes in the Critical Zone

Mariasilvia Giamberini and Antonello Provenzale

Abstract The terrestrial carbon cycle is rooted in the Critical Zone, as its main processes involve matter and energy exchanges between vegetation, soil, water, rock and the atmosphere. The biological carbon cycle is, together with the water cycle, a crucial element cycle for the regulation of life and climate on Earth. Soils and the terrestrial carbon cycle are intimately linked, as none of them would exist without the other: primitive soils were formed by the action of the first terrestrial photosynthetically active organisms, and today soils support vegetation on Earth and represent the greatest carbon pool of the terrestrial biosphere, containing much more carbon than the atmosphere. Carbon is exchanged mainly through CO_2 fluxes, which take place at the soil–vegetation–atmosphere interface and whose magnitude is influenced by the characteristics of the soil, of climate and of ecosystems. Photosynthesis and respiration from terrestrial vegetation and soils represent the two major components of carbon fluxes between the atmosphere and the biosphere. Their balance (or imbalance) regulates the evolution of climate on Earth and is in turn influenced by the climatic conditions. The disturbances caused by human activities (e.g. fossil fuels emissions, land use changes, fires) are now altering this balance. Measuring and modelling this cycle in different ecosystem types and under different levels of human pressures is necessary to deepen our understanding of the dynamics of the Earth life support system and preserve the services it provides to the benefit of humankind.

1 Introduction: Carbon, the Globetrotter of Life

Carbon is perhaps the most impressive element in the periodic table. It can form hundreds of thousands of different natural molecules, travelling from the highest atmosphere to the deeper layers of the planetary mantle and back, passing through oceans and volcano's mouths, from the breath of animals to rocks, from leaf stomata to soils. Residence time in one of its forms can be a fraction of a second or hundreds of millions of years. In this travel, carbon goes from the most oxidised form (CO_2)

M. Giamberini (✉) · A. Provenzale
Institute of Geosciences and Earth Resources, National Research Council of Italy, Pisa, Italy
e-mail: mariasilvia.giamberini@cnr.it

T. White and A. Provenzale (eds.), *Critical Zone and Ecosystem Dynamics*,
Advances in Critical Zone Science, https://doi.org/10.1007/978-3-031-69076-1_10

to reduced ones (organic molecules) up to the most reduced carbon molecule, CH_4, and back through the metabolism, life cycle and food web of all living organisms.

Carbon also contributes to regulating the climate of the Earth through its journey between the living and the unliving realms. Looking at the wider picture, the work of geologists, biologists, oceanographers, soil scientists, climatologists and palaeoclimatologists has made clear that the carbon cycle regulates the Earth climate through carbon exchanges between the atmosphere, oceans, soils, living organisms and rocks. In particular, the sequestration of carbon in vegetation, and its underground burial, is a crucial control factor of the Earth's climate at large scale [1–4]. Palaeoclimatologists have been able to reconstruct the concentration of CO_2 in the atmosphere in the last 800,000 years from the analysis of air bubbles trapped in the ice cores of the Antarctica ice sheet; relating them to temperatures showed the strong link of the two variables on the scale of thousands of years [5, 6].

At the same time, climate affects the carbon cycle: air temperature regulates both the CO_2 exchange rate and absorption capacity of the oceans (among other drivers such as acidity and CO_2 fugacity) and the rate of metabolic processes in soil and vegetation, thus providing positive feedback in the temperature—CO_2 loop.

The buffering capacity of vegetation, soil and oceans to store and release CO_2 (and greenhouse gases, GHGs, in general), acting as mitigators of climate changes, has been ascertained for the past and present climate. Estimates of today's CO_2 emissions and measurements of atmospheric concentrations reveal that about half of the CO_2 anthropogenic emissions have been absorbed by marine and terrestrial ecosystems and by ocean waters (fossil fuel and land use change emissions from 1750 to 2019: 700 ± 75 PgC, [7]). Indeed, the absorption capacity of terrestrial ecosystems and oceans has "spared" Earth from the experience of an even faster warming, limiting the growth of the atmospheric CO_2 concentration by anthropogenic emissions.

Apparently, the CO_2 buffering capacity of land and oceans has remained roughly proportional to CO_2 atmospheric concentration and the reasons for this behaviour have been widely investigated [3]. In terrestrial ecosystems, the higher temperatures, the longer vegetative season, nutrient fertilisation and the so-called "fertilization effect" (according to which the higher the CO_2 air concentration, the higher the CO_2 uptake if nutrients are available) are accounted among the main drivers of the observed increase in vegetation productivity [8], but the role of CO_2 fertilisation in the different biomes and in the long run as a main driver of the increase of Net Biome Productivity (NBP) is uncertain. Injecting CO_2-enriched air in greenhouses is used to enhance the production of vegetables, but estimates on the carbon budget subject to artificial CO_2 fumigation revealed that also respiration and litterfall decomposition are boosted, not leading to a substantial increase in carbon sequestration in the long run [9, 10]. Such conclusions have been obtained mainly for temperate forests (the Free Air CO_2 Enrichment (FACE) experiment) and it is not possible to generalise them to other biomes. In addition, nutrient deficiency, common in some environments, can limit the effect of fertilisation.

Trends in the capacity of terrestrial ecosystems to buffer CO_2 emissions are not constant in time and are not evenly distributed in space. They depend on a variety of factors, such as the intrinsic characteristics of the ecosystems' microbial food web, oscillations in temperature and precipitations associated to "El Niño", availability of nutrients and water, and anthropogenic factors. These latter include changes in land use (in both directions: deforestation and intensive farming inducing carbon loss, and abandonment, which causes trees encroachment and afforestation) and disturbances, like fires and pests, making it difficult to obtain accurate assessments and future forecasts.

How long such absorption processes will continue, and if and how it will change are questions of paramount importance, as the answer will determine the future increase rate of atmospheric CO_2 and the time remaining to limit the growth of surface temperature below the limit of 1.5 °C indicated in the Paris Agreement in 2015 [3]. For these reasons, efforts to increase monitoring and modelling for an accurate assessment of the natural carbon pools and fluxes and a better understanding of the processes regulating the carbon cycle of land and oceans and the related feedbacks is necessary.

In the following of this chapter, we shall consider the land surface–atmosphere carbon cycle; specific examples will be made about the Arctic tundra, because its relative simplicity allows to focus on the basic processes and because the carbon cycle perturbations that are likely to occur in the Arctic can greatly influence the Earth's climate. We focus on the CO_2 cycle as one of the major factors that couples the biosphere and the atmosphere in controlling climate processes. We acknowledge that CO_2 fluxes are coupled with energy and water vapour fluxes and that other gases are also involved and important, as CH_4, N_2O and VOCs.

Figure 1 illustrates the basic fluxes used in describing the terrestrial carbon cycle, including anthropogenic perturbations.

The terminology of the terrestrial natural carbon process and fluxes is well known, but a recap is worthy for the sake of completeness and because such terms will be recalled further on. Here, the convention used is that fluxes from the atmosphere to the soil are negative and fluxes from the soil to the atmosphere are positive. In this sense, Gross Primary Productivity (GPP) and Net Primary Productivity (NPP) are always negative, Respiration terms (AR, HR and ER) are always positive, while the Net Ecosystem Exchange (NEE) can be either positive or negative, depending on the prevailing process (respiration or photosynthesis, respectively).

Gross Primary Productivity (GPP) is the flux of carbon uptaken by plants through photosynthesis (flux: mass per unit of time and surface). Part of this carbon will be respired by plants for their metabolism.

Net Primary Productivity (NPP) is the net flux of carbon that is effectively used to produce plants tissues, once the carbon used by plants metabolism, i.e. consumed in autotrophic respiration, is subtracted from the GPP. It therefore represents the carbon that enters the food web in terrestrial ecosystems, and is given by the algebraic sum of GPP (negative by definition) and autotrophic respiration (AR, positive by definition but smaller than the absolute value of GPP), resulting in a negative value.

Autotrophic Respiration (AR) is the flux of carbon respired by plants for their metabolism.

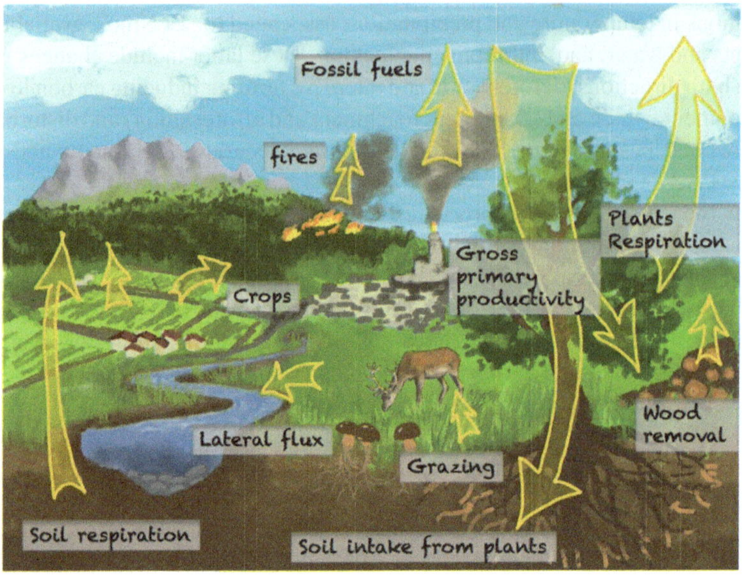

Fig. 1 The terrestrial carbon fluxes, including anthropogenic perturbations as fossil fuels emissions, fires, crops and wood removal (courtesy of Irene Sebastiani ©)

Heterotrophic Respiration (HR) is the flux of carbon respired by microorganisms in the soil during their metabolic activities, consuming molecules produced by plants or decomposing dead organic matter.

Ecosystem Respiration (ER or Reco) is the sum of HR and AR.

Net Ecosystem Exchange (NEE) is the algebraic sum between GPP and ER and represents the net flux of carbon in a given amount of time and a given area.

A **Carbon pool** is a medium (atmosphere, above ground vegetation, soil, oceans…), or a portion of it, that stores and exchanges carbon with other pools.

A carbon pool is a **carbon sink** at a given time if its net ecosystem exchange is negative, NEE <0, i.e. if it stores more carbon than it emits. It is a **carbon source** if the NEE is positive, i.e. carbon emissions at a given time exceed the carbon uptake.

Besides the major fluxes and pools, carbon can enter or leave an ecosystem also because of other factors, such as the flow of dissolved and particulate organic matter into surface and subsurface water, the production of VOCs (especially from forests), disturbances (pests, fires, droughts), crops and grazing, and changes in land use [7]. Inland waters are carbon pools themselves, with their incoming/outgoing fluxes and carbon balance, although quantities are much less with respect to vegetation and soil. Carbon fluxes from soil to water and production of VOCs are natural processes, while fires, pests, etc. can be both of natural or anthropogenic origin, or exacerbated by human influence (e.g. from climate change and land use). Such fluxes may not have a high magnitude in themselves, but all together they can turn a carbon sink into a source at local or regional scale, with significant impacts in case of sudden and large events (e.g. droughts, pests or fires) (Fig. 2).

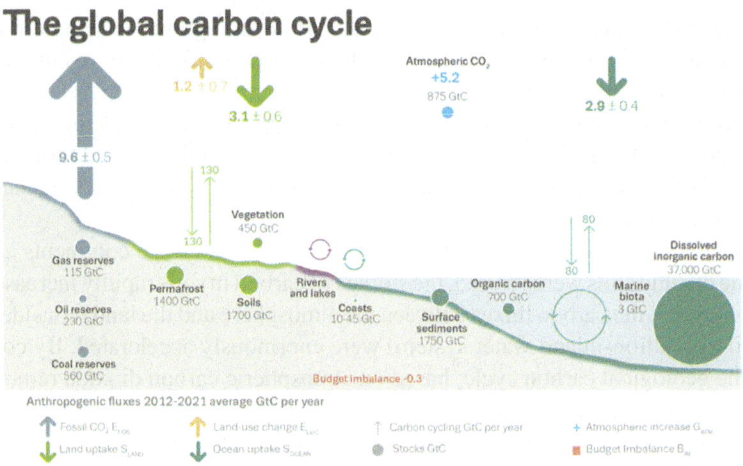

Fig. 2 Schematic representation of the overall perturbation of the global carbon cycle caused by anthropogenic activities averaged globally for the decade 2012–2021. See legends in the image for the corresponding arrows and units. The uncertainty in the atmospheric CO_2 growth rate is very small (± 0.02 GtC year^{-1}) and is neglected in the figure. The anthropogenic perturbation occurs on top of an active carbon cycle, with fluxes and stocks represented in the background and taken from [2] for all numbers, except for the carbon stocks in coasts, which are from a literature review of coastal marine sediments [11]. Figure reproduced from [7] under the Creative Commons Attribution 4.0 License

2 Soil Functions and Carbon Cycle Processes

Soils exert the most important function in the biological terrestrial carbon cycle as they host the vegetation and the decomposing micro-organisms, and store the organic matter produced by photosynthesis.

Soils are formed by the weathering of rock parent material, where soil microorganisms feed on nutrients and produce organic acids and proteins that contribute to further rock alteration and form the first organic substrates, as we can observe in recent volcanic depositions or periglacial areas exposed to the atmosphere after glacier melting. Algae, fungi, lichens, pioneering vegetation then start to colonise the rocks, accumulating and releasing carbon compounds with the interplay of photosynthesis and respiration. Regulating the terrestrial carbon cycle through the two processes of photosynthesis and respiration is one of the most important soil functions.

In the Earth's past, some of the first evidences of structured photosynthetic communities come from the fossil records of "stromatolites", formed about 3.5 billion years ago in the Precambrian period and found in Australia, North America, Greenland and Africa [12]. Such structures are made of mats of layered photosynthetic cyanobacteria and single cell algae, formed by capturing sediment particles then cemented to form mineral microbial mats. Stromatolites were able to capture, store and release carbon, performing the processes forming the basics of the carbon cycle later occurring in terrestrial ecosystems, such as uptake of CO_2 from the atmosphere, synthesis and release of organic substances, carbon storage in the sediment and in living organisms, release of CO_2 back to the atmosphere owing to respiration and decomposition.

However, the living material of stromatolites is poorly structured, and it does not lead to the formation of a proper "soil". Soils as we know today indeed formed much later, in the Devonian period (around 400 M years bp), more or less together with plants firmly establishing on land [13].

Once life (and primarily, terrestrial plants) established on the continents and the first true organic soils were formed, the storage of carbon in soils rapidly increased. As a consequence, the carbon fluxes between the atmosphere and the land (considered as the soil-vegetation-inland water system) were enormously accelerated. By comparison, the geological carbon cycle, based on atmospheric carbon dioxide removal by surface rock weathering, transport to and deposition in coastal sediments, subduction and possibly melting of the sediments into the mantle and carbon dioxide emission by volcanic activity, is much slower and it exchanges probably less than 1 Pg (C) per year [14].

Today, soil is the largest terrestrial carbon pool, storing carbon either in the topsoil and in the deep soil (>3 m). Estimates of carbon pools in the various biomes rely both on interpolation from direct surveys and models; their complexity and uncertainty are related to the high spatial heterogeneity and to the difficulties met in providing a detailed ground-based data coverage.

Soil carbon can be very old (millions of years) as it remained buried, and its bioavailability to decomposition varies depending on its chemical composition, soil structure, temperature and water availability [15]. The soil pools are not evenly distributed in the different biomes.

The latest estimates from various sources report a global carbon stock in soils of around 2300–2900 Pg (C) (1200 in permafrost soil), by far larger than the carbon stock in above ground vegetation (estimated around 450 Pg (C)). The atmosphere currently contains around 590 Pg (C) with an expected average increase of about 5 Pg(C)/year [2, 16]. Further detailed data regarding carbon in soils, also at regional/ecosystem level, can be found in many reports from the UN FAO (www.fao.org), in [4] and in [16].

Indeed, photosynthesis is the key process that links the atmosphere and the terrestrial biosphere through biogeochemical cycles [17]. Around half of the CO_2 captured through photosynthesis (Gross Primary Productivity) is returned to the atmosphere through the respiration of the vegetation (Autotrophic Respiration) and the other half contributes to the Net Primary Productivity, which fuels of energy all the other living organisms in the trophic web and is the engine and the key regulator of ecosystems.

The exchanges of carbon across the atmosphere, vegetation and soil in photosynthesis-dominated ecosystems are ultimately driven by the energy captured from solar radiation and transformed into chemical energy, transferring it to all forms of life in the biosphere. Even the energy from fossil fuels that we humans have been using since the industrial revolution comes from solar radiation that was transferred to plants and animals, and then trapped in non-fully decomposed organisms that formed coal, oil and natural gas (Tables 1 and 2).

Table 1 Natural and anthropogenic (for comparison) land–atmosphere carbon fluxes. Data are expressed as Pg(C) per year. When available, more than one value is given for the same parameter. In these cases, different authors may report different values, depending on the evaluation method used, and different definitions

Flux	Pg (C)/year	Uncertainty Pg(C)/year	References
Natural fluxes			
PRIMARY PRODUCTIVITY			
Gross primary production before anthropogenic influence	123		Ciais et al. [18] and Jansson et al. [19] (in: Ussiri and Lal [4])
Gross primary production (GPP)	115–190		Cai and Prentice [8] (in: Crisp et al. [3])
Net primary production (NPP)	50	44–57	Ciais, Yao [20] (in: Crisp et al. [3])
Net primary productivity prior to human influence	60		Ussiri and Lal [4]
Net primary productivity included human influence	63 (average)	56.4–63.6	Jansson et al. [19], Ruimy et al. [21], Field et al. [22], Ito [23], Running [24] (in: Ussiri and Lal [4])
RESPIRATION and other natural carbon losses to the atmosphere			
Autotrophic respiration (AR)	60		Lloyd, Farquar [25]; Waring et al. [26] (in: Ussiri and Lal [4])
Autotrophic respiration (AR)	64	±12	[27] (in: Crisp et al. [3])
Decomposition of litter	55		Ussiri and Lal [4]
Soil heterotrophic respiration (SHR)	39	33–46	Ciais, Yao [20] (in: Crisp et al. [3])
Wildfires	4		Ussiri and Lal [4]
Fires	1.6	±0.1	Ciais, Yao [20] (in: Crisp et al. [3])
Other (grazing, outgassing of inner water, biogenic reduction, decay of wood)	3.3	±0.8	Ciais, Yao [20] (in: Crisp et al. [3])
Net flux into the biosphere			
Storage of new SOC and plant material = Net Biome Production	−1.4	±0.7	[28] (in: Ussiri and Lal [4])
Net Ecosystem Exchange	−2.4	±0.7	Ciais, Yao [20]
Anthropogenic fluxes			
Fossil fuels burning/cement (2021)	9.6	±0.5	Friedlingstein [7]
Consumption of harvested crops	1.5	±0.2	Ciais, Yao [20] (in: Crisp et al. [3])

(continued)

Table 1 (continued)

Flux	Pg (C)/year	Uncertainty Pg(C)/year	References
Land use change	1.1	±0.5	Ciais, Yao [20] (in: Crisp et al. [3])
Emissions from land use (deforestation/carbon loss)	1.2	±0.7	Friedlingstein [7]

Table 2 Carbon stocks [Pg(C)] in Earth's global pools (updated to 2014). (*) = estimated natural pool at 1750 AD; (**) = anthropogenic contribution after the industrial revolution by 2014; Black = total by 2014. ± = uncertainty; references: (1) Prentice et al. [28], (2) = Ussiri and Lal [4], (3) = Ciais et al. [18], (4) = Bridgham et al. [29] (5): Tarnocai et al. [30], (6): Friedlingstein [7]

Pool	Pg (C)	Uncertainty Pg (C)	Notes and references
Atmosphere	590 (*) + 260 (**) = 850	± 10	Average increase: 4 Pg year; (1 2)
Vegetation	450 ÷ 650 (*) –30 (**)	± 45	Deforestation; uncertainty refers to anthropogenic contribution (3, 1)
Soils + litter (SOC) <1 m	1,500–2,400 (*)		SOM + litter for top 1 m; (3)
Soil (exc. Permafrost)	1,700		(6)
Permafrost	1,400		(6)
Permafrost	1,700		Permafrost region including deep permafrost (5)
Wetlands	300–700		(4)
Surface ocean	1,000 + 170 (**)	±20 (**)	Acidification; uncertainty refers to anthropogenic contribution (2, 3)
Intermediate and deep ocean	37,100 (DIC)		(2, 3)
sediment	1,750		(2, 3)
Fossil fuels	>6,000–400 (**)	± 20 (**)	Extraction (2)

3 Mechanisms of CO_2 Uptake and Release in Plants and Soils

Photosynthesis transfers solar energy into ecosystems: it is the engine of life. It allows carbon to enter the life cycle of plants and forming all the natural substances needed for building living organisms. Through the food web, energy and organic matter are transferred from plants to animals. They will be finally released back to the atmosphere through respiration and decomposition of dead tissues. In terrestrial ecosystems, photosynthesis happens in vascular and non-vascular vegetation, lichens, and photosynthetic bacteria such as cyanobacteria. The cell organ deputed to photosynthesis is the chloroplast, and it is found in leaves and in the other green or orange parts of plants. Roots, trunk, branches do not perform photosynthesis. Respiration, instead, happens in all cells of plants, both aboveground and belowground.

The rate and efficiency of photosynthesis depend on physical and climate factors (light, *in primis*) and on plant physiology. Carbon dioxide diffuses into the leaf cells from pores called *stomata*, whose aperture regulates the flux rates of water vapour and gas molecules, as CO_2 and molecular oxygen. Stomatal aperture determines the "stomatal conductance", i.e., the rate of gas exchange through a unit of leaf surface, and links the photosynthesis rate to transpiration of water molecules. As the stomatal conductance depends on the CO_2 gradient at the leaf surface and the availability of light and water, air moisture influences the photosynthetic activity through plant transpiration. Indeed, as the exchange of water and CO_2 molecules happens through stomata, there is a trade-off between the uptake of CO_2 and the loss of water, especially in arid climates. When stomata are open, CO_2 can enter the leaves, but water will be lost. During the hottest hours of summer, plants may reduce the stomatal opening to preserve water, at the cost of decreasing the photosynthesis rate. In arid climates, this trade-off between the need to perform photosynthesis and to preserve water resulted in a different photosynthetic mechanism: it is the case of "crassulacean" plants, which capture the light during the day but open the stomata and complete the photosynthesis at night, allowing CO_2 to enter the stomata when the consequent loss of water is less stressful for the plant. The Water Use Efficiency (WUE), defined as the ratio between the mass of CO_2 fixed and the mass of water lost, represents the ability of plants to cope with water scarcity. CO_2 fertilization increases the water use efficiency: there is evidence that the density of stomata has decreased with CO_2 fertilization from the industrial revolution [31]. An indirect measure of WUE can be derived from δ13-C isotope ratio in plant material. In fact, as 12-C is lighter than 13-C, it diffuses faster into plant stomata, and this difference in plant uptake is greatest when the stomatal conductance is high. Instead, the closer stay the stomata, the less is isotope fractionation.[1]

The net productivity of plants is influenced by the abundance of nitrogen: nitrogen deficiency is a limiting factor to plant growth, mainly because it is an essential constituent of proteins and enzymes, and it is also present in chlorophyll [33]. In CO_2 fertilization, as the increase of CO_2 atmospheric concentration boosts the water

[1] THE isotopic ratio refers to the ratio of the atomic abundances of two isotopes of the same element, e.g., 13C/12C. The notation δ13C indicates the way this relative abundance is measured, i.e. as the [(isotopic ratio in the sample towards the isotopic ratio in a reference material) -1] × 1000. In natural systems, the isotope ratio of stable isotopes varies with climate and environmental conditions, as physical–chemical as well as biological processes are selective towards isotopes through kinetic fractionation. This makes the isotopic ratio a powerful tool to investigate natural and climate processes. The isotopic fractionation is the relative partitioning of the heavier and lighter isotopes between two coexisting phases in a natural system (Definition from: [32]. https://doi.org/10.1016/B978-0-12-419968-2.00003-0).

use efficiency of plants and enhances the C/N ratio in plant tissues, nitrogen availability in plants may decrease as a consequence of the rise in atmospheric CO_2 concentration, with consequences on the whole trophic web and on the long-term capacity of ecosystems to buffer CO_2 emissions [34, 35].

Carbon compounds are ultimately used by all living organisms to produce energy through cellular respiration, either autotrophic (in photosynthetic organisms) or heterotrophic (by macro and microfauna, fungi and the microbiota). Aerobic respiration is the most common mechanism when oxygen is present; in this case, CO_2 and H_2O are produced. Where oxygen is scarce, e.g. in waterlogged environments such as peatbogs, decomposition is reduced and fermentation can occur with emissions of CH_4 [36].

Most of the heterotrophic and part of the autotrophic respiration (e.g. from plant roots) take place below ground and constitute the "soil respiration". Roots (autotrophic) respiration accounts for a large part of soil respiration, which can range from 10% up to 90% depending on the root biomass and respiration rate, as well as on other characteristics of the soil [37, 38].

Soil respiration is regulated by a number of abiotic factors such as temperature, moisture and availability of nutrients. As different groups (cohorts) of decomposing micro-organisms are usually present in the soil, the range of temperatures within which soil respiration is possible is broad and can encompass temperatures above 40 °C as well as a few Celsius degrees below zero, making soil respiration possible in harsh environments such as deserts and the polar regions.

Microbial respiration in soil is linked to temperature via an exponential Arrhenius type relationship [39]. Thus, a future increase in CO_2 emissions from microbial respiration is foreseen in all biomes. As said in the premises of this chapter, this could translate into a net emission if primary productivity, which is also expected to increase, cannot balance the higher respiration [40–43].

Soil moisture is the other important control factor of soil respiration. Respiration is greatest at intermediate moisture levels, while it is lowest in dry and in very wet soils. It is observed that moisture acts as a limiting factor in arid or semiarid ecosystems, while it blocks aerobic respiration in water-logged environments, unless the specialized wetland vegetation is present. In arid or semiarid soils, the re-wetting of soil caused by rainfalls generates pulses of CO_2 production as it mobilises and re-hydrates dry organic matter, soil enzymes and micro-organisms, making it possible their encounter, also disrupting soil aggregates. In this sense, micro-organisms act as a catalyst of the decomposition of dry organic matter, and water is the reaction medium [44].

The processes governing the CO_2 release from soil to the atmosphere are as important as those governing CO_2 formation in determining the flux rate and temporal trends. Once CO_2 is produced, it is dissolved in the intracellular water. Its release outside the cell is driven by gas diffusion, i.e. by the concentration gradient between the cell and the external environment. Outside the cell, CO_2 is dissolved in the water contained in soil pores either as dissolved gas or as carbonate ions, and it is ultimately released to the atmosphere as a gas. This flux is driven by a number of abiotic factors, including diffusion and convection at the soil-air interface. The CO_2-carbonates equilibrium in water and the ratio between CO_2 in gaseous form and that dissolved in water at the air–water surface is influenced by the pore water pH and by the presence of carbonates and other ions of abiotic origin (e.g. Ca and Mg). Diffusion and convection to the atmosphere are influenced by pressure and wind. In particular, wind gusts induce convective motions that mechanically remove the CO_2-enriched air from the soil pores. Besides controlling the kinetics of biochemical reactions, soil moisture also controls the amount of CO_2 dissolved in the droplets in soil pores: wet soils trap CO_2 dissolved in water, that can be suddenly released when the soil dries. Also snow pores trap CO_2. Air bubbles at the bottom of the snowpack can reach CO_2 concentrations of a few thousands of ppm and if a sudden change occurs in the snowpack, e.g. owing to wind gusts, changes in temperature or pressure or a rain event, CO_2 can be suddenly released, producing pulse CO_2 emissions. Snow cover also affects CO_2 respiration by insulating the soil and reducing the winter decrease of temperature in comparison with uncovered soil. A steady snowpack may then buffer the CO_2 flux [45–48].

For all these reasons, the time and spatial variability of CO_2 fluxes from soil and vegetation cannot be described as a function of one or a few variables only, e.g. radiation for primary productivity and temperature for respiration, as atmospheric and environmental conditions, biotic and physical–chemical processes in the soil all contribute to determine CO_2 fluxes.

If any of such factors vary in space and time, CO_2 fluxes will vary accordingly, and their mutual interplay will make the flux variation complex to model and predict, at any scale of observation, from the point scale of a single plant-soil-atmosphere interaction to the landscape scale.

Nevertheless, when modelling CO_2 fluxes at larger scales, e.g. at ecosystem or landscape scale by using remote sensing data and time-series spanning several years, some assumptions must be made about the relevant drivers, and the variability at local scale is averaged. In this case, the number of ground-based observations needed for a sound ground validation of the remote sensing products has to be as large as possible to average the local variability. The scale of observation of CO_2 fluxes determines the modelling approach and the techniques needed for performing the observations.

4 Estimating CO_2 Fluxes at Different Scales

Understanding and modelling climate variability require knowledge of the natural carbon cycle, which is linked with the atmospheric CO_2 concentration. The feedbacks to the increase of temperature can be positive (increase of respiration, which enhances CO_2 emissions and so on) or negative (increase of vegetation biomass, which subtracts CO_2 from the atmosphere and has a cooling effect through evapotranspiration). The mechanisms involved are complex and interrelated, as the carbon cycle implies exchanges between the several pools constituted by soil, microbiota, vegetation, animals, atmosphere. The complete picture involves also lateral fluxes towards water and oceans and bio-geochemical reactions that link the biological cycle to the geological carbon cycle of rock weathering and volcanic – geological emissions.

The first step for an accurate modelling is the availability of high-quality and comprehensive data of the carbon pools and fluxes at global, regional, biome and ecosystem level, allowing to identify the main drivers and their influence on the various components of the carbon cycle at all scales.

To these aims, an array of methods for measuring carbon fluxes and pools have been developed and are continuously upgraded for various applications that include, besides climate studies, soil science, agriculture, and the management of forest, wetlands and other biotic resources as well as land use planning.

The improvement of instrument precision, handling, automation and affordability for measuring and monitoring CO_2 fluxes in terrestrial ecosystems has made it possible to increase the number and the type of observations. More and more accurate and spatially dense ground-based and remote-sensed datasets are now available, to be used for emission and stock inventories and for scientific research, such as investigating processes in soil and vegetation at plot or ecosystem scales and developing diagnostic and forecasting modelling at regional and global scales. At the same time, an array of different modelling techniques has been implemented, for extracting estimates of carbon stocks, gain and losses, and in particular gross and net primary productivity and ecosystem respiration at all scales. However, some regions and biomes of the planet, such as tropical forests, mountain and arctic regions are still underrepresented, mainly because they are difficult to reach and because of the high costs needed for operating monitoring stations in remote locations.

Measurements techniques for CO_2 fluxes at plot scale were originally set up in the'70 for measuring CO_2 respiration in agricultural soils. These consisted in closed flux chambers posed on the soil and containing a CO_2 absorbing system based on soda lime. The absorbed CO_2 was analysed in the lab through titration of the soda-lime solution. This system, not expensive and easy to use but time demanding for the lab analysis, provided an accurate time-averaged measurement of the CO_2 emissions over the exposure time of the soda-lime solution in the flux chamber. Thanks to the progress in analytical instrumentation, such method was quickly replaced by gas chromatography of air sampled inside the flux chamber at different times, and, finally, by in situ gas analysers that rely on laser-based infrared light spectroscopic determination of CO_2 concentration (IRGA, i.e. Infra-Red Gas Analyser). This method is still widely used today, and it is known as the "flux chamber method" [49].

The flux chamber method is based on the general principle of isolating the atmosphere above the soil to detect temporal changes in the CO_2 concentration inside the chamber, which are influenced by the exchange processes between the soil, the vegetation and the atmosphere. The method can be used either on bare soil or on the soil-vegetation system to respectively measure heterotrophic (from soil microorganisms) or ecosystem (autotrophic + heterotrophic) respiration (ER) by using a dark chamber that inhibits photosynthesis. When a transparent chamber is used over the soil-vegetation system, the Net Ecosystem Exchange (NEE) is measured. Gross Primary Productivity is calculated by subtracting ER from NEE.

Several instrumental set-ups coupling an IRGA and chambers are possible, the common rationale being that the flux can be derived by analysing the CO_2 concentration in the chamber air, either from the time-dependent variation of the CO_2 concentration inside the chamber (in the dynamic closed-chamber method) or from the concentration of CO_2 at the inlet and the outlet of the stationary flux (in the steady-state chamber method). The instrumentation set-up is affordable, is portable and easy to use, and the method requires a relatively simple calculation to estimate the fluxes [50]. A desired feature for high-resolution and small-scale investigations is the possibility to use a portable instrumentation to capture the spatial heterogeneity of the fluxes. This method, however, is time demanding when a large spatial coverage is required, and it does not provide continuous time series of data. Fixed, automated chambers are sometimes installed to obtain continuous monitoring. Precautions need to be implemented to avoid instrumental bias and perturbation in the soil-vegetation system where the chambers are installed.

At ecosystem scale, CO_2 fluxes are measured by the Eddy Covariance micrometeorological method, basically consisting in correlating the simultaneous measurement of the air CO_2 concentration and the air turbulent velocity [51, 52]. While more expensive, as well as technologically and computational more demanding than the flux chamber method, the Eddy Covariance technique has the advantage of a larger footprint for flux detection and of continuous monitoring in time. However, Eddy Covariance allows the direct measure of NEE only, while ER and GPP during daytime are derived applying partitioning methods that rely on various modelling assumptions [53].[2] Validation with flux chambers for the day-time computation of GPP and ER is sometimes applied. Coordinated international networks of instrumented sites equipped with Eddy Covariance towers which share the same methodology, such as FLUXNET at global scale and ICOS in Europe, make it possible to compare values and to use them for calibrating models at the biome or regional scale [54, 55]. However, the spatial coverage of the network of ground-based flux towers and chambers is not uniform, and it is not yet sufficient to cover all biomes and to reduce uncertainties for a global assessment of CO_2 fluxes from terrestrial ecosystems (Figs. 3 and 4). The regional coverage is poor in the Global South and in scarcely populated areas or in regions which are more difficult to reach, such as in Africa, in the Arctic, in Central Asia, in the Andes, in Central Australia and in Oceania [56]. The Arctic and Subarctic, as well as the steppe and the desert biomes are especially underrepresented, making it challenging to reduce estimate uncertainties. When looking at the "temperature vs precipitation" climate envelope, we see that some climatic regions are not sufficiently represented, and in particular the cold and arid regions (steppe and cold deserts, Arctic, mountains), the hot and humid regions (pluvial regions) and the hot and arid regions (hot deserts) [57]. A geographic and climatic representation of the ground-based CO_2 flux measurement coverage is reported in Figs. 3 and 4.

[2] Some insights on ground-based instrumentation for measuring CO_2 fluxes is given in Chapter 14 of this volume.

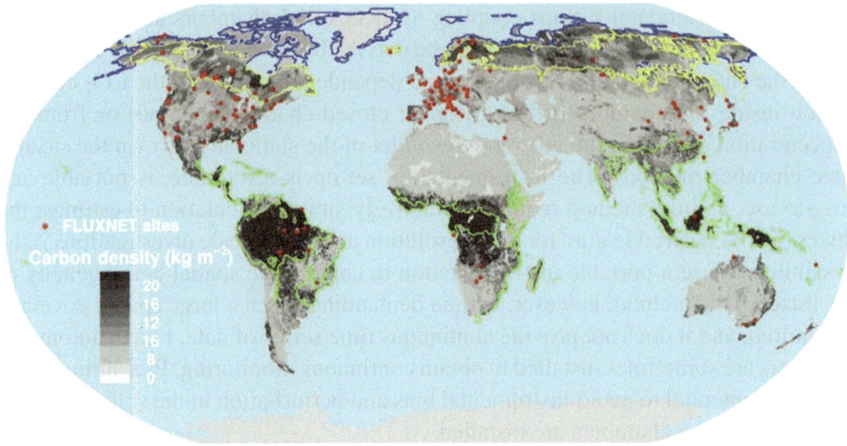

Fig. 3 Spatial distribution of terrestrial vegetation and soil carbon storage with the three 'tipping element' regions identified by Lenton et al. [58]. The tipping elements coincide with regions of high storage and hence high potential for carbon losses able to influence the atmospheric CO_2 concentration. Red points show the distribution of carbon flux field observations, showing that this network has sparse coverage in the tipping element regions. Figure from [57]: Observing terrestrial ecosystems and the carbon cycle from space. Glob Change Biol 21: 1762–1776. https://doi.org/https://doi.org/10.1111/gcb.12822. © 2014 John Wiley and Sons Ltd

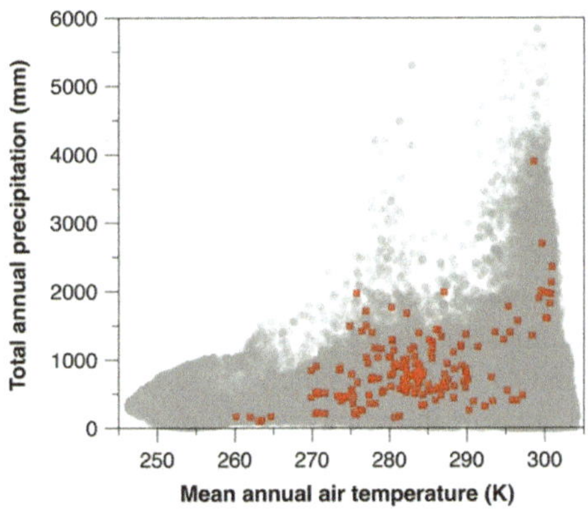

Fig. 4 FLUXNET Eddy Covariance sites plotted in climate space (red points), with the distribution of ecosystems in the same climate space shown in grey. Temperature and precipitation are annual means for 1° latitude–longitude cells. This figure shows the climate-based coverage of the ntwork. Figure from [57]: Observing terrestrial ecosystems and the carbon cycle from space. Glob Change Biol 21: 1762–1776. https://doi.org/https://doi.org/10.1111/gcb.12822. © 2014 John Wiley and Sons Ltd

Remote sensing (RS) observations are meant to provide continuous and spatially broad data enabling to estimate variables related to CO_2 fluxes (GPP in particular), although the issues regarding scarce ground-based data, needed for validation of accurate RS-based products, are still present.

The first attempts to evaluate Gross Primary Productivity and the Above Ground biomass from remote sensing started at the very beginning of the era of Earth Observation in the'70. These estimates were based on the empirical observation that vegetation production is related to the absorbed photosynthetically active radiation (PAR) when temperature, water and nutrients are not a limiting factor [59].

The NDVI (Normalised Differential Vegetation Index) and LAI (Leaf Area Index) have been widely used as indicators of vegetation productivity and biomass. The overall idea is that the light reflectance (or absorbance) at specific bands in the electromagnetic spectrum due to the presence of vegetation is proportional to the vegetation biomass and productivity. Efforts have been made to design space-born sensors devoted to vegetation monitoring, and today a number of specific vegetation indices, related to the activity of Chlorophyll and photosynthetic pigments, are used, such as NIRv, CSI and R_{chl} (e.g. see [60–62]). Another effort goes in the direction of enhancing the resolution of the sensors, for improving the discrimination ability between the spectral signature of vegetation and that of other surface covers such as bare soil, litter, rocks. A database of remote sensing indices can be found at www.indexdatabase.de/. Queries can be done per sensor, per index and per application, providing a useful tool for selecting the appropriate remote sensing index and sensor for an array of specific monitoring requirements.

Remote-sensing-based "light use efficiency" models for measuring GPP are the most common approaches; they rely on the idea that the efficiency to convert solar energy into biomass, i.e. the vegetation productivity, is proportional to the photosynthetically active incident radiation (PAR), to the "fraction of Absorbed Photosynthetically Active Radiation" (fAPAR) and to the "Light Use Efficiency" (LUE), i.e. the efficiency of chloroplasts to use the energy of the absorbed radiation to perform photosynthesis [63, 64], in a general equation:

$$\mathrm{GPP} = \mathrm{PAR} \times \mathrm{fAPAR} \times \mathrm{LUE} \tag{1}$$

PAR can be directly measured in field surveys or derived from climate products. fAPAR and LUE are estimated in various ways using indices derived from measured remote sensing reflectance spectra and the resulting models need to be calibrated on ground-based data.

Some examples of models derived from the light use efficiency equations can be seen in a number of scientific works aimed to improve the assessment of GPP at various scales [8, 65, 66].

Another approach to quantify GPP from remote sensing uses the measurement of SIR (Sun Induced Fluorescence) which has been demonstrated to be proportional to the radiation effectively used in photosynthesis, thus to GPP [67, 68]. Many applications of this method have been proposed so far. The usefulness of this method is that it provides a direct measure of the response of vegetation to light, discriminating from other sources of scattered radiation and constraining the GPP estimates [57, 69].

5 Example: Quantifying Carbon Fluxes in the Arctic Tundra

The High Arctic is the realm of the tundra biome, where trees cannot grow owing to the low temperature of the ground and the thin thickness of the "active layer", the unfrozen upper portion of the arctic permafrost soil, and where the vegetation consists of shrubs and grass or, more north, of low prostrate vascular vegetation mixed with mosses, lichens and a biocrust composed of cyanobacteria, heterotrophic bacteria and fungi [70, 71].

In the High Arctic, a large part of the ground is continuously frozen for most of the year. It is the permafrost, "the part of the ground that remains frozen for at least two consecutive years". Today, at least a quarter of the soil in the northern hemisphere is occupied by permafrost, either in the tundra or in the taiga. Estimates of how much carbon is trapped in permafrost indicate between 1460 and 1600 Pg of carbon [72], which counts for around half of the organic carbon in the soils of all the warmest regions, and about twice the amount currently contained in atmospheric CO_2 molecules.

During the Holocene, the tundra has acted as a carbon sink, subtracting CO_2 from the atmosphere, mainly due to the formation of peatlands [73–76]. But as the Arctic is warming up to four times as fast as the average on Earth [77], this picture is rapidly changing and the question whether the tundra (and more in general permafrost areas) is still a carbon sink or has turned into a carbon source, and how the system will behave in the next decades, has become crucial, as the answer will affect the global carbon emissions budget and the prediction of global warming in the coming years.

In fact, on the one hand, growing temperatures are increasing the active layer thickness, exposing the Soil Organic Matter (SOM) to stronger biodegradation by soil microorganisms, and resulting in a greater production of CO_2 (under conditions of aerobic respiration) or methane (in the absence of oxygen, as in peat bogs).

On the other hand, milder temperatures and the unfrozen soil foster the growth of plants and will allow species with greater biomass to shift further north; the greater number of snow-free days might lengthen the period in which the vegetation is able to perform photosynthesis and thus absorb carbon dioxide; the higher concentration of CO_2 in the atmosphere also exerts a fertilising effect, further increasing the plant biomass and the associated primary productivity.

On top of this, multiple feedbacks in different directions are foreseen, but are difficult to quantify: less snow means darker soil and then warmer conditions. More vegetation means more water transpiration with a cooling effect on the surface soil temperature.

The quantification of these processes is a current matter of investigation and debate across the scientific community. Records from circumpolar measurements of the Thermal State of Permafrost confirm that permafrost temperatures are increasing, potentially leading to an estimated loss of carbon of 120 ± 85 Pg of carbon emissions by 2100 [78]. Whether these emissions will be offset by the increase of primary productivity is still unclear, as both observed and modelled prevailing trends of primary productivity and soil respiration show high uncertainty [79–81].

For example, time series analysis of vegetation indices from Remote Sensing over the Arctic reveals an increasing greenness trend [82] but also a mixed greening—browning tendency depending on the latitude and the region [83–85].

Uncertainties in current and future estimates of the global Arctic carbon fluxes are mainly due to the poor data and estimates, especially for what concerns winter measurements [86], and to regional differences in CO_2 emission and absorption rates. An effort is necessary to reduce such uncertainties and to include the effects of the permafrost climate feedbacks in the future climate projections [78].

A special effort is necessary to better quantify winter carbon fluxes, which is one of the "great unknown" of the Arctic: measurements of Arctic CO_2 fluxes often refer to the short summer season, when the vegetation is photosynthesising and the 'active layer' is thawed. However, recent assessments have led to the conclusion that even in winter months, thanks to the 'protective shield' of the snowpack, the metabolism of micro-organisms does not cease to function, and that therefore CO_2 emissions, albeit of a lesser intensity, continue and make up a significant part of the annual balance. For this reason, probably the CO_2 balance must be revised upwards [86]

For improving the quantification and the understanding of such processes all year round and their evolution with time, it is necessary to acquire data, perform experiments and improve models, both diagnostic, i.e. which explain current and past data, and predictive, i.e. which try to estimate what will happen in the future.

Diagnostic modelling, whether 'empirical' (trying to explain experimental data, without going into the details of the processes that generate the observed phenomena) or 'deterministic' (trying to explain the processes underlying the observed phenomena) is fundamental to understand which variables most influence a given behaviour, and to quantify their contribution. Obviously, empirical models require data, whose acquisition is challenging and expensive because of the harsh environmental conditions and the remoteness of the places: the lack of continuous and coherent pan-Arctic data is an issue to be solved to reach reliable carbon budgeting and forecast. Efforts to cover the data and modelling gaps are done by the scientific institutions working in the Arctic regions. Many research projects have been or are devoted to measure the Arctic carbon fluxes, the vegetation growth, the active layer depths and the microbial processes therein at local and pan-Arctic scale and using an array of techniques, including flux chambers, Eddy Covariance and Remote Sensing (e.g. see the EU funded projects: PERCCOM (https://cordis.europa.eu/project/id/333796); Page21 (https://cordis.europa.eu/project/id/282700) and the US funded project: Arctic-Boreal Vulnerability Experiment (ABoVE: https://daac.ornl.gov/cgi-bin/dataset_lister.pl?p=34); a database of CO_2 fluxes in the Arctic can be found at: https://daac.ornl.gov/ABOVE/guides/Arctic_Boreal_CO2_Flux.html).

Among those, the Critical Zone Observatory at the research base of Ny Ålesund (Svalbard, NO), managed by the Italian National Research Council, is devoted to measure and model CO_2 fluxes [87], active layer depth and vegetation cover. There, intense summer field work together with the recent installation of new instruments, such as the Eddy Covariance tower installed in 2019, will lead to a better quantification of both summer and winter CO_2 fluxes. A new set of sensors to measure winter CO_2 fluxes from the snowpack has been installed in 2024.

The improvement of remote sensing technologies can partially overcome the difficulties of monitoring pan-Arctic regions, providing better estimates of the greenness and contributing to improve modelling of the permafrost status [84–90]. However, two challenges need to be addressed: (a) the scarcity of cloud-free summer days results in very few available optical images, and (b) models that relate primary productivity and ecosystem respiration to remote sensing observations still need ground-based calibration and validation data, which are sparse and discontinuous. Hence, only a coordinated effort to build a network of ground-based measurements and satellite constellations with a frequent revisit time can overcome these issues.

In addition to CO_2 emissions, another major unknown in permafrost thaw is the formation and release of methane (or natural gas), a climate-altering gas twenty times stronger than CO_2. Methane emission from permafrost to the atmosphere happen in "escape points" such as lakes, ponds and "pingoes"; in this last case, the release of methane is due to pockets of gas formed in the past and accumulated in 'lenses' within the permafrost or under the glacier beds, ready to escape into the atmosphere as a result of glacier melting and fracturing, or the opening of pores in the thawed soil [91–93]. Lakes constitute an escape route of methane through groundwater discharge [94]. The 'thermokarsting', i.e. the collapse of the soil due to thawing and run-off of water, generates methane: the melting of ice in the soil leaves voids over which the overlying soil collapses. In the depressions thus formed, new wetlands are created in which methane-producing anaerobic bacteria prevail [95]. These emissions are localised, and can be intense but discontinuous, thus difficult to predict, quantify and intercept, contributing to the overall uncertainty of GHG emissions from the Arctic [96].

Unfortunately, the lack of a comprehensive quantification of winter emissions and methane release makes it likely that GHG emissions are currently underestimated, and that the Arctic greening will not be able to compensate for their increase. The question is therefore not whether the net CO_2 (and more general, GHG) emissions from the tundra will increase due to global warming, but how much and at what pace, and what will be the consequences of this positive feedback on both Arctic climate and ecosystems and global climate change.

The IPCC "Special Report on the Ocean and Cryosphere in a Changing Climate" allows us to understand what to expect in the future for the permafrost areas with respect to various scenarios of anthropogenic GHG emissions, through an accurate synthesis of the results of scientific publications in the field: it concludes that permafrost thaw will continue throughout this century and beyond, as an effect of the 'positive feedback' of the Arctic amplification. In the most restrictive anthropogenic emission scenario, a decrease between 10 and 40 per cent in the permafrost area is expected by 2100, extending to 50–90 per cent in the worst case, with total emissions estimated to be half of the anthropogenic emissions in the worst emission scenario considered in the climate models [97].

Against this dramatic backdrop, the positive news is that effective anthropogenic abatement policies, by limiting global warming, can also significantly curb the rate at which permafrost thaws and mitigate the resulting CO_2 or methane emissions. It is therefore up to us to take action to avoid future climate instabilities.

6 Summary and Conclusions

The biological carbon cycle is deeply rooted in biotic and abiotic ecosystem processes, and it is driven by the exchanges of matter and energy across the food web and ecosystem components. The exchanges of gases such as CO_2 and methane (CH_4) between soil, vegetation and atmosphere in turn influence (and are influenced by) climate through the feedbacks over the atmospheric GHG concentrations. Carbon is contained in all compartments of the Critical Zone, that is vegetation, fungi, microbiota and animals as well as rocks, soil and water, and is exchanged across all of them and between them and the atmosphere. Thus, the carbon cycle has a fundamental place in the dynamics of the Critical Zone. The vegetation primary productivity fuels the food web as it is the gateway for letting matter and energy enter into ecosystems, while the Net Biome Productivity (NBP) determines the magnitude and direction of the overall climate feedback of terrestrial ecosystems. The importance of the underlying regulatory mechanisms of the carbon cycle cannot be overestimated, and a special effort is thus necessary to understand the associated mechanisms and to assess carbon pools and fluxes. The first step is measuring and modelling the carbon (CO_2 and CH_4) fluxes for the different biomes and understanding their drivers at all scales, together with an assessment of the carbon pools. To this end, a coordinated global effort is done with both ground measurement networks and space-born sensors. Improvements to fill data gaps in the most remote regions and in less developed countries, as well as in measurement techniques and modelling tools, will allow to increase of the accuracy and representativeness of measurements. Understanding the temporal changes of the fluxes and pools, and the magnitude of their feedbacks, is the next step for evaluating the response and the future of the Critical Zone as a life support system of ecosystems and human societies.

Acknowledgements We thank prof. Michele Freppaz from the University of Turin for careful reading of this chapter and the very useful suggestions on how to improve it.

References

1. Barral A, Gomez B, Fourel F et al (2017) CO_2 and temperature decoupling at the million-year scale during the Cretaceous Greenhouse. Sci Rep 7:8310. https://doi.org/10.1038/s41598-017-08234-0
2. Canadell JG, Monteiro PMS, Costa MH, Cotrim da Cunha L, Cox PM et al (2021) Global carbon and other biogeochemical cycles and feedbacks. In: Masson-Delmotte V, Zhai P, Pirani A, Connors SL, Péan C, Berger S, Caud N, Chen Y, Goldfarb L, Gomis MI, Huang M, Leitzell K, Lonnoy E, Matthews JBR, Maycock TK, Waterfield T, Yelekçi O, Yu R, Zhou B (eds.) Climate change 2021: the physical science basis. Contribution of working group I to the sixth assessment report of the intergovernmental panel on climate change. Cambridge University Press, pp 673–816 https://doi.org/10.1017/9781009157896.007
3. Crisp D, Dolman H, Tanhua T, McKinley GA, Hauck J, Bastos A et al (2022) How well do we understand the land-ocean-atmosphere carbon cycle? Rev Geophys 60:e2021RG000736. https://doi.org/10.1029/2021RG000736
4. Ussiri DAN, Lal R (2017) Carbon sequestration for climate change mitigation and adaptation. Publisher, Springer Cham. https://doi.org/10.1007/978-3-319-53845-7
5. Baldovin M, Cecconi F, Provenzale A et al (2022) Extracting causation from millennial-scale climate fluctuations in the last 800 kyr. Sci Rep 12:15320. https://doi.org/10.1038/s41598-022-18406-2
6. Lüthi D, Le Floch M, Bereiter B et al (2008) High-resolution carbon dioxide concentration record 650,000–800,000 years before present. Nature 453:379–382. https://doi.org/10.1038/nature06949
7. Friedlingstein P et al (2022) Global carbon budget 2022. Earth Syst Sci Data 14:4811–4900. https://doi.org/10.5194/essd-14-4811-2022
8. Cai W, Prentice IC (2020) Recent trends in gross primary production and their drivers: analysis and modelling at flux-site and global scales. Environ Res Lett 15:4050. https://doi.org/10.1088/1748-9326/abc64e
9. Jiang M, Medlyn BE, Drake JE et al (2020) The fate of carbon in a mature forest under carbon dioxide enrichment. Nature 580:227–231. https://doi.org/10.1038/s41586-020-2128-9
10. Schlesinger W, Lichter J (2001) Limited carbon storage in soil and litter of experimental forest plots under increased atmospheric CO_2. Nature 411:466–469. https://doi.org/10.1038/35078060
11. Price JT, Warren R (2016) Literature review of the potential of "blue carbon" activities to reduce emissions. 1104872/AVOID2 WPE.2 Report 1. https://docslib.org/doc/10123770/blue-carbon-activities-to-reduce-emissions
12. Brasier MD, Green O, Jephcoat AP, Grassineau NV et al (2002) Questioning the evidence for Earth's oldest fossils. Nature 416(6876):76–81. https://doi.org/10.1038/416076a
13. Ponomarenko AG (2015) Early evolutionary stages of soil ecosystems. Biol Bull Rev 5:267–279. https://doi.org/10.1134/S207908641503007X
14. Orcutt BN, Daniel I, Dasgupta R (eds) (2019) Deep carbon: Past to present. Cambridge University Press. ISBN: 9781108677950. https://doi.org/10.1017/9781108677950
15. Shi Z, Allison SD, He Y et al (2020) The age distribution of global soil carbon inferred from radiocarbon measurements. Nat Geosci 13:555–559. https://doi.org/10.1038/s41561-020-0596-z
16. Jackson RB, Lajtha K, Crow SE, Hugelius G, Kramer MG, Piñeiro G (2017) The ecology of soil carbon: pools, vulnerabilities, and biotic and abiotic controls. Annu Rev Ecol Evol Syst 48(1):419–445
17. Schlesinger WH, Bernhardt ES (2013) Biogeochemistry—an analysis of global change, 3rd edn. Academic Press, Elsevier. https://doi.org/10.1016/C2010-0-66291-2
18. Ciais P, Sabine CL, Govindasamy B, Bopp L, Brovkin V, Canadell J et al (2013). Carbon and other boigeochemical cycles. In: Stocker TF, Qin D, Plattner G-K et al (eds) Climate change 2013: physical science basis. Contribution of working group I to the fifth assessment report of

the intergovernmental panel on climate change. Cambridge University Press, Cambridge, UK, New York, USA, p 465–570

19. Jansson C, Wullschleger SD, Kalluri UC, Tuskan GA (2010) Phytosequestration: carbon biosequestration by plants and the prospects of genetic engineering. Bioscience 60(9):685–696. https://doi.org/10.1525/bio.2010.60.9.6
20. Ciais P, Yao Y, Gasser T, Baccini A, Wang Y, Lauerwald R et al (2020) Empirical estimates of regional carbon budgets imply reduced global soil heterotrophic respiration. Natl Sci Rev 8(2). https://doi.org/10.1093/nsr/nwaa145
21. Ruimy A, Saugier B, Dedieu G (1994) Methodology for the estimation of terrestrial net primary production from remotely sensed data. J Geophys Res-Atmos 99(D3):5263–5283. https://doi.org/10.1029/93jd03221
22. Field CB, Behrenfeld MJ, Randerson JT, Falkowski P (1998) Primary production of the biosphere: integrating terrestrial and oceanic components. Science 281(5374):237–240. https://doi.org/10.1126/science.281.5374.237
23. Ito (2011) A historical meta-analysis of global terrestrial net primary productivity: are estimates converging? Global Change Biol 17(10):3161–3175. https://doi.org/10.1111/j.1365-2486.2011.02450.x
24. Running SW (2012) A measurable planetary boundary for the biosphere. Science 337(6101):1458–1459. https://doi.org/10.1126/science.1227620
25. Lloyd J, Farquhar GD (1996) The CO_2 dependence of photosynthesis, plant growth responses to elevated atmospheric CO_2 concentrations and their interaction with plant nutrient status. Funct Ecol 10:4–32. https://doi.org/10.2307/2390258
26. Waring RH, Landsberg JJ, Williams M (1998) Net primary production of forests: a constant fraction of gross primary production? Tree Physiol 18(2):129–134
27. Ito A (2020) Constraining size-dependence of vegetation respiration rates. Sci Rep 10(1):4304. https://doi.org/10.1038/s41598-020-61239-0
28. Prentice IC, Farquhar GD, Fasham MJR, Goulden ML, Heimann M, Jaramillo VJ, Kheshgi HS, Quéré CL, Scholes RJ, Wallace DWR (2001) The carbon cycle and atmospheric carbon dioxide. In: Houghton JT, Ding Y, Griggs DJ et al (eds) Climate change 2001: the scientific basis. Contribution of working group I to the third assessment report of the intergovernmental panel on climate change. Cambridge University Press, Cambridge, United Kingdom, New York, USA, pp 183–237
29. Bridgham SD, Megonigal JP, Keller JK et al (2006) The carbon balance of North American wetlands. Wetlands 26:889–916. https://doi.org/10.1672/0277-5212(2006)26[889:TCBONA]2.0.CO;2
30. Tarnocai C, Canadell JG, Schuur EAG, Kuhry P, Mazhitova G, Zimov S (2009) Soil organic carbon pools in the northern circumpolar permafrost region. Global Biogeochem Cycles 23.https://doi.org/10.1029/2008GB003327
31. Woodward FI (1987) Stomatal numbers are sensitive to increases in CO_2 from pre-industrial levels. Nature 327:617–618. https://doi.org/10.1038/327617a0
32. Tiwari M, Singh AK, Sinha DK (2015) Chapter 3 - stable isotopes: tools for understanding past climatic conditions and their applications in chemostratigraphy. In: Ramkumar M (ed) Chemostratigraphy. Elsevier, pp 65–92. https://doi.org/10.1016/B978-0-12-419968-2.00003-0
33. Novoa R, Loomis RS (1981) Nitrogen and plant production. Plant Soil 58:177–204. https://doi.org/10.1007/BF02180053
34. Drake BG, Gonzàlez-Meler MA (1997) MORE EFFICIENT PLANTS: a consequence of rising atmospheric CO_2? Annu Rev Plant Physiol Plant Mol Biol 48(1):609–639. https://doi.org/10.1146/annurev.arplant.48.1.609
35. Reich PB, Hungate BA, Luo Y (2006) Carbon-nitrogen interactions in terrestrial ecosystems in response to rising atmospheric carbon dioxide. Annu Rev Ecol Evol Syst 37(1):611–636. https://doi.org/10.1146/annurev.ecolsys.37.091305.110039
36. Kotsyurbenko OR, Glagolev MV, Merkel AY, Sabrekov AF, Terentieva IE (2019) Methanogenesis in Soils, Wetlands, and Peat. In: Stams A, Sousa D (eds) Biogenesis of hydrocarbons. handbook of hydrocarbon and lipid microbiology. Springer Cham. https://doi.org/10.1007/978-3-319-78108-2_9

37. Andrews J, Harrison K, Matamala R, Schlesinger W (1999) Separation of root respiration from total soil respiration using carbon-13 labeling during free-air carbon dioxide enrichment (FACE). Soil Sci Soc Am J 63:1429–1435. https://doi.org/10.2136/sssaj1999.6351429x
38. Raich JW, Mora G (2005) Estimating root plus rhizosphere contributions to soil respiration in annual croplands. Soil Sci Soc Am J 69(3):0361–5995. https://doi.org/10.2136/sssaj2004.0257
39. Lloyd J, Taylor JA (1994) On the temperature dependence of soil respiration. Funct Ecol 8(3):315–23. JSTOR. https://doi.org/10.2307/2389824
40. Davidson E, Janssens I (2006) Temperature sensitivity of soil carbon decomposition and feedbacks to climate change. Nature 440:165–173. https://doi.org/10.1038/nature04514
41. Karhu K, Auffret M, Dungait J et al (2014) Temperature sensitivity of soil respiration rates enhanced by microbial community response. Nature 513:81–84. https://doi.org/10.1038/nature13604
42. Schlesinger WH, Andrews JA (2000) Soil respiration and the global carbon cycle. Biogeochemistry 48:7–20. https://doi.org/10.1023/A:1006247623877
43. Stocker TF et al (eds) (2013). Climate change 2013: the physical science basis. Contribution of working Group I to the fifth assessment report of the intergovernmental panel on climate change. Cambridge Univ. Press
44. Barnard RL, Blazewicz SJ, Firestone MK (2020) Rewetting of soil: Revisiting the origin of soil CO_2 emissions. Soil Biol Biochem 147:s107819, ISSN 0038-0717.https://doi.org/10.1016/j.soilbio.2020.107819
45. Graham L, Risk D (2018) Explaining CO_2 fluctuations observed in snowpacks. Biogeosciences 15:847–859. https://doi.org/10.5194/bg-15-847-2018
46. Musselman RC, Massman WJ, Frank JM, Korfmacher JL (2005) The temporal dynamics of carbon dioxide under snow in a high elevation rocky mountain subalpine forest and meadow. Arct, Antarct, Alp Res 37(4):527–538. http://www.jstor.org/stable/4095872
47. Seok B, Helmig D, Williams MW et al (2009) An automated system for continuous measurements of trace gas fluxes through snow: an evaluation of the gas diffusion method at a subalpine forest site, Niwot Ridge, Colorado. Biogeochemistry 95:95–113. https://doi.org/10.1007/s10533-009-9302-3
48. Sullivan BW, Dore S, Montes-Helu MC, Kolb TE, Hart SC (2012) Pulse emissions of carbon dioxide during snowmelt at a high-elevation site in Northern Arizona, U.S.A. Arct, Antarct, Alp Res 44(2):247–254. http://www.jstor.org/stable/23252818
49. Livingston GP, Hutchinson GL (1995) Enclosure-based measurement of trace gas exchange: applications and sources of error. In: Matson PA, Harris RC (eds) Biogenic trace gases: measuring emissions from soil and water. Blackwell Science, Oxford, pp 14–51
50. Baneschi I, Raco B, Magnani M, Giamberini MS, Lelli M, Mosca P, Provenzale A, Coppo L, Guidi M (2022) Non-steady-state closed dynamic chamber to measure soil CO_2 respiration: a protocol to reduce uncertainty. Front Environ Sci 10.https://doi.org/10.3389/fenvs.2022.1048948
51. Baldocchi DD (2003) Assessing the eddy covariance technique for evaluating carbon dioxide exchange rates of ecosystems: past, present and future. Glob Change Biol 9:479–492. https://doi.org/10.1046/j.1365-2486.2003.00629.x
52. Baldocchi DD (2008) 'Breathing' of the terrestrial biosphere: lessons learned from a global network of carbon dioxide flux measurement systems. Aust J Bot 56(1):26. https://doi.org/10.1071/BT07151
53. Reichstein M, Falge E, Baldocchi D, Papale D, Aubinet M et al (2005) On the separation of net ecosystem exchange into assimilation and ecosystem respiration: review and improved algorithm. Global Change Biol 11:1424–1439. https://doi.org/10.1111/j.1365-2486.2005.001002.x
54. Pastorello G, Trotta C, Canfora E et al (2020) The FLUXNET2015 dataset and the ONEFlux processing pipeline for eddy covariance data. Sci Data 7:225. https://doi.org/10.1038/s41597-020-0534-3

55. Rebmann C, Aubinet M, Schmid H, Arriga N, Aurela M, Burba G, Clement R et al (2018) ICOS eddy covariance flux-station site setup: a review. Int Agrophys 32(4):471–494. https://doi.org/10.1515/intag-2017-0044
56. Kim DG, Bond-Lamberty B, Ryu Y, Seo B, Papale D (2022) Ideas and perspectives: enhancing research and monitoring of carbon pools and land-to-atmosphere greenhouse gases exchange in developing countries. Biogeosciences 19:1435–1450. https://doi.org/10.5194/bg-19-1435-2022
57. Schimel D, Pavlick R, Fisher JB, Asner GP, Saatchi S, Townsend P, Miller C, Frankenberg C, Hibbard K, Cox P (2015) Observing terrestrial ecosystems and the carbon cycle from space. Glob Change Biol 21:1762–1776. https://doi.org/10.1111/gcb.12822
58. Lenton TM, Held H, Kriegler E, Hall JW, Lucht W, Rahmstorf S, Schellnhuber HJ (2008) Tipping elements in the Earth's climate system. PNAS 105(6):1786–1793. https://doi.org/10.1073/pnas.0705414105
59. Lieth H, Whittaker RH (Eds) (1975) Primary productivity of the biosphere. Springer, New York.https://doi.org/10.1007/978-3-642-80913-2
60. Badgley G et al (2017) Canopy near-infrared reflectance and terrestrial photosynthesis. Sci Adv 3:e1602244. https://doi.org/10.1126/sciadv.1602244
61. Zhang H, Li J, Liu Q, Lin S, Huete A, Liu L, Croft H et al (2022) A novel red-edge spectral index for retrieving the leaf chlorophyll content. Methods Ecol Evol 13:2771–2787. https://doi.org/10.1111/2041-210X.13994
62. Wu W, Epstein HE, Guo H, Li X, Gong C (2021) A pigment ratio index based on remotely sensed reflectance provides the potential for universal gross primary production estimation. Environ Res Lett 16:054065. https://doi.org/10.1088/1748-9326/abf3d
63. Monteith JL (1972) Solar radiation and productivity in tropical ecosystems. J Appl Ecol 9(3):747–766. https://www.jstor.org/stable/2401901
64. Monteith JL (1977) Climate and the efficiency of crop production in Britain. Phil Trans R Soc Lond B 28:1277–1294. https://doi.org/10.1098/rstb.1977.0140
65. Cheng YB, Zhang Q, Lyapustin AI, Wang Y, Middleton EM (2014) Impacts of light use efficiency and fPAR parameterization on gross primary production modeling. Agric For Meteorol 189–190:187–197. https://doi.org/10.1016/j.agrformet.2014.01.006
66. Liang S, Wang J (eds) (2020) Estimate of vegetation production of terrestrial ecosystem, Chapter 15. In: Advanced remote sensing, 2nd edn. Academic Press, p 581–620. https://doi.org/10.1016/B978-0-12-815826-5.00015-5
67. Frankenberg C et al (2011) New global observations of the terrestrial carbon cycle from GOSAT: Patterns of plant fluorescence with gross primary productivity. Geophys Res Lett 38.https://doi.org/10.1029/2011GL048738
68. Joiner J, Yoshida Y, Vasilkov AP, Yoshida Y, Corp LA, Middleton EM (2011) First observations of global and seasonal terrestrial chlorophyll fluorescence from space. Biogeosciences 8:637–651. https://doi.org/10.5194/bg-8-637-2011
69. Norton AJ, Rayner PJ, Koffi EN, Scholze M, Silver JD, Wang Y-P (2019) Estimating global gross primary productivity using chlorophyll fluorescence and a data assimilation system with the BETHY-SCOPE model. Biogeosciences 16:3069–3093. https://doi.org/10.5194/bg-16-3069-2019
70. Lee YK (2020) The arctic plants of Svalbard, what we learn from the green in the treeless white world. Springer Cham. https://doi.org/10.1007/978-3-030-34560-0
71. Nabe-Nielsen J, Normand S, Hui FKC et al (2017) Plant community composition and species richness in the High Arctic tundra: From the present to the future. Ecol Evol 7:10233–10242. https://doi.org/10.1002/ece3.3496
72. Schuur EAG et al (2015) Climate change and the permafrost carbon feedback. Nature 520(7546):171–179. https://doi.org/10.1038/nature14338
73. Billings WD (1987) Carbon balance of Alaskan tundra and taiga ecosystems: past, present and future. Quatern Sci Rev 6(2):165–177. https://doi.org/10.1016/0277-3791(87)90032-1
74. He Y, Jones MC, Zhuang Q, Bochicchio C, Felzer BJ, Mason E, Yu Z (2014) Evaluating CO_2 and CH_4 dynamics of Alaskan ecosystems during the Holocene thermal maximum. Quatern Sci Rev 86:63–77. https://doi.org/10.1016/j.quascirev.2013.12.019

75. Lindgren A, Hugelius G, Kuhry P (2018) Extensive loss of past permafrost carbon but a net accumulation into present-day soils. Nature 560:219–222. https://doi.org/10.1038/s41586-018-0371-0
76. Loisel J, Yu Z, Beilman DW, Camill P, Alm J et al (2014) A database and synthesis of northern peatland soil properties and Holocene carbon and nitrogen accumulation. The Holocene 24(9):1028–1042. https://doi.org/10.1177/0959683614538073
77. Rantanen M, Karpechko AY, Lipponen A et al (2022) The Arctic has warmed nearly four times faster than the globe since 1979. Commun Earth Environ 3:168. https://doi.org/10.1038/s43247-022-00498-3
78. Schaefer K, Lantuit H, Romanovsky VE, Schuur EAG, Witt R (2014) The impact of the permafrost carbon feedback on global climate. Environ Res Lett 9(8). https://doi.org/10.1088/1748-9326/9/8/085003
79. Belshe EF, Schuur EAG, Bolker BM (2013) Tundra ecosystems observed to be CO_2 sources due to differential amplification of the carbon cycle. Ecol Lett 16:1316–1324. https://doi.org/10.1111/ele.12164
80. López-Blanco E, Exbrayat JF, Lund M, Christensen TR, Tamstorf MP, Slevin D, Hugelius G, Bloom AA, Williams M (2019) Evaluation of terrestrial pan-Arctic carbon cycling using a data-assimilation system. Earth Syst Dyn 10:233–255. https://doi.org/10.5194/esd-10-233-2019
81. McGuire AD, Christensen TR, Hayes D, Heroult A et al (2012) An assessment of the carbon balance of Arctic tundra: comparisons among observations, process models, and atmospheric inversions. Biogeosciences 9:3185–3204. https://doi.org/10.5194/bg-9-3185-2012
82. Potter C (2014) Regional analysis of NASA satellite greenness trends for ecosystems of arctic Alaska. Int J Geosci 5:997–1006. https://doi.org/10.4236/ijg.2014.59085
83. Berner LT, Massey R, Jantz P et al (2020) Summer warming explains widespread but not uniform greening in the Arctic tundra biome. Nat Commun 11:4621. https://doi.org/10.1038/s41467-020-18479-5
84. Fu D, Su F, Wang J, Sui Y (2019) Patterns of arctic tundra greenness based on spatially downscaled solar-induced fluorescence. Remote Sens 11(12):1460. https://doi.org/10.3390/rs11121460
85. Li ZL, Mu C, Chen X, Wang XY (2021) Changes in net ecosystem exchange of CO_2 in Arctic and their relationships with climate change during 2002–2017. Adv Clim Chang Res 12(4):475–481. https://doi.org/10.1016/j.accre.2021.06.004
86. Natali SM, Watts JD, Rogers BM et al (2019) Large loss of CO_2 in winter observed across the northern permafrost region. Nat Clim Chang 9:852–857. https://doi.org/10.1038/s41558-019-0592-8
87. Magnani M, Baneschi I, Giamberini M et al (2022) Microscale drivers of summer CO_2 fluxes in the Svalbard High Arctic tundra. Sci Rep 12:763. https://doi.org/10.1038/s41598-021-04728-0
88. Nelson PR, Maguire AJ, Pierrat Z, Orcutt EL, Yang D, Serbin S, Frost GV et al (2022) Remote sensing of tundra ecosystems using high spectral resolution reflectance: opportunities and challenges. J Geophys Res: Biogeosci 127(2). https://doi.org/10.1029/2021JG006697
89. Trofaier AM, Westermann S, Bartsch A (2017) Progress in space-borne studies of permafrost for climate science: towards a multi-ECV approach. Remote Sens Environ 203:55–70. https://doi.org/10.1016/j.rse.2017.05.021
90. Ma D, Wu X, Ma X, Wang J, Lin X, Mu C (2021) Spatial, phenological, and inter-annual variations of gross primary productivity in the arctic from 2001 to 2019. Remote Sens 13(15):2875. https://doi.org/10.3390/rs13152875
91. Hodson AJ, Nowak A, Hornum MT et al (2020) Sub-permafrost methane seepage from open-system pingos in Svalbard. Cryosphere 14:3829–3842. https://doi.org/10.5194/tc-14-3829-2020
92. Kleber GE, Hodson AJ, Magerl L et al (2023) Groundwater springs formed during glacial retreat are a large source of methane in the high Arctic. Nat Geosci 16:597–604. https://doi.org/10.1038/s41561-023-01210-6

93. Kohnert K, Serafimovich A, Metzger S et al (2017) Strong geologic methane emissions from discontinuous terrestrial permafrost in the Mackenzie Delta. Canada. Sci Rep 7:5828. https://doi.org/10.1038/s41598-017-05783-2
94. Olid C, Rodellas V, Rocher-Ros G et al (2022) Groundwater discharge as a driver of methane emissions from Arctic lakes. Nat Commun 13:3667. https://doi.org/10.1038/s41467-022-31219-1
95. Li L, Xue B (2021) Methane emissions from northern lakes under climate change: a review. Springer Nat Appl Sci 3:883. https://doi.org/10.1007/s42452-021-04869-x
96. Schuur EAG et al (2022) Permafrost and climate change: carbon cycle feedbacks from the warming arctic. Annu Rev Environ Resour 47(1):343–371. https://doi.org/10.1146/annurev-environ-012220-011847
97. Meredith M, Sommerkorn M, Cassotta S et al (2019) Polar Regions. In: Pörtner H-O, Roberts DC, Masson Delmotte V, Zhai P, Tignor M, Poloczanska E, Mintenbeck K, Alegría A, Nicolai M, Okem A, Petzold J, Rama B, Weyer NM (eds) IPCC special report on the ocean and cryosphere in a changing climate. Cambridge University Press, Cambridge, UK and New York, NY, USA, pp 203–320. https://doi.org/10.1017/9781009157964.005

Models, Digital Twins and Virtual Research Environments

Marta Magnani and Antonello Provenzale

Abstract The intrinsic complexity of the Critical Zone (CZ) poses unique challenges to modelling. The high number of components and scales involved in the CZ dynamics allows to describe this zone from different perspectives, thus resulting in a variety of modelling methods. This chapter provides an overview of the issues encountered when modelling the CZ, and the main approaches applied to date, spanning from conceptual and classical models to machine learning techniques and digital twins. Future perspectives to advance our knowledge of the CZ are then discussed. These includes: (i) the creation of open and accessible datasets that assist in data assimilation within models; (ii) global alliance of scientists sharing objectives, vocabulary, knowledge, and possibly coding languages; (iii) platforms fostering collaborations and interactive modelling, such as virtual research environments. Such efforts would lead to the synthesis of knowledge from different disciplines, working groups and networks to finally achieve a "whole-system" approach capable of reproducing most CZ processes at once.

The synergistic use of modeling, mapping, and monitoring is a promising strategy to address the complexity of the Critical Zone (CZ) [1, 2]. While mapping provides snapshots of spatial patterns, monitoring records the temporal changes of such patterns, together informing model inputs, validating the outputs, and supplying information for the representation of the processes at work. In turn, models are useful tools for the quantitative assessment, understanding and prediction of CZ dynamics. These can for instance convey crucial information for the design of new observatories [3]. Hence, modeling-mapping-monitoring is an inseparable triad, in which each of the components is essential. Keeping in mind such a framework, this chapter provides an overview of the characteristics (Sect. 1), approaches (Sect. 2) and challenges (Sect. 3) of modeling the CZ.

M. Magnani (✉) · A. Provenzale
Institute of Geosciences and Earth Resources, National Research Council, Pisa, Italy
e-mail: marta.magnani@igg.cnr.it

T. White and A. Provenzale (eds.), *Critical Zone and Ecosystem Dynamics*,
Advances in Critical Zone Science, https://doi.org/10.1007/978-3-031-69076-1_11

1 The "Organized Complexity" of the CZ

The Critical Zone is an open system, where the balance of energy and matter is influenced by lateral transports and vertical fluxes [2–4]. At its lower limit, the chemical and physical weathering of the bedrock produces the regolith that enters the CZ layer (e.g., [5]). At the upper limit of the CZ, vegetation-soil-atmosphere exchanges of moisture and gasses, molecular deposition on soils, solar irradiance and heat fluxes are some examples of how CZ interacts with external reservoirs [3, 6]. In addition, water runoff and soil and regolith sliding along hill and mountain slopes transport the elements in and out of the CZ (e.g., [7, 8]).

Nonetheless, the CZ is not a passive constituent of Earth's climate system. This is the "living reactor" of our planet [9, 10], that links the subsurface with the atmosphere, through which many compounds are cycled, transformed, or produced [11]. Hence, CZ modelling mostly addresses local processes, whose connection with large scale circulations and regional (or global) climate is twofold. On the one hand, environmental response to global changes often takes place at local scale [12]. On the other, local feedbacks may propagate to regional and continental scale through cross-scale effects [13, 14]. The vision of scale emergence suggests that if the whole complex of feedbacks is positive on all scales, then critical transitions in the large-scale climate system may occur [15, 16]. It is worth recalling here that large-scale transitions due to cross-scale emergence may, in turn, feedback to local systems. From this, one can appreciate the importance of modeling CZ processes for local dynamics and possibly also for planetary-scale dynamics.

Processes occurring within the CZ are coupled with the environment outside the CZ and also across themselves. The CZ has been often conceptualized as a "layer cake" made of atmosphere, vegetation, root zone, deeper vadose, saturated aquifer, and bedrock, with general trends of increasing density and response time (i.e., the time needed to reach a new steady state after perturbation) with depth [2]. The vertical layers, which are intrinsically heterogeneous because include both biotic and abiotic components, are separated by interfaces and interact among them through the cycles of water, mass, and energy [10, 17]. In turn, these cycles regulate the formation, transformation and functioning of the layers [18]. Probably the most intuitive example is the water cycle, which involves all layers, thus linking subsurface processes with surface dynamics, influencing soil geochemistry, determining the resources for living organisms, and contributing to the physical transformation of the whole system [19]. In wider terms, the combination of all the cycles makes the CZ a strongly coupled environment [20], in which water plays a crucial role.

The cycles occurring in the CZ, and the related processes, span across several space and time scales, from seconds to eons and from molecular to global dimensions (NRC [17, 21, 22]), thus requiring a multiscale framework [23]. Most of these cycles lay in the range between the slow geological processes, involving rock weathering, erosion, transport, and deposition, sedimentation in the oceans, lithification, tectonic and volcanic activities, and the fast biological cycles, involving plant productivity, litterfall, accumulation and decomposition of organic materials, and trophic chain processes [20]. Most cycles include both the fast and the slow components (e.g., [17]) and tie the dynamics of an element to another, as in the case of carbon and water [24].

The scale chosen to represent the CZ system dictates the degree of complexity of the model. Depending on the scale, matter can be considered as either uniform and continuous or heterogeneous and discrete [1, 9, 25]. The intrinsic heterogeneity and patchiness of some surface processes can result in "hot spots" and/or "hot moments" [26], which are respectively areas or periods featuring abrupt changes in the characteristic rates of the processes, such as biogeochemical reaction rates, relative to the surrounding environment or relative to the typical behavior over long time periods. Although the definition of hot spots and hot moments has been originally formulated for biogeochemical cycling, this can be easily generalized to any process or variable. The rate shift of hot spots and hot moments can be occasional and episodic (i.e., intermittent) or long-lasting, as a consequence of the intrinsic nonlinearity of surface processes, that often involve feedbacks between the different components. Permanent changes may also occur either gradually or abruptly, when a threshold is exceeded [20]. Clearly, weather a phenomenon is a hot moment, or a permanent state shift depends on the reference domain or period. Apparent irreversible changes can be such only on a limited time span, and they could be part of a more complex temporal dynamics over much longer time scales. And on longer times, the system is usually non-stationary. An example of a process that appears as long-lasting one over human time scales, but that may be an occasional hot moment over geologic time scales, is the 'great acceleration' of climate change since 1950. In a span of a few decades, the changes of the climatic indicators that describe the structure and functioning of the Earth System, such as surface temperature and atmospheric concentration of carbon dioxide [27], have significantly accelerated and we do not know whether such behavior will lead to a "new state", probably not particularly pleasant for human societies. Hence, the same process could be perceived and interpreted from different perspectives, depending on the scale of observation.

In addition, processes can appear as stochastic or having a "noise" component at certain scales. Lin [2] suggested the distinction of three regimes featuring low, intermediate, and high degree of randomness that is associated to the number of constituents or degrees of freedom. At low randomness, systems of *organized simplicity* allow for analytical, mechanistic tractability. At high randomness, systems of *unorganized complexity* are characterized by large populations of constituents that are sufficiently regular to enable statistical approaches, which capture the ensemble behavior while disregarding small fluctuations. The most challenging regime is the intermediate complexity range, where both deterministic and statistical approaches essentially fail owing to the high complexity of the interactions. This is the domain of the so-called *organized complexity*.

As a final remark, a distinctive characteristic of the CZ framework is the focus on the planetary components that are 'critical' to supporting life on Earth. In addition to climate change effects, the anthropic demand for food, water and goods is degrading soil structure and functions, which in turn broadcast this stress to the whole CZ [10, 23]. Hence, CZ models should represent, or at least inform on ecosystem services, human activity, risk management and societal impact [11, 28–33]. This would assist in mitigation policies and restoration practices [10].

All the above characteristics are to be accounted in the representation of the whole CZ. The variety of scales, elements and processes is the origin of the several approaches that are used to address the modeling of the CZ. These approaches are the subject of the next section, which has not to be intended as an exhaustive review of the published models, but rather as an overview of the possible techniques and rationale to address the complex task of modelling the CZ.

2 Modeling Approaches

Given the complexity of the CZ, different modeling approaches are typically adopted. The most intuitive and probably the most common approach is conceptual modeling. Conceptual models are problem-oriented abstractions of a real process, often addressing specific aspects of the system, whose primary objective is to convey the fundamental principles and basic functionality of the real processes [34, 35]. The reader is invited to search for the term 'conceptual model' in the Giardino and Houser book [36], to see examples of CZ conceptual models. These models use simplifications to reduce complexity and possibly assumptions where there are uncertainties, thus leading to a schematic visual representation of natural processes. Conceptual models can also be combined with quantitative models that are implemented in numerical codes, for instance solving classical model equations or building neural networks. An overview of quantitative model characteristics is provided in the next sections.

2.1 Classical Modeling Approaches

Classical models are usually based on governing equations, i.e., on mathematical relationships. Table 1 reports seven of the main classifications of quantitative classical models that correspond to different modeling approaches. This table also provides few examples of such approaches, focusing on models applied in various CZOs across the globe, in the spirit of the modelling-mapping-monitoring strategy. Although the approaches presented in Table 1 are usually considered as antithetic, they would rather be complementary to achieve a whole-system representation of the CZ. Of these dichotomies, the first group is linked to the type of mathematical description, while the second group is identified by the type of theoretical, i.e., conceptual, approach. This list does not pretend to be exhaustive of the whole array of possibilities, it rather introduces the reader to some of the major concepts in CZ modeling.

One of the goals of CZ models is the study of processes that are either poorly understood or underrepresented. Exploratory models are often based on empirical relationships between measured variables, relying on the possibility to obtain information

Table 1 Main types of classical models and illustrative references for each approach applied in CZOs

	Classification	Description	Examples
Mathematical representation	Data-driven	Based on relationships among measured data, often heuristic. Regression models	[37–42]
	Process-based	Based on consolidated knowledge of process characteristics and drivers	[33, 43–49]
	Diagnostic	Describe the response variables at the same time of the drivers	[38–42, 50, 51]
	Prognostic	Forecast the future evolution of the system based on present relationships between the drivers and the response variable	[33, 43– 47, 49, 52, 53]
	Spatially implicit	Also called one-dimensional models, such as multiple-box models or column models. They account for vertical dynamics only	[45, 53]
	Spatially explicit	Account for lateral dynamics and transport. Suitable for studying patterns	[33, 44, 46–49, 52, 54–56]
	Deterministic	The system is described by the dynamics of state variables, usually disregarding smaller-scale fluctuations	[33, 47, 48]
	Stochastic	Include one or more data-driven stochastic components. The system is described in terms of average behavior and probability distributions of variables	[50]

(continued)

Table 1 (continued)

	Classification	Description	Examples
Theoretical representation	Bottom-up	Upscaling. The models infer large-scale relationships from small-scale processes	[3, 51, 52]
	Top-down	Downscaling. The models drive small-scale dynamics from large-scale behavior	[3, 33, 57]
	Cycle-oriented	Follow one compound through the whole CZ, including several layers and processes	[45, 46, 48, 49, 53, 55]
	Interface-oriented	Describe the characteristics and dynamics of one or a few compounds at the interface between different constituents or layers	[39, 40, 52]
	Reductionist	Focused on one or a few specific aspects of the system, aiming at reducing the complexity of the problem	[38, 43, 49, 50, 53, 55]
	Whole-system	Include all the main components of the system. The complexity of model description can depend on the target scale	[33, 46, 47]

directly from the data without, or with just a few, assumptions. The purpose of this approach is to select the best model in representing the data and process, according to some index [58], and the main outcome is a regression or multi-regression model, called *data-driven model*. Beyond helping in representing new processes, data-driven models can also serve as parametrization within process-based models, for instance helping to reduce the computational time. *Process-based models* are based on a priori information on system dynamics and use physically-based equations whose solution may require computer simulations and long computations [59]. As a world of caution, the term data-driven model is sometimes used to describe a process-based model, as defined here, in which the inputs and parameter values are provided by data.

Another goal of CZ modelling is the projection of current processes to future conditions, the so-called "Earthcast" [60], a term coined by Chris Paola at the

National Center for Earth Dynamics in analogy to weather forecasts [61]. Process-based models are usually preferred for this aim. Earthcasting models generally use *prognostic* differential equations that describe the time evolution of a variable of interest in response to the forcing. Since we entered the Anthropocene [62], where human activities are a relevant geologic, hydrologic, atmospheric, and geomorphologic forcing, such an exercise can be particularly useful for management decisions and adaptation strategies [10]. Similar approaches can also be used for the backward extrapolation of the present CZ into the past, based on historical records and proxies [61, 63]. Even parameters, i.e., constant values, used in process-based model equations, may change over long simulation times. For instance, soil structure has been shown to be a dynamic property of the CZ because of climatic and anthropogenic forcings [23]. Parameterizations may thus assist in describing the evolution of dynamic CZ properties, which can be considered as constant over short time scale but evolve over long times. Parameterizations have often the form of *diagnostic* equations, which describe the target variable as a function of the drivers at a certain time. Hence, diagnostic equations are descriptive rather than predictive, but can enter predictive models when coupled with prognostic equations. Of course, it remains to be seen whether a given diagnostic relationship, derived from current data, can still hold under different climatic and environmental conditions.

It is worth noting that the intrinsically local essence of CZOs allows for modelling a single vertical column, that is, a one-dimensional (1D) domain, which represents a prototype system. This corresponds to an (horizontally) *implicit space* model, in which only vertical fluxes are dynamically modeled, while lateral transports are represented by source or sink terms. In contrast, when landscape heterogeneities are relevant, an *explicit space* (2D or 3D) model is preferred. This approach aims at describing both the temporal evolution and the spatial distribution of the system variables; therefore, it is especially relevant for modeling water flows in complex geomorphology, plant self-organization owing to interactions between individuals [64], and more generally pattern formation processes and macroscopic organization [3, 65]. These features are connected to the spontaneous emergence of spatial macroscopic structures owing to nonlinear dynamics, such as feedbacks between processes, compartments, and elements of the system [66]. Hence, the identification and modeling of heterogeneities associated with (or conducive to) pattern formation are a relevant issue for upscaling, downscaling and cross-scale emergence of CZ processes.

Depending on the chosen scale, the process at hand, and even our knowledge about the system, a random component may be included in the model. Examples of random processes can be fire ignition [67], extreme weather events, insect pest outbreaks (which could indeed be described also by a dynamical system [68]). If the random process has general characteristics that can be described using, for instance, probability distributions and statistical moments, the model can be considered as *stochastic*. In most cases, CZ models are *deterministic*, i.e., the target variable or its derivative are univocally defined when the values of the drivers are known. In these models, the parameters in the equations represent average macroscopic characteristics of the system, which subsume microscopic heterogeneities (e.g., soil water diffusivity in Richards' equation corresponds to an average behavior, which varies at the microscale depending on the specific 3D random distribution of soil pore size; [30]). Nonetheless, a stochastic behavior of the simulated variables can arise even in deterministic models from the randomness of an input variable, such as precipitation (which is instead deterministic in a meteorological model). In this sense, the stochastic or deterministic nature of the model is often the choice of the modeler, based on the scales and the processes to be resolved in detail.

Beyond empirical data-driven models, new models used in CZOs can originate from the coupling and modification of existing codes [23, 69]. An example of this approach is the evolution of the Coupled Carbon, Aggregation and Structure Turnover (CAST) model, which mostly focus on soil processes, into the integrated critical zone (1D-ICZ) model, which also includes hydrological and biological processes, as described in [10]. The issue of multiscale processes in the CZ and the interaction of small-scale dynamics with large scale circulations/transports has been addressed by arrays of nested models (see also the multiscale modelling review by [70]). This has been done by building cascades of models, in which the output of one component that simulate a process at a certain scale is used as input to other components that simulate processes at smaller scales, possibly requiring downscaling techniques, not dissimilarly from what is done in climate impact studies. In this framework, a cascade model may consist of a climate model that feeds a dynamic global vegetation model, which in turn is used to drive the weathering processes of a geochemical model. This is a *top-down approach*. The deep interconnection of modelling and observation in CZ science also provides a fertile ground for the inverse, *bottom-up* approach, in which field (plot) studies are upscaled to hillslope, catchment, watershed and regional scales. Clearly, simple averaging is often impossible owing to the nonlinearity of the system (see for instance the fractal property of basalt weathering rate at different scales, [71]). Upscaling approaches can be informed by the intercomparison of studies performed at different CZOs and mapping or can arise from the coupling of models working at different scales and used in distinct disciplines (e.g., [42, 72, 73]).

Representing the CZ structure not only involves multiscale issues but also a potentially large number of compounds and climatic components. Reducing the system complexity is sometimes achieved by either focusing on specific elements, following their cycling through the whole depth and extent of the CZ, or considering only the processes occurring at the interface between two CZ layers or components, such as soil-roots, surface-atmosphere, vadose-ground water, soil–bedrock. These correspond to *cycle-oriented* and *interface-oriented* approaches, respectively. Although many processes can be represented in cycle-oriented and interface-oriented models, these are usually considered as *reductionist* approaches, because they do not consider the CZ in its entirety. In contrast, modeling the structure and functioning of the whole CZ (from the canopy top, throughout the soil and groundwater to the bedrock limit) is the aim of the *whole-system* approach. Whole-system models include all the main CZ components, such as vegetation, microbiota, gases, water, soils, soil organisms and rocks, possibly to the detriment of model detail. Their comprehensive approach is needed to study the CZ as 'one physical system', fundamentally corresponding to

Tansley's idea of the ecosystem, including both the biotic component and the abiotic matrix that sustain and is modified by living organisms [74].

2.2 *Digital Twin and Machine Learning Approaches*

The effort of thinking and representing the CZ as "one physical system" is in the spirit of the Digital Twin (DT) vision. This approach, which was born in industry to check, control and drive the state of an industrial plant and its products, or of a moving vehicle, combines machine learning (ML) with the physics-based knowledge of the system to mirror the dynamics and characteristics of the real system–the physical twin-at both micro- and macroscopic level [75, 76]. Currently, the European Union and the European Space Agency (ESA) are supporting projects and programs to develop DTs of specific components of the Earth System (e.g., the ocean) and of the planet as a whole, largely using as input data the large quantity of Earth observations provided by satellite constellations.

The DT approach involves an array of modeling techniques, and mixes well-established equations and statistical techniques, including correlations, decision trees and neural networks, for the ultimate purpose of reproducing the functioning of the physical twin. As an essential ingredient of DTs, the ML can be thought of as a data-driven approach which does not necessarily produce an equation; it rather identifies patterns within data using different possible mathematical algorithms based on optimization techniques. To this end, the ML learns from data and from the differences between the predicted and actual state of the system. Data are often split into a training and a testing set, which are respectively used to build the network and assess the network performance. After being tested, the network can provide future projections, and can be updated in time using new data to capture possible changes in the system functioning. Similarly, DTs are built on and fed by qualitative and quantitative real-time observations to achieve an accurate description of a real system, the CZ in our case. Such a data-based approach goes beyond the classical concept of modeling that relies on equations, and therefore on assumptions and approximations, in the spirit of "all models are wrong, but some are useful" [77, 78]. The continuous exchange of information between the physical and the virtual twins is the signature of the DT approach. Data assimilation is needed for building the DT, updating its state and optimizing its structure, to provide real-time state estimations, as well as past and future predictions of system trajectories. Notably, the main purpose of DTs until now has been risk assessment, decision making and management (e.g., [79, 80, 81]).

The CZ poses two main challenges to the DT archetype: (i) the physical twin, i.e. the CZ, undergoes continuous changes and adaptations (i.e., CZ is in a non-steady state), thus complicating long term predictions since the DT description is based on the current features of the system; (ii) the continuous data series needed by the DT would be provided by open-air laboratories involving a high number of constituents and degrees of freedom, thus complicating the identification of the most appropriate information to be provided to the DT, possibly requiring huge investments. In this sense, the identification of "CZ Essential Variables", in the same spirit of those already identified for Climate, Ocean and Biodiversity, could be useful (see e.g. https://www.earthdata.nasa.gov/learn/backgrounders/essential-variables). For the CZ, the quest of the DT holy grail is still ongoing, despite few steps forward have been achieved in specific disciplines [42, 82, 83]. As a final remark, efforts in this direction can also be found in ML-based studies, despite often focused on specific aspects of the CZ [84] and specific sites [85].

3 Open Challenges

If CZ is the science of the Anthropocene [86, 87], DT is the modelling approach of the Anthropocene, fully benefitting from advances in information and communication technologies and cloud infrastructures. The DT framework poses new challenges to the scientific community that range from the creation of open and accessible datasets to a global alliance of scientists and the development of huge virtual infrastructures. More generally, these are the challenges posed by the development of a model of any kind (whether classical or ML-based) that aims at representing and forecasting the whole CZ.

First, data assimilation is a crucial point of the DT-modeling paradigm. This will require a common effort in merging data catalogues, identifying prioritized (essential) variables, and gathering harmonized and standardized measurements in observatories across the world [88]. Such a goal should encourage synergies in the observational networks, such as the CZO network [69, 89], the International Long-Term Ecosystem Research network (ILTER) [90], the US National Ecological Observatory Network (NEON, [91]), the Chinese Ecosystem Research Network (CERN) [92, 93], the Australian Terrestrial Ecosystem Research Network (TERN) and SuperSite Network [94], Integrated Carbon Observation System (ICOS) [95], etc. Some of these collaborations are already ongoing, as in the case of ILTER and CZOs, and from there connections should progressively spread out.

Second, the creation of a global alliance of scientists will not only involve measurements, but also common approaches, by sharing methods and knowledges across scales, processes, and disciplines [96]. To this purpose, models (either classical or DTs) offer a unique opportunity for scientists of different disciplines to develop a common vocabulary [9], as in the spirit of CZ transdisciplinary, and aim at a shared description of the system. Beyond terminologies, common coding languages between disciplines and working groups would also foster the representation of the whole CZ as "one system" and the sharing of knowledge. Desired characteristics of shared codes would be portability trough different hardware architectures (e.g., CPUs and GPUs), readability for scientists of any discipline, flexibility for the application at different scales and for the inclusion of different modules, and efficiency in both time and energy consumption [97]. One option to achieve such a standard is the component-based programming, supporting a linking framework that provides a plug-and-play environment, as proposed by the Community Surface Dynamics Modeling System (CSDMS) [98, 99]. The key idea is to have a compilation of interfaced objects (i.e., components), possibly using different coding languages, which represent either models or constituents of the systems or processes that can be coupled together to create new composite models. A few applications of this idea have been proposed (e.g., [30, 100–102]), and the same exercise could be done also including ML-based components to build the DT of the CZ. This context may help to address the main gaps of classical modeling approaches: the transdisciplinary representation of deeply coupled processes; and the scale bridging or multiscale modeling (e.g., [2, 60, 61]). A common library of interfaced components (or models) that can be declined to specific case studies as well as account for regional to continental heterogeneities would help in simulating the coupled processes of the CZ across the globe, instead of having one single supermodel. This effort is needed to understand what factors are dominant at different scales and address cross-scale feedback and scale emergence (see Sect. 1).

Third, common libraries of data and models will require infrastructures, in terms of both hardware, (i.e., supercomputer and storage servers, or cloud approaches) and software (i.e., platforms, efficient interfaces and toolkits) that allow end-users to easily and intuitively apply their expertise. Virtual research environments (VRE) are promising frameworks to serve this purpose, offering catalogues of data and model as well as toolkits for online analysis through an intuitive graphical user interface (GUI) built on community platforms. These services will allow users to retrieve information on specific aspects of the system, test, modify, improve, and configure new toolkits and interfaces. These interactive environments require a multi-layered software structure [79, 97]. The VRE structure would thus appear as an iceberg: at the top there would be services that hide the complexity to the user and do not require expert intervention, for instance using visual representation of model components in a GUI, then complexity would increase diving into the structure. VREs would thus become incubators of methodologies and innovation, as well as neutral grounds for merging information and knowledge and joining the efforts across different disciplines. Examples of platforms linked to the CZ community that offer collaboration environments already exists, still these are often oriented to specific disciplines (e.g., HydroShare, https://www.hydroshare.org/). A hierarchical environment connecting such initiatives and extending the existing networks with new platforms is therefore needed. In Europe, the LifeWatch ERIC e-infrastructure (https://www.lifewatch.eu/) is devoted to this task, and it has the goal of creating a system of VREs for ecosystems and biodiversity. A specific extension of this vision is currently ongoing in the Italian European-funded ITINERIS infrastructure project (https://itineris.cnr.it/), which will support the creation of a specific CZ VRE, with the aim of later extending it to the international community.

All the above operational, methodological, and technical efforts will hopefully lead to the synthesis of knowledge from different disciplines, working groups and networks, which is still largely lacking at global scale, to support the development of a unified theoretical framework of CZ evolution and functions (Fig. 1). Until now, hard work has been done to explore processes at the catchment scale in CZOs [96],

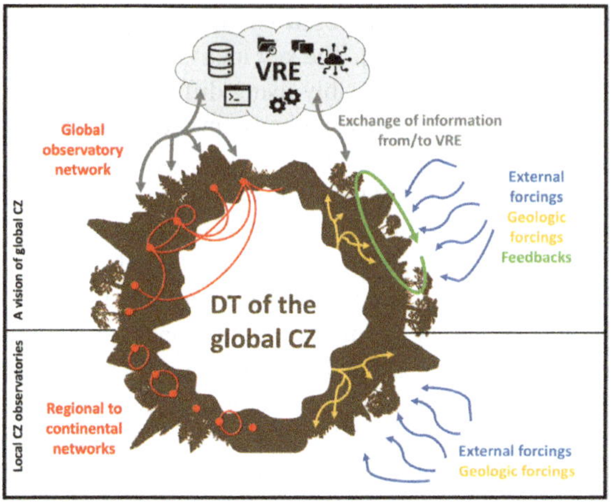

Fig. 1 Perspective of the global CZ framework. Present and future (suggested) frameworks are represented in the lower and upper part of the panel, respectively. In the upper part, the dynamics of the CZ Digital Twin (DT) is driven by external forcings (blue arrows), geological forcings (yellow arrows) and local-to-large scale feedbacks (green arrow). A global observatory network (red edges and nodes) exchanges information (grey two-way arrows) with virtual research environments (VRE, grey cloud). Same color conventions in the lower part of the panel

and time is now ripe for advancing towards a more comprehensive and large-scale representation of CZ processes. Still, the modeling approaches discussed in Sect. 2.1 should not be abandoned, but lessons learned from these efforts should concur to a coherent characterization of the CZ across the whole globe.

The global CZ perspective is summarized in Fig. 1. The framework involves a worldwide CZO network, where exchanges of knowledges and data across observatories are supported by dedicated VREs that provide storage facilities and toolkits for statistical analyses. This allows for the study of the CZ dynamics including climatic, geologic, and anthropic forcings, as well as non-linear processes and feedbacks acting at differing scales. The VRE also provide platforms for the interfacing of classical models and ML techniques to create a global DT of the CZ (CZ-DT). The CZ-DT will be continuously informed by field and remote measurements conveyed by the global CZO network. The use of flexible coding techniques within the VRE permits the declination of the DT across scales, spanning from the local to the global level. The outcomes are expected to provide information for the establishment of new observatories, contribute to the understanding of the CZ dynamics, and assist in the development of climate change adaptation strategies. Such a framework would create a fertile ground for a global synthesis of measurements, knowledges, and expertise, which is at the heart of the CZ interdisciplinary science.

Acronyms

CZ	Critical zone
CZO	Critical zone observatory
ESM	Earth system model
DT	Digital twin
ML	Machine learning
VRE	Virtual research environment
GUI	Graphical user interface

References

1. Kirchner JW (2006) Getting the right answers for the right reasons: linking measurements analyses and models to advance the science of hydrology. Water Resour Res 42(3)
2. Lin H (2010) Earth's critical zone and hydropedology: concepts characteristics and advances. Hydrol Earth Syst Sci 14(1):25–45
3. Chorover J, Troch PA, Rasmussen C, Brooks PD, Pelletier JD, Breshears DD, Durcik M (2011) How water carbon and energy drive critical zone evolution: the Jemez-Santa Catalina critical zone observatory. Vadose Zone J 10(3):884–899
4. Rasmussen C, Troch PA, Chorover J, Brooks P, Pelletier J, Huxman TE (2011) An open system framework for integrating critical zone structure and function. Biogeochem 102(1):15–29
5. Brantley SL (2010) Rock to regolith. Nat Geosci 3(5):305–306
6. Pelletier JD, Barron-Gafford GA, Gutiérrez-Jurado H, Hinckley ELS, Istanbulluoglu E, McGuire LA, Tucker GE (2018) Which way do you lean? Using slope aspect variations to understand critical zone processes and feedbacks. Earth Surf Proc Land 43(5):1133–1154
7. Anderson SP, von Blanckenburg F, White AF (2007) Physical and chemical controls on the critical zon. Elements 3(5):315–319
8. Braun J, Mercier J, Guillocheau F, Robin C (2016) A simple model for regolith formation by chemical weathering. J Geophys Res Earth Surf 121(11):2140–2171
9. Brantley SL, Eissenstat DM, Marshall JA, Godsey SE, Balogh-Brunstad Z, Karwan DL, Weathers KC (2017a) Reviews and syntheses: on the roles trees play in building and plumbing the critical zone. Biogeosciences 14(22):5115–5142
10. Banwart SA, Nikolaidis NP, Zhu YG, Peacock CL, Sparks DL (2019) Soil functions: connecting earth's critical zone. Annu Rev Earth Planet Sci 47:333–359
11. Amundson R, Richter DD, Humphreys GS, Jobbágy EG, Gaillardet J (2007) Coupling between biota and earth materials in the critical zone. Elements 3(5):327–332
12. Parmesan C (2006) Ecological and evolutionary responses to recent climate change. Annu Rev Ecol Evol Syst 637–669
13. Fan Y (2015) Groundwater in the Earth's critical zone: relevance to large-scale patterns and processes. Water Resour Res 51(5):3052–3069
14. Heffernan JB, Soranno PA, Angilletta MJ Jr, Buckley LB, Gruner DS, Keitt TH, Weathers KC (2014) Macrosystems ecology: understanding ecological patterns and processes at continental scales. Front Ecol Environ 12(1):5–14
15. Rietkerk M, Brovkin V, van Bodegom PM, Claussen M, Dekker SC, Dijkstra HA, Seneviratne SI (2011) Local ecosystem feedbacks and critical transitions in the climate. Ecol Complex 8(3):223–228
16. Soranno PA, Cheruvelil KS, Bissell EG, Bremigan MT, Downing JA, Fergus CE, Webster KE (2014) Cross-scale interactions: quantifying multi-scaled cause–effect relationships in macrosystems. Front Ecol Environ 12(1):65–73

17. Arènes A, Latour B, Gaillardet J (2018) Giving depth to the surface: an exercise in the Gaia-graphy of critical zones. Anthropocene Rev 5(2):120–135
18. Chorover J, Kretzschmar R, Garcia-Pichel F, Sparks DL (2007) Soil biogeochemical processes within the critical zone. Elements 3(5):321–326
19. Brooks PD, Chorover J, Fan Y, Godsey SE, Maxwell RM, McNamara JP, Tague C (2015) Hydrological partitioning in the critical zone: recent advances and opportunities for developing transferable understanding of water cycle dynamics. Water Resour Res 51(9):6973–6987
20. Guo L, Lin H (2016) Critical zone research and observatories: current status and future perspectives. Vadose Zone J 15(9)
21. Brantley SL, Goldhaber MB, Ragnarsdottir KV (2007) Crossing disciplines and scales to understand the critical zone. Elements 3(5):307–314
22. National Research Council (2001) Basic research opportunities in earth science. National Academy Press, Washington D.C.
23. Sullivan PL, Billings SA, Hirmas D, Li L, Zhang X, Ziegler S, Wen H (2022) Embracing the dynamic nature of soil structure: a paradigm illuminating the role of life in critical zones of the anthropocene. Earth Sci Rev 225:103873
24. Yu X, Duffy C, Kaye J, Crow W, Bhatt G, Shi Y (2014) Watershed reanalysis of water and carbon cycle models at a critical zone observatory. Remote Sens Terr Water Cycle 493–509
25. Young IM, Crawford JW (2004) Interactions and self-organization in the soil-microbe complex. Science 304(5677):1634–1637
26. McClain ME, Boyer EW, Dent CL, Gergel SE, Grimm NB, Groffman PM, Pinay G (2003) Biogeochemical hot spots and hot moments at the interface of terrestrial and aquatic ecosystems. Ecosys 301–312
27. Steffen W, Broadgate W, Deutsch L, Gaffney O, Ludwig C (2015) The trajectory of the anthropocene: the great acceleration. Anthropocene Rev 2(1):81–98
28. Richter DD Jr (2007) Humanity's transformation of Earth's soil: pedology's new frontier. Soil Sci 172(12):957–967
29. Richter DD Jr, Mobley ML (2009) Monitoring Earth's critical zone. Science 326(5956):1067–1068
30. Vereecken H, Schnepf A, Hopmans JW, Javaux M, Or D, Roose T, Young IM (2016b) Modeling soil processes: Review key challenges and new perspectives. Vadose zone J 15(5)
31. Wilkinson BH (2005) Humans as geologic agents: a deep-time perspective. Geol 33(3):161–164
32. Wilkinson BH, McElroy BJ (2007) The impact of humans on continental erosion and sedimentation. Geol Soc Am Bull 119(1–2):140–156
33. Zipper SC, Soylu ME, Kucharik CJ, Loheide SP II (2017) Quantifying indirect groundwater-mediated effects of urbanization on agroecosystem productivity using MODFLOW-AgroIBIS (MAGI) a complete critical zone model. Ecol Model 359:201–219
34. Robinson S (2008a) Conceptual modelling for simulation part I: definition and requirements. J Oper Res Soc 59(3):278–290
35. Robinson S (2008b) Conceptual modelling for simulation Part II: a framework for conceptual modelling. J Oper Res Soc 59(3):291–304
36. Giardino JR, Houser C (2015) Principles and dynamics of the critical zone. Elsevier, Amsterdam
37. Coughlan MR, Nelson DR, Lonneman M, Block AE (2017) Historical land use dynamics in the highly degraded landscape of the Calhoun critical zone observatory. Land 6(2):32
38. Glossner KL, Lohse KA, Appling AP, Cram ZK, Murray E, Godsey SE, Pierson FB (2022) Long-term suspended sediment and particulate organic carbon yields from the Reynolds creek experimental watershed and critical zone observatory. Hydrol Process 36(2):e14484
39. Magnani M, Baneschi I, Giamberini M, Mosca P, Raco B, Provenzale A (2020) Drivers of carbon fluxes in Alpine tundra: a comparison of three empirical model approaches. Sci Total Environ 732:139139
40. Magnani M, Baneschi I, Giamberini M, Raco B, Provenzale A (2022) Microscale drivers of summer CO2 fluxes in the Svalbard high Arctic tundra. Sci Rep 12(1):1–14

41. Regelink IC, Stoof CR, Rousseva S et al (2015) Linkages between aggregate formation, porosity and soil chemical properties. Geoderma 247–248:24–37. https://doi.org/10.1016/j.geoderma.2015.01.022
42. Song XD, Wu HY, Liu F, Tian J, Cao Q, Yang SH, Zhang GL (2019) Three-dimensional mapping of organic carbon using piecewise depth functions in the red soil critical zone observatory. Soil Sci Soc Am J 83(3):687–696
43. Bastola S, Dialynas YG, Bras RL, Noto LV, Istanbulluoglu E (2018) The role of vegetation on gully erosion stabilization at a severely degraded landscape: a case study from Calhoun experimental critical zone observatory. Geomorphol 308:25–39
44. Duffy C, Shi Y, Davis K, Slingerland R, Li L, Sullivan PL, Brantley SL (2014) Designing a suite of models to explore critical zone function. Procedia Earth Planet Sci 10:7–15
45. Giannakis GV, Nikolaidis NP, Valstar J, Rowe EC, Moirogiorgou K, Kotronakis M, Banwart SA (2017) Integrated critical zone model (1D-ICZ): a tool for dynamic simulation of soil functions and soil structure. Adv Agron 142:277–314
46. Le PV, Kumar P (2017) Interaction between ecohydrologic dynamics and microtopographic variability under climate change. Water Resour Res 53(10):8383–8403
47. Shi Y, Davis KJ, Duffy CJ, Yu X (2013) Development of a coupled land surface hydrologic model and evaluation at a critical zone observatory. J Hydrometeorol 14(5):1401–1420
48. Woo DK, Kumar P (2017) Role of micro-topographic variability on the distribution of inorganic soil-nitrogen age in intensively managed landscape. Water Resour Res 53(10):8404–8422
49. Yu X, Bhatt G, Duffy C, Shi Y (2013) Parameterization for distributed watershed modeling using national data and evolutionary algorithm. Comput Geosci 58:80–90
50. Abban B, Papanicolaou AN, Cowles MK, Wilson CG, Abaci O, Wacha K, Schnoebelen D (2016) An enhanced Bayesian fingerprinting framework for studying sediment source dynamics in intensively managed landscapes. Water Resour Res 52(6):4646–4673
51. Rousseva S, Kercheva M, Shishkov T, Lair GJ, Nikolaidis NP, Moraetis D, Banwart SA (2017) Soil water characteristics of European SoilTrEC critical zone observatories. Adv Agron 142:29–72
52. Banwart SA, Bernasconi SM, Blum WE, de Souza DM, Chabaux F, Duffy C, Zhang B (2017) Soil functions in Earth's critical zone: key results and conclusions. Adv Agron 142:1–27
53. Stamati FE, Nikolaidis NP, Banwart SA, Blum WE (2013) A coupled carbon aggregation and structure turnover (CAST) model for topsoils. Geoderma 211:51–64
54. Lepore C, Arnone E, Noto LV, Sivandran G, Bras RL (2013) Physically based modeling of rainfall-triggered landslides: a case study in the Luquillo forest Puerto Rico. Hydrol Earth Syst Sci 17(9):3371–3387
55. Papanicolaou AN, Wacha KM, Abban BK, Wilson CG, Hatfield JL, Stanier CO, Filley TR (2015) From soilscapes to landscapes: a landscape-oriented approach to simulate soil organic carbon dynamics in intensively managed landscapes. J Geophys Res Biogeosci 120(11):2375–2401
56. Shi Y, Baldwin DC, Davis KJ, Yu X, Duffy CJ, Lin H (2015) Simulating high-resolution soil moisture patterns in the Shale hills watershed using a land surface hydrologic model. Hydrol Process 29(21):4624–4637
57. Sullivan PL, Goddéris Y, Shi Y, Gu X, Schott J, Hasenmueller EA, Brantley SL (2019) Exploring the effect of aspect to inform future earthcasts of climate-driven changes in weathering of shale. J Geophys Res Earth Surf 124(4):974–993
58. Höge M, Wöhling T, Nowak W (2018) A primer for model selection: the decisive role of model complexity. Water Resour Res 54(3):1688–1715
59. Hipsey MR, Hamilton DP, Hanson PC, Carey CC, Coletti JZ, Read JS, Brookes JD (2015) Predicting the resilience and recovery of aquatic systems: a framework for model evolution within environmental observatories. Water Resour Res 51(9):7023–7043
60. Goddéris Y, Brantley SL (2013) Earthcasting the future critical zone. Elementa: Science of the Anthropocene 1

61. Sullivan PL, Li L, Goddéris Y, Brantley SL (2020) Poised to hindcast and earthcast the effect of climate on the critical zone. In: Dontsova K, Balogh-Brunstad Z, Le Roux G (eds) Biogeochemical Cycles. Sanchez-Guzman G, Velasquez W, Alvarez-Alvarado MS (2022) Modeling a simulated forest to get burning times of tree species using a digital twin. In: 2022 IEEE 12th annual computing and communication workshop and conference (CCWC). IEEE, pp 0639–0643. https://doi.org/10.1002/9781119413332.ch10
62. Lewis SL, Maslin MA (2015) Defining the anthropocene. Nature 519(7542):171–180
63. Heidari P, Li L, Jin L, Williams JZ, Brantley SL (2017) A reactive transport model for Marcellus shale weathering. Geochim Cosmochim Acta 217:421–440
64. Gilad E, von Hardenberg J, Provenzale A, Shachak M, Meron E (2004) Ecosystem engineers: from pattern formation to habitat creation. Phys Rev Lett 93:098105
65. Quijano J, Lin H (2014) Entropy in the critical zone: a comprehensive review. Entropy 16(6):3482–3536
66. Vereecken H, Pachepsky Y, Simmer C, Rihani J, Kunoth A, Korres W, Shao Y (2016a) On the role of patterns in understanding the functioning of soil-vegetation-atmosphere systems. J Hydrol 542:63–86
67. Baudena M, D'Andrea F, Provenzale A (2009) An idealized model for tree–grass coexistence in savannas: the role of life stage structure and fire disturbances. J Ecol 98:74–80. https://doi.org/10.1111/j.1365-2745.2009.01588.x
68. Vissio G, Provenzale A (2022) On-off intermittency and irruptions in host-parasitoid dynamics. J Theor Biol 546:111174. https://doi.org/10.1016/j.jtbi.2022.111174
69. Brantley SL, McDowell WH, Dietrich WE, White TS, Kumar P, Anderson SP, Gaillardet J (2017b) Designing a network of critical zone observatories to explore the living skin of the terrestrial Earth. Earth Surf Dyn 5(4):841–860
70. Scheibe TD, Murphy EM, Chen X, Rice AK, Carroll KC, Palmer BJ, Wood BD (2015) An analysis platform for multiscale hydrogeologic modeling with emphasis on hybrid multiscale methods. Groundwater 53(1):38–56
71. Navarre-Sitchler A, Brantley SL (2007) Basalt weathering across scales. Earth Planet Sci Lett 261(1–2):321–334
72. Clark MP, Fan Y, Lawrence DM, Adam JC, Bolster D, Gochis DJ, Zeng X (2015) Improving the representation of hydrologic processes in Earth system models. Water Resour Res 51(8):5929–5956
73. McDonnell JJ, Sivapalan M, Vaché K, Dunn S, Grant G, Haggerty R, Weiler M (2007) Moving beyond heterogeneity and process complexity: a new vision for watershed hydrology. Water Resour Res 43(7)
74. Richter DD Jr, Billings SA (2015) 'One physical system': Tansley's ecosystem as Earth's critical zone. New Phytol 206(3):900–912
75. Jones D, Snider C, Nassehi A, Yon J, Hicks B (2020) Characterizing the digital twin: a systematic literature review CIRP. J Manuf Sci Technol 29:36–52
76. Semeraro C, Lezoche M, Panetto H, Dassisti M (2021) Digital twin paradigm: a systematic literature review. Comput Ind 130:103469
77. Box GE (1979) Robustness in the strategy of scientific model building. In: Robustness in statistics. Academic Press, Cambridge, Massachusetts, pp 201–236
78. Box GE, Draper NR (1987) Empirical model-building and response surfaces. John Wiley and Sons, Hoboken, New Jersey
79. Bauer P, Stevens B, Hazeleger W (2021b) A digital twin of Earth for the green transition. Nat Clim Chang 11(2):80–83
80. Moshrefzadeh M, Machl T, Gackstetter D, Donaubauer A, Kolbe TH (2020) Towards a distributed digital twin of the agricultural landscape. J Digit Landsc Arch 5:173–118
81. Pillai UPA, Pinardi N, Alessandri J, Federico I, Causio S, Unguendoli S, Staneva J (2022) A digital twin modelling framework for the assessment of seagrass nature based solutions against storm surges. Sci Total Environ 847:157603
82. Mukhtar H, Wunderlich RF, Lin YP (2022) Digital twins of the soil microbiome for climate mitigation. Environments 9(3):34

83. Skobelev P, Laryukhin V, Simonova E, Goryanin O, Yalovenko V, Yalovenko O (2020) Multi-agent approach for developing a digital twin of wheat. In: 2020 IEEE international conference on smart computing (SMARTCOMP). pp 268–273
84. Bui EN (2016) Data-driven critical zone science: a new paradigm. Sci Total Environ 568:587–593
85. Zhou Q, Fellows A, Flerchinger GN, Flores AN (2019) Examining interactions between and among predictors of net ecosystem exchange: a machine learning approach in a semi-arid landscape. Sci Rep 9(1):1–11
86. Minor J, Pearl JK, Barnes ML, Colella TR, Murphy PC, Mann S, Barron-Gafford GA (2020) Critical zone science in the anthropocene: opportunities for biogeographic and ecological theory and praxis to drive earth science integration. Prog Phys Geogr: Earth Environ 44(1):50–69
87. Latour B (2014) Some advantages of the notion of "Critical Zone" for geopolitics. Proc Earth Planet Sci 10:3–6
88. Haase P, Tonkin JD, Stoll S, Burkhard B, Frenzel M, Geijzendorffer IR, Schmeller DS (2018) The next generation of site-based long-term ecological monitoring: linking essential biodiversity variables and ecosystem integrity. Sci Total Environ 613:1376–1384
89. Gaillardet J, Braud I, Hankard F, Anquetin S, Bour O, Dorfliger N, Zitouna R (2018) OZCAR: the French network of critical zone observatories. Vadose Zone J 17(1):1–24
90. Mirtl M, Borer ET, Djukic I, Forsius M, Haubold H, Hugo W, Haase P (2018) Genesis goals and achievements of long-term ecological research at the global scale: a critical review of ILTER and future directions. Sci Total Environ 626:1439–1462
91. Loescher HW, Kelly EF, Lea R (2017) National ecological observatory network: beginnings programmatic and scientific challenges and ecological forecasting. In: Terrestrial ecosystem research infrastructures. CRC Press, Boca Raton, Florida, pp 27–52
92. Fu B, Li S, Yu X, Yang P, Yu G, Feng R, Zhuang X (2010) Chinese ecosystem research network: progress and perspectives. Ecol Complex 7(2):225–233
93. Shenggong L, Guirui Y, Xiubo Y, Honglin H, Xuebing G (2015) A brief introduction to Chinese ecosystem research network (CERN). J Resour Ecol 6(3):192–196
94. Karan M, Liddell M, Prober SM, Arndt S, Beringer J, Boer M, Wardlaw T (2016) The Australian SuperSite network: a continental long-term terrestrial ecosystem observatory. Sci Total Environ 568:1263–1274
95. Heiskanen J, Brümmer C, Buchmann N, Calfapietra C, Chen H, Gielen B, Kutsch W (2022) The integrated carbon observation system in Europe. Bull Am Meteor Soc 103(3):E855–E872
96. Baatz R, Sullivan PL, Li L, Weintraub SR, Loescher HW, Mirtl M, Van Looy K (2018) Steering operational synergies in terrestrial observation networks: opportunity for advancing Earth system dynamics modelling. Earth Syst Dyn 9(2):593–609
97. Bauer P, Dueben PD, Hoefler T, Quintino T, Schulthess TC, Wedi NP (2021a) The digital revolution of Earth-system science. Nat Comput Sci 1(2):104–113
98. Peckham SD, Hutton EW, Norris B (2013) A component-based approach to integrated modeling in the geosciences: the design of CSDMS. Comput Geosci 53:3–12
99. Tucker G, Hutton E, Piper M, Campforts B et al (2021) Numerical modeling of Earth's dynamic surface: a community approach. https://doi.org/10.31223/X51615
100. David O, Ascough JC II, Lloyd W, Green TR, Rojas KW, Leavesley GH, Ahuja LR (2013) A software engineering perspective on environmental modeling framework design: the object modeling system. Environ Model Softw 39:201–213
101. Goodall JL, Robinson BF, Castronova AM (2011) Modeling water resource systems using a service-oriented computing paradigm. Environ Model Softw 26(5):573–582
102. Jiang P, Elag M, Kumar P, Peckham SD, Marini L, Rui L (2017) A service-oriented architecture for coupling web service models using the basic model interface (BMI). Environ Model Softw 92:107–118

Critical Zone Science Informs Landscape Management Through Quantification of Slow Episodic Constraints on Ecosystem Services

Kathleen A. Lohse, Amy S. Commendador, Kayla L. Glossner, Sarah A. Stalder, Carolyn Macek, and Ruth B. MacNeille

Abstract The critical zone supports many ecosystem services from which humans derive both direct and indirect benefits as a society, yet these processes that form over long periods of time are often neglected and are rarely quantified. Thus linking them to ecosystem services has lagged behind other ecosystem service assessment studies. Critical zone services extends the concept of ecosystem services to understand the maintenance of ecosystem services for future generations through measurement of processes operating at longer time scales and their interaction with fast processes. In this chapter, we ask how improved understanding of the critical zone and its services can be used to inform land management. We posit that integrating critical zone science into land management will assist in understanding and predicting the sustainability and/or disruption of natural services. To this end, we review critical zone science conducted at the NSF Critical Zone Observatories (CZOs) over the past 7 years and identify key areas in which critical zone science has advanced our understanding of ecosystem services. We find that intermediate scale (plot to landscape) measurements and modeling provided by CZOs are enhancing our understanding of climate, topography, and bedrock controls on fine-scale variability in ecosystem services. Measurements of slow variables (at deeper spatial and temporal scales) inform management of baseline rates that are key to predicting responses to extreme events, climate and land use change, and to sustaining services as the frequency of extreme events and disturbances increase.

Key points

- Most critical zone services are intermediate regulating services that have received less attention in the ecological literature and enhance our understanding of the

K. A. Lohse (✉) · A. S. Commendador · K. L. Glossner · S. A. Stalder · C. Macek · R. B. MacNeille
Department of Biological Sciences, Idaho State University, Pocatello, ID 83209, USA
e-mail: klohse@isu.edu

T. White and A. Provenzale (eds.), *Critical Zone and Ecosystem Dynamics*, Advances in Critical Zone Science, https://doi.org/10.1007/978-3-031-69076-1_12

maintenance of ecosystem services through measurement of longer time scale processes.
- Intermediate scale measurements are enhancing understanding of climate, topography, and bedrock controls on ecosystem services.
- Measurements at deeper spatial and temporal scales inform management of baseline rates that are key to predicting responses to extreme events.

1 Introduction

We live in a time where unprecedented pressures are being placed on the critical zone, defined as the thin veneer of Earth's surface from the top of the vegetation canopy to the bottom of groundwater [69]. This critical zone provides many goods and services from which humans derive benefit as a society. These benefits have been termed ecosystem goods and services and are considered final provisioning services (e.g., food, water, shelter) or intermediate regulating services (e.g., carbon sequestration, soil formation) that contribute indirectly to human well-being [8]. The availability of these ecosystem goods and services are typically assessed on a year-to-decadal basis, and highly contingent upon local to global climate, land use, and geographic factors. Within the ecosystem services framework, "slow variables", such as soil production, within these intermediate regulating services have been identified as key to supporting ecosystem structure and function [6, 19], yet research quantifying these slow intermediary processes and linking them to ecosystem services has lagged behind other ecosystem service assessments. In part, this lag can be explained by the physical and monetary costs of quantifying these slow variables. In addition, this lag may in part be explained by the slow percolation of ideas across disciplines, and therefore only recent incorporation of the ecosystem services framework into the Earth sciences.

Critical Zone (CZ) Science has emerged as an interdisciplinary field of study with a mission to study how Earth's living skin is structured, evolves, and provides critical functions that sustain life [22] and how this zone is influenced by climate, land use, and other disturbance events. To address these challenges, CZ science frequently examines the interactions of fast and slow processes, often occurring belowground, and how these processes and products manifest themselves at the catchment scale. It is through the interactions among ecologists, hydrologists, geoscientists, and increasingly social scientists conducting place-based science on specific areas, defined in space and time, that the notion of "critical zone services" has emerged.

Critical zone services embrace the concept of ecosystem services to include understanding the maintenance of ecosystem services for future generations, typically through the measurement of slow processes operating at longer time scales and their interaction with faster ones. Field et al. [32] posited that these slow variables act as key constraints (rate limiting steps) in the provisioning and regulating aspects of ecosystem services. Provisioning and intermediate regulating services are defined here to include all the direct products obtained from the critical zone (e.g., food, water) and those intermediary processes that support services provided by the critical zone (e.g., nutrient cycling, primary production) and those that act to regulate significant processes, cycles and flows that impact societal health in myriad ways (e.g., disease, water quality and control, soil erosion, local to regional climate), respectively (Table 1 and Fig. 1). As such, sustaining an ecosystem service such as forest productivity for future generations requires (1) foundational knowledge and understanding of critical zone processes such as soil production and erosion rates in the absence of human perturbations and (2) an understanding how human activities and management actions such as forestry alter these processes rates. In addition, understanding how changes in extreme events such as floods, fires, and droughts and altered disturbance regimes alter or disrupt ecosystem services requires long-term observations to identify sensitivities and thresholds of these slow processes to extreme events.

In this chapter, we conduct an exemplary review of CZ science conducted across the NSF Critical Zone Observatory (CZO) network over the past 7 years (2013–2020) and explore how its science has advanced our understanding of the supply of ecosystem services. We constrain this review to these years because the Critical Zone Collaborative Network (CZCN) succeeded this network in 2020. Other reviews have focused on select research advances in the critical zone [38, 86] and future directions [13, 14]. Here we posit that integrating CZ science into land management will

Table 1 Critical zone services, constraints, and currencies. Constraints are rate-limiting processes that control availability of critical zone services, and currencies are the societal valuation (market or non-market) of benefit from those services. Adapted from [32]

Critical zone services	CZ constraints	Currencies
Final provisioning services		
Water storage	Surface water, groundwater storage	Water supply
Food supply	Soil quality, erosion, extreme events	Food security
Intermediate regulating services		
Primary production	Soil formation, fertility, erosion, extreme events-fire	Sustainable crops, timber, forage, livestock
Soil formation	Soil production, weathering	Water storage, carbon sequestration
Nutrient cycling	Long-range dust transport, rock derived, N fixation	Nutrient availability carbon storage
Water quality regulation	Hydrological supply rate	Filtration and buffering
Water supply regulation	Water supply management (flood control, irrigation) change in rain to snow	Regulation of hydrological flows
Climate regulation	Vegetation and soil responses	Regulation of global temperature and precipitation
	Soil formation and weathering	

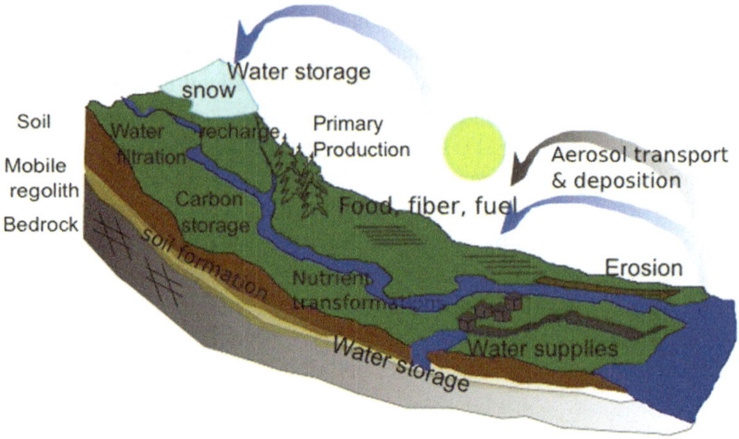

Fig. 1 Critical zone and critical zone services

assist in understanding and predicting the sustainability and/or disruption of natural services. We organize the review around critical zone services (Table 1) and synthesize these findings at the end in an attempt to address the following questions: How can improved understanding of the critical zone and its services be used to inform land management on: (1) its resiliency to climate change and associated extreme events –droughts, floods, fires, and (2) sensitivities, thresholds, and interactions that may sustain or restore the critical zone? We focus this review on final provisioning and intermediary regulating services; cultural services were beyond the scope of this review.

2 Brief Background on Critical Zone Observatories

The U.S. NSF Critical Zone Observatory (CZO) network consisted of nine sites in 2020 (Fig. 2).

Three sites were initially selected as the first CZOs (Boulder Creek (BC), Southern Sierras (SS), and Shale Hills (SH)) based on a proposal competition in 2007. Additional sites (Santa Catalina-Jemez River (SC-JR), Luquillo (LUQ), and Christina (CR)) were added in 2009. In 2013, another four sites (Reynolds Creek (RC), Eel River (Eel), Intensively Managed Landscape (IML), and Calhoun (C)) were established and one discontinued (CR) to form what was the CZO network of nine sites in 2020. CR is not included in this review hereafter. This network also coincided with the establishment of the Critical Zone Observatory National Office [106]. In 2020, the CZO network was succeeded by the CZCN.

The sites span a range of geologic, climatic, ecologic, and physiographic environments (Table 2). Geology varies from igneous-felsic intrusive and extrusive,

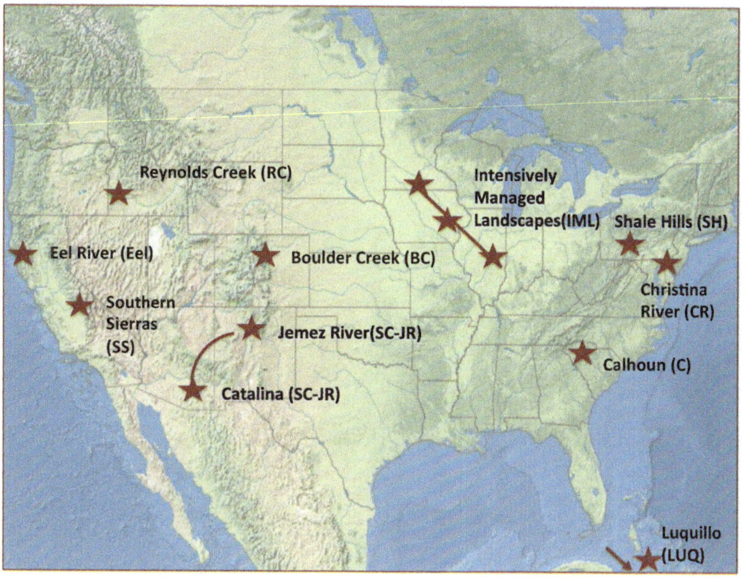

Fig. 2 Location of critical zone observatory sites

igneous ultramafic extrusive (BC, RC, SC-JR, LUQ, SS), sandstones and shales (Eel, SH) to unconsolidated materials (IML, C). Mean annual precipitation ranges from 235 mm yr^{-1} at RC to 5000 mm yr^{-1} at LUQ. Vegetation varies from grasslands to forests. Most sites are in relatively protected areas, though many CZOs have land use legacies (BC, SH, SS, SC-JR) and some of the recent CZOs include more extensive (RC) to intensive (IML) human activities and past human mediated disturbance histories (C).

3 Critical Zone Science Contributions to Advancing Our Understanding of Constraints on the Supply of Ecosystem Services

In the following sections, we review contributions of CZ science to our understanding of final provisioning and intermediate regulating ecosystem services.

3.1 Final Provisioning Services

Final provisioning services consist of all the products obtained from the critical zone, such as food, water, fiber, and energy. Provisioning services have received

Table 2 Critical zone observatory attributes. Geology abbreviated as follows: igneous-felsic intrusive (Igneous-FI) and extrusive (Igneous-FE), igneous ultramafic extrusive (Igneous UME), igneous mafic extrusive (Igneous ME), metamorphic (Met), unconsolidated materials (Uncon), sandstone (Sand), sedimentary (Sed). Land cover abbreviated as follows: agriculture (Ag), grassland-herbacious (Grass-Herb), shrubland (Shrub), evergreen forest (E Forest), mixed forest (M Forest), deciduous forest (D Forest). Soils are abbreviated as follows: inceptisol (Incept), Mollisol (Moll), Alfisol (Alf), Ultisol (Ult), Entisol (Ent), Aridisol (Arid), Oxisol (Oxi), and Histisol (Hist)

CZO	Area (km^2)	Elevation range (m)	MAP (mm/yr)	MAT (°C)	Geology	Soils	Land cover	Climate
Boulder creek	1160	1480–4120	500–1000	−3.8–8	Igneous-FI and FE, Igneous UME, other Met	Incept, Moll, Alf	Alpine Tundra, M Forest, Grass-Herb, Ag-Pasture, developed-low	Temperate subhumid
Calhoun	180–200	100–200	1270	16	Uncon, Igneous-FI, Meta-igneous	Ult, Incept, Ent	Grass-Herb, M Forest	Humid subtropical
Catalina-Jemez	3.1	830–2780	400–800	11–18	Igneous-FI, other met	Ent, Moll	Barren Land, E Forest, Grass-Herb	Temperate Subhumid
	446.8	1700–3432	400–800	3–11	Igneous-FE	Arid, Alf, Moll, Incept	M Forest, Grass-Herb	
Eel river	9540	0–2464	2160	11	Sand, Other Sed, Other Met	Ult, Incept	Grass-Herb, M Forest, Developed-High, Ag-Crops	Mediterranean Hyperhumid
IML	3700	180–291	1000	11	Uncon, Other Sed	Moll, Alf, Ent, Incept	Ag-Crops, Grass-Herb, Shrub	Temperate Humid
	270	197–248	889	9	Same as above	Same as above	Same as above	
	44,000	210–294	747	5	Same as above	Same as above	Same as above	

(continued)

Table 2 (continued)

CZO	Area (km^2)	Elevation range (m)	MAP (mm/yr)	MAT (°C)	Geology	Soils	Land cover	Climate
Luquillo	300	0–1075	1650–5000	21	Igneous-FI, Igneous-UME, Meta-igneous	Incept, Ult, Oxi	D Forest, Developed-Low	Tropical Humid
Reynolds creek	239	1099–2244	235–1100	4.7–8.9	Igneous-FI and FE, Igneous ME, Igneous-ME, Uncon	Arid, Alf, Moll, Ent	Grass-Herb, E Forest, Shrub	Mediterranean Dry
Shale hills	165	256–310	1050	9.5	Shale, Sand, Carbonate	Incept, Alf, Ult	D Forest, Ag-Pasture	Temperate Humid
Southern Sierra	4400	400–2700	510–1100	4–16.5	Igneous-FI	Ent, Incept, Alf, Hist	E Forest, Shrub, Grass-Herb	Mediterranean Humid

the most attention from ecosystem services assessments, yet knowledge gaps still remain with regard to our understanding of water storage and food security. CZ science has focused on addressing these gaps through better quantification of both rock and snowpack water storage potentials and flows and their effect on ecosystem resiliency, as well as evaluating different production systems and associated costs (loss processes including erosion and leaching) and benefits in intensively managed systems.

3.1.1 Water Storage

Studies in the critical zone have been exploring the interactions between ground and surface water and how each of these water sources are involved in water storage. Recent advancements in this arena include improved quantification of exchangeable rock moisture and its contribution to the timing and magnitude of runoff [83] as well as advances in quantifying snowpack water storage and recharge [10, 78].

Exchangeable Rock Water Storage

Water storage is typically associated with the thin veneer of soils at the surface of the critical zone. However, recent studies have associated significant water storage in weathered bedrock, also termed rock moisture [83, 87], and that this represents a significant component of the water budget in all critical zone systems. Holbrook et al. [47] indicated that the saprolite is a significant reservoir of water in the SS CZO, possibly storing an average of 3 m^3/m^2 of water along the forested slope. Similarly, Kormos et al. [52] found that the storage potential in bedrock due to infiltration in the RC CZO was approximately 34% of the precipitation. In the BC CZO, Langston et al. [55] showed that water storage in the upper unsaturated zone was dictated more by the duration of a precipitation event than the magnitude, allowing more storage of water from several pulse events rather than one large event, with recharge occurring across all seasons. Studies at the Eel CZO pointed to water available in the rock water reservoir mediating the timing and magnitude of runoff [71, 82].

Snowpack Water Storage

Quantifying the spatial distribution in snowpack water storage and the timing and path that snow melt takes to recharge groundwater and feed streams is of great interest as climate change has made the timing and amount of water supply less predictable. This has long reaching effects on other critical zone services, such as water regulation (see Sect. 3.3). Kormos et al. [52] indicated that southwest facing slopes in Idaho drain more often throughout the snow season while northeast facing slopes contribute more to soil drainage over the whole year, and that the peaks in catchment soil drainage and bedrock infiltration are synchronized with rain or snow events. However, as shown at the BC CZO, a changing climate will reduce the amount of infiltration as slow snowmelt shifts to shorter duration, larger magnitude events, thus decreasing water storage in the unsaturated zone [10, 78]. In this way, consistently earlier snowmelt from a warming climate could decrease near-surface water storage and availability to plants and soil biota [10]. Warming temperatures may also alter atmospheric humidity, exerting a major control on how seasonal snow responds to a changing climate [43]. Strategies in snow-dominated elevations to attenuate snowmelt as well as reduce evapotranspiration have been proposed to combat these long reaching consequences of climate change [49].

Better quantification of the distribution of snowpack has been key to executing these objectives. Light Detection and Ranging (LiDAR)-derived snowpack data sets are increasingly being used to address this issue and estimate regional snow water storage and sensitivites to climate change in the critical zone [42, 100]. For example, Tennant et al. [100] documented trends in snow-elevation storage using LiDAR and identified sensitivities of snow storage to climate change across the western CZOs (BC, SS, SC-JR, RC). In particular, they showed a relatively short distance between elevations at freezing levels and elevations of significant storage, indicating greater sensitivity at RC and SS CZOs to changes from snow to rain. They also indicated that changes in vegetation cover, due to drought or wildfire, could drive changes in the spatial distribution of snow storage. These findings collectively point to the need for advancements in water regulation (see Sect. 3.3).

3.1.2 Food Supply

Studies in the critical zone seek to understand how land use and other anthropogenic changes alter soil quality and soil organic matter content and characteristics, which are key constraints on food supply. Soil organic matter content is the balance between biomass inputs and losses via heterotrophic respiration, erosion and leaching. In highly managed systems, outputs also include timber harvesting and crop yields. An early study by Montgomery [63] highlighted that erosion rates from conventionally plowed agricultural fields are 1–2 orders of magnitude greater than erosion under native vegetation and long-term geological processes. CZ science is contributing to better estimates of geological rates of erosion using studies based on cosmogenic ^{10}Be and ^{26}Al (see Sect. 3.2.2) and whole-catchment erosion rates to compare against contemporary values.

One of the only CZOs studying intensive agriculture, the IML CZO, has focused on understanding interactions between geology and past and present legacies of agriculture (Fig. 1) on ecosystems services such as food supply, water quality and erosion. By coupling erosion and carbon (C) models to account for the role of selective entrainment and deposition on soil organic carbon (SOC) redistribution, Papanicolaou et al. [73] developed spatiotemporal predictions of SOC stocks at the hillslope scale under different hydrologic and land use scenarios. Their model simulations suggested that conservation tillage reversed the downward trend in SOC losses, resulting in nearly 26% of the lost SOC being regained. Integrative modeling by

Richardson and Kumar [84] further assessed the cumulative impacts of provisioning services associated with feed and energy-based production systems on regulating and supporting service bundles, and pointed to higher costs than previously estimated associated with water quality and atmospheric stabilizing services, measured as SOC storage, respiration, and nitrate leaching [84]. Other CZOs (RC, SS, Eel, SH, LUQ, BC, SC-JR, C) are less intensively managed and focus on working rangeland to forest ecosystems. These systems provide important intermediate regulating services such as forage, fiber, and timber through sustained soil formation and nutrient cycling, as well as water quality protection and water regulation to society discussed hereafter.

3.2 *Intermediate Regulating Services*

Intermediate regulating services, previously termed supporting services [54], are processes that allow for the maintenance of services provided by the critical zone. These processes have received less attention by the ecological community but form the basis of many Earth system science studies. For example, rangelands and pastures are managed through grazing and specialized farming practices geared towards the production of forage for livestock, whereas tree crops and forests are managed for timber used to produce fiber, shelter and other products. Forage and timber are products of primary productivity, and primary production is maintained by soil formation processes such as soil development, weathering and nutrient cycling, that in turn control currencies such as water storage and carbon sequestration (Table 1). Significant contributions of CZ science include understanding fine-scale topographic and geologic controls on productivity and feedbacks on erosion, advances in measurement of biomass in complex terrain, understanding the role of soil thickness in controlling water storage and ecohydrologic responses, impacts of fire and extreme events on soil formation processes, and recovery of soil C sequestration following disturbances. Other key contributions of CZ science include a greater understanding of the impacts of extreme events on water quality, and improved flood regulation through LiDAR snow flights that allow for data collection on spatial distributions.

3.2.1 Primary Production

Studies in the critical zone have been venturing into more complex terrain to understand the role of topography and bedrock composition and depths in modulating primary productivity, a supporting service provided by the critical zone. Key contributions of CZ science have been improved measurement of productivity at fine to intermediate scales, enhanced understanding of topographic and bedrock composition controls on productivity, and improved predictions of productivity to responses to drought and fire.

Fine Scale Controls of Topography on Primary Productivity

Numerous studies within the CZ network have focused on understanding the role of topography such as aspect, slope, and curvature on fine scale spatial patterning of primary productivity and biomass using LiDAR data as well as quantifying C uptake and losses through the measurement of net ecosystem exchange (NEE) using eddy covariance towers. Although topography has long been acknowledged as an important state factor shaping ecosystem structure and function (e.g. [16, 19]) as well as soil development [50], prior research has been primarily conducted along toposequences (e.g. [89]) or at plot scales. Intermediate scale measurements of productivity spanning plot to landscape scale are still relatively sparse.

High resolution LiDAR digital elevation models and other airborne cloud point data are filling this gap in intermediate scale measurements of biomass and productivity. Using multi-scale data of field-measured biomass, and terrestrial (TLS) and airborne laser scanning (ALS) data, Li et al. [57] developed a hierarchical modeling framework for biomass prediction in sagebrush-steppe, a low stature vegetation community that is challenging to quantify using spectral-based aerial and satellite remote sensing. Li et al. [58] further incorporated vegetation vertical structure obtained from LiDAR with ground-measured biomass data to scale shrub biomass from small field sites (1 m subplots and 1 ha plots) to a larger landscape. TLS techniques have also been used in other hot desert shrub environments to estimate biovolumes [41]. In the BC and SC-JR CZOs, Swetnam et al. [99] used LiDAR to estimate forest aboveground biomass C and incorporated these estimates into a net ecosystem carbon budget [76]. In particular, Swetnam et al. [99] showed that topographic position was more important than climate in explaining the distribution of aboveground biomass C. Vegetation estimates were driven more by the ridge-top or valley bottom position than by aspect, elevation, or precipitation. These findings have important implications for modeling forest sensitivity to climate change and predicting carbon reservoirs.

Bedrock Composition and Depth Controls on Productivity

Work in the critical zone is also showing critical interconnections between landscape evolution and plant productivity. Bedrock minerals provide important nutrients needed for tree growth that can place first order constraints on primary productivity. Typically nitrogen (N) is a limiting nutrient in terrestrial ecosystems [56], but release of phosphorus (P) from primary minerals during soil formation is controlled by bedrock composition. At the SS CZO, Hahm et al. [39] demonstrated that bedrock composition and geochemistry strongly influenced landscape evolution through differences in primary productivity. They showed that vegetation cover was highly correlated to bedrock P when elevation and precipitation were constrained. The availability of nutrients from bedrock resulted in a positive feedback loop in landscape evolution; increases in nutrients resulted in more vegetation, and this enhanced rock weathering and soil formation processes, which subsequently released more nutrients and increased water retention. These bedrock differences in soil development rates resulted in differences in erosion, where soil mantled terrain had two times higher erosion than bare rock.

Building off of this work, Uhlig et al. [102] examined the dynamics between plant and microbial nutrient demand and chemical weathering rates of parent material in a controlled weathering regime in the southern Sierra Nevadas. They found P, and to a lesser extent potassium (K), were the only elements heavily recycled, whereas calcium (Ca), magnesium (Mg) and silicon (Si) were not recycled (i.e., the regolith is the only source of plant uptake of these elements). Indeed, they found that trees obtain Mg to deep depths (6 m) via roots or associated mycorrhizae suggesting a tight link between weathering rates and nutrient uptake and erosion.

Improved Predictions of Productivity to Responses to Drought and Fire

Other studies in the critical zone are linking measurements of C uptake and loss using eddy covariance (EC) towers with remotely sensed and experimental data to identify sensitivities to drought and other perturbations. Previous work in the SS CZO by [101], for example, linked remotely sensed normalized difference vegetation indices (NDVI) to variations in carbon uptake as measured by gross ecosystem exchange (GEE) along an elevation gradient. They identified mid-elevation mountain forest ecosystems as particularly sensitive to future increases in temperature and concurrent changes in snow accumulation and melt. More recent research by Bales et al. [7] identified four main mechanisms that potentially offset each other to determine the impact of drought, in this case the 2012–2015 California drought, on forests in the Sierras; (1) partitioning of precipitation to evapotranspiration (ET) and access to water storage at depth, (2) warmer temperatures than the previous decade, (3) runoff from higher elevations offsetting ET demands, and (4) drought associated dieback and a wildfire thinned forest that reduced ET. Bales et al. [7] estimated usable water storage of at least 1300 mm yr^{-1} as root accessible water, conferring resiliency to drought. Finally, other studies within the CZO network are identifying temperature sensitivities to drought-induced mortality and investigating causal mechanisms. Adams et al. [2], for example, showed 5% proportional mortality per degree C temperature increase, with hydraulic failure as the dominant process explaining mortality across multiple species. These findings collectively highlight the sensitivities of forest to drought and the need to develop strategies to maximize forest resiliency and services provided by them [21].

Other CZOs are linking remotely sensed and experimental data to examine woody encroachment and prescribed fires at landscape scales. Woody plant cover, especially as western juniper (*Juniperus occidentalis*) has increased by 10 times over the past 100 years in the western USA [61]. Junipers are thought to alter the timing and amount of water use across a landscape but the mechanisms have remained poorly quantified. Sankey et al. [88] fused Landsat 5 Thematic Mapper and LiDAR–based data to develop juniper classifications and quantify expansion patterns in the RC

CZO. They showed an 85% juniper expansion based on comparisons to a 1965 vegetation cover map. Niemeyer et al. [66] showed that juniper reduced surface water input through interception by an average of 138 mm or 24% of the total site water budget compared to low and mountain sagebrush. Using coupled continuous soil moisture probe data and spatially distributed electrical magnetic induction (EMI) surveys, Niemeyer et al. [67, 68] also found that juniper extracted soil moisture well into the saprolite, as deep as 12 m below the surface. Niemeyer et al. [67, 68] further modeled ET associated with juniper conversion back to grassland across the western US, finding that only 29% of the juniper distribution had the potential for meaningful increases in streamflow. The development of these fused products and other geophysical tools offer new methods for management to quantify changes in canopy cover at larger scales.

In response to woody encroachment, prescribed fire is increasingly used as a management tool to reduce woody material and increase forage for grazing. In an attempt to determine how prescribed fire impacts water and carbon dynamics, a 26-ha mixed aspen sagebrush subcatchment was burned in 2007. Evapotranspiration as measured by EC recovered within 2 years on low sagebrush and aspen areas, and within 4 years for deeper-rooted mountain big sagebrush, and only marginally increased streamflow (20%) for 6 years following the fire [33]. Consistent with these findings, Fellows et al. [30] showed rapid recovery of gross ecosystem production (GEP) and similar, not increased, rates of ecosystem respiration following fire. They associated this rapid recovery with increased herbaceous and grass productivity. Interestingly, Fellows et al. [30] correlated annual GEP to NDVImax ($r^2 = 0.75$) suggesting the possible utility in extending GEP measurement to the landscape scale. Collectively, findings indicated that prescribed fire in more mesic (>500 MAP) montane sagebrush steppe ecosystems might be an effective tool to enhance forage for grazing. Prescribed fires in lower elevations need further evaluation to determine risk of invasion by *Bromus tectorum* (cheatgrass) and state-transition to homogeneous grasslands and loss of other services.

3.2.2 Soil Formation

Prior to the CZO program, soil production had not received much attention as a process by ecosystem service assessments, but with increasing pressures placed on soils, understanding the factors maintaining soil quantity and quality has become imperative. Recent studies have shown that 75% of actively cycling carbon (C), 1300–1500 Pg C, is held within the top meter of soil [40], however, this is constrained by long-term processes occurring at depth, in particular, the weathering of bedrock to soils. As this ultimately serves as the foundation for food, fiber, and energy production, water availability, biodiversity, and cycling of carbon and other essential nutrients, various CZOs have devoted much attention quantifying soil production, long term erosion, and the effects of extreme events on erosional processes.

Regolith Formation Mechanisms

Much research has been focused on the mechanisms of regolith formation, the transformation of bedrock to regolith (broken or unconsolidated rock). Geophysical surveys were conducted across a number of the CZOs to evaluate hypotheses related to topographical stresses. Specifically, St. Clair et al. [94] hypothesized that elastic stresses induced by surface topography could be substantial enough to fracture rocks in some landscapes, with implications for slope stability, erosion rates, and bedrock hydrologic properties. They showed considerable differences in stress profiles with depth beneath ridgetops and valley floors consistent with this hypothesis. St. Clair et al. [96] further examined topographic stresses across a range of CZOs (C, CR, BC) and showed that stress modifies the abundance of open fractures in bedrock. With topographic stress initiating bedrock fracturing, the second rate determining factor in transforming bedrock to regolith was water availability [31]. Anderson et al. [3] went on to demonstrate the importance of frost wedge cracking in the formation of mobile regolith at the BC CZO. The fastest rate of regolith production thus far has been observed on volcaniclastic bedrock in the LUQ CZO. This production rate is 5 to 30 times greater than previous estimates for granite and shale bedrock systems, providing insight that not all lithologies produce regolith at comparable rates [29].

Source-Sink Dynamics of Soil Organic Carbon in Complex Terrain

Soil C sequestration and storage is a key currency supplied by soil production and formation. Much uncertainty in soil C budgets stems from distributing soil C across complex terrain where soil thickness is largely unknown. To date, soil C models in complex terrain have used local controls such as vegetation cover, slope, elevation, hillslope position and soil properties to distribute soil C. Patton et al. [74] discovered a simple empirical model to predict soil thickness at *any* location within a catchment using high-resolution digital elevation models and a limited number of soil thickness measurements. Combining this method with soil C, the total amount of soil profile C can be accurately predicted to saprolite at 3 m resolution, and the entire watershed C and nitrogen (N) stores can be quantified. Findings from Patton et al. [75] showed that a significant amount of C was stored deep in the critical zone and that sampling at fixed depths between 30 and 100 cm (as is typically done) can vastly underestimate total soil C stores on complex terrain.

Several CZOs are also addressing how topographic complexity controls recovery of SOC in degraded systems as well as investigating the legacies of forest management and erosion on soil carbon sequestration. In the C CZO, a degraded critical zone where past agricultural practices decapitated surface horizons (specifically O and A horizons) and eroded 32% of its SOC, Dialynas et al. [26, 27] showed that topography exerted a strong control on the landscape's capacity to serve as a C source or a sink. Specifically, only 34% of eroded C has been replaced by C sequestration because of fine scale source-sink dynamics driven by erosion. In the LUQ CZO, Dialynas et al. [26, 27] used the same model to show that lithology largely constrains the rate of recovery in tropical environments. Specifically, they showed that landscapes underlain by volcaniclastic rock approached a state close to geomorphic equilibrium whereas a quartz diorite dominated landscape experienced greater rates of denudation and therefore slower rates of recovery. These findings collectively are useful in informing current management of legacies of the past.

Roles of Plants in Regolith Formation

Other contributions by CZOs to understanding regolith formation are testing hypotheses related to plants as builders and plumbers of the critical zone [13]. Specifically, Hasenmueller et al. [44] examined the role of roots in transforming rock to regolith and how rapidly soil is produced. They showed root-rock interactions were important for regolith weathering in the SH CZO where there was a high correlation between root density and bedrock fracturing. Brantley et al. [13] came up with 9 hypotheses to extend our understanding of the role of plants as plumbers in the critical zone. These hypotheses can be used in future cross CZO and other network related activities.

Roles of Extreme Events Such as Fire in Driving Denudation Rates

Extreme events (floods, fires, hurricanes) are accelerating rates of erosion, and several CZOs have been in a position to measure the responses to these episodic, extreme events. All five CZOs (100%) in the Western US (SS, SC-JR, RC, BC, Eel) have experienced wildfires in the last funding cycle (2013–2018). The SC-JR CZO experienced two fires in 2011 and 2013 and studied the effects of fire on long-term denudation rates in the Valles Caldera, NM, USA. Relative to an unburned catchment, a burned catchment experienced 3 orders of magnitude higher erosion rates [70]. The immense amount of suspended sediment flushed over one year following a fire explained 95% of long term suspended sediment erosion. Others have documented the redistribution of pyrogenic carbon (PyC) following the Thompson Ridge Fire in New Mexico [36]. Comparing post burn soil to soil collected prior to the fire, Galanter et al. [36] showed that Thompson Ridge Fire had the effect of homogenizing initial PyC differences among soil horizons and watersheds within the burn perimeter. Other CZOs such as the BC CZO have been evaluating the geomorphic, hydrologic and geochemical responses to extreme events, including extreme floods and intense rainfall events. Anderson et al. [4], for example, documented that the Boulder Creek September 2013 flood resulted in over 1100 landslides and debris flows and moved the equivalent of hundreds to thousands of years of hillslope weathering products in this single event, accounting for the majority of sediment transport and channel erosion [4]. These studies collectively point to the value and need of long-term monitoring to understand and predict the resistance and resiliency of ecosystems to extreme events.

3.2.3 Nutrient Cycling

Some nutrients are derived from the atmosphere through wet and dry (dust, aerosol) deposition [18] as well as biological nitrogen fixation, whereas others are thought to be derived primarily from rocks. Internal cycling of nutrients is thought to provide >90% of essential nutrients such as N and P to ecosystems [20] and take place in the upper surface horizons where SOC and minerals are more readily available. However, limited studies exist examining microbial-rock dynamics, depth-dependent zonation in nutrient cycling, and the relative contributions of atmospheric and rock derived nutrient to internal cycling. These patterns of input may vary across ecosystems depending on their climate and weathering environment.

CZ science is adding more details to the role of short and long-range dust inputs in providing nutrients to ecosystems, digging deeper into questions related to distribution of microbial processes to understand biologically (microbes and roots) mediated weathering, the impacts of grazing on rates of biological N fixation and the microbial communities responsible for it, as well as exploring deeper spatial extents in the critical zone as a new habitat frontier for different microbial communities. The ability of ecosystems to sequester carbon can be constrained by nutrients such that understanding nutrient inputs and losses as controlled by fire or land management are critical to predicting carbon storage and other services.

Long Range Dust Deposition

The importance of dust-derived nutrients is generally assumed to be low compared to nutrient supplies resulting from in situ bedrock weathering and soil formation, with the greatest impact predicted for heavily weathered systems, or those with naturally low bedrock-P. However, recent research by Aciego et al. [1] found that while longer-term P fluxes are dominated by bedrock weathering, over the short term P fluxes from exogenic dust are actually greater than bedrock-derived P, across a range of bedrock P content, in the Californian Southern Sierra Nevada montane forest ecosystem. Arvin et al. [5] went on to show that dust contributed most of the neodymium in vegetation to a site in the Sierras confirming the ecological significance of dust in eroding mountain landscapes. McClintock et al. [60] also found spatial variability in exogenic dust accumulation across the Luquillo Mountain ecosystem in Puerto Rico depending on bedrock type, with volcanic rocks having higher concentrations than quartz diorite. They argue this may be the result of soil residence time, or edaphic properties such as Fe or Al concentrations that might impact dust retention. Interestingly however, the greater dust accumulation in volcanic soils had little effect on soil P fractions, whereas dust content was significantly correlated with soil P in the more P-poor quartz diorite, suggesting exogenic dust inputs are more important in the P cycle in those systems. Overall, consistent deposition of exogenic dust over the long term with increasingly significant contributions over the short term suggests dust-derived nutrients has and continues to play a critical role in the evolution of both tropical and temperate ecosystems, which may be considerably enhanced with current projections of climate and land use change.

Nitrogen Inputs and Microbial Communities Controlled by Climate and Disturbance

Atmospherically deposited N and new N derived from biological N fixation are being altered by human activities. In drylands, soil biological crust communities provide vital services by bringing in 'new' N and C via the process of biological N_2 fixation and photosynthesis, respectively, but these rates are poorly constrained. Schwabedissen et al. [91] quantified rates of nitrogen cycling, nitrogenase activity and the soil biological crust bacterial communities associated with these rates along an elevation gradient in the RC CZO. They showed that warmer, drier climates at lower elevations hosted greater biological soil crust cover and higher nitrogenase activity (N inputs) and higher rates of nitrogen cycling compared with colder, wetter climates at higher elevations during the wet spring growing season. Dramatic shifts in the abundance of individual N_2-fixing groups from cyanobacteria to symbiotic were observed as elevation increased [11, 92]. In contrast to expectations, Schwabedissen et al. [91] found no significant grazing effect in those areas excluded from grazing for > 40 yrs and slight shifts in microbial communities after 40 + yr of exclusion except at the lowest elevation sites where fire legacies may have influenced patterns [11, 91]. The minimal impacts observed at higher elevations may be explained by relatively low rates of grazing as well as seasonal grazing patterns (movement from lower to higher elevations with warming and drying of soils).

In the Colorado Front Range at higher alpine areas receiving atmospherically deposited N, changing patterns of snowmelt may influence the amount of N inputs that enters groundwater and streams draining the area. At the rain-snow transition, Hinckley et al. [45] showed transport of anthropogenic N to streams was primarily governed by hillslope aspect via snowmelt patterns, attendant soil moisture and associated microbial responses. During summer rain events, Hinckley et al. [46] used isotope tracers to show that differences in soil properties on different aspects controlled the timing, movement, and the fate of N deposited into different soil pools. In a meta-analysis of microbial responses to anthropogenic N additions, Carey et al. [17] identified that abundances of bacteria (AOB) increased in response to N additions (329%) relative to ammonia-oxidizing archaea (AOA) (27%) suggesting that elevated N supply increases soil nitrification potential through increases in AOB.

Other critical zone studies have been documenting patterns of nutrient cycling following fire. In the SC-JR CZO, Weber et al. [105], for example, showed convergence of microbial communities in Ponderosa Pine and Mixed Conifer even 3 months after fire. These results suggest that the presence of specific taxa may be more important in predicting community compositional shifts after exposure to high burn severity than overall community composition or physical and chemical parameters of unburned soils, which were quite distinct. Collectively, these findings suggest that the maintenance of nutrient cycling services depends on the microbial communities and their responses to disturbances.

Depth Associated Carbon and Nutrient Stabilization

In addition to understanding inputs to the critical zone, CZ science has dug into understanding the stability of deep soil C and other elements as well as the microbial communities responsible for transforming these materials. Early work by Fontaine et al. [35] showed that adding fresh plant-derived C to the subsoil (0.6–0.8 m depth) resulted in microbial mineralization of old (~2,500 yr) C. These findings suggested that the lack of fresh SOC –the energy source for microbes–promoted the stability of deep C. More recently, Stone et al. [97, 98] explicitly examined soil organic matter (SOM) stability and microbial activity as a function of soil depth. While their results supported the idea that reduced respiration in subsoils increased SOM stability, they also found that these patterns were mostly driven by declines in energy availability and the size of the microbial biomass pool. They indicated that on a per-unit biomass basis, subsoil OM was as sensitive to destabilization as surface SOM.

New research is changing our understanding of SOM persistence in soils, previously explained largely as a result of molecular recalcitrance. Models are now shifting towards a more ecosystem-based approach, with SOM stabilization viewed as a possible product of soil biogeochemical properties and interactions rather simply relying on intrinsic characteristics of the OM itself [90], CZ studies are also contributing to this debate. Cyle et al. [23] examined the effect of substrate quality, given as the ratio of neutral detergent fiber to crude protein (NDF/CP), on SOM stability downline. They found that despite higher initial microbial mineralization rates, higher quality substrates resulted in more silt and clay C and N fractions, where there is more potential for mineral adsorption and thus greater residence time and stabilization. While this is based on soils with initially low C concentrations, it provides more evidence that SOM stabilization is more complex than previously thought.

Studies in the CZ have also been evaluating SOC sensitivity to decomposition as a function of temperature and other ecosystem properties. In a review, Rasmussen et al. [80] indicated that clay content was a poor predictor of SOC decomposition. Rather, exchangeable Ca strongly predicted SOM content and likely decomposition in water-limited soils. In contrast, iron- and aluminum-oxyhydroxides were better predictors in wetter, more acidic environments. Laboratory incubations of soils collected along an elevation gradient at the RC CZO lend supporting evidence that clay is a poor predictor of SOC decomposition. This study showed greater quantity of CO_2 released from higher elevations compared to lower elevation sites owing to greater soil C concentrations. However, the sensitivity of soil C decomposition to increasing temperatures appears to be greater at lower elevations [25], more C was lost as CO_2 on a per gram soil C basis at lower sites compared to higher ones. In this environment, sand content was a better predictor of SOC decomposition, suggesting availability rather than quality of SOC determined decomposition.

The role of aging exo-enzymes and changes in microbial communities in controlling the stabilization or destabilization of SOC has also been explored. Billings et al. [9] from C CZO found that exo-enzyme aging contributed to variation in apparent exo-enzyme activities. Wang et al. [104] investigated the distribution of enzymes in aggregates and showed that the distribution of enzyme activity indicates a higher contribution of fast-growing microorganisms in the micro- compared to macro-aggregates size fractions. Nitrogen additions and increased water availability altered these activities and their distribution suggesting that altered inputs and precipitation regimes may change the stability of this C.

Soil C also comes in inorganic forms as carbonates, and these soil inorganic carbon (SIC) stores are significant in arid and semi-arid regions. Pedogenic carbonate often follows a morphogenetic development sequence where carbonates coat rocks, form masses and nodules and eventually may become engulfed and cemented [28]. The inorganic carbon associated with these carbonate coatings on rocks and masses are not typically quantified by soil scientists because they are greater than 2 mm in size, the operational definition of soil, and they are difficult to measure owing to the heterogeneity and scale. Consequently, SIC may be underestimated in arid and semi-arid regions. The RC CZO has been contributing to a gap in the C cycle through the quantification of inorganic carbon associated with rocks. Rocks with carbonate coatings were pulverized to quantify bulk inorganic carbon, and analyses showed that in some cases rock carbonate coating can represent as much as 43% of the total inorganic carbon in soil profiles. These findings suggested that studies only examining the fine fraction of soils (<2 mm) are overlooking a significant store of carbon in gravelly parent materials [95].

Long-term (20 year) experimental rainfall and vegetation manipulations in the sagebrush steppe ecosystems associated with the RC CZO indicate the critical need to quantify both forms of carbon when considering dryland soils, and also the incorporation of deep measurements. Huber et al. [48] showed that extrapolation from surface soils may be inappropriate due to differential response of surface versus deep C pools to altered rainfall and plant communities. Dryland soil organic C pools were susceptible to changing rainfall and altered plant community composition. Accumulation of pedogenic carbonates were also susceptible to these treatment factors. Predicting future responses to climate and vegetation will require consideration of changes in both organic and inorganic C pools and the full vertical dimension of soils.

3.3 Regulating Services

Regulating services are those services that act to regulate significant processes, cycles and flows within an ecosystem, such as disease, waste mediation, water quality and control, soil erosion, local to regional climate, biological life cycles and production, and so on.

3.3.1 Water Quality Regulation

Current water quality regulation in the U.S. is largely focused on land use as a major contributor to water degradation, however, there is little research on how land use and climate change interact to impact water quality. Loecke et al. [59] recently examined how changes in climate extremes may interact with agricultural land use practices, and the potential impacts this may have on water quality in the Midwestern U.S. They found that extreme changes—dry years followed by wet years—resulted in "weather whiplash" whereby higher than predicted concentrations of nutrients were flushed from soils and entered water bodies. These findings indicate that management may need to anticipate greater nutrient loading (and subsequent reduction in water quality) in water bodies with projected increases in alternating weather extremes, and incorporate mediation measures for flood events, water quality regulation, and water supply forecasting.

In addition to changes in climate patterns, more extreme disturbance events such as fires and floods will likely alter water quality regulation. After catastrophic wildfires, for example, increases in discharge following storm events were observed in a part of the greater Rio Grande River watershed [81]. These increases were due to increases in overland flow and resulted in increased erosion and associated increased turbidity that lasted long after storm events. Turbidity also altered stream metabolism because it impacted light availability and decreased in-stream rates of gross primary production, but also provided pulses of nutrients to stream systems [81]. In the SC-JR CZO, Murphy et al. [64] reported that water quality was still impaired even 3 years after a fire owing to intense storms (30 min rainfall intensity > 10 mm h^{-1}). In other systems, such as the Eel CZO with elevated nutrients, toxic algal blooms have been a concern in waters because of their rapid spread and ability to harm ecosystem and human health. A recent study found that benthic cyanobacteria use trapped oxygen bubbles produced during photosynthesis to float downstream. When these bubbles are dislodged in ripples or areas of backwater, the healthy mass sinks back to the bottom where it can colonize a new area [12]. By understanding the dispersal techniques of cyanobacteria or the water quality impacts of wildfire, water managers may be able to make better-informed decisions on how to deal with current problems and be better prepared to deal with future problems as disturbance events become increasingly normal.

3.3.2 Water Supply Regulation

Water management (flood control, irrigation) places key constraints on water supply regulation as a critical zone service and therefore on currencies of hydrological flow. Changing climate and extremes are forcing water management agencies to rethink how they manage water. Key contributions of CZ science to management has been the application of repeat snow-on LIDAR to forecast water supplies and manage flooding, and use of microtopography from LIDAR to advance our understanding of the impacts of tile drainage in low relief regions.

Repeat Airborne Snow Observations in Management of Water Supplies

Several CZOs are investigating the role of infrastructure and its interaction with climate change in altering water supply regulation. This knowledge is essential to informing decision making and land management. In particular, several CZOs are asking questions about how snow pack storage will change with climate change and how this will influence water supplies. Taking advantage of Airborne Snow Observations (ASO) with LiDAR, investigators at RC CZO are coupling CZO related models (iSNOBAL) to LiDAR to forecast water supply for California and the Boise River Basin [72]. In 2017, the Boise River experienced spring flooding from snow melt. One of the main issues was that most of the SNOWTEL sites were reporting no snow, but 0.25–0.5 m of snow was still heterogeneously distributed across the landscape. By coupling spatial distributions of snow mapped via aerial LiDAR with iSNOBAL, a physically-based spatially-distributed snow distribution and melt model, they were able to manage the release of water from the dam above Boise, Idaho to minimize extreme flooding. The National Resources Conservation Service in the US has since adopted this as an approach for forecasting water supplies in California.

LiDAR Enhanced Measurement of Microtopography in Low Relief Areas

The impacts of agricultural drainage on hydrology and biogeochemistry have long been appreciated [93]. Drainage improvements when combined with agricultural land use change can lead to increased peak runoff rates, sediment losses, and nutrient losses. However, the influence of spatially resolved micro-topographic variability, such as depressions and roadside ditches, on tile flow and their accumulated impact on hydrology and nutrient dynamics remain poorly understood. Wagner et al. [103] incorporated tile flow as an explicit parameter into an integrated ecohydrologic-flow model at the IML CZO to examine the influence of tile drains on ecohydrologic and nutrient dynamics in intensively managed agricultural fields at LiDAR-resolution scales. Advances in coupled model and fine resolution topography may help to improve precision agriculture and reduce or minimize nutrient and sediment losses beyond agriculture fields.

3.3.3 Climate Regulation

Vegetation responses (e.g., respiration and photosynthesis) to climate change may be constrained by geochemical processes such as weathering and soil formation [32]. Geophysical measurements in the CZOs are pointing to deep coupling between climate and weathering extent. In several Western US cases, these average depths increased with decreasing elevations. In the case of the SS and RC CZOs, there is an increase in weathering extent along elevation gradients that coincides with the rain-snow transition [24, 65, 79, 85]. The largest difference in weathering depth between two aspects at the RC CZO occurred where the difference in snow accumulation was the greatest [65]. Findings collectively suggest that elevation and dominant form of precipitation are major controls on the depth to unweathered bedrock. Given the documented changes in the rain to snow transition over the past 30 years [37, 51, 52] and projected changes [51], these findings point to likely changes in weathering that will come in the future as the transition to snow free zones increases. Changes in vegetation cover, due to climate change, drought or wildfire, could also drive changes in the spatial distribution of snow storage and hence weathering and weathering products (with links to climate regulation).

4 Synthesis

In this chapter, we examined the contributions of CZ science to our understanding of critical zone services. We identified key contributions to the supply of services, and end by returning to the questions posed at the beginning:

(1) How can improved understanding of the critical zone be used to inform land management on resiliency to climate change and associated extreme events (droughts, floods, fires) that may disrupt these services? Our review highlights the following key findings, among others not discussed, that may be beneficial in management planning:

- Rock moisture is important for water storage and resiliency of forests to drought.
- Extreme precipitation events will likely increase erosion and lead to declines in soil formation and productivity.
- Increasing fire frequency in the western US is likely to increase erosion rates and reduce productivity and soil formation because soil loss will be greater than inputs of materials to sustain them.
- Predicting productivity and soil carbon storage responses to climate change and plant community shifts will require quantifying both surface and deep pools of soil carbon and both organic and inorganic forms in semi-arid regions.
- New tools and models are emerging that can improve prediction of soil thickness and therefore soil carbon storage across the landscape.
- Management may need to anticipate weather whiplash with more extreme weather–drier followed by wetter years–for flood events, water quality regulation, and water supply forecasting.

(2) How can improved understanding of the critical zone be used to inform land management on sensitivities, uncertainties, and thresholds in response to climate change and interactions with land use to sustain or restore ecosystem services?

- *Water storage*–Sierra Nevada and the northern basin and range have high sensitivity to climate change in terms of snow water storage. Management can be aided by snow observatory LiDAR and CZO related models of snow distribution.
- *Water supply*–Snow-on LiDAR spearheaded by CZOs and physically based models are becoming management norms to forecast water supplies.
- *Productivity*–Recovery time of soil organic carbon is determined by management, topographic complexity, and lithological constraints.
- *Nutrient cycling and biodiversity*–We need to evaluate shifts in rates of N inputs from dust and alterations in microbial communities in response to climate that constrain productivity.

Critical zone science and its services are already receiving attention within management and policy frameworks. Recently, critical zone services have been adopted as a framework for soil protection policy in Europe because it extends the definition and discussion of soils beyond the traditional boundaries of soils to incorporate other processes [62]. Indeed, the European Union identifies seven functions that it wants to maintain,biomass production, buffering and filtering of water, biodiversity pool, source of raw materials, support for housing and infrastructure, carbon sink and archive of cultural heritage [62]. These similar functions could be adopted into US policy frameworks to inform soil conservation and land management in the U.S. and ultimately assist in sustaining and increasing resiliency of critical resources.

5 Conclusions

Critical zone services expands the notion of ecosystem services to understand the maintenance of ecosystem services for future generations, typically through the measurement of slow processes operating at longer time scales and their interaction with faster ones. Numerous advancements have been made by critical zone science that can contribute to landscape management and predicting the sustainability and/or disruption of natural services. Some of the most exciting advancements in the CZOs have been and likely will be at the interface between disciplines and among different scientists, rather than within a single discipline. Key contributions of critical zone science to management are mainly in the area of supporting services to intermediate regulating services to a lesser extent provisioning services. In particular numerous advancements in intermediate scale (plot to landscape) measurements and modeling are enhancing our understanding of climate, topography, and bedrock controls on fine-scale variability in ecosystem services. Measurements of slow variables (at deeper spatial and temporal scales) provide baseline rates that managers can monitor and sustain. They are also key to predicting responses to extreme events, climate and land use change, as well as to sustaining services as the frequency of extreme events and disturbances increase.

Acknowledgements Support for this research was provided by the NSF via RC CZO Cooperative agreement NSF EAR-1331872. We thank Tim White for a friendly review of the manuscript. The data referred to in this review are contained in a published source (not a repository) cited in the References.

Author Contributions KAL conceived the idea and organization, KAL, AC, KG, SS, CM, RM contributed to initial review of literature and annotated bibliographies in graduate seminar course, KAL wrote initial manuscript, and KAL and AC prepared manuscript for submission.

Competing Interests The authors declare that they have no conflict of interest but one is a principal investigator of a Critical Zone Observatory (K. A. Lohse).

References

1. Aciego SM, Riebe CS, Hart SC, Blakowski MA, Carey CJ, Aarons SM et al (2017) Dust outpaces bedrock in nutrient supply to montane forest ecosystems. Nat Commun 8:14800. https://doi.org/10.1038/ncomms14800
2. Adams HD, Zeppel MJ, Anderegg WR, Hartmann H, Landhäusser SM, Tissue DT et al (2017) A multi-species synthesis of physiological mechanisms in drought-induced tree mortality. Nat Ecol & Evol 1(9):1285. https://doi.org/10.1038/s41559-017-0248-x
3. Anderson RS, Anderson SP, Tucker GE (2013) Rock damage and regolith transport by frost: an example of climate modulation of the geomorphology of the critical zone. Earth Surf Proc Land 38(3):299–316. https://doi.org/10.1002/esp.3330
4. Anderson SW, Anderson SP, Anderson RS (2015) Exhumation by debris flows in the 2013 Colorado front range storm. Geol 43(5):391–394. https://doi.org/10.1130/G36507.1
5. Arvin LJ, Riebe CS, Aciego SM, Blakowski MA (2017) Global patterns of dust and bedrock nutrient supply to montane ecosystems. Sci Adv 3(12):eaao1588. https://doi.org/10.1126/sciadv.aao1588
6. Millennium Ecosystem Assessment (2005) Ecosystems and human well-being: synthesis. Island Press, Washington, DC. https://www.millenniumassessment.org/documents/document.356.aspx.pdf
7. Bales RC, Goulden ML, Hunsaker CT, Conklin MH, Hartsough PC, O'Geen AT et al (2018) Mechanisms controlling the impact of multi-year drought on mountain hydrology. Sci Rep 8(1):690. https://doi.org/10.1038/s41598-017-19007-0
8. Banwart S, Black H, Cai Z, Gicheru P, Joosten H, Victoria R et al (2014) Benefits of soil carbon: report on the outcomes of an international scientific committee on problems of the environment rapid assessment workshop. Carbon Manag 5(2):185–192. https://doi.org/10.1080/17583004.2014.913380
9. Billings S, Min K, Ballantyne F, Chen Y, Sellers M (2016) Aging exo-enzymes can create temporally shifting, temperature-dependent resource landscapes for microbes. Biogeochem 131(1–2):163–172. https://doi.org/10.1007/s10533-016-0273-x
10. Blankinship JC, Meadows MW, Lucas RG, Hart SC (2014) Snowmelt timing alters shallow but not deep soil moisture in the Sierra Nevada. Water Resour Res 50(2):1448–1456. https://doi.org/10.1002/2013WR014541
11. Blay ES, Schwabedissen SG, Magnuson TS, Aho KA, Sheridan PP, Lohse KA (2017) Variation in biological soil crust bacterial abundance and diversity as a function of climate in cold steppe ecosystems in the intermountain West, USA. Microb Ecol 74(3):691–700. https://doi.org/10.1007/s00248-017-0981-3
12. Bouma-Gregson K, Power ME, Bormans M (2017) Rise and fall of toxic benthic freshwater cyanobacteria (Anabaena spp.) in the Eel river: Buoyancy and dispersal. Harmful Algae 66:79–87. https://doi.org/10.1016/j.hal.2017.05.007

13. Brantley SL, Eissenstat DM, Marshall JA, Godsey SE, Balogh-Brunstad Z, Karwan DL et al (2017a) Reviews and syntheses: on the roles trees play in building and plumbing the critical zone. Biogeosciences 14(22):5115. https://doi.org/10.5194/bg-14-5115-2017
14. Brantley SL, McDowell WH, Dietrich WE, White TS, Kumar P, Anderson SP et al (2017b) Designing a network of critical zone observatories to explore the living skin of the terrestrial Earth. Earth Surf Dyn 5(4):841–860. https://doi.org/10.5194/esurf-5-841-2017
15. Brooks JR, Barnard HR, Coulombe R, McDonnell JJ (2010) Ecohydrologic separation of water between trees and streams in a Mediterranean climate. Nat Geosci 3(2):100–104. https://doi.org/10.1038/ngeo722
16. Burke IC, Reiners WA, Olson RK (1989) Topographic control of vegetation in a mountain big sagebrush steppe. Vegetatio 84(2):77–86. https://doi.org/10.1007/BF00036508
17. Carey CJ, Dove NC, Beman JM, Hart SC, Aronson EL (2016) Meta-analysis reveals ammonia-oxidizing bacteria respond more strongly to nitrogen addition than ammonia-oxidizing archaea. Soil Biol Biochem 99:158–166. https://doi.org/10.1016/j.soilbio.2016.05.014
18. Chadwick OA, Derry LA, Vitousek PM, Huebert BJ, Hedin LO (1999) Changing sources of nutrients during four million years of ecosystem development. Nature 397(6719):491. https://doi.org/10.1038/17276
19. Chapin FS, Matson PA, Mooney HA (eds) (2002) Principles of terrestrial ecosystem ecology, 1st edn. Springer-Verlag, New York
20. Chapin III FS, Matson PA, Vitousek PM (eds) (2012) Principles of terrestrial ecosystem ecology, 2nd Edn. Springer-Verlag, New York. https://doi.org/10.1007/978-1-4419-9504-9
21. Cobb RC, Ruthrof KX, Breshears DD, Lloret F, Aakala T, Adams HD et al (2017) Ecosystem dynamics and management after forest die-off: a global synthesis with conceptual state and transition models. Ecosphere 8(12). https://doi.org/10.1002/ecs2.2034
22. Critical Zone Strategic Plan Summary (2016) https://criticalzone.org/images/national/associated-files/1National/Critical_Zone_Strategic_Plan_Summary_2016-02-11.pdf
23. Cyle KT, Hill N, Young K, Jenkins T, Hancock D, Schroeder PA, Thompson A (2016) Substrate quality influences organic matter accumulation in the soil silt and clay fraction. Soil Biol Biochem 103:138–148. https://doi.org/10.1016/j.soilbio.2016.08.014
24. Dahlgren RA, Boettinger JL, Huntington GL, Amundson RG (1997) Soil development along an elevational transect in the western Sierra Nevada, California. Geoderma 78(3–4):207–236. https://doi.org/10.1016/S0016-7061(97)00034-7
25. Delvinne HH (2016) Temperature impacts on soil organic carbon decomposition across an environmental gradient in a semi-arid ecosystem (Doctoral dissertation). Boise State University Theses and Dissertations. Boise State University, Boise. https://scholarworks.boisestate.edu/td/1162
26. Dialynas YG, Bastola S, Bras RL, Billings SA, Markewitz D (2016a) Topographic variability and the influence of soil erosion on the carbon cycle. Global Biogeochem Cycles 30(5):644–660. https://doi.org/10.1002/2015GB005302
27. Dialynas YG, Bastola S, Bras RL, Marin-Spiotta E, Silver WL, Arnone E, Noto LV (2016b) Impact of hydrologically driven hillslope erosion and landslide occurrence on soil organic carbon dynamics in tropical watersheds. Water Resour Res 52(11):8895–8919. https://doi.org/10.1002/2016WR018925
28. Ditzler C, Scheffe K, Monger HC (eds) (2017) Soil survey manual (USDA Handbook 18). Government Printing Office, Washington, D.C. https://www.nrcs.usda.gov/wps/portal/nrcs/detail/soils/scientists/?cid=nrcs142p2_054262
29. Dosseto A, Buss HL, Suresh PO (2012) Rapid regolith formation over volcanic bedrock and implications for landscape evolution. Earth Planet Sci Lett 337:47–55. https://doi.org/10.1016/j.epsl.2012.05.008
30. Fellows AW, Flerchinger GN, Lohse KA, Seyfried MS (2018) Rapid recovery of gross production and respiration in a mesic mountain big sagebrush ecosystem following prescribed fire. Ecosyst 1–12. https://doi.org/10.1007/s10021-017-0218-9
31. Ferrier KL, Kirchner JW, Finkel RC (2012) Weak influences of climate and mineral supply rates on chemical erosion rates: measurements along two altitudinal transects in the Idaho Batholith. J Geophys Res: Earth Surf 117(F2). https://doi.org/10.1029/2011JF002231

32. Field JP, Breshears DD, Law DJ, Villegas JC, López-Hoffman L, Brooks PD et al (2015) Critical zone services: expanding context, constraints, and currency beyond ecosystem services. Vadose Zone J 14(1). https://doi.org/10.2136/vzj2014.10.0142
33. Flerchinger GN, Seyfried MS, Hardegree SP (2016) Hydrologic response and recovery to prescribed fire and vegetation removal in a small rangeland catchment. Ecohydrol 9(8):1604–1619. https://doi.org/10.1002/eco.1751
34. Flerchinger GN, Fellows AW, Seyfried MS, Clark PE, Lohse KA (In Review) Water and carbon fluxes along a climate gradient in sagebrush ecosystems. Ecosyst
35. Fontaine S, Barot S, Barré P, Bdioui N, Mary B, Rumpel C (2007) Stability of organic carbon in deep soil layers controlled by fresh carbon supply. Nature 450(7167):277. https://doi.org/10.1038/nature06275
36. Galanter A, Cadol D, Lohse K (2018) Geomorphic influences on the distribution and accumulation of pyrogenic carbon (PyC) following a low severity wildfire in northern New Mexico. Earth Surf Process Landforms. https://doi.org/10.1002/esp.4386
37. Godsey SE, Marks D, Kormos PR, Seyfried MS, Enslin CL, Winstral AH, McNamara JP, Link TE (2018) Eleven years of mountain weather, snow, soil moisture and stream flow data from the rain-snow transition zone–the Johnston draw catchment, reynolds creek experimental watershed and critical zone observatory, USA. Earth Syst Sci Data Discuss. https://doi.org/10.5194/essd-2017-112
38. Guo L, Lin H (2016) Critical zone research and observatories: current status and future perspectives. Vadose Zone J 15(9). https://doi.org/10.2136/vzj2016.06.0050
39. Hahm WJ, Riebe CS, Lukens CE, Araki S (2014) Bedrock composition regulates mountain ecosystems and landscape evolution. Proc Natl Acad Sci 111(9):3338–3343. https://doi.org/10.1073/pnas.1315667111
40. Harden JW, Hugelius G, Ahlström A, Blankinship JC, Bond-Lamberty B, Lawrence CR et al (2018) Networking our science to characterize the state, vulnerabilities, and management opportunities of soil organic matter. Glob Change Biol 24(2):e705–e718. https://doi.org/10.1111/gcb.13896
41. Harman CJ, Lohse KA, Troch PA, Sivapalan M (2014) Spatial patterns of vegetation, soils, and microtopography from terrestrial laser scanning on two semiarid hillslopes of contrasting lithology. J Geophys Res Biogeosci 119(2):163–180. https://doi.org/10.1002/2013JG002507
42. Harpold AA, Guo Q, Molotch N, Brooks PD, Bales R, Fernandez-Diaz JC et al (2014) LiDAR-derived snowpack data sets from mixed conifer forests across the Western United States. Water Resour Res 50(3):2749–2755. https://doi.org/10.1002/2013WR013935
43. Harpold AA, Brooks PD (2018) Humidity determines snowpack ablation under a warming climate. Proc Natl Acad Sci 201716789. https://doi.org/10.1073/pnas.1716789115
44. Hasenmueller EA, Gu X, Weitzman JN, Adams TS, Stinchcomb GE, Eissenstat DM et al (2017) Weathering of rock to regolith: the activity of deep roots in bedrock fractures. Geoderma 300:11–31. https://doi.org/10.1016/j.geoderma.2017.03.020
45. Hinckley ELS, Barnes RT, Anderson SP, Williams MW, Bernasconi SM (2014) Nitrogen retention and transport differ by hillslope aspect at the rain-snow transition of the Colorado front range. J Geophys Res Biogeosci 119(7):1281–1296. https://doi.org/10.1002/2013JG002588
46. Hinckley ELS, Ebel BA, Barnes RT, Murphy SF, Anderson SP (2017) Critical zone properties control the fate of nitrogen during experimental rainfall in montane forests of the Colorado front range. Biogeochem 132(1–2):213–231. https://doi.org/10.1007/s10533-017-0299-8
47. Holbrook WS, Riebe CS, Elwaseif M, Hayes LJ, Basler-Reeder K, Harry LD et al (2014) Geophysical constraints on deep weathering and water storage potential in the Southern Sierra critical zone observatory. Earth Surf Proc Land 39(3):366–380. https://doi.org/10.1002/esp.3502
48. Huber DP, Lohse KA, Commendador AS, Joy S, Finney B, Aho K, Germino M (2019) Vegetation and precipitation shifts interact to alter organic and inorganic carbon storage in cold desert soils. Ecosphere 10(3):e02655. https://doi.org/10.1002/ecs2.2655

49. Hunsaker CT, Whitaker TW, Bales RC (2012) Snowmelt runoff and water yield along elevation and temperature gradients in California's southern Sierra Nevada. JAWRA J Am Water Resour Assoc 48(4):667–678. https://doi.org/10.1111/j.1752-1688.2012.00641.x
50. Jenny H (1941) Factors of soil formation; a system of quantitative pedology. McGraw-Hill Book Company Inc., New York
51. Klos PZ, Link TE, Abatzoglou JT (2014) Extent of the rain-snow transition zone in the western US under historic and projected climate. Geophys Res Lett 41(13):4560–4568. https://doi.org/10.1002/2014GL060500
52. Kormos PR, McNamara JP, Seyfried MS, Marshall HP, Marks D, Flores AN (2015) Bedrock infiltration estimates from a catchment water storage-based modeling approach in the rain snow transition zone. J Hydrol 525:231–248. https://doi.org/10.1016/j.jhydrol.2015.03.032
53. Kormos PR, Marks DG, Seyfried MS, Havens SC, Hedrick A, Lohse KA, Sandusky M (2018) 31 years of hourly spatially distributed air temperature, humidity, and precipitation amount and phase from Reynolds critical zone observatory. Earth Syst Sci Data Discuss. https://doi.org/10.5194/essd-2017-82
54. Lamothe KA, Sutherland IJ (2018) Intermediate ecosystem services: the origin and meanings behind an unsettled concept. Int J Biodivers Sci, Ecosyst Serv & Manag 14(1):179–187
55. Langston AL, Tucker GE, Anderson RS, Anderson SP (2015) Evidence for climatic and hillslope-aspect controls on vadose zone hydrology and implications for saprolite weathering. Earth Surf Proc Land 40(9):1254–1269. https://doi.org/10.1002/esp.3718
56. LeBauer DS, Treseder KK (2008) Nitrogen limitation of net primary productivity in terrestrial ecosystems is globally distributed. Ecol 89(2):371–379. https://doi.org/10.1890/06-2057.1
57. Li A, Glenn NF, Olsoy PJ, Mitchell JJ, Shrestha R (2015) Aboveground biomass estimates of sagebrush using terrestrial and airborne LiDAR data in a dryland ecosystem. Agric for Meteorol 213:138–147. https://doi.org/10.1016/j.agrformet.2015.06.005
58. Li A, Dhakal S, Glenn NF, Spaete LP, Shinneman DJ, Pilliod DS et al (2017) Lidar aboveground vegetation biomass estimates in shrublands: prediction, uncertainties and application to coarser scales. Remote Sensing 9(9):903. https://doi.org/10.3390/rs9090903
59. Loecke TD, Burgin AJ, Riveros-Iregui DA, Ward AS, Thomas SA, Davis CA, Clair MAS (2017) Weather whiplash in agricultural regions drives deterioration of water quality. Biogeochem 133(1):7–15. https://doi.org/10.1007/s10533-017-0315-z
60. McClintock MA, Brocard G, Willenbring J, Tamayo C, Porder S, Pett-Ridge JC (2015) Spatial variability of African dust in soils in a montane tropical landscape in Puerto Rico. Chem Geol 412:69–81. https://doi.org/10.1016/j.chemgeo.2015.06.032
61. Miller RF, Bates JD, Svejcar TJ, Pierson FB, Eddleman LE (2005) Biology, ecology, and management of western juniper (Juniperus occidentalis). Tech Bull 152. Agricultural Experiment Station, Oregon State University, Corvallis, Oregon. http://ir.library.oregonstate.edu/concern/technical_reports/cz30pv075
62. Montanarella L, Panagos P (2015) Policy relevance of critical zone science. Land Use Policy 49:86–91. https://doi.org/10.1016/j.landusepol.2015.07.019
63. Montgomery DR (2007) Soil erosion and agricultural sustainability. Proc Natl Acad Sci 104(33):13268–13272. https://doi.org/10.1073/pnas.0611508104
64. Murphy SF, Writer JH, McCleskey RB, Martin DA (2015) The role of precipitation type, intensity, and spatial distribution in source water quality after wildfire. Environ Res Lett 10(8):084007. https://doi.org/10.1088/1748-9326/10/8/084007
65. Nielson T (2017) Application of hydrogeophysical imaging in the Reynolds creek critical zone observatory (Doctoral dissertation). Boise State University Theses and Dissertations. 1354. Boise State University, Boise, Idaho. https://doi.org/10.18122/B22X42
66. Niemeyer RJ, Link TE, Seyfried MS, Flerchinger GN (2016) Surface water input from snowmelt and rain throughfall in western juniper: potential impacts of climate change and shifts in semi-arid vegetation. Hydrol Process 30(17):3046–3060. https://doi.org/10.1002/hyp.10845
67. Niemeyer RJ, Heinse R, Link TE, Seyfried MS, Klos PZ, Williams CJ, Nielson T (2017a) Spatiotemporal soil and saprolite moisture dynamics across a semi-arid woody plant gradient. J Hydrol 544:21–35. https://doi.org/10.1016/j.jhydrol.2016.11.005

68. Niemeyer RJ, Link TE, Heinse R, Seyfried MS (2017b) Climate moderates potential shifts in streamflow from changes in pinyon-juniper woodland cover across the western US. Hydrol Process 31(20):3489–3503. https://doi.org/10.1002/hyp.11264
69. National Research Council (2001) Basic research opportunities in Earth science. The National Academies Press, Washington, DC. https://doi.org/10.17226/9981
70. Orem CA, Pelletier JD (2016) The predominance of post-wildfire erosion in the long-term denudation of the Valles Caldera, New Mexico. J Geophys Res Earth Surf 121(5):843–864. https://doi.org/10.1002/2015JF003663
71. Oshun J, Dietrich WE, Dawson TE, Fung I (2016) Dynamic, structured heterogeneity of water isotopes inside hillslopes. Water Resour Res 52(1):164–189. https://doi.org/10.1002/2015WR017485
72. Painter TH, Berisford DF, Boardman JW, Bormann KJ, Deems JS, Gehrke F et al (2016) The airborne snow observatory: fusion of scanning lidar, imaging spectrometer, and physically-based modeling for mapping snow water equivalent and snow albedo. Remote Sens Environ 184:139–152. https://doi.org/10.1016/j.rse.2016.06.018
73. Papanicolaou AN, Wacha KM, Abban BK, Wilson CG, Hatfield JL, Stanier CO, Filley TR (2015) From soilscapes to landscapes: a landscape-oriented approach to simulate soil organic carbon dynamics in intensively managed landscapes. J Geophys Res Biogeosci 120(11):2375–2401. https://doi.org/10.1002/2015JG003078
74. Patton NR, Lohse KA, Seyfried M, Crosby BT, Godsey SE (2018) Predicting soil thickness on soil mantled hillslopes. Nat Commun 9:3329. https://doi.org/10.1038/s41467-018-05743-y
75. Patton NR, Lohse KA, Seyfried MS, Parsons S (2019) Topographic controls on soil organic carbon on soil mantled landscapes. Sci Rep 9:6390. https://doi.org/10.1038/s41598-019-42556-5
76. Perdrial J, Brooks PD, Swetnam T, Lohse KA, Rasmussen C, Litvak M et al (2018) A net ecosystem carbon budget for snow dominated forested headwater catchments: linking water and carbon fluxes to critical zone carbon storage. Biogeochem 138(3):225–243. https://doi.org/10.1007/s10533-018-0440-3
77. Poley A (2017) Deriving landscape-scale vegetation cover and aboveground biomass in a semi-arid ecosystem using imaging spectroscopy (Doctoral dissertation). Boise State University Theses and Dissertations. 1360. Boise State University, Boise, Idaho. https://doi.org/10.18122/B25Q72
78. Rasouli K, Pomeroy JW, Marks DG (2015) Snowpack sensitivity to perturbed climate in a cool mid-latitude mountain catchment. Hydrol Process 29(18):3925–3940. https://doi.org/10.1002/hyp.10587
79. Rasmussen C, Dahlgren RA, Southard RJ (2010) Basalt weathering and pedogenesis across an environmental gradient in the southern Cascade Range, California, USA. Geoderma 154(3–4):473–485. https://doi.org/10.1016/j.geoderma.2009.05.019
80. Rasmussen C, Heckman K, Wieder WR, Keiluweit M, Lawrence CR, Berhe AA et al (2018) Beyond clay: towards an improved set of variables for predicting soil organic matter content. Biogeochemistry 137(3):297–306. https://doi.org/10.1007/s10533-018-0424-3
81. Reale JK, Van Horn DJ, Condon KE, Dahm CN (2015) The effects of catastrophic wildfire on water quality along a river continuum. Freshw Sci 34(4):1426–1442. https://doi.org/10.1086/684001
82. Rempe DM, Dietrich WE (2014) A bottom-up control on fresh-bedrock topography under landscapes. Proc Natl Acad Sci 111(18):6576–6581. https://doi.org/10.1073/pnas.1404763111
83. Rempe DM, Dietrich WE (2018) Direct observations of rock moisture, a hidden component of the hydrologic cycle. Proc Natl Acad Sci 115(11):2664–2669. https://doi.org/10.1073/pnas.1800141115
84. Richardson M, Kumar P (2017) Critical zone services as environmental assessment criteria in intensively managed landscapes. Earth's Future 5(6):617–632. https://doi.org/10.1002/2016EF000517

85. Riebe CS, Kirchner JW, Finkel RC (2004) Sharp decrease in long-term chemical weathering rates along an altitudinal transect. Earth Planet Sci Lett 218(3–4):421–434. https://doi.org/10.1016/S0012-821X(03)00673-3
86. Riebe CS, Hahm WJ, Brantley SL (2017) Controls on deep critical zone architecture: a historical review and four testable hypotheses. Earth Surf Proc Land 42(1):128–156. https://doi.org/10.1002/esp.4052
87. Salve R, Rempe DM, Dietrich WE (2012) Rain, rock moisture dynamics, and the rapid response of perched groundwater in weathered, fractured argillite underlying a steep hillslope. Water Resour Res 48(11). https://doi.org/10.1029/2012WR012583
88. Sankey TT, Glenn N, Ehinger S, Boehm A, Hardegree S (2010) Characterizing western juniper expansion via a fusion of landsat 5 thematic mapper and lidar data. Rangel Ecol Manage 63(5):514–523. https://doi.org/10.2111/REM-D-09-00181.1
89. Schimel D, Stillwell MA, Woodmansee RG (1985) Biogeochemistry of C, N, and P in a soil catena of the shortgrass steppe. Ecol 66(1):276–282. https://doi.org/10.2307/1941328
90. Schmidt MW, Torn MS, Abiven S, Dittmar T, Guggenberger G, Janssens IA et al (2011) Persistence of soil organic matter as an ecosystem property. Nature 478(7367):49. https://doi.org/10.1038/nature10386
91. Schwabedissen SG, Lohse KA, Reed SC, Aho KA, Magnuson TS (2017) Nitrogenase activity by biological soil crusts in cold sagebrush steppe ecosystems. Biogeochem 134(1–2):57–76. https://doi.org/10.1007/s10533-017-0342-9
92. Schwabedissen SG (2016) Nitrogen fixation by biological soil crusts and their communities: climatic and grazing controls (Master's Thesis). Idaho State University, Pocatello, Idaho
93. Skaggs RW, Breve MA, Gilliam JW (1994) Hydrologic and water quality impacts of agricultural drainage. Crit Rev Environ Sci Technol 24(1):1–32. https://doi.org/10.1080/10643389409388459
94. Slim M, Perron JT, Martel SJ, Singha K (2015) Topographic stress and rock fracture: a two-dimensional numerical model for arbitrary topography and preliminary comparison with borehole observations. Earth Surf Proc Land 40(4):512–529. https://doi.org/10.1002/esp.3646
95. Stanbery CA, Pierce JL, Benner SG, Lohse K (2017) On the rocks: Quantifying storage of inorganic soil carbon on gravels and determining pedon-scale variability. CATENA 157:436–442. https://doi.org/10.1016/j.catena.2017.06.011
96. St. Clair J, Moon S, Holbrook WS, Perron JT, Riebe CS, Martel SJ et al (2015) Geophysical imaging reveals topographic stress control of bedrock weathering. Science 350(6260):534–538. https://doi.org/10.1126/science.aab2210
97. Stone MM, DeForest JL, Plante AF (2014) Changes in extracellular enzyme activity and microbial community structure with soil depth at the luquillo critical zone observatory. Soil Biol Biochem 75:237–247. https://doi.org/10.1016/j.soilbio.2014.04.017
98. Stone MM, Plante AF (2015) Relating the biological stability of soil organic matter to energy availability in deep tropical soil profiles. Soil Biol Biochem 89:162–171. https://doi.org/10.1016/j.soilbio.2015.07.008
99. Swetnam TL, Brooks PD, Barnard HR, Harpold AA, Gallo EL (2017) Topographically driven differences in energy and water constrain climatic control on forest carbon sequestration. Ecosphere 8(4):e01797. https://doi.org/10.1002/ecs2.1797
100. Tennant CJ, Harpold AA, Lohse KA, Godsey SE, Crosby BT, Larsen LG et al (2017) Regional sensitivities of seasonal snowpack to elevation, aspect, and vegetation cover in western North America. Water Resour Res 53:6908–6926. https://doi.org/10.1002/2016WR019374
101. Trujillo E, Molotch NP, Goulden ML, Kelly AE, Bales RC (2012) Elevation-dependent influence of snow accumulation on forest greening. Nat Geosci 5(10):705. https://doi.org/10.1038/ngeo1571
102. Uhlig D, Schuessler JA, Bouchez J, Dixon JL, von Blanckenburg F (2017) Quantifying nutrient uptake as driver of rock weathering in forest ecosystems by magnesium stable isotopes. Biogeosciences 14(12):3111. https://doi.org/10.5194/bg-14-3111-2017
103. Wagner DC (2016) Modeling and measurement of tile drain controls in intensively managed landscapes (Doctoral dissertation). IDEALS, University of Illinois at Urbana-Champaign, Urbana, Illinois. http://hdl.handle.net/2142/90823

104. Wang R, Dorodnikov M, Yang S, Zhang Y, Filley TR, Turco RF et al (2015) Responses of enzymatic activities within soil aggregates to 9-year nitrogen and water addition in a semi-arid grassland. Soil Biol Biochem 81:159–167. https://doi.org/10.1016/j.soilbio.2014.11.015
105. Weber CF, Lockhart JS, Charaska E, Aho K, Lohse KA (2014) Bacterial composition of soils in ponderosa pine and mixed conifer forests exposed to different wildfire burn severity. Soil Biol Biochem 69:242–250. https://doi.org/10.1016/j.soilbio.2013.11.010
106. White T, Brantley S, Banwart S, Chorover J, Dietrich W, Derry L et al (2015) The role of critical zone observatories in critical zone science. In: Giardino J, Houser C (eds) Developments in Earth surface processes, vol 19. Elsevier, pp 15–78

Water, Carbon, and Weathering in the Catalina-Jemez CZO

Jon Chorover, Greg Barron-Gafford, Marcy Litvak, and Craig Rasmussen

Abstract The Catalina-Jemez Critical Zone Observatory (CZO) was designed to test the hypothesis that critical zone structure evolves predictably in response to climatic forcing, and that the latter can be quantified on the basis of effective energy and mass transfer (EEMT) deriving from effective precipitation and net ecosystem production. Several findings support the predictive power of EEMT for CZ structure, including those pertaining to soil thickness, above ground biomass, valley density, clay content, and mineral weathering rates. To unravel the mechanisms underpinning these relationships, the Catalina-Jemez CZO was established along a climate gradient in Arizona and New Mexico and instrumented to quantify land-atmosphere exchanges of water and carbon, subsurface translocation of fluids and solutes, and surface water dynamics. Here we report on selected results that highlight how pulsed delivery of water into the weathering profile stimulates biota to directly impact in the geochemical transformation of rock to soil.

1 Introduction

From the perspective of thermodynamics, the critical zone (CZ) is an open system that is subject to a continuous through-flux of matter and energy. As such, it is predicted to evolve over time to a structure that is most effective for energy dissipation [19]. Energy and matter fluxes through the CZ are governed by climate and lithology [6, 9], which can be considered as "top-down" and "bottom-up" forcings, respectively. For a given climate, different rock types weather at different rates, leading to variations in biological colonization and soil formation [1].

J. Chorover (✉) · C. Rasmussen
Department of Environmental Science, University of Arizona, Tucson, AZ, USA
e-mail: chorover@arizona.edu

G. Barron-Gafford
School of Geography, Development and Environment, University of Arizona, Tucson, AZ, USA

M. Litvak
Department of Biology, University of New Mexico, Albuquerque, NM, USA

T. White and A. Provenzale (eds.), *Critical Zone and Ecosystem Dynamics*,
Advances in Critical Zone Science, https://doi.org/10.1007/978-3-031-69076-1_13

To capture top-down forcing, the effective energy and mass transfer (EEMT) model [14] quantifies climate-driven energy input to the CZ for the purposes of predicting rates and trajectories of CZ evolution (Fig. 1). This model, which incorporates net ecosystem production and precipitation in excess of evapotranspiration as drivers for soil formation, served as the conceptual framework for hypothesis development and design of the Catalina-Jemez Critical Zone Observatory [4]. Derivation and application of the model is described in detail elsewhere [15, 14]. Briefly, EEMT is the sum of energy input to the subsurface via effective precipitation (E_{ePPT}, energy input associated with precipitation, PPT, in excess of evapotranspiration, ET) and net ecosystem production (E_{BIO}) components [14]:

$$\mathrm{EEMT} = E_{e\mathrm{PPT}} + E_{\mathrm{BIO}} \tag{1}$$

where

$$E_{e\mathrm{PPT}} = F c_{\mathrm{w}} \Delta T \tag{2}$$

$$\mathrm{E}_{BIO} = \mathrm{NEP}\ h_{\mathrm{bio}} \tag{3}$$

F is the mass influx of water to the subsurface weathering profile (= PPT − ET, kg m^{-2} s^{-1}), ΔT is ambient air temperature −273 (K), c_w is the specific heat of water (J kg^{-1} K^{-1}), ET is evapotranspiration, NEP is net ecosystem production

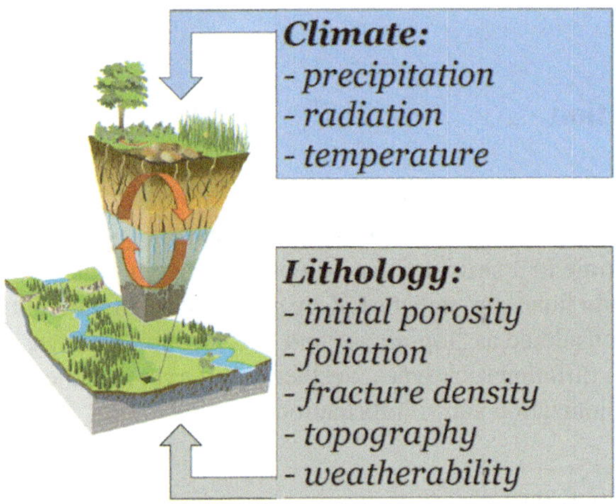

Fig. 1 CZ evolution is the result of top-down climatic and bottom-up lithologic forcings. Internal CZ fluxes and transformations of matter and energy (red arrows) occur on hydrologic event time scales. Examples include wetting front propagation, photosynthesis, microbial activity, gas exchange, mineral dissolution and precipitation. These internal fluxes and transformations—cumulative over pedogenic and geologic time scales—lead to the evolution of CZ structure and associated attributes, such as mineral composition, porosity, microbial communities, and mineral-stabilized and occluded carbon (Modified from [5])

of biomass and organic matter, and h_{bio} is specific biomass energy content (22×10^6 J kg^{-1}). Hence, our CZO team proposed that calculation of the landscape-scale distribution of EEMT offers a powerful approach to explore the climate forcing space of CZ evolution and its heterogeneity along environmental gradients [4]. To account for variation introduced by complex terrain, the calculations of P_{eff} and NEP were refined by Rasmussen et al. [15] by considering the topographic effects of slope and aspect on lateral water and carbon subsidies, local vapor pressure deficit, associated evapotranspirative flux, and point-scale vegetation controls on surface resistance and primary production.

Therefore, in conjunction with a bare earth digital elevation map, normalized difference vegetation index (NDVI) and meteorological data, the EEMT model can be used to calculate, at square meter spatial resolution, top-down energy inputs to the CZ subsurface across complex landscapes that feed into the weathering zone (Fig. 2). Our CZO team hypothesized that increasing EEMT would result in increasing weathering extent, increased soil and regolith depth, increased mobile element depletion and immobile element enrichment, and enhancement of weathering-induced CZ structure. While the EEMT model provides predictions of the extent of pedogenic weathering at a given location in the landscape, it does not elucidate the specific mechanisms of CZ evolution. In this way, it is like any other thermodynamic model. However, by designing an observatory to measure the fluxes and transformations directly, our CZO team also sought to identify the mechanisms whereby water and carbon inputs interact to affect weathering dynamics. Such interactions include the coupled reactions among water, gases, microbes, minerals, and organic matter, as they ensue in response to the wetting and drying that occurs throughout the year in these water limited systems.

2 Siting the Catalina-Jemez CZO Along a Gradient in Climatic Forcing

In water-limited CZ systems—such as those in the arid to sub-humid southwestern U.S. and over 40% of Earth's land area—EEMT increases with increasing precipitation, which, in turn, increases with elevation above sea level. In addition, throughout the Basin and Range Province of the western U.S., EEMT increases from south-facing to north-facing aspect, because greater insolation on south-facing slopes leads to higher ET and, therefore, reduced effective precipitation. The higher precipitation occurring in mountain regions of the Southwest serves as the principal source of potable water sustaining basin cities. For all of these reasons, the CJ CZO is situated along elevation gradients in three lithologies (granite, schist, and rhyolite) that sample multiple aspects in the Santa Catalina (AZ, granite and schist) and Jemez (NM, rhyolite) mountain ranges (Figs. 2 and 3). For the remainder of this chapter, examples of key results are presented from selected sites within (and in one case outside) the Catalina-Jemez CZO (Fig. 3). This is not intended to be an exhaustive review of our

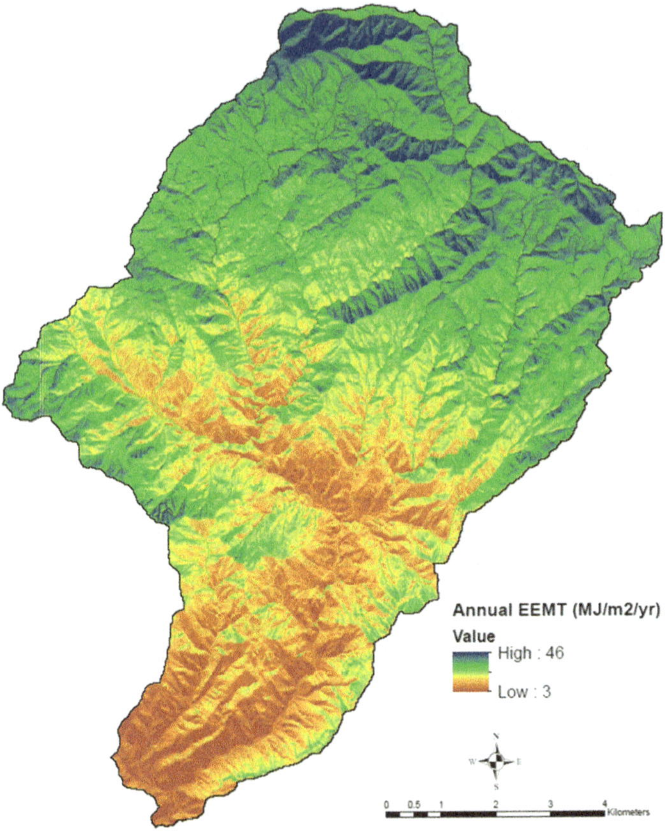

Fig. 2 Map of calculated effective energy and mass transfer (EEMT) flux across the upper CZ boundary for Sabino Canyon watershed in the Catalina CZO. EEMT is shown to increase with elevation (from bottom to top of figure) and with northness of aspect. Increase in EEMT is the result of increasing effective precipitation (PPT-ET) resulting from local water balance and calculated increase in net primary production of water-limited ecosystems along the elevation gradient. Data from Rasmussen et al. [15]

CZO results, but rather to highlight specific aspects of new understanding that results from interdisciplinary collaboration in CZ science.

3 Observatory Infrastructural Design

The objective of our CZO design was to develop a mechanistic understanding of how carbon, water, and lithogenic element cycles are coupled throughout the critical zone and how they promote long-term evolution of the weathering profile. Therefore, our infrastructural approach involved deployment of co-located eddy covariance flux

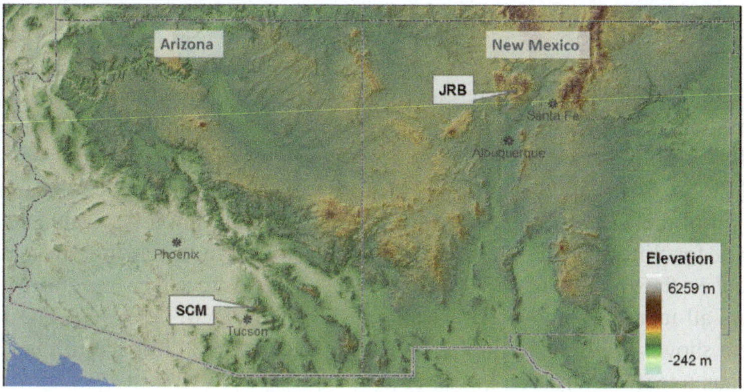

Fig. 3 Map showing location of the Santa Catalina Mountains (SCM) and Jemez River Basin (JRB) components of the Catalina-Jemez CZO. The SCM is located north of Tucson and University of Arizona and the JRB is located north of Albuquerque and the University of New Mexico

towers, instrumented pedons, groundwater monitoring wells, and instrumented weirs or flumes for surface water discharges (Fig. 4). We refer to this set up as our "CJ CZO sensor-sampler network array" and our CZO team instrumented several of our primary focal CZO sites (zero order basins, or ZOBs) in this way to enable concurrent measurements of land–atmosphere exchange of water and carbon, subsurface biogeochemistry, surface water dynamics, and associated landscape evolution [4].

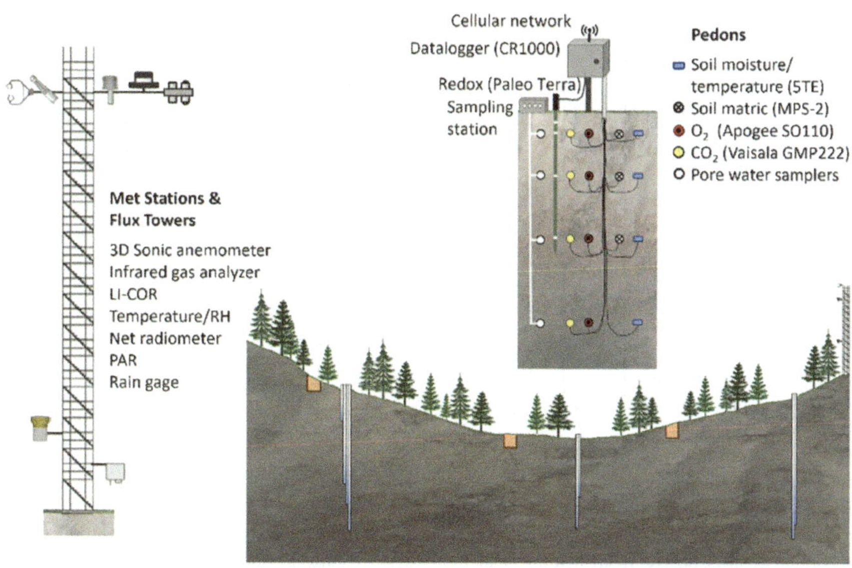

Fig. 4 Catalina-Jemez CZO infrastructure includes met stations and eddy covariance flux towers, instrumented pedons, groundwater monitoring wells, and surface water weirs or flumes

4 Testing EEMT-Based Predictions

Several findings from CJ CZO research support the contention that EEMT is predictive of a range of CZ structural and functional attributes within a given lithology, including soil thickness, relief, valley density, soil texture, and mineral weathering rates [13, 20, 22]. Pelletier et al. [13] showed that EEMT increases linearly with elevation in the Santa Catalina Mountains, such that elevation is a valuable direct surrogate for climate in this region. By combining field, remote sensing, and modeling approaches, they found that soil thickness, mean distance to valley, and above ground biomass all increased non-linearly with elevation and EEMT (Fig. 5). The EEMT gradient shown in Fig. 5 is similar to that depicted in EEMT space in Fig. 2 (i.e., more than a tenfold variation in EEMT).

The rate and extent of geochemical weathering is reflected in the composition of waters percolating through the porous subsurface and discharged into surface waters [8]. Zapata et al. [22] sampled spring discharge points through multiple field seasons in the mixed conifer zone of the Jemez CZO to assess the impacts of relatively small differences in upgradient EEMT—deriving primarily from variation in catchment aspect—on the extent of geochemical weathering. They found maximum EEMT to be an effective predictor of aqueous Na^+ concentration (a non-reactive tracer of silicate mineral dissolution) and the sum of base cation ($Na^+ + K^+ + Mg^{2+} + Ca^{2+}$) concentrations (Fig. 6). Base cation release was strongly correlated with dissolved inorganic carbon (DIC) concentration (dominantly HCO_3^- at the measured

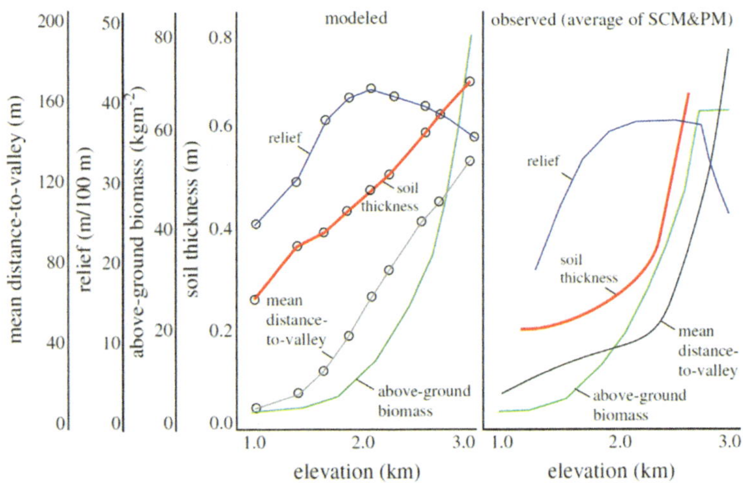

Fig. 5 Comparison of model-predicted (left panel) and measured (right panel) trends in landscape-scale parameters in the Santa Catalina Mountains (SCM) and nearby Pineleño Mountains (PM). Modeling results derive from a numerical landscape evolution code that utilizes EEMT as the climate-forcing parameter [13]

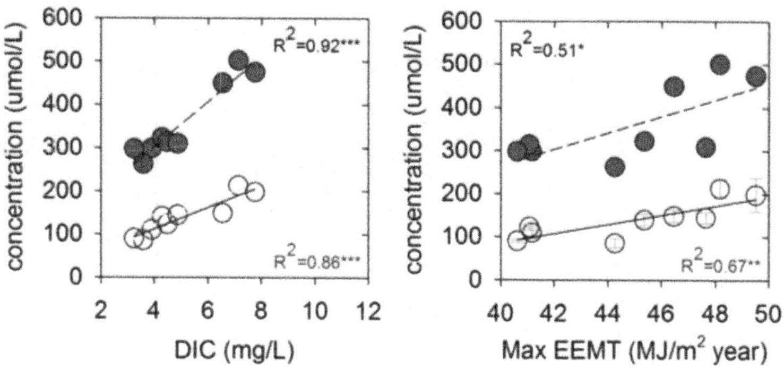

Fig. 6 Correlations among weathering derived solutes and EEMT values as the vary across different aspects within the mixed conifer vegetation zone of the Jemez CZO. Open symbols are Na^+ concentrations and filled symbols are the sum of base cation concentrations [22]

pH values), consistent with CO_2-driven silicate dissolution reactions increasing with EEMT.

Whereas solute loads are one indication of primary mineral weathering, another is the production of secondary minerals that accompanies the incongruent dissolution of rock during soil formation [3]. Secondary mineral products accumulate in the clay (<2 μm) size fraction of soils, and the clay fraction of bulk soil is, therefore, a useful measure of weathering profile alteration. Shepard et al. [20] conducted a meta-analysis of pedogenic chronosequence studies in the literature and found strong probabilistic relations between EEMT and both soil clay content and soil thickness. In particular, soil clay content increased predictably with increasing time-integrated EEMT—i.e., cumulative EEMT over the residence time of soil on the landscape (Fig. 7). This reflects the internal soil forming processes of mineral weathering and clay production responding to energy and mass fluxes.

These results provide examples whereby EEMT is a predictive independent variable for CZ structure evolution for constant lithology, and that relatively small variation in EEMT associated with north versus south aspect can result in significant differences in bedrock weathering and soil development rates. However, Leone et al. [10] showed that there are cases when variations in rock fabric within a given lithology can trump small differences in EEMT. Using geophysical seismic refraction methods, these authors found that the south facing slopes of a schist catchment in mixed conifer forest in the upper elevation range of the Catalina Mountains had deeper weathered regolith than the corresponding north facing slopes. The differences in regolith thickness were deduced from seismic velocities whose values correlate with the weathering-derived porosity distribution of subsurface material. In Fig. 8, highest seismic velocities (ca. 4000–5000 m s^{-1}) correspond to intact schist bedrock, and lower velocities (<1250 m s^{-1}) are interpreted to correspond to soil and saprolite.

It should not be surprising that increasing foliation angle (as defined in Fig. 8) correlates with regolith thickness. With increasing angle, the more weatherable layers

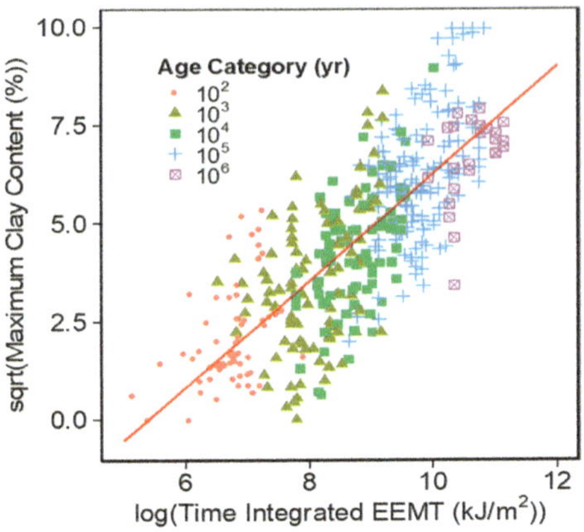

Fig. 7 The bivariate relationship of maximum clay content and time integrated EEMT across a synthesis of soil chronosequence studies. Points indicate individual soil profiles where color and marker for each point vary by logarithmic age category. The clay and EEMT data were transformed to approximate normal distributions and the red line is the best fit line among all points [20]

are increasingly exposed to the vertical infiltration of meteoric waters. As metamorphic layering becomes closer to normal to the slope, water is able to attack preferentially the weatherable layers, whereas propagation of regolith through layering parallel to the slope is limited by the persistence of least weatherable layers.

5 Ecosystem Metabolism and Resilience to Disturbance

While the EEMT model has been shown to be predictive of chemical denudation [14, 22], the precise nature of patterns linking water and carbon inputs to the CZ with associated solute effluxes and mineral transformations remain unclear. In the Catalina-Jemez CZO, our CZO team sought to investigate these linkages by direct, concurrent measurements of aqueous and gas phase variation through the weathering profile from the vegetated surface, through soil and weathered bedrock to stream water discharges, and their response to hydrologic events. Precipitation in excess of evapotranspiration, which gives rise to F (Eq. 2), is positive on an annual basis, and drives subsurface moisture propagation, with interannual variation depending especially on snowpack accumulation during the winter months (Fig. 9). Similarly, gross primary production (GPP) exceeds ecosystem respiration (RE) throughout most of the year, generating a negative net ecosystem exchange of carbon (–NEE), indicating that carbon is sequestered in plant, soil or the deeper subsurface. NEE exhibits positive values on a monthly time scale only in the winter months when photosynthetic CO_2 uptake is temperature limited, and soil respiration persists under the snowpack (Litvak et al. unpublished data). Interestingly, despite a large, stand replacing wildfire in June 2013, which killed a majority of the mixed conifer vegetation and resulted in

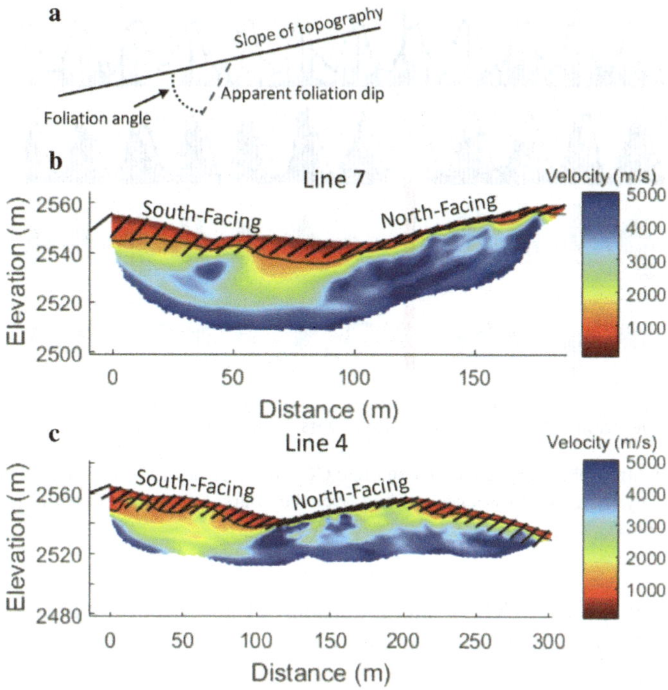

Fig. 8 Seismic refraction tomograms showing variation in regolith thickness across north and south facing hillslopes in the Bigelow Zero Order Basin, at 2500 m elevation in the Santa Catalina Mountains. Results show that foliation angle exerts strong control on weathering depth in schist bedrock. **a** Foliation angle is defined as that observed between topography slope and foliation dip. As shown in **b** and **c**, regolith depth increases with foliation angle such that south facing hillslopes have deeper weathering profiles than north–facing slopes [10]

the site being a net source of CO_2 to the atmosphere in 2013 and 2014, gross primary production and net carbon sequestration rebounded quickly in the subsequent years as a result of rapid regrowth (Fig. 9). This suggests a high degree of ecosystem resiliency (in terms of carbon partitioning) to this form of disturbance.

6 Partitioning of Soil Respiration

The instrumented pedons included sensors at four depths down to the soil–bedrock interface for continuous monitoring at each depth of soil moisture, matric potential, temperature, redox potential, and partial pressures of CO_2 and O_2. In particular, the co-location of sensors for P_{CO_2} and P_{O_2} enabled the estimation of vertical fluxes (F) of CO_2 and O_2 using the gradient method [17], which is necessary for calculation of the apparent respiratory quotient (ARQ) [2]. The ARQ provides an independent assessment of total soil respiration (Rsoil_ARQ) on the basis of O_2 influx, and

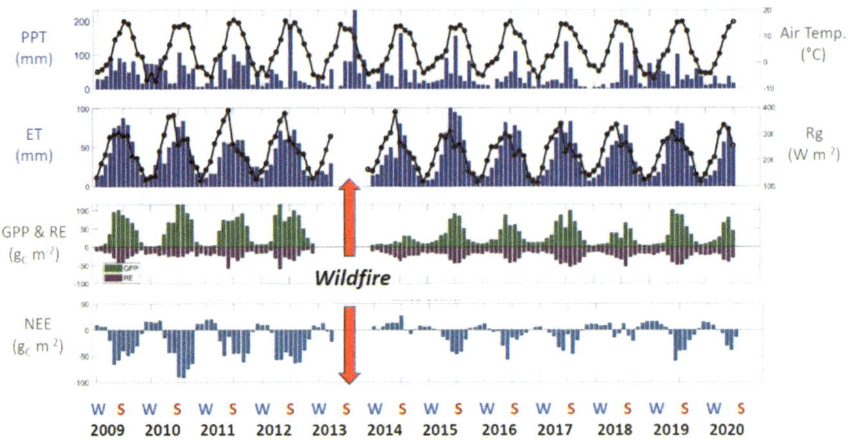

Fig. 9 Land–atmosphere exchange of water and CO_2 at the Jemez (NM USA) mixed conifer site as measured by the eddy covariance method over the course of the CZO project (Litvak et al. unpublished data)

therefore any "loss" of CO_2 to lateral transport or geochemical reaction relative to that released vertically back to the atmosphere. The ARQ is the ratio of soil CO_2 efflux to soil O_2 influx, and it is given by [2]:

$$\mathrm{ARQ} = \frac{-D_{S,\mathrm{CO}_2}}{-D_{S,\mathrm{O}_2}} \frac{\partial \mathrm{CO}_2}{\partial \mathrm{O}_2} = 0.76 \frac{\partial \mathrm{CO}_2}{\partial \mathrm{O}_2} \tag{4}$$

where Ds_{CO_2} and Ds_{O_2} are the diffusion coefficients of CO_2 and O_2 in the soil medium the ratio of which gives rise to the 0.76 constant on the right side, ∂CO_2 and ∂O_2 are the mole fraction gradients of CO_2 and O_2, respectively, with depth The moles of O_2 consumed per mole of CO_2 produced during soil respiration, referred to as the *oxidative ratio*, is a function of the types of biomolecules undergoing oxidative degradation, and it typically ranges from 0.7 for organic acids to 1.4 for lipids. A mean value of 1.1 is commonly used.

Interestingly, we found that, in fact, most of the CO_2 generated by soil respiration in the Bigelow ZOB mixed conifer forest site (Catalina CZO) was subject to lateral transport and consumption as the vertical efflux accounted for less than half of total soil respiration (Fig. 10). This result indicated that quantifying total soil respiration on the basis of time-dependent accumulation of CO_2 in closed chambers installed at the soil surface (as is commonly done) likely results in a significant underestimation of total respiratory CO_2 production. Furthermore, it raised the question as to the fate of the CO_2 not captured by the vertical efflux. It highlighted the importance of accounting for that "lost" CO_2 and discerning the extent to which it may drive subsurface geochemical reactions including silicate weathering and soil formation.

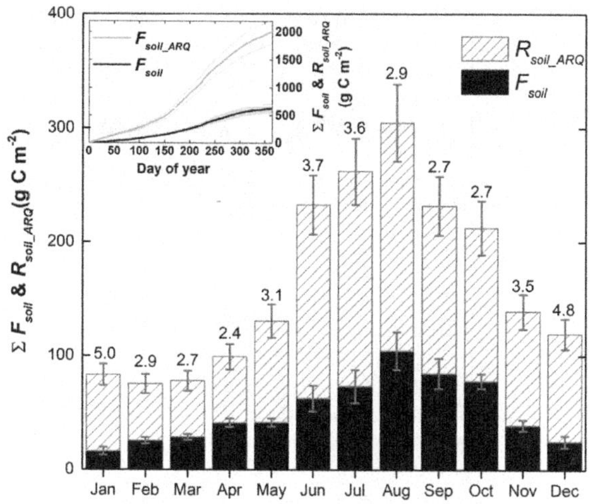

Fig. 10 Monthly cumulative vertical soil CO_2 efflux (F_{soil}) (with uncertainty represented as the standard error) and monthly cumulative soil CO_2 production (R_{soil_ARQ}) when accounting for the additional amount of CO_2 that is subsequently removed from total soil respiration prior to vertical efflux. The values above of each bar indicate the ratio R_{soil_ARQ}/F_{soil}. The inset figure shows the annual cumulative of F_{soil} and R_{soil_ARQ} and its uncertainty [18]

7 Coupling of Gas and Water Fluxes in the Vadose Zone

Time series data from the sensor array in the Bigelow ZOB show that gross GPP and ET are highest during the warm summer months when rains from the North American Monsoon (blue hyetographs, top panel) provide essential soil moisture for photosynthesis (Fig. 11). Temperature (T), moisture (θ), and gas (P_{CO_2} and P_{O_2}) sensors installed in a north facing instrumented pedon (soil over low permeability schist bedrock at 1 m depth) show that rain and snowmelt events produce wetting front propagation to depth and that this leads to periodic appearance and growth of a saturated zone at the soil–bedrock interface (red color in Θ data, Fig. 11).

This periodic subsoil saturation is accompanied by oxygen depletion and likely reductive transformation of alternative terminal electron acceptors (e.g., NO_3^-, Mn^{IV}, Fe^{III}) as a result of heterotrophic microbial respiration of newly solubilized organic matter. There is clear evidence of a P_{CO_2} response to moisture infusion during the summer monsoons (especially see July–August 2017). Importantly, however, wetting front propagation is also frequently accompanied by a *decrease* in P_{CO_2} (Fig. 11), rather than an increase, as might be expected to occur with a respiration pulse. This occurs most often during snowmelt and fall rain infusions. This pattern is opposite to the commonly invoked *Birch effect* [11, 21], which is attributed to rapid oxidative degradation of organic matter made bioavailable by solubilization in pore waters.

This decrease in P_{CO_2} with downward migration of the wetting front is most evident from the pattern, particularly during cooler months, where slow and steady accumulation of CO_2 in the unsaturated zone is suddenly wiped away as a result of infusion of meteoric waters (Fig. 11, e.g., Oct–Dec 2015, Jan–Mar 2017). This process is attributed to the fact that incoming waters, pre-equilibrated with the relatively low values of atmospheric P_{CO_2} suddenly encounter the 10–100 fold higher P_{CO_2} contained within soil pores, which leads to dissolution of gaseous CO_2 into the

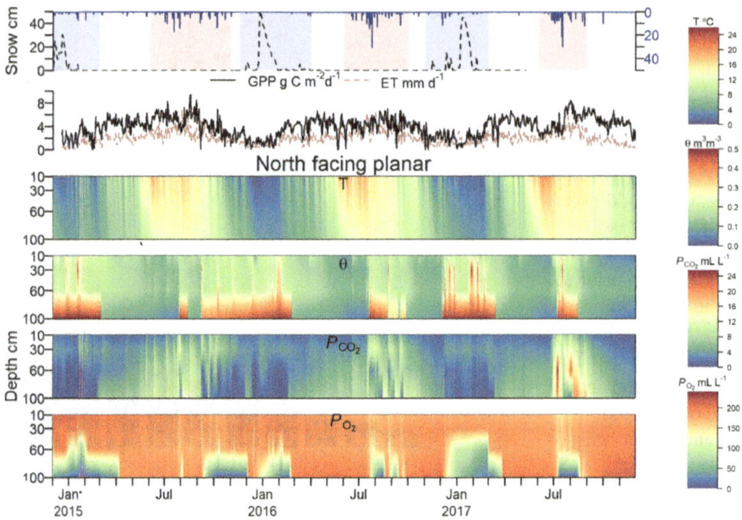

Fig. 11 Coupled water, energy, CO_2 and O_2 dynamics on a north-facing slope of the Bigelow ZOB of the Catalina CZO [12]

percolating aqueous phase in accordance with Henry's law:

$$\left[H_2CO_3^*\right] = K_H P_{CO_2} \tag{5}$$

$$\left[H^+\right] = K_1 K_H P_{CO_2}/[HCO_3^-] \tag{6}$$

where K_H is the Henry's constant for CO_2 (34.2 mol m^{-3} atm^{-1} at 25 °C) and $\left[H_2CO_3^*\right]$ is the concentration of dissolved CO_2 plus carbonic acid. The acidity of soil solution is mediated by P_{CO_2} and the proportional dissociation of carbonic acid as indicated in Eq. 6, where K_1 is the proton dissociation constant for $H_2CO_3^*$ ($10^{-6.3}$ at 25 °C). These H^+ ions then drive silicate mineral weathering via the proton-promoted dissolution mechanism. Hence, the prevalence of proton-promoted silicate dissolution in the weathering zone is strongly dependent on the primary production of biomass and the ensuing heterotrophic plus autotrophic subsurface respiration.

8 Pore Water Chemistry and Mineral Transformations

Soil solution chemistry from instrumented pedons at four locations in the Bigelow ZOB plot consistently within the kaolinite stability field (Fig. 12), which is consistent with the prevalence of kaolinite in the soil clay fraction as indicated by quantitative X-ray diffraction (XRD) [12]. These same XRD data also indicated a prevalence of plagioclase and orthoclase inherited from the parent rock. Hence, the geochemical weathering reactions mediated by respiratory CO_2 for plagioclase and orthoclase, respectively, weathering to form kaolinite, can be written as:

$$2Na_{0.5}Ca_{0.5}Al_{1.5}Si_{2.5}O_8 + 3CO_{2(g)} + 8.5H_2O_{(l)} \leftrightarrow 1.5Si_2Al_2O_5(OH)_{4(s)} + Na^+_{(aq)} + Ca^{2+}_{(aq)} + 3HCO^-_{3(aq)} + 2Si(OH)^0_{4(aq)} \quad (7)$$

$$2KAlSi_3O_{8(s)} + 2CO_{2(g)} + 11H_2O_{(l)} \leftrightarrow Si_2Al_2O_5(OH)_{4(s)} + 2K^+_{(aq)} + 2HCO^-_{3(aq)} + 4Si(OH)^0_{4(aq)} \quad (8)$$

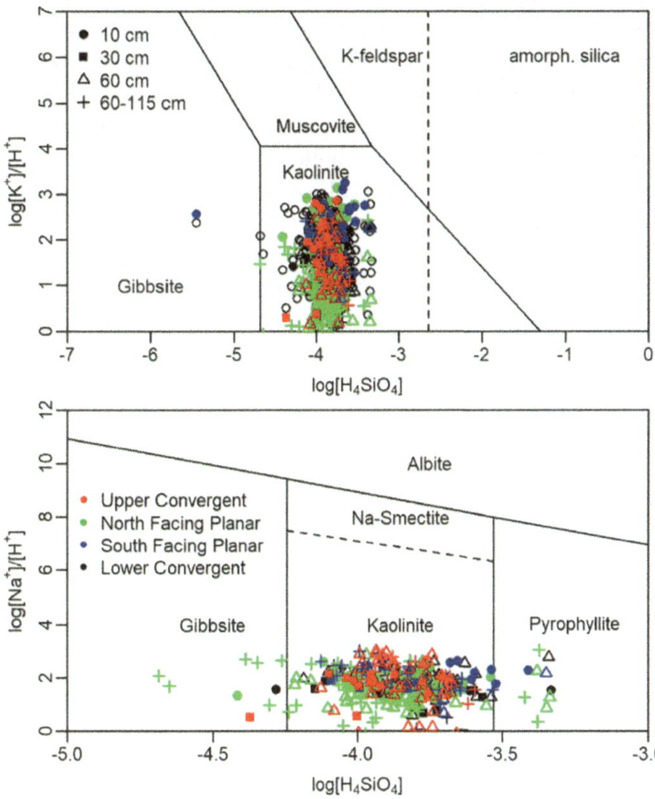

Fig. 12 Pore water chemistry from four depths in instrumented pedons at four landscape locations in the Bigelow ZOB of the Catalina CZO (upper convergent, lower convergent, north facing planar, and south facing planar instrumented hillslopes all included) plotted within stability diagrams for selected aluminosilicate solid phases [12]. Data indicate that soil solution consistently plots within the kaolinite stability field, consistent with the incongruent weathering of feldspars to form kaolinite in the weathering profiles

The formation of secondary clays such as kaolinite promotes the development of soil structure, the retention of plant available water, natural organic matter, and nutrients. In this way, ecosystem metabolism—as quantified directly using the CZO monitoring systems put in place (e.g., Fig. 9)—promotes soil formation and sustains life in the CZ.

9 Surface Water Dynamics and CO_2 Evasion

The appearance of a saturated zone on the hillslope (Fig. 11, and note that this is the same north facing hillslope with shallow regolith depth to high velocity bedrock as shown in Fig. 6) signals the initiation of lateral subsurface hydrologic flow toward the convergent zone with discharge into the ephemeral stream along the central axis of the ZOB. Since CO_2 partitions from gas to aqueous phase during these vadose zone wetting events (Fig. 13, top), lateral flows become charged with carbonic acid. This lateral subsurface flow occurs at the soil–bedrock interface, promoting the weathering reactions shown in Eqs. (7) and (8), forming soil from bedrock and transmitting rock-derived solutes into the stream (Fig. 13, bottom). Using similar event-based hydrochemical data for Si, Olshansky et al. [12] estimated a reaction rate (*R*) for feldspar weathering to kaolinite that ranged from $10^{-13.5}$ to $10^{-12.7}$ mol m^{-2} s^{-1}, depending on event, which is intermediate between laboratory and previously reported field-based reaction rates (the latter being integrated over longer than event time scales).

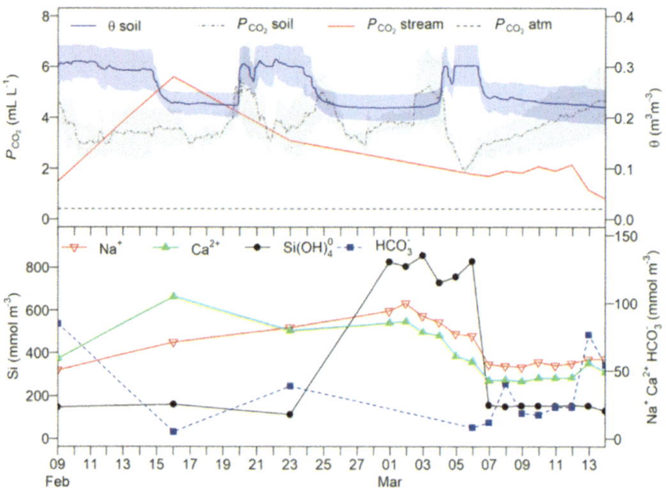

Fig. 13 High resolution view of the temporal interplay between soil moisture (θ), P_{CO_2} and discharge of soluble weathering products into the stream for Feb–Mar 2017 in the Bigelow ZOB [12]

Despite the consumption of CO_2 by silicate weathering along hydrologic flow paths (Eqs. 7 and 8), lateral flow was observed to discharge from hillslopes to the stream charged with CO_2 concentrations that are significantly higher than that in equilibrium with the atmosphere (Fig. 13 top). This results in evasion of streamwater CO_2 along the surface water flow path, a mechanism of land–atmosphere exchange of CO_2 that is under-represented in Earth surface models [7, 16, 18]. To the extent that this evasion occurs outside of the footprint of eddy covariance towers, it may lead to under estimation of net ecosystem exchange (release of CO_2 from land to atmosphere).

10 Summary and Mechanisms of EEMT-Driven Weathering

The EEMT conceptual model asserts that weathering of rock to form soil is driven by inputs of water, carbon, and energy to the CZ. Several lines of evidence—including geomorphological studies, spring hydrochemical analyses, and literature surveys—supported this contention, but the mechanisms remained poorly resolved. Therefore, the Catalina-Jemez CZO was designed, not only with this conceptual framework in mind, but also with the specific goal of opening the "black box" of the CZ to reveal the specific processes whereby water, carbon and energy interact to drive CZ weathering.

Our CZO sensor/sampler infrastructure was built to provide 3D information on water, gas, and solute fluxes across time in ZOBs, including at the time scale of hydrologic events. The role of ecosystem metabolism was considered central, as gross primary production, ecosystem respiration, and evapotranspiration all exert strong controls over the amount of water and carbon that makes its way into the CZ subsurface to fuel biogeochemical weathering and hillslope hydrologic fluxes.

Our team observed that a large fraction of total soil respiration is partitioned into vertical and lateral hydrologic flows, where it drives silicate weathering reactions at the bedrock-soil interface. We found that gas production and consumption was tightly coupled to moisture dynamics in our water-limited systems, but not always in the way that these interactions are predicted to occur for water-limited systems. Specifically, respiratory CO_2 dissolution into meteoric waters resulted in large decreases in subsurface P_{CO_2} that, in turn, drove proton-promoted incongruent weathering of primary framework silicates to form secondary phyllosilicates. These weathering reactions were not sufficiently rapid to consume the elevated total dissolved CO_2 and carbonic acid present in hillslope waters, and a significant amount of respiratory CO_2 was found to be re-emitted to the atmosphere from streams receiving this CO_2-charged groundwater discharge.

Acknowledgements Funding for the research reported in this chapter and for the Catalina-Jemez Critical Zone Observatory was provided by the US National Science Foundation Grant Numbers EAR-0724958 and EAR-1331408.

References

1. Amundson R (2003) Soil formation. Treatise Geochem 1–35
2. Angert A et al (2015) Using O_2 to study the relationships between soil CO_2 efflux and soil respiration. Biogeosciences 12(7):2089–2099
3. Chadwick OA, Chorover J (2001) The chemistry of pedogenic thresholds. Geoderma 100(3–4):321–353
4. Chorover J et al (2011) How water carbon and energy drive critical zone evolution: the Jemez-Santa Catalina critical zone observatory. Vadose Zone J 10(3):884–899
5. Chorover J, Kretzschmar R, Garcia-Pichel F, Sparks DL (2007) Soil biogeochemical processes within the critical zone. Elements 3:321–326
6. Dixon JL, Chadwick OA, Vitousek PM (2016) Climate-driven thresholds for chemical weathering in postglacial soils of New Zealand. J Geophys Res Earth Surf 121(9):1619–1634
7. Duvert C, Butman DE, Marx A, Ribolzi O, Hutley LB (2018) CO_2 evasion along streams driven by groundwater inputs and geomorphic controls. Nat Geosci 11(11):813–8182
8. Gaillardet J, Dupré B, Louvat P, Allègre CJ (1999) Global silicate weathering and CO_2 consumption rates deduced from the chemistry of large rivers. Chem Geol 159(1–4):3–30
9. Hahm WJ, Riebe CS, Lukens CE, Araki S (2014) Bedrock composition regulates mountain ecosystems and landscape evolution. Proc Natl Acad Sci USA 111(9):3338–3343
10. Leone JD, Holbrook WS, Riebe CS, Chorover J, Ferré TP, Carr BJ, Callahan R (2020) Strong slope-aspect control of regolith thickness by bedrock foliation. Earth Surf Proc Land 45(12):2998–3010. https://doi.org/10.1002/esp.4947
11. Moyano FE, Manzoni S, Chenu C (2013) Responses of soil heterotrophic respiration to moisture availability: an exploration of processes and models. Soil Biol Biochem 59:72–85
12. Olshansky Y, Knowles JF, Barron-Gafford G, Rasmussen C, Abramson N, Chorover J (2019) Soil fluid biogeochemical response to climatic events. J Geophys Res: Biogeosci 124(9):2866–2882. https://doi.org/10.1029/2019JG005216
13. Pelletier JD et al (2013) Coevolution of nonlinear trends in vegetation soils and topography with elevation and slope aspect: a case study in the sky islands of southern Arizona. J Geophys Res Earth Surf 118(2):741–758
14. Rasmussen C et al (2011) An open system framework for integrating critical zone structure and function. Biogeochemistry 102(1):15–29
15. Rasmussen C, Pelletier JD, Troch PA, Swetnam TL, Chorover J (2015) Quantifying topographic and vegetation effects on the transfer of energy and mass to the critical zone. Vadose Zone J 14(11)
16. Raymond PA et al (2013) Global carbon dioxide emissions from inland waters. Nature 503(7476):355–359
17. Sánchez-Cañete EP, Scott RL, van Haren J, Barron-Gafford GA (2017) Improving the accuracy of the gradient method for determining soil carbon dioxide efflux. J Geophys Res Biogeosci 122(1):50–64
18. Schelker J, Singer GA, Ulseth AJ, Hengsberger S, Battin TJ (2016) CO_2 evasion from a steep high gradient stream network: importance of seasonal and diurnal variation in aquatic pCO_2 and gas transfer. Limnol Oceanogr 61(5):1826–1838
19. Schneider ED, Sagan D (2005) Into the cool: energy flow thermodynamics and life. University of Chicago Press, Chicago, IL
20. Shepard C, Schaap MG, Pelletier JD, Rasmussen C (2017) A probabilistic approach to quantifying soil physical properties via time-integrated energy and mass input. Soil 3(1):67–82
21. Unger S, Máguas C, Pereira JS, David TS, Werner C (2010) The influence of precipitation pulses on soil respiration—assessing the "Birch effect" by stable carbon isotopes. Soil Biol Biochem 42(10):1800–1810
22. Zapata-Rios X et al (2015) Climatic and landscape controls on water transit times and silicate mineral weathering in the critical zone. Water Resour Res 51(8):6036–6051

Alpine Critical Zone Observations at the Gran Paradiso National Park, Italy

Ilaria Baneschi, Mariasilvia Giamberini, Marta Magnani, Pietro Mosca, Brunella Raco, Gianna Vivaldo, and Antonello Provenzale

Alpine ecosystems are typically cold and partially glaciated environments located in high mountains, where altitudinal, geographical and temporal gradients occurring within small space and short time scales offer a compact framework for the study of Critical Zone (CZ) processes. These ecosystems are widely distributed across the world and in total cover more than 3 Mkm^2 [111, 164]. Strictly speaking, Alpine ecosystems are present in mountain belts between the treeline at the lower limit (e.g., [108]) and permanent snowfields, glaciers or exposed rocks at the upper boundary. Below the lower boundary we find sub-alpine or mountain forests, whereas above the upper boundary the so-called nival band extends up to the mountain summit [111]. The absence of trees, mostly owing to the low average temperature (typically, the mean annual temperature is below 6 °C, [19, 110, 164]), is a common feature of Alpine ecosystems. On the other hand, the local bioclimatic characteristics are significantly different across different geographical regions [109, 164], a variability which is associated with a rich biodiversity (e.g., about 10,000 plant species, [99]). The dynamics of Alpine ecosystems is coupled with the evolution of both sub-alpine and nival areas, whose boundaries mainly occur as smooth ecotones rather than as sharp boundaries [31, 77, 110]. For such reasons, in the following we will focus on the European Alps, and in addition to the Alpine features, we will also address aspects of the sub-alpine and nival areas. This chapter is structured as follows: Sect. 1 provides a general overview of the main Alpine features; Sect. 2 discusses the observed and expected changes in the mountain environment; Sect. 3 describes the Gran Paradiso National Park (GPNP) in the Italian Alps; Sect. 4 is devoted to discussing the CZ Observatories (CZO) established at GPNP, including the main scientific results obtained till now; and Sect. 5 considers future perspectives.

I. Baneschi (✉) · M. Giamberini · B. Raco · G. Vivaldo · A. Provenzale
Institute of Geosciences and Earth Resources, National Research Council of Italy, Pisa, Italy
e-mail: i.baneschi@igg.cnr.it

M. Magnani · P. Mosca
Institute of Geosciences and Earth Resources, National Research Council of Italy, Torino, Italy

T. White and A. Provenzale (eds.), *Critical Zone and Ecosystem Dynamics*,
Advances in Critical Zone Science, https://doi.org/10.1007/978-3-031-69076-1_14

1 Overview of Alpine Features

The Alps are arc-shaped mountains in Europe, in map view about 1200 km long and between 100 and 400 km wide, extending from Colle di Cadibona (Italy) to Austria and Slovenia. The Alps are a complex orogenic system structured throughout the Cenozoic in the framework of collision between the Europe and Adriatic plates following the complete closure of their intervening Piedmont-Ligurian ocean (e.g. [38]). During the Pleistocene, the Alps were repeatedly glaciated ([2] and references therein) with ice caps that covered almost the entire mountain belt and substantial parts of the northern foreland. The morphological setting of the Alps led to a variety of microclimates, habitat types and soils. Given the great variety of soil forming conditions, such as a wide range of climatic regimes [157], distinct vegetation belts, very diverse geological substrata, the high variability of topography [13], and microtopography that dominates the soil pattern above the timberline, Alpine soils are highly variable over short spatial ranges, leading to complex patterns of soil characteristics and humus properties [46, 83]. In particular, the dynamics of the soil-forming processes changes with altitude, with additional differentiation generated by exposure, climate and vegetation [48, 50].

For instance, Egli et al. [47] found that the rates of soil organic carbon stock and the chemical weathering increase up to the timberline (about 1600 m a.s.l.) and then decrease in alpine meadows, especially over north-facing slopes. Further, multilayered soil profiles and fossil soils are quite common in the Alps, reflecting a sequence of different processes (such as periglacial dynamics after glacier retreat, nunatak-like areas or sediment deposition events) in the same spot as a result of morphodynamic changes ([13] and references therein).

As the elevation increases, precipitation usually increases and evapotranspiration decreases, leading to a water surplus. In turn, this results into pseudogleyification, element leaching and podsolization on siliceous substrates [47, 51]. Usually, pH values decrease with elevation owing to the higher leaching and acidification. At the same time, the biological activity and the mineralization rate of organic material decrease. Therefore, soils have high organic matter content, and marked organic horizons are frequently found [144]. Soils on north facing slopes have higher organic matter content than those on southern slopes [49]. In addition, "time" must be considered as an especially important soil forming condition. Owing to the last glaciation, Alpine soils are often young (<10,000 years BP) and they are strongly influenced by the periglacial environment (Fig. 1).

Thanks to the high elevation, the Alpine chain features several cryospheric components, which are an integral element of the mountain landscape and include lingering snow cover, glaciers, permafrost, and lake and river ice [93]. From autumn to spring, precipitation falls mainly as snow, which accumulates on the soil and provides soil thermal insulation [68]. On exposed slopes, the mountain hydrologic cycle is characterized by a strong interannual variability [124, 149] and the snow cover can be rather heterogeneous over short distances owing to wind erosion and microtopography, thus influencing the patterns of ground temperature, permafrost soils and

Fig. 1 An example of a typical high-altitude Alpine environment above the tree line, here at the Nivolet Plain, Gran Paradiso National Park. In the background, the Gran Paradiso massif, composed mainly of gneiss. Close to the lakes, in the foreground, are areas with exposed carbonatic rocks

runoff [151, 175]. Overall, the cryosphere provides important ecosystem services, such as freshwater for downstream regions during late spring and summer [173]. For this reason, the Alps are often considered Europe's water towers thanks to their large water resources [55].

Mountain ecosystems provide goods and services which include benefits to watersheds, such as slope stability and discharge regulation, food and energy production, recreational services and tourism [146]. In particular, carbon stocking is an important ecosystem service provided by Alpine soils and biomass [30]. Carbon forms about 55–60% of soil organic matter (SOM, [145]). The SOM acts as a stabilizer of soil structure, supporting soil aggregation, aeration and water infiltration, and contributes to the resistance to erosion and to water retention. It also provides nutrients for microbes and the soil fauna.

The spring snowmelt triggers a period of rapid microbial growth [68], even though the microbial biomass can grow even under snow [121] and in frozen soils [45]. Plant litter provides nutrients for microbial communities [95], including bacteria, fungi and archaea (e.g., [10]). Microbial activity persists throughout the whole year, though the community undergoes seasonal shifts, with fungi dominating during winter and bacteria during summer [11, 21, 133]. During summer, microbes compete with plants for nutrient resources, with minimum microbial biomass occurring in summer and minimum plant biomass occurring in or at the end of the cold season [11]. At the same time, mycorrhizal and ectomycorrhizal fungi that live in symbiosis with plants can accelerate the biochemical weathering of the parental material, which is especially relevant in the early stage of soil formation (e.g., in the foreland of retreating glaciers), thus providing organic, and not mineralized nutrients to plants [37].

The annual patterns of temperature, precipitation, nutrients and photoperiod (i.e., the duration of daily illumination by solar irradiance) drive vegetation growth and microbial activity, and therefore carbon dioxide (CO_2) fluxes. Photosynthesis is associated with a flux of CO_2 from the atmosphere to the plants, while respiration (both by plants and soil microbiota) leads to an emission of CO_2 (or CH_4 if anaerobic conditions prevail) to the atmosphere. Alpine ecosystems typically are CO_2 sinks during summer, when net sequestration from the atmosphere occurs, and CO_2 sources during winter, when a net (mostly microbial) emission is observed [54]. The amount of net carbon storage—another important ecosystem service—is controversial [42, 93, 142, 155].

The heterogeneity of microclimatic conditions experienced by mountain ecosystems results in a mosaic of habitats and high biodiversity [177]. Alpine vegetation above the treeline consists of mountain shrublands, grasslands and tundra, populated by dwarf shrubs, herbs, algae, lichens and mosses [79, 109], and mostly including endemic and specialized species [148]. The Alpine Critical Zone (CZ) has a thickness that is at the lower bound of the range recorded at mid latitudes [184]. In Alpine environments, vegetation mostly includes species that are productive during summer, whereas in winter only belowground biomass and/or seeds can persist [19]. Such physical stressful conditions favor facilitation, i.e., positive interactions, between species rather than competition [4, 29, 43, 129]. Spring snowmelt releases large amounts of water in soils and causes soil carbon and nutrient mineralization (e.g., [92, 121, 143]), thus ultimately benefitting plant biomass build-up as nitrogen and phosphorus are re-absorbed by the plants in the long term [11, 37, 45, 88].

2 Past and Future Mountain Ecosystem Changes

Mountain ecosystems are considered as sentinels of change, as they are highly sensitive to the impacts of both climate and land-use change [146]. European Alps are seminatural environments that have sustained agro-pastoral practices for centuries [107]. In the subalpine horizon, land abandonment is currently leading to forest regrowth [72, 73], changing the habitat characteristics and impacting biodiversity [41, 171], with potential consequences on carbon sequestration and soil stabilization. In such environments, the effects of land and pasture abandonment are mixed with those of climate change, further complicating local studies [60].

A significant temperature increase is currently observed in the Alps, and it is expected to continue in the future [93]. High mountains are experiencing a larger temperature increase compared to low-elevation sites, owing to the so-called "elevation dependent warming" (EDW, [136, 139, 146]). The EDW consists in the amplification of the rate of warming (i.e., temperature change in degrees per decade) across elevation bands, owing to factors that preferentially increase the net flux of energy to the surface and therefore the surface temperature. The feedback effects that mostly contribute to this warming are linked to the decrease of surface albedo, i.e. surface reflectivity, owing to the reduction of the snow cover, and the nonlinear sensitivity of radiative fluxes to variations in the atmospheric composition. At global scale, the EDW has been observed above 500 m a.s.l. (e.g., [147, 174]). At regional to local scales, observations can be contradictory [168], despite a general warming trend being observed, usually more pronounced in summer and spring across the Alps [7, 34, 93].

As a consequence of the temperature rise and of the lowering solid/liquid ratio in precipitation [93], the Alpine cryosphere is strongly affected [18]. In the European Alps, the duration and depth of mountain snow cover has generally reduced since the middle of the twentieth century [100, 124–126], and warming, degradation and destabilization of permafrost at individual sites has been observed [20, 25, 27, 87, 89, 118, 119]. Note, also, that Alpine permafrost observations are still rare and, in many mountain regions, permafrost behavior is inferred from regional modeling studies [26, 81, 82, 87, 126, 135]. Glacier area, length and mass are strongly declining [186], and a large interannual and regional variability is also observed [165, 182]. Summer and annual runoff have changed owing to both intensified ice and snow melt and modified rainfall regimes, producing contrasting effects at different elevations, locations, and in different seasons [9, 32].

Major and trace ions as well as acid rock drainage have been released in surface freshwater by permafrost degradation [36, 103]. The drainage of chemical compounds and the observed changes in river regimes have also affected plankton phenology and productivity, as well as benthos communities (e.g., [58, 134, 152]).

Range shift of plants and animals towards higher altitudes (e.g., [66, 67, 169]) and changes in community composition [138, 163] were observed in response to habitat changes, with cold-adapted species being the most threatened [78, 114, 154, 170]. Changes in snow cover duration and accumulation shifted plant phenology [97, 169, 183], with potential effects on primary productivity (e.g., [160]). In turn, changes in the Alpine vegetation may alter the quantity and quality of forage for herbivores, which can be critical for the survival of mammal newborns [104, 130], and this could feed back to vegetation changes as herbivores contribute to limit the upward expansion of shrubs [181]. These observations agree with patterns observed in mountain regions across the world [93], although topographic heterogeneity can cause intra-site variability [152].

The above changes are expected to be exacerbated in the future. Further glacier mass loss, snowfall decrease and rainfall increase, snow cover and permafrost decline are projected in the Alps [18, 75, 93]; for an updated review of expected climate changes in the Alps see Kotlarski et al. [112]. A general decrease of summer river discharge, an increase of winter discharge and a shift towards more rain-driven regimes are expected [1, 22, 106]. In addition, for glacier-fed rivers it has been predicted that the persistent glacier shrinkage for several years or decades, drawing from long-term ice storage, would lead to a turning point beyond which the annual glacier runoff would start decreasing [101]. Small glaciers, such as those in the Alps, are most affected by glacier retreat and runoff regime shift [102]. On the other hand, upslope migration and possible collapse of the Alpine flora may be buffered by microrefugia (a localized environment different from surrounding areas that provides suitable conditions [86, 159, 187], thus favoring mountain species compared to species leaving in flat and homogeneous landscapes [80].

Many of the projected changes will alter ecosystem services and human activities either directly or depending on the ecological response [98]. To name a few examples, temperature increase and changes in water regimes may affect crop and livestock production in mountain sites and valley floors [69]; permafrost degradation can lead to land destabilization [82] and rockfalls [3] also affecting the accessibility of hiking trails and tourisms; changes of mountain landscapes may impact recreation activities such as skiing [161]; glacier retreat can lead to debris destabilization [52]; and extreme temperature events at high altitudes can trigger avalanches from steep mountain faces [33, 44, 56, 61, 141], with potential devastating impacts on people and ecosystems. Hydropower facilities, which are a significant source of electricity and revenue in mountain regions (e.g., [70]), would be affected by catchment-dependent changes in runoff [71, 85, 158]. Freshwater quality may be affected by the melting glaciers that are subject to pollutant loads of anthropogenic origin [24, 94, 131]. These side effects of climate change pose potential hazard to human well-being, communities, and infrastructures, thus dictating the need of conservation policies and adaptation measures [16, 17, 53, 57, 69, 90].

3 The Gran Paradiso National Park

The Alps include the oldest Italian protected area: the Gran Paradiso National Park (GPNP; [189]). The GPNP was created in 1922 for the conservation of the Alpine ibex, *Capra ibex*, and its habitat in general (www.pngp.it). The park area includes five valleys (i.e., Rhêmes, Valsavarenche, Cogne, Soana and Orco) located in north-western Italy, in an area between Aosta Valley (52% of the park area) and Piedmont (48%), at the border with France. In the GPNP, the main Alpine features are present within a surface of 720 km^2 (between 45° 25′ and 45° 45′ N, and 7° and 7° 30′ W) with an elevation range between 800 and 4061 m a.s.l., providing an ideal framework to study mountain CZ processes. Regular surveys and monitoring of the endemic flora and fauna are carried out by the park surveillance and technical personnel, often in collaboration with research centers and universities, also in the frame of international collaborations. Many studies address the ecology of Alpine environments, including the behavior, composition, and population dynamics of the local terrestrial, avian and aquatic fauna, with a special focus on herbivores such as the Alpine ibex (*Capra ibex*), the chamois (*Rupicapra rupicapra*) and the marmot (*Marmota marmota*), community composition and biodiversity changes, and Alpine lake ecosystems (see www.pngp.it for a list of publications, and also [14, 23, 30, 40, 74, 96, 113, 137, 166]). The GPNP promotes many actions for sustainable tourism [15] and in 2014 it was included in the International Union for the Conservation of Nature (IUCN) Green List, in recognition of the effectiveness of the protected area management.

The park area is covered with 54% grasslands, 18% forests and shrubs, 20% bare rocks, and 7% permanent snow and ice [28, 59]. Most plant species are endemic [39]. Forests includes mixed-broadleaved woods (*Castanea sativa, Quercus* spp., *Betula alba, Populus* spp., *Fraxinus* spp., *Acer* spp.) in the valley bottom and coniferous woods (*Larix decidua, Picea abies, Pinus cembra, Abies alba*) from about 1000 m a.s.l. to the treeline, generally occurring at 1600–2200 m a.s.l [113]. Grasslands are composed of typical Alpine species, such as *Nardus* sp., *Festuca* sp. and *Carex curvula* [39], while pioneer species, such as *Saxifraga oppositifolia, Silene acaulis* and *Poa alpina,* are common in glacier forelands [120]. Abandonment of rural areas peaked after the Second World War and shrubs such as *Berberis, Rhododendron, Juniperus, Alnus, Salix* spp., bracken *Pteridium aquilinum* and European larch are encroaching previously open grasslands at the sub-alpine ecotone [113]. Low intensity grazing by sheep and cows is maintained from July to September in some areas of the park. A recent assessment showed that grasslands carbon stock resulted to be the second most intensive one among the different mountain habitats in the GPNP, after the broadleaf deciduous forest, with an average content of 5.7 and 6.3 kg/m^2, depending on the grassland land cover category [30].

From November to early-mid June, at elevations above 2000 m a.s.l. the soil is usually covered with snow (even though a significant snow reduction was observed in the last years), with maximum snow depth larger than 250 cm in the years 2017–2019 as measured by the meteorological station at Lake Serrù (2450 m a.s.l.), on the road climbing to the Nivolet Plain. Daily records of precipitation, temperature and snow depth at the Serrù station are available from 1962. At this station, the mean annual precipitation over the whole record is 1185 mm/year, with average precipitation during the snow-free season (June–October) of 491.79 mm/year. The mean daily temperature during June–October ranges from 3 to 12 °C. (5th and 95th quantiles, respectively).

4 The GPNP Critical Zone Observatory

The GPNP hosts a "diffused" Critical Zone Observatory, where open air measurements, monitoring and surveys are carried out in different valleys and various elevations within the park, with a specific focus on carbon dioxide and water fluxes between soil, vegetation and atmosphere. Figure 2 shows the instruments installed at the Nivolet Plain, snapshots of Noaschetta Valley, where we measure the differences between grazed and non-grazed plots, and of Levionaz, where we measure the differences between irrigated and non-irrigated plots. Ongoing work is reported in Table 1.

The metamorphic bedrock of the sampling sites belongs to the Gran Paradiso Unit, composed of abundant gneisses and minor quartzites, dolostones and marbles, and to the oceanic Piedmont-Ligurian zone, consisting of calcschists with serpentinites and metabasites [140]. Quaternary glacial and alluvial deposits variably cover these rocks.

To fix ideas and terminology, in these studies we focus on the estimate of the Gross Primary Production (GPP), related to the absorption of CO_2 by photosynthesis and the associated building of organic material, and the Ecosystem Respiration (ER), given by the combination of the autotrophic respiration by plants and the heterotrophic respiration from soil microbiota. Typically, GPP is negative (i.e., a flux from the atmosphere to the soil/vegetation) and ER is positive (a flux to the atmosphere). The

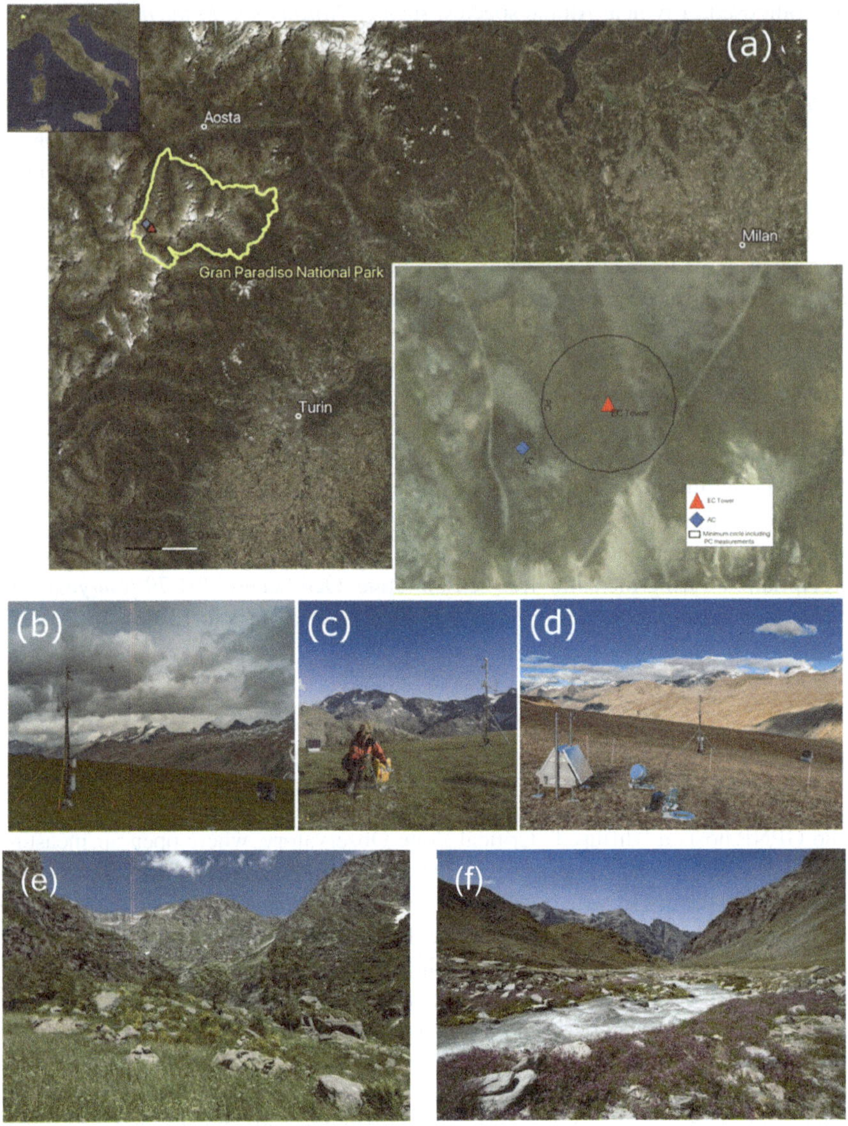

Fig. 2 Panels **a–d** (from [172]): Fixed and portable instruments installed at the Nivolet Plain (about 2700 m a.s.l.), as a part of the GPNP Critical Zone Observatory. **a** Location of the Gran Paradiso National Park, of the Eddy Covariance station (EC, red triangle) and of the automated flux chambers (AC, blue diamond) for measuring CO_2 and water fluxes between the surface and the atmosphere. The site is an ICOS ERIC associated station. **b** The Eddy Covariance station; **c** measurement of fluxes with the portable flux chamber; **d** the automated flux chambers. Panels **e** and **f** show pictures of the Noaschetta Valley (about 1500 m a.s.l.) and Levionaz (about 2600 m a.s.l.), respectively

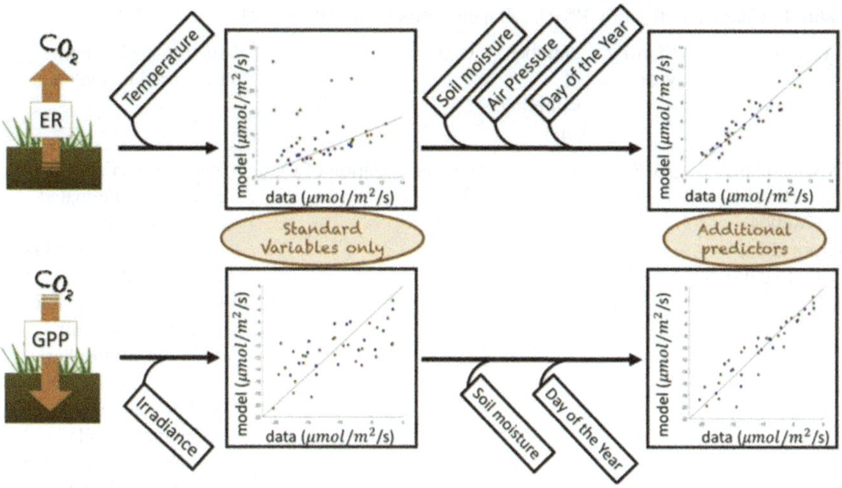

Fig. 3 Graphical abstract from the work of Magnani et al. [122], showing how the ER and GPP fluxes measured at site scale cannot be reproduced solely by the standard drivers (temperature for ER and solar irradiance for GPP), but require further drivers such as soil moisture and day-of-the-year, this latter being a proxy for the phenological state of the vegetation

Net Ecosystem Exchange (NEE) is the sum of GPP and ER, i.e., NEE = GPP + ER, and it can be positive or negative depending on whether ER or GPP dominates. In addition, evapotranspiration (ET), i.e., the release of water vapor from plants and soil, provides further information on ecosystem functioning and a necessary ingredient to close the water balance.

Since 2017, periodic field surveys have been held each summer during the vegetative season (typically from early July to the end of September) at Nivolet plain and in the Noaschetta Valley. Within the Nivolet plain, NEE, ER, ET, meteoclimatic and environmental variables are measured in four sites, which differ for bedrock type, aspect and vegetation composition. One of such sites has been equipped with an Eddy Covariance station in 2019 and it is now an Associated ICOS site (Integrated Carbon Observation System, a European Research Infrastructure, www.icos.eu), and with two couples of fixed flux chambers working during the snow-free season since 2021. In the Noaschetta Valley, a long-term measurement effort aims at evaluating the differences in CO_2 fluxes from grazed areas versus ungrazed plots (where large herbivore exclusion is obtained by electrified fences). In summer 2022, CO_2 and water flux measurements started in Levionaz (Valsavarenche), where a manipulation experiment has been launched for evaluating the differences between natural and irrigated plots.

Table 1 Characteristics of GPNP CZO measurements and their goals

Location	Starting year	Number of sampling sites	Goals	Instrument/method: measured factors
Nivolet plain (45.5019181, 7.1786064) (from about 2500 to 2700 m asl)	2017	5	Characterization of CO_2 fluxes	Portable accumulation chambers, automated chambers, eddy covariance tower: CO_2 and water fluxes Portable weather station: air temperature, air moisture, atmospheric pressure, solar irradiance Soil probes: soil temperature, soil volumetric water content Soil pits: soil chemistry
Noaschetta (45.4635473, 7.3197014) (about 1500 m asl)	2017	2 + 2	Effects of grazing (comparison of carbon fluxes from grazed and fenced, ungrazed areas at two sites)	Portable accumulation chambers: CO_2 and water fluxes Portable weather station: air temperature, air moisture, atmospheric pressure, solar irradiance Soil probes: soil temperature, soil volumetric water content
Levionaz (45.573704, 7.2435368) (about 2500 m asl)	2022	2 + 2	Effect of irrigation (comparison of irrigated and natural grasslands co-located)	Portable accumulation chambers: CO_2 and water fluxes Portable weather station: air temperature, air moisture, atmospheric pressure, solar irradiance Soil probes: soil temperature, soil volumetric water content

The first column reports the name of the location, its coordinates (lat-long) and the approximate elevation. The second column reports the year when the measurements started, the third column the number of different measurement sites/plots, the fourth column the main goals of the studies, and the last column some characteristics of the instrumentation. 2 + 2 indicates 2 measurement sets at each site where the experiment involves the comparison of two different plots

4.1 Scientific Aims

All field measurements currently aim at determining the dependence of CO_2 and water fluxes (i.e., GPP and ER for CO_2, and ET for water) on climate, environmental parameters, ecological drivers and human management actions, at both inter- and intra-annual time scale. At the Nivolet plain, also the spatial variability at the valley scale is taken into account. Field data have been used to build data-driven models that identify such dependences using a statistical approach.

At the Nivolet plain, the drivers of CO_2 fluxes are studied at different spatial scales, from the size of the individual point measurement (with diameter of about 20 cm) to the plot scale (20 m), to the ensemble of plots at the scale of the valley (that we could define as "landscape scale" for this type of highly variable environment). Intercomparison campaigns have been designed, to build a robust spatial and temporal upscaling framework for the estimation of CO_2 and water fluxes at different spatial scales and provide relevant parameterizations to process-based models, ultimately aiming at predicting the carbon and water flux response to climate change. A further step in the upscaling chain would be to build GPP models for broader areas of Alpine grasslands using remote sensing observations (from drones and satellite) validated by the in-situ measurements.

At Noaschetta and Levionaz, the long-term measurements are focused on the impact of management practices (grazing or irrigation) on the CO_2 and water fluxes. At Noaschetta, the measurements were designed to monitor the effect of land abandonment on the regrowth of wild vegetation and the possible differences in CO_2 exchanges between grazed and ungrazed grassland. At Levionaz, periodic grassland irrigation in a limited area (using pre-existing, historical water storage and distribution facilities) is adopted to prevent grassland from drying out during summer drought events, which are becoming more frequent in the Alps and expected to increase further in the near future [12].

4.2 *Measurement Facilities*

In all sites, a portable apparatus was used to measure fluxes and meteoclimatic variables. Such instrumentation consists in:

- A portable Infrared Gas Analyser (IRGA) equipped with sensors for CO_2 and H_2O determination;
- An accumulation chamber in polycarbonate that allows to measure NEE and, when shaded, ER;
- Meteorological stations equipped with radiometer, barometer and thermo-hygrometer;
- Soil sensors for measuring soil temperature and volumetric water content.

Using this portable apparatus, CO_2 fluxes were measured using the non-steady-state closed dynamic chamber method [35]. Flux estimates were obtained by means of the transparent cylindrical polycarbonate accumulation chamber connected to the LI-COR LI-840 IR spectrophotometer. In each site and for each measurement campaign, 20 to more than 30 randomly selected sample points were measured during a temporal window of about two hours. For each sample point, each flux measurement was performed first with the transparent chamber and then with the chamber shaded by a dark cover, placing the chamber over a stainless-steel collar inserted into the soil to a depth of about 2 cm [122]. The temporal variation of the CO_2 concentration in the chamber was monitored in real time and recorded by a palmtop computer, with a record every second for about 90 s of total measurement. Between the different measurements, the apparatus was purged with ambient air in order to clean it from the excess/deficiency of CO_2.

To obtain flux estimates, either parametric or non-parametric models were used to estimate the CO_2 concentration derivative (dC/dt in ppmV/s) at time $t = 0$. As discussed by Baneschi et al. [8], the best performing method was a linear regression, provided an appropriate range of concentration and time values is selected. For each field campaign, we followed a standardized procedure, starting with laboratory calibration as an essential part of the CO_2 flux measurements, calibrating the apparatus within the range of fluxes expected in the field. To this end, in the laboratory we imposed a sequence of flux values and measured them with the apparatus to be used in the field. A linear calibration curve was thus obtained as the fitting line of the estimated concentration derivative (dC/dt) versus the imposed flux. We finally convert the temporal variation of the CO_2 concentration inside the chamber (in ppmV/s) into a CO_2 flux (μmol $CO_2/m^2/s$), after correction for in-situ pressure and temperature.

The Net Ecosystem Exchange (NEE) and the Ecosystem Respiration (ER) were then recorded in a short time interval. Assuming negligible variations of the environmental variables between the two consecutive measurements of NEE and ER at the same sample point, this procedure allows to obtain the Gross Primary Production (GPP), i.e. the net carbon fixation due to photosynthesis, as the difference between these two consecutive measurements (GPP = NEE − ER). Note that, given the altitude, the typical vegetation height never exceeded the collar level, and thus no volumetric correction was needed for our sampling method and area [132]. The raw dataset was quality controlled by statistical analysis of the frequency distribution and by specific tests (Rosner test, [153]) to identify potential outliers.

Together with the flux sampling, basic meteo-climatic and environmental variables were recorded. Air humidity, temperature and pressure and solar irradiance were measured at about 1.5 m above the ground at the center of the plot with a thermo-hygrometer and a pyranometer. Soil temperature and soil volumetric water content, expected to have significant spatial heterogeneity, were measured at the base of the collar at a depth of about 5 cm.

The Nivolet site is also equipped with an eddy covariance tower (installed in 2019) and two automated flux chambers (installed in 2021). To respect the policy of protected areas, the fixed instruments are powered by solar panels, and therefore work only during the summer season, when the panels are not covered with snow.

The automated chambers consist of two couples each composed by a transparent methyl methacrylate (for NEE) and steel (for ER) chamber, with the same size and less than 2 m far from each other. The two couples are at a distance of about 10 m. All chambers were connected to the same LI-COR LI-850 IRGA. Once every hour, one of the chambers alternatively closed and measured the carbon flux for 5 min, excluding an initial cleaning transient of 20 s. The closure frequency was dictated by the available power supply provided by the solar panels. The automated chambers followed the same measurement method of the portable chamber.

The eddy covariance station is equipped with a LI-COR sensor for CO_2/H_2O measurement, a 3D sonic anemometer, a thermometer, a barometer, a hygrometer, a pluviometer, and a sensor for photosynthetically active radiation (PAR). The eddy covariance station is part of the ICOS network (Station ID: IT-Niv) and the Nivolet site is part of the eLTER program. The flux calculations using the raw signals recorded by the gas analyzer and the sonic anemometer are performed by EddyPro® [65]; LI-COR [116] using default settings. Data quality control is performed for the energy and mass fluxes following Mauder and Foken [127] 0–1–2 policy [128]. The final quality control flags are obtained as a combination of partial statistical flags based on two tests: (i) steady state, and (ii) developed turbulent conditions [63, 64, 76]. The flux random uncertainty due to sampling errors, computed according to the Finkelstein and Sims [62] method implemented in EddyPro®, was used as an estimate of the measurement error. The storage term is expected to be negligible for short canopy and was estimated from one single CO_2 concentration measurement at eddy covariance flux measurement height [6, 185].

The stable isotopes and chemistry laboratories at IGG-CNR (Institute of Geosciences and Earth Resources - National Research Council of Italy) support and complete the field surveys providing analysis of solid, liquid and gaseous matrices. Datasets related to the Gran Paradiso CZO are published in Parisi et al. [188] and in the "IGG-CNR Critical Zone Observatories" community of the Zenodo repository (www.zenodo.org/communities/?p=IGG-CNR).

4.3 Insights from the Field Measurements

An exponential dependence of ER on the environmental temperature [117] and a response of GPP to light intensity through a Michaelis–Menten function [156] are commonly assumed. Nonetheless, a large residual variance was observed in the measured fluxes after accounting for those two factors. For such reasons, different multi-regression models were tested, thus including additional drivers in the model [122]. Different modeling approaches were tested to find the most efficient mathematical formulation and the most relevant set of drivers. In the selected formulation, additional variables were introduced as perturbations of the standard exponential and rectangular hyperbolic functions for ER and GPP, respectively. The optimum set of variables explaining the inter- and intra-annual flux variability included solar irradiance, soil moisture and day of the year for GPP, and air temperature, soil moisture, air pressure and day of the year for ER, in agreement with other studies. In this study, the day of the year was interpreted as a proxy of phenology, which is especially pronounced in high mountains (see Sect. 1). Such a model resulted in large explained variances (83–94%), when calibrated for each of the Nivolet sites and sampling years [122]. Figure 3 shows the graphical abstract of this work, which reports the comparison between the measured data and the model outputs.

A similar study was performed in a High-Arctic tundra, in the Ny-Ålesund area in Spitzbergen Island [123]. The study showed that in the Arctic tundra air temperature and solar irradiance well reproduce the daily variations of ER and GPP, respectively, at a fixed measurement point and over short periods. However, the classical models failed in reproducing the spatial and temporal flux variations at the site scale. Using the same model developed for the Alps, the authors obtained an analogue set of predictors of ER and GPP for the Arctic dataset, namely air temperature, soil moisture and vegetation green fractional cover for ER, and solar irradiance, soil moisture and vegetation green fractional cover for GPP. Such similar results may suggest a common functioning of high-altitude and high-latitude ecosystems, especially concerning summer CO_2 fluxes at the site scale.

Recent research focused on the study of the year-to-year flux variability compared to the site-to-site variability and on the comparison of different co-located measurement techniques. Understanding the relevance of flux dynamics variability between years and between sites provides information about the limits to the application of the estimated model parameterizations for spatial extrapolations and temporal projections. The results of the analysis [115] indicate that, in this high-altitude area, the effects of interannual variability on the parameters of the empirical models for GPP and ER are stronger with respect to the parameter variability across the different sites. This result indicates that (a) spatial upscaling of the empirical flux models from site to landscape scale is presumably feasible, for example by using remote sensing estimates of the drivers, and (b) interannual variability, especially in the hydrological variables, leads to significant impact on the CO_2 fluxes. In turn, this suggests the need for long-term monitoring and measurement programs.

Another important point concerns the relationships between automated chambers, portable chambers and eddy covariance measurements, and the question on whether these different approaches provide a consistent view of the carbon dioxide fluxes [172]. The complex topography, the presence of winter snow cover, the sharp seasonal variation of plant phenology and the sudden wind changes in high mountains pose several challenges to the eddy covariance methods, thus requiring dedicated data quality controls [84, 150] and jeopardizing the measurements of nighttime emissions [91, 179]. For such reasons, eddy covariance systems installed on mountain slopes have been sometimes validated against independent measurements obtained by flux chambers [84, 179]. The results of the analysis of the Nivolet CZO measurements indicate that daytime net fluxes from the average of several point measurements by portable chambers and the estimates provided by eddy covariance are largely consistent with each other. On the other hand, automated chambers provide a pointwise estimate of the fluxes and the consistency with averaged fluxes depends on the degree of heterogeneity of the soil-vegetation system (for example, generated by highly spatially variable soil moisture) and, in such a case, on the exact location of the automated chambers. In this case, the installation of an ensemble of several automated chambers would be ideal. Finally, the extrapolation of nighttime respiration to daytime conditions (as estimated by partitioning methods) does not fully agree with the direct estimate of daytime respiration obtained with darkened portable chambers. This points to a relevant role of plant physiological differences between daytime and nighttime respiration, which cannot be explained by considering only the temperature difference between night and day.

5 Future Perspectives

At the Gran Paradiso CZO, a relevant question concerns the role and dynamics of groundwater and soil moisture. Besides carbon fluxes, vertical exchanges of water across the surface portion of the Critical Zone (which roughly corresponds to the vadose or unsaturated zone in groundwater studies) and soil moisture itself affect many CZ processes and the carbon fluxes. Soil moisture emerged as an important control variable for high-altitude ecosystems [122], which are typically rich in water at the beginning of summer (for example by snowmelt), but could become water-limited as summer progresses, owing to the strong evapotranspiration. In the last few years, summers have been quite dry and several portions of high-altitude grasslands were affected by droughts, also at Gran Paradiso. Currently, we are measuring and modeling water fluxes at the soil–vegetation–atmosphere interface, considering in particular their relationships with the carbon fluxes. A specific issue concerns the effects of a simple manipulation experiment, where specific high-altitude areas of the grasslands in the Levionaz Valley (at about 2500 m a.s.l.) and in other locations are irrigated by using a small portion of the water collected by existing reservoirs. A study is now ongoing where the differences between irrigated and non-irrigated plots are determined, including water and carbon fluxes, soil characteristics, vegetation properties and species distribution, and invertebrate biodiversity.

At the time of writing, we are planning to explore in detail the changes in proglacial environments, which are often characterized by high microscale climate variability. Glaciers are rapidly retreating in the Alps, and Gran Paradiso is no exception. This leaves wide areas open to the establishment of Critical Zone processes, soil formation and colonization by vascular and non-vascular vegetation. The proglacial area of Lauson, in Valnontey (between about 2600 and 3000 m a.s.l.) is ideal for this type of analysis as it has been already studied from the point of view of vegetation chrono-sequences and biodiversity [120]. A measure of the carbon stocks and of water and carbon fluxes at different distances from the retreating glaciers is an interesting endeavor planned for the coming years.

Another issue concerns the impact of changes in land use, and in particular the effects of pasture abandonment. In the Noaschetta Valley, also in the framework of the ABRESO project of the Belmont Forum, the differences between pastured and non-pastured mid-altitude grasslands (at about 1500 m a.s.l.) are monitored since 2017. This is a long-term study which should provide information on how the mountain environment is changing in face of temperature rise, bush and tree encroachment, and land-use modification.

Broadening the scale, it is becoming of central importance to upscale the results obtained at individual sites (by closed chamber or eddy covariance methods) to a larger scale, either that of the valley or even to the whole mountain region considered (clearly for the same type of ecosystem). The upscaling challenge can be addressed by resorting to remote sensing observations, which however should have a detailed enough spatial resolution to properly cope with the small-scale variability of mountain environments. Besides commercial satellites, currently the highest resolution for monitoring ecosystems dynamics is provided by the European Sentinel-2 constellation, which convey important information on the drivers of carbon and water fluxes, as well as on vegetation properties. One should then delve into the issue of combining site measurements, satellite observations and drone surveys (with various types of sensors) to develop simple data-driven models (based on correlations or on machine learning) to relate the fluxes and the variables of interest to the environmental variables (e.g., [5, 105, 162, 167]).

A further step is related to the establishment of other CZOs in nearby locations, to provide an even broader view of the high-altitude Critical Zone in a larger mountain environment. Other existing research sites are currently involved in the effort of establishing a wide "regional" CZO in the north-western Alps, complementing the existing facilities with novel instrumentation that can provide similar information at all sites. Along these lines, it becomes crucial to determine a minimal set of "Critical Zone variables", or indicators, to be measured at all sites in the same type of environment (clearly, high-mountain variables will be different from those needed for coastal wetlands). Similar efforts have been undertaken by other communities under the framework of the so-called "Essential Variables" (e.g., [178]), and are currently ongoing for example in the eLTER Research Infrastructure in Europe ("eLTER Standard Observations"). The difficulty, of course, resides in the competing needs of properly covering the main components of the CZ and, at the same time, of not multiplying the variables to be measured beyond what reasonably can be done.

A final point concerns the dissemination of the results and re-use of the data, information and knowledge gained through studies at a given CZO, such as that at GPNP. Often, the results are published in scientific journals but the data and knowledge, although openly available, can be difficult to find and used by other researchers. Instead, data and research products should be FAIR: Findable, Accessible, Interoperable, and Reusable [176]. It can be difficult for single research groups to set up an operational framework for achieving this goal. For this reason, in Europe there are several Research Infrastructures that can help in the effort of making the research findings FAIR. LifeWatch ERIC (www.lifewatch.eu), in particular, is an e-science infrastructure that aims at establishing an ensemble of Virtual Research Environments (VRE) where researchers can find data, knowledge links, analysis methods and models and combine them as they need, also adding their own contribution in terms of data or software. The eLTER Research Infrastructure is also developing data integration portals where the long-term data from their sites across all Europe will be made available. A start has already been made in the Site and Metadata Portal DEIMS-SDR [180]. A true community endeavor, which is open to non-expert users such as park wardens, land managers and planners, or farmers. In Italy, the large national project ITINERIS is combining several different European Research Infrastructures to create a system of VREs able to tackle—as test cases—some relevant environmental issues. One of such VREs is explicitly devoted to establish a virtual system based on a set of Critical Zone Observatories, including the GPNP CZO.

Acknowledgements We thank Thomas Dirnböck for careful reading of this chapter and very useful suggestions on how to improve it. We acknoweldge support by the H2020 projects ECOPOTENTIAL (grant number: 641762), e-shape (grant number: 820852), eLTER PLUS (grant number: 871128), by the Italian National Biodiversity Future Center (NBFC), National Recovery and Resilience Plan (NRRP; mission 4, component 2, investment 1.4 of the Ministry of University and Research, funded by the European Union– NextGenerationEU; project code CN00000033), and by the ITINERIS NRRP Italian infrastructure project (project code No. IR0000032 - ESFRI Environment). This work benefited from the help of Francesca Avogadro di Valdengo, Alice Baronetti, Andrea Berton, Virginia Boiani, Maurizio Catania, Simona Gennaro, Silvana Goiran, Sara Lenzi, Silvio Marta, Angelica Parisi. We thank Bruno Bassano, Ramona Viterbi, Andrea Mainetti and the surveillance personnel of GPNP for their continued assistance and support.

References

1. Addor N et al (2014) Robust changes and sources of uncertainty in the projected hydrological regimes of Swiss catchments. Water Resour Res 50(10):7541–7562
2. Allard JL, Hughes PD, Woodward JC (2021) A radiometric dating revolution and the quaternary glacial history of the Mediterranean mountains. Earth Sci Rev 223:10384
3. Allen S, Huggel C (2013) Extremely warm temperatures as a potential cause of recent high mountain rockfall. Glob Planet Chang 107:59–69
4. Anthelme F, Cavieres LA, Dangles O (2014) Facilitation among plants in alpine environments in the face of climate change. Front Plant Sci 5:387
5. Argenti G, Chiesi M, Fibbi L, Maselli F (2022) Use of remote sensing and bio-geochemical models to estimate the net carbon fluxes of managed mountain grasslands. Ecol Model 474:110152
6. Aubinet M, Vesala T, Papale D (eds) (2012) Eddy covariance: a practical guide to measurement and data analysis. Springer Science and Business Media
7. Auer I et al (2007) HISTALP—historical instrumental climatological surface time series of the Greater Alpine Region. Int J Climatol 27(1):17–46
8. Baneschi I et al (2023) Non-steady-state closed dynamic chamber to measure soil CO_2 respiration: a protocol to reduce uncertainty. Front Environ Sci 10:1048948. https://doi.org/10.3389/fenvs.2022.1048948
9. Bard A et al (2015) Trends in the hydrologic regime of Alpine rivers. J Hydrol 529:1823–1837
10. Bardelli T et al (2017) Effects of slope exposure on soil physico-chemical and microbiological properties along an altitudinal climosequence in the Italian Alps. Sci Total Environ 575:1041–1055
11. Bardgett RD et al (2005) A temporal approach to linking aboveground and belowground ecology. Trends Ecol Evol 20(11):634–641
12. Baronetti A et al (2022) Future droughts in northern Italy: high-resolution projections using EURO-CORDEX and MED-CORDEX ensembles. Clim Chang 172:22 https://doi.org/10.1007/s10584-022-03370-7
13. Baruck J et al (2016) Soil classification and mapping in the Alps: the current state and future challenges. Geoderma 264:312–331
14. Baumann E et al (2016) Diversity and functional composition of alpine grasslands along an elevational transect in the Gran Paradiso National Park (NW Italy). Tuexenia 36:337–358
15. Beltramo R, Duglio S Cappelletti GM (2022) Should I stay or can I go? Accessible tourism and mountain huts in Gran Paradiso National Park. Sustainability 14(5):2936

16. Beniston M, Stoffel M, Hill M (2011) Impacts of climatic change on water and natural hazards in the Alps: can current water governance cope with future challenges? Examples from the European "ACQWA" project. Environ Sci Policy 14(7):734–743
17. Beniston M, Stoffel M (2014) Assessing the impacts of climatic change on mountain water resources. Sci Total Environ 493:1129–1137
18. Beniston M et al (2018) The European mountain cryosphere: a review of its current state, trends, and future challenges. Cryosphere 12(2):759–794
19. Billings WD (1973) Arctic and alpine vegetations: similarities, differences, and susceptibility to disturbance. Bioscience 23(12):697–704
20. Biskaborn BK et al (2019) Permafrost is warming at a global scale. Nat Commun 10(1):1–11
21. Björk RG et al (2008) Temporal variation in soil microbial communities in Alpine tundra. Soil Biol Biochem 40(1):266–268
22. Bliss A, Hock R, Radić V (2014) Global response of glacier runoff to twenty-first century climate change. J Geophys Res Earth Surf 119(4):717–730
23. Boitani L, Barrasso P, Grimod I (1984) Ranging behavior of the red fox in the Gran Paradiso National Park (Italy). Italian J Zool 51(3–4):275–284
24. Bogdal C et al (2010) Release of legacy pollutants from melting glaciers: model evidence and conceptual understanding. Environ Sci Technol 44(11):4063–4069
25. Bodin X et al (2017) The 2006 collapse of the Bérard rock glacier (Southern French Alps). Permafrost Periglac Process 28(1):209–223
26. Boeckli L et al (2012) A statistical approach to modelling permafrost distribution in the European Alps or similar mountain ranges. Cryosphere 6(1):125–140
27. Buchli T et al (2013) Characterization and monitoring of the Furggwanghorn rock glacier, Turtmann Valley, Switzerland: results from 2010 to 2012. Vadose Zone J 12(1)
28. Buchhorn M et al (2019) Copernicus global land service: land cover 100m: collection 2: epoch 2015: Globe
29. Callaway et al (2002) Positive interactions among alpine plants increase with stress. Nature 417:844–848
30. Canedoli C et al (2020) Evaluation of ecosystem services in a protected mountain area: soil organic carbon stock and biodiversity in alpine forests and grasslands. Ecosyst Serv 44:101135
31. Carlson BZ et al (2013) Working toward integrated models of alpine plant distribution. Alp Bot 123(2):41–53
32. Casassa G et al (2009) Detection of changes in glacial run-off in alpine basins: examples from North America, the Alps, central Asia and the Andes. Hydrol Process 23(1):31–41
33. Castebrunet H et al (2014) Projected changes of snow conditions and avalanche activity in a warming climate: the French Alps over the 2020–2050 and 2070–2100 periods. Cryosphere 8(5):1673–1697
34. Ceppi P et al (2012) Revisiting Swiss temperature trends 1959–2008. Int J Climatol 32(2):203–213
35. Chiodini G et al (1998) Soil CO_2 flux measurements in volcanic and geothermal areas. Appl Geochem 13:543–552
36. Colombo N et al (2018) Impacts of permafrost degradation on inorganic chemistry of surface fresh water. Glob Planet Chang 162:69–83
37. D'Amico M, Almeida JP, Barbieri S, Castelli F, Sgura E, Sineo G, Celi L (2020) Ectomycorrhizal utilization of different phosphorus sources in a glacier forefront in the Italian Alps. Plant Soil 446:81–95
38. Dal Piaz GV, Bistacchi A, Massironi M (2003) Geological outline of the Alps. Episodes 26(3):175–180
39. Danna C et al (2022) Ethnomedicinal and ethnobotanical survey in the Aosta Valley Side of the Gran Paradiso National Park (Western Alps, Italy). Plants 11(2):170
40. Debernardi P, Patriarca E, Reutter B (2003) Contribution to the knowledge of Apodemus genus in the Gran Paradiso National Park. Hystrix 14(1–2)
41. Dirnböck T, Dullinger S, Grabherr G (2003) A regional impact assessment of climate and land-use change on alpine vegetation. J Biogeogr 30(3):401–417

42. Dirnböck T, Kraus D, Grote R, Klatt S, Kobler J, Schindlbacher A, Kiese R (2020) Substantial understory contribution to the C sink of a European temperate mountain forest landscape. Landscape Ecol 35:483–499
43. Dullinger S, Kleinbauer I, Pauli H, Gottfried M, Brooker R, Nagy L, Grabherr G (2007) Weak and variable relationships between environmental severity and small-scale co-occurrence in alpine plant communities. J Ecol 95(6):1284–1295
44. Eckert N et al (2013) Temporal trends in avalanche activity in the French Alps and subregions: from occurrences and runout altitudes to unsteady return periods. J Glaciol 59(213):93–114
45. Edwards AC, Scalenghe R, Freppaz M (2007) Changes in the seasonal snow cover of alpine regions and its effect on soil processes: a review. Quatern Int 162:172–181
46. Egli M et al (2005) Modellierung von Bodentypen und Bodeneigenschaften im Oberengadin (Schweiz) mit Hilfe eines Geographischen Informationssystems (GIS). Geographica Helvetica 60(2):87–96
47. Egli M, Mirabella A, Sartori G, Zanelli R, Bischof S (2006) Effect of north and south exposition on weathering and clay mineral formation in Alpine soils. Catena 67:155–174
48. Egli M, Mirabella A, Sartori G (2008) The role of climate and vegetation in weathering and clay mineral formation in late quaternary soils of the Swiss and Italian Alps. Geomorphology 102:307–324
49. Egli M, Sartori G, Mirabella A, Favilli F, Giaccai D, Delbos E (2009) Effect of north and south exposure on organic matter in high Alpine soils. Geoderma 149:124–136
50. Egli M, Sartori G, Mirabella A, Giaccai D (2010) The effects of exposure and climate on the weathering of late Pleistocene and Holocene Alpine soils. Geomorphology 114:466–482
51. Egli M, Dahms D, Norton K (2014) Soil formation rates on silicate parent material in alpine environments: different approaches—different results? Review article. Geoderma 213:320–333
52. Eichel J, Draebing D, Meyer N (2018) From active to stable: paraglacial transition of Alpine lateral moraine slopes. Land Degrad Dev 29(11):4158–4172
53. Elsasser H, Bürki R (2002) Climate change as a threat to tourism in the Alps. Climate Res 20(3):253–257
54. Ernakovich JG et al (2014) Predicted responses of arctic and alpine ecosystems to altered seasonality under climate change. Glob Change Biol 20(10):3256–3269
55. European Environment Agency (ed) (2009) Regional climate change and adaptation: the Alps facing the challenge of changing water resources. Office for Official Publications of the European Communities, Luxembourg
56. Faillettaz J, Funk M, Vincent C (2015) Avalanching glacier instabilities: review on processes and early warning perspectives. Rev Geophys 53(2):203–224
57. Farinotti D, Pistocchi A, Huss M (2016) From dwindling ice to headwater lakes: could dams replace glaciers in the European Alps? Environ Res Lett 11(5):054022
58. Fell SC et al (2018) Declining glacier cover threatens the biodiversity of alpine river diatom assemblages. Glob Chang Biol 24(12):5828–5840
59. Filippa G et al (2022) On the distribution and productivity of mountain grasslands in the Gran Paradiso National Park, NW Italy: a remote sensing approach International. J Appl Earth Obs Geoinf 108:102718
60. Filippa G et al (2019) Climatic drivers of greening trends in the Alps. Remote Sens 11(21):2527
61. Fischer L et al (2013) Slope failures and erosion rates on a glacierized high-mountain face under climatic changes. Earth Surf Proc Land 38(8):836–846
62. Finkelstein L, Sims F (2001) Sampling error in eddy correlation flux measurements. J Geophys Res Atmos 106(D4):3503–3509
63. Foken T, Wichura B (1996) Tools for quality assessment of surface-based flux measurements. Agric For Meteorol 78(1–2):83–105
64. Foken T et al (2004) Post-field data quality control. Handbook of micrometeorology. Springer, Dordrecht, pp 181–208
65. Fratini G, Mauder M (2014) Towards a consistent eddy-covariance processing: an intercomparison of EddyPro and TK3. Atmos Meas Tech 7:2273–2281. https://doi.org/10.5194/amt-7-2273-2014

66. Freeman BG et al (2018) Expanding, shifting and shrinking: the impact of global warming on species' elevational distributions. Glob Ecol Biogeogr 27(11):1268–1276
67. Frei E, Bodin J, Walther GR (2010) Plant species' range shifts in mountainous areas—all uphill from here? Bot Helv 120(2):117–128
68. Freppaz M et al (2018) Topsoil and snow: a continuum system. Appl Soil Ecol 123:435–440
69. Fuhrer J, Smith P, Gobiet A (2014) Implications of climate change scenarios for agriculture in alpine regions—a case study in the Swiss Rhone catchment. Sci Total Environ 493:1232–1241
70. Gaudard L, Gilli M, Romerio F (2013) Climate change impacts on hydropower management. Water Resour Manage 27(15):5143–5156
71. Gaudard L et al (2014) Climate change impacts on hydropower in the Swiss and Italian Alps. Sci Total Environ 493:1211–1221
72. Gehrig-Fasel J, Guisan A, Zimmermann NE (2007) Tree line shifts in the Swiss Alps: climate change or land abandonment? J Veg Sci 18(4):571–582
73. Gellrich M, Zimmermann NE (2007) Investigating the regional-scale pattern of agricultural land abandonment in the Swiss mountains: a spatial statistical modelling approach. Landsc Urban Plan 79(1):65–76
74. Giuliano D, Battisti A, Bassano B (2019) Feeding ecology of the red fox in the Soana Valley, Gran Paradiso National Park. J Mt Ecol 12:1–18
75. Gobiet A et al (2014) 21st century climate change in the European Alps—a review. Sci Total Environ 493:1138–1151
76. Göckede M et al (2008) Quality control of CarboEurope flux data—Part 1: coupling footprint analyses with flux data quality assessment to evaluate sites in forest ecosystems. Biogeosciences 5(2):433–450
77. Gottfried M, Pauli H, Grabherr G (1998) Prediction of vegetation patterns at the limits of plant life: a new view of the alpine-nival ecotone. Arct Alp Res 30(3):207–221
78. Gottfried M et al (2012) Continent-wide response of mountain vegetation to climate change. Nat Clim Chang 2(2):111–115
79. Grabherr G, Nagy L, Thompson DBA (2003) An outline of Europe's alpine areas. In: Alpine biodiversity in Europe. Springer, Berlin, Heidelberg, pp 3–12
80. Graae BJ et al (2018) Stay or go—how topographic complexity influences alpine plant population and community responses to climate change. Perspect Plant Ecol Evol Syst 30:41–50
81. Gruber S (2012) Derivation and analysis of a high-resolution estimate of global permafrost zonation. Cryosphere 6(1):221–233
82. Gruber S, Haeberli W (2007) Permafrost in steep bedrock slopes and its temperature-related destabilization following climate change. J Geophys Res Earth Surf 112(F2)
83. Hagedorn F, Mulder J, Jandl R (2010) Mountain soils under a changing climate and land use. Biogeochemistry 97:1–5
84. Hammerle A, Haslwanter A, Schmitt M, Bahn M, Tappeiner U, Cernusca A, Wohlfahrt G (2007) Eddy covariance measurements of carbon dioxide, latent and sensible energy fluxes above a meadow on a mountain slope. Bound-Layer Meteorol 122:397–416
85. Hänggi P, Weingartner R (2012) Variations in discharge volumes for hydropower generation in Switzerland. Water Resour Manag 26(5):1231–1252
86. Hannah L et al (2014) Fine-grain modeling of species' response to climate change: holdouts, stepping-stones, and microrefugia. Trends Ecol Evol 29(7):390–397
87. Harris C et al (2009) Permafrost and climate in Europe: monitoring and modelling thermal, geomorphological and geotechnical responses. Earth-Sci Rev 92(3–4):117–171
88. Haselwandter K, Hofmann A, Holzmann HP, Read DJ (1983) Availability of nitrogen and phosphorus in the nival zone of the Alps. Oecologia 57:266–269
89. Hilbich C et al (2008) Monitoring mountain permafrost evolution using electrical resistivity tomography: a 7-year study of seasonal, annual, and long-term variations at Schilthorn, Swiss Alps. J Geophys Res Earth Surf 113(F1)
90. Hill M (2013) Adaptive capacity of water governance: cases from the Alps and the Andes. Mt Res Dev 33(3):248–259

91. Hiller R, Zeeman MJ, Eugster W (2008) Eddy-covariance flux measurements in the complex terrain of an Alpine valley in Switzerland. Bound-Layer Meteorol 127:449–467
92. Hiltbrunner E, Schwikowski M, Körner C (2005) Inorganic nitrogen storage in alpine snow pack in the Central Alps (Switzerland). Atmos Environ 39(12):2249–2259
93. Hock R et al (2019) High mountain areas. In: Pörtner H-O et al (eds) IPCC special report on the ocean and cryosphere in a changing climate, pp 131–202
94. Hodson AJ (2014) Understanding the dynamics of black carbon and associated contaminants in glacial systems. Wiley Interdiscip Rev Water 1(2):141–149
95. Hofmann K et al (2016) Distribution of prokaryotic abundance and microbial nutrient cycling across a high-alpine altitudinal gradient in the Austrian Central Alps is affected by vegetation, temperature, and soil nutrients. Microb Ecol 72(3):704–716
96. Hoffmann S et al (2019) Data on alpine grassland diversity in Gran Paradiso National Park, Italy. Data Brief 24:103942
97. Huelber K et al (2006) Phenological responses of snowbed species to snow removal dates in the Central Alps: implications for climate warming. Arct Antarct Alp Res 38(1):99–103
98. Huggel C et al (2019) Loss and damage in the mountain cryosphere. Reg Environ Chang 19(5):1387–1399
99. Hughes CE, Atchison GW (2015) The ubiquity of alpine plant radiations: from the Andes to the Hengduan Mountains. New Phytol 207(2):275–282
100. Hüsler F et al (2014) A satellite-based snow cover climatology (1985–2011) for the European Alps derived from AVHRR data. Cryosphere 8(1):73–90
101. Huss M, Fischer M (2016) Sensitivity of very small glaciers in the Swiss Alps to future climate change. Front Earth Sci 4:34
102. Huss M, Hock R (2018) Global-scale hydrological response to future glacier mass loss. Nat Clim Chang 8(2):135–140
103. Ilyashuk BP et al (2018) Rock glaciers in crystalline catchments: hidden permafrost-related threats to alpine headwater lakes. Glob Chang Biol 24(4):1548–1562
104. Jacobson AR et al (2004) Climate forcing and density dependence in a mountain ungulate population. Ecology 85(6):1598–1610
105. Jägermeyr J, Gerten D, Lucht W, Hostert P, Migliavacca M, Nemani R (2014) A high-resolution approach to estimating ecosystem respiration at continental scales using operational satellite data. Glob Chang Biol 20(4):1191–1210
106. Jenicek M, Seibert J, Staudinger M (2018) Modeling of future changes in seasonal snowpack and impacts on summer low flows in alpine catchments. Water Resour Res 54(1):538–556
107. Kaplan JO, Krumhardt KM, Zimmermann N (2009) The prehistoric and preindustrial deforestation of Europe. Quatern Sci Rev 28(27–28):3016–3034
108. Körner C (2007) Climatic treelines: conventions, global patterns, causes (Klimatische Baumgrenzen: Konventionen, globale Muster, Ursachen). Erdkunde 316–324
109. Körner C (2021) Alpine plant life: functional plant ecology of high mountain ecosystems. Springer Nature
110. Körner C, Paulsen J (2004) A world-wide study of high altitude treeline temperatures. J Biogeogr 31(5):713–732
111. Körner C, Paulsen J, Spehn EM (2011) A definition of mountains and their bioclimatic belts for global comparisons of biodiversity data. Alp Bot 121(2):73–78
112. Kotlarski S et al (2023) 21st Century alpine climate change. Clim Dyn 60:65–86. https://doi.org/10.1007/s00382-022-06303-3
113. Laiolo P et al (2004) Consequences of pastoral abandonment for the structure and diversity of the alpine avifauna. J Appl Ecol 41(2):294–304
114. Lamprecht A et al (2018) Climate change leads to accelerated transformation of high-elevation vegetation in the central Alps. New Phytol 220(2):447–459
115. Lenzi S et al (2023) Spatial and temporal variability of carbon dioxide fluxes in the Alpine Critical Zone: the case of the Nivolet Plain, Gran Paradiso National Park, Italy. PLoS One 18(5):e0286268

116. LI-COR Biosciences (2017) Eddy covariance processing software (Version 706) [Software]. www.licor.com/EddyPro
117. Lloyd J, Taylor JA (1994) On the temperature dependence of soil respiration. Funct Ecol 315–323
118. Luethi R, Phillips M, Lehning M (2017) Estimating non-conductive heat flow leading to intra-permafrost Talik formation at the Ritigraben Rock Glacier (Western Swiss Alps). Permafrost Periglac Process 28(1):183–194
119. Lugon R, Stoffel M (2010) Rock-glacier dynamics and magnitude–frequency relations of debris flows in a high-elevation watershed: Ritigraben, Swiss Alps. Glob Planet Chang 73(3–4):202–210
120. Mainetti A et al (2021) Successional herbaceous species affect soil processes in a high-elevation alpine proglacial chronosequence. Front Environ Sci 8:615499
121. Magnani A et al (2017) Interannual variability of soil N and C forms in response to snow—cover duration and pedoclimatic conditions in alpine tundra, northwest Italy. Arct Antarct Alp Res 49(2):227–242
122. Magnani M et al (2020) Drivers of carbon fluxes in Alpine tundra: a comparison of three empirical model approaches. Sci Total Environ 732:139139
123. Magnani M et al (2022) Microscale drivers of summer CO_2 fluxes in the Svalbard High Arctic tundra. Sci Rep 12:763 https://doi.org/10.1038/s41598-021-04728-0
124. Marty C, Meister R (2012) Long-term snow and weather observations at Weissfluhjoch and its relation to other high-altitude observatories in the Alps. Theoret Appl Climatol 110(4):573–583
125. Marty C, Tilg AM, Jonas T (2017) Recent evidence of large-scale receding snow water equivalents in the European Alps. J Hydrometeorol 18(4):1021–1031
126. Matiu M et al (2021) Observed snow depth trends in the European Alps: 1971 to 2019. Cryosphere 15(3):1343–1382
127. Mauder M, Foken T (2004) Documentation and instruction manual of the eddy covariance software package TK2. Univ Bayreuth, Abt Mikrometeorol 26–42, ISSN 161489166
128. Mauder M et al (2013) A strategy for quality and uncertainty assessment of long-term eddy-covariance measurements. Agric For Meteorol 169:122–135, ISSN 0168-1923
129. Michalet R et al (2006) Do biotic interactions shape both sides of the humped-back model of species richness in plant communities? Ecol Lett 9(7):767–773
130. Mignatti A et al (2012) Sex- and age-structured models for Alpine ibex Capra ibex ibex population dynamics. Wildl Biol 18(3):318–332
131. Milner AM et al (2017) Glacier shrinkage driving global changes in downstream systems. Proc Natl Acad Sci 114(37):9770–9778
132. Morton A, Heinemeyer A (2018) Vegetation matters: correcting chamber carbon flux measurements using plant volumes. Sci Total Environ 639:769–772
133. Nemergut DR et al (2005) Structure and function of alpine and arctic soil microbial communities. Res Microbiol 156(7):775–784
134. Nevalainen L et al (2014) Zooplankton (Cladocera) species turnover and long-term decline of Daphnia in two high mountain lakes in the Austrian Alps. Hydrobiologia 722(1):75–91
135. Obu J et al (2019) Northern Hemisphere permafrost map based on TTOP modelling for 2000–2016 at 1 km^2 scale. Earth Sci Rev 193:299–316
136. Palazzi E et al (2019) Elevation-dependent warming in global climate model simulations at high spatial resolution. Clim Dyn 52(5):2685–2702
137. Palmegiani I, Gazzola A, Apollonio M (2013) Wolf diet and its impact on the ungulates community in a new recolonized area of Western Alps: Gran Paradiso National Park. Folia Zool 62(1):59–66
138. Pauli H et al (2012) Recent plant diversity changes on Europe's mountain summits. Science 336(6079):353–355
139. Pepin N et al (2015) Elevation-dependent warming in mountain regions of the world. Nat Clim Chang 5:424–430

140. Piana F et al (2017) Geology of Piemonte region (NW Italy, Alps–Apennines interference zone). J Maps 13(2):395–405. https://doi.org/10.1080/17445647201713162 18
141. Pielmeier C et al (2013) Wet snow avalanche activity in the Swiss Alps–trend analysis for mid-winter season. In: Proceedings of the international snow science workshop, grenoble and chamonix, pp 1240–1246
142. Pintaldi E, D'Amico ME, Colombo N, Colombero C, Sambuelli L, De Regibus C, Freppaz M (2021) Hidden soils and their carbon stocks at high-elevation in the European Alps (North-West Italy). Catena 198:105044
143. Pintaldi E et al (2022) Snowbed communities and soil C and N dynamics during a four-year investigation in the NW-Italian Alps. Arct Antarct Alp Res 54(1):368–385
144. Ponge JF et al (2014) The impact of parent material, climate, soil type and vegetation on Venetian forest humus forms: a direct gradient approach. Geoderma 226–227:290–299
145. Pribyl DW (2010) A critical review of the conventional SOC to SOM conversion factor. Geoderma 156(3–4):75–83
146. IPCC (2022) Climate Change 2022: impacts, adaptation and vulnerability. Contribution of working group II to the sixth assessment report of the intergovernmental panel on climate change [Pörtner H-O, Roberts M, Tignor ES, Poloczanska K, Mintenbeck A, Alegría M, Craig S, Langsdorf S, Löschke V, Möller A, Okem BR (eds)]. Cambridge University Press, Cambridge, UK and New York, NY, USA, 3056p. https://doi.org/10.1017/9781009325844
147. Qixiang W, Wang M, Fan X (2018) Seasonal patterns of warming amplification of high-elevation stations across the globe. Int J Climatol 38(8):3466–3473
148. Quinn JA (2008) Arctic and Alpine biomes, vol 6. Greenwood Publishing Group
149. Rabatel A et al (2013) Changes in glacier equilibrium-line altitude in the western Alps from 1984 to 2010: evaluation by remote sensing and modeling of the morpho-topographic and climate controls. Cryosphere 7(5):1455–1471
150. Reba ML, Link TE, Marks D, Pomeroy J (2009) An assessment of corrections for eddy covariance measured turbulent fluxes over snow in mountain environments. Water Resour Res 45(4)
151. Rödder T, Kneisel C (2012) Influence of snow cover and grain size on the ground thermal regime in the discontinuous permafrost zone. Swiss Alps Geomorphol 175:176–189
152. Rogora M et al (2018) Assessment of climate change effects on mountain ecosystems through a cross-site analysis in the Alps and Apennines. Sci Total Environ 624:1429–1442
153. Rosner B (1983) Percentage points for a generalized ESD many-outlier procedure. Technometrics 25:165–172
154. Rumpf SB et al (2018) Range dynamics of mountain plants decrease with elevation. Proc Natl Acad Sci 115(8):1848–1853
155. Rumpf SB, Gravey M, Brönnimann O, Luoto M, Cianfrani C, Mariethoz G, Guisan A (2022) From white to green: snow cover loss and increased vegetation productivity in the European Alps. Science 376(6597):1119–1122
156. Ruimy A et al (1995) CO_2 fluxes over plant canopies and solar radiation: a review. Adv Ecol Res 26:1–68 (Academic Press)
157. Schär et al (1998) Current Alpine climate. In: Cebon P et al (eds) Views from the Alps Regional perspectives on climate change. The MIT Press, Cambridge, London, pp 21–72
158. Schaefli B et al (2019) The role of glacier retreat for Swiss hydropower production. Renew Energy 132:615–627
159. Scherrer D, Körner C (2011) Topographically controlled thermal-habitat differentiation buffers alpine plant diversity against climate warming. J Biogeogr 38(2):406–416
160. Scholz K et al (2018) Analyzing the effects of growing season length on the net ecosystem production of an alpine grassland using model-data fusion. Ecosystems 21(5):982–999
161. Spandre P et al (2019) Winter tourism under climate change in the Pyrenees and the French Alps: relevance of snowmaking as a technical adaptation. Cryosphere 13(4):1325–1347
162. Spinosa A, Fuentes-Monjaraz MA, El Serafy G (2023) Assessing the use of Sentinel-2 data for spatio-temporal upscaling of flux tower gross primary productivity measurements. Remote Sens 15(3):562

163. Steinbauer MJ et al (2018) Accelerated increase in plant species richness on mountain summits is linked to warming. Nature 556(7700):231–234
164. Testolin R, Attorre F, Jiménez-Alfaro B (2020) Global distribution and bioclimatic characterization of alpine biomes. Ecography 43(6):779–788
165. Thibert E et al (2018) Causes of glacier melt extremes in the Alps since 1949. Geophys Res Lett 45(2):817–825
166. Tiberti R et al (2013) Ecological dynamics of two remote Alpine lakes during ice-free season. J Limnol 72(3):401–416
167. Tonolli S, Dalponte M, Neteler M, Rodeghiero M, Vescovo L, Gianelle D (2011) Fusion of airborne LiDAR and satellite multispectral data for the estimation of timber volume in the Southern Alps. Remote Sens Environ 115(10):2486–2498
168. Tudoroiu M et al (2016) Negative elevation-dependent warming trend in the Eastern Alps. Environ Res Lett 11(4):044021
169. Vitasse Y et al (2021) Phenological and elevational shifts of plants, animals and fungi under climate change in the European Alps. Biol Rev 96(5):1816–1835
170. Viterbi R et al (2020) Effects of temperature rise on multi-taxa distributions in mountain ecosystems. Diversity 12:210. https://doi.org/10.3390/d12060210
171. Vittoz P et al (2009) Low impact of climate change on subalpine grasslands in the Swiss Northern Alps. Glob Chang Biol 15(1):209–220
172. Vivaldo G et al (2023) Carbon dioxide exchanges in an alpine tundra ecosystem (Gran Paradiso National Park, Italy): a comparison of results from different measurement and modelling approaches. Atmos Environ 305:119758
173. Viviroli D et al (2011) Climate change and mountain water resources: overview and recommendations for research, management and policy. Hydrol Earth Syst Sci 15(2):471–504
174. Wang Q, Fan X, Wang M (2016) Evidence of high-elevation amplification versus Arctic amplification. Sci Rep 6(1):1–8
175. Warscher M, Strasser U, Kraller G, Marke T, Franz H, Kunstmann H (2013) Performance of complex snow cover descriptions in a distributed hydrological model system: a case study for the high Alpine terrain of the Berchtesgaden Alps. Water Resour Res 49(5):2619–2637
176. Wilkinson MD et al (2016) The FAIR guiding principles for scientific data management and stewardship. Sci Data 3(1):1–9
177. Winkler M et al (2016) The rich sides of mountain summits—a pan-European view on aspect preferences of alpine plants. J Biogeogr 43(11):2261–2273
178. WMO—World Meteorological Organization (2022) Essential climate variables. https://library.wmo.int/records/item/58111-the-2022-gcos-ecvs-requirements-gcos-245?offset=143
179. Wohlfahrt G, Anfang C, Bahn M, Haslwanter A, Newesely C, Schmitt M, Cernusca A (2005) Quantifying nighttime ecosystem respiration of a meadow using eddy covariance, chambers and modelling. Agric For Meteorol 128(3–4):141–162
180. Wohner C, Peterseil J, Poursanidis D, Kliment T, Wilson M, Mirtl M, Chrysoulakis N (2019) DEIMS-SDR—a web portal to document research sites and their associated data. Eco Inform 51:15–24
181. Wookey A et al (2009) Ecosystem feedbacks and cascade processes: understanding their role in the responses of Arctic and alpine ecosystems to environmental change. Glob Chang Biol 15(5):1153–1172
182. Wouters B, Gardner AS, Moholdt G (2019) Global glacier mass loss during the GRACE satellite mission (2002–2016). Front Earth Sci 7:96
183. Xie J et al (2018) Relative influence of timing and accumulation of snow on alpine land surface phenology. J Geophys Res Biogeosci 123(2):561–576
184. Xu X, Liu W (2017) The global distribution of Earth's critical zone and its controlling factors. Geophys Res Lett 44(7):3201–3208
185. Xu L, Baldocchi DD (2004) Seasonal variation in carbon dioxide exchange over a Mediterranean annual grassland in California. Agric For Meteorol 123(1–2):79–96
186. Zemp M et al (2019) Global glacier mass changes and their contributions to sea-level rise from 1961 to 2016. Nature 568(7752):382–386

187. Marta S et al (2023) Heterogeneous changes of soil microclimate in high mountains and glacier forelands. Nat Commun 14:5306
188. Parisi A et al (2024) Carbon dioxide fluxes in Alpine grasslands at the Nivolet Plain, Gran Paradiso National Park, Italy 2017–2023. Sci Data 11:652
189. Lovari S, Cassola F (1975) Nature conservtion in Italy: the existing national parks and otherprotected areas. Biol Conserv 8(2):127–142

Introducing the Avon River Critical Zone Observatory—Part of the Australian Critical Zone Observatories Network

Matthias Leopold, Louise Barton, Jason Beringer, Deirdre B. Gleeson, Caitlin E. Moore, Andrew Rate, Megan H. Ryan, Talitha Santini, and Sally E. Thompson

1 Introduction

Critical Zone Observatories (CZOs) are a global attempt to understand the function of the vertical 'geo-bio-atmosphere boundary down to the bedrock' critical zone (CZ) profile by characterising its physical, chemical and biological structure, learning how matter and energy are stored, transported and transformed within the CZ, and elucidating how this storage, movement and transformation contribute to the evolution of CZ structure [28, 54, 133]. These processes and interactions between structure and function can only be observed in contemporary CZs, which form today's terrestrial landscape.

Contemporary CZs are the outcome of interactions between biological, physical and chemical weathering and pedogenic, erosive and depositional processes acting throughout the life of the CZ (e.g. [38, 89]). The structure and function of contemporary CZs are determined by this history, including land management (e.g., [41]). Therefore, CZ history has important implications for land and water management and environmental stewardship today [84, 87, 95, 136]. The structure of contemporary CZs also preserves a record of the CZ history [8, 58, 105], providing important insights into the trajectories along which CZs evolve and the factors influencing this evolution.

The growing global network of CZOs is situated overwhelmingly in the Northern Hemisphere, primarily in North America and Europe (see Fig. 1). These observatories offer insights where the age of the CZ is constrained by the relatively recent formation

M. Leopold (✉) · L. Barton · J. Beringer · D. B. Gleeson · C. E. Moore · A. Rate · M. H. Ryan · T. Santini
School of Agriculture and Environment, Institute of Agriculture, The University of Western Australia, Crawley, WA, Australia
e-mail: matthias.leopold@uwa.edu.au

S. E. Thompson
School of Engineering, The University of Western Australia, Crawley, WA, Australia

T. White and A. Provenzale (eds.), *Critical Zone and Ecosystem Dynamics*, Advances in Critical Zone Science, https://doi.org/10.1007/978-3-031-69076-1_15

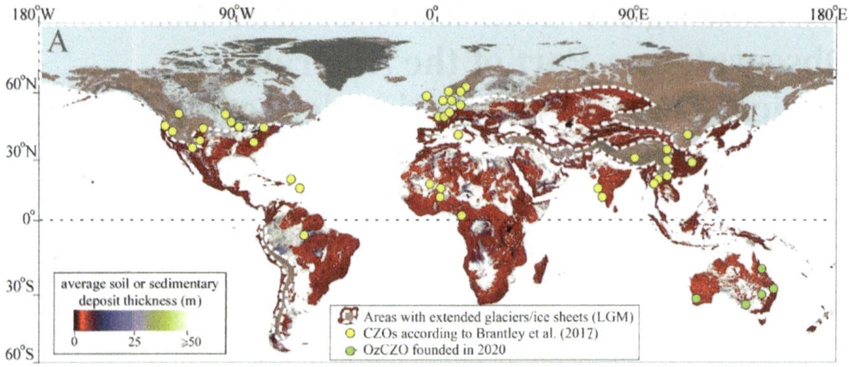

Fig. 1 Global Critical Zone Observatories (CZOs) and Australian Critical Zone Observatories (OzCZO) showing how sites are located relative to ice coverage during the last glacial maximum (dashed lines) and contemporary regolith depth (shading, after [108]). The Northern Hemisphere sites are biased towards previously glaciated locations, extended periglacial rims and areas of shallow regolith, while OzCZO adds to the deep regolith and unglaciated locations

of parent materials (e.g. recent uplift on the US West Coast, [62]), or glaciation during the last ice age that substantially removed (e.g. over the Archaean rocks of the Canadian shield) or reworked the CZ (e.g., in the US Midwest, [136]). The distribution of fresh non-weathered minerals via eolian processes and the subsequent incorporation via periglacial processes extend the spatial significance of the ice ages for CZ formation far beyond the boundaries of the former ice sheets and glaciers, especially in the Northern Hemisphere.

While older parent materials are represented in the Northern Hemisphere CZOs, e.g. Eocene parent materials in the Sierra Nevada [92, 106] and Mesozoic sites, such as Boulder Creek in the Rocky Mountains and Luquillo in Puerto Rico [9, 33], existing sites are biased towards parent materials from the Paleozoic or younger, and have often experienced significant recent disturbance through glaciation. Thus, the periods when the CZs at most CZOs evolved are constrained relative to the range of CZ ages on the planet (see Fig. 1).

One opportunity to broaden the length of the geological history sampled by CZOs is to expand the global CZO network into the Southern Hemisphere, which contains geologic parent materials spanning the Archaean to the Cenozoic. The Southern Hemisphere has experienced little glaciation relative to the Northern Hemisphere, offering the opportunity to learn about CZ development across the full extent of those timescales. Of the 45 CZOs worldwide mapped by Brantley et al. [28], only one is in the Southern Hemisphere (in Brazil). However, in 2020, a new foundational network of five CZOs was established in Australia [61].

The initial Australian Critical Zone Observatory Network (OzCZO) contains five sites (see Figs. 1 and 2): (1) Avon River, a mixed, rainfed agricultural site on the Archaean parent materials of the Yilgran Craton in the south-west of the continent; (2) Calperum, a dry woodland on Quaternary eolian dunes bordering the floodplain of the

Murray River in South Australia; (3) Wellington, a basaltic site sustaining a mixture of woodland and pasture located in the Barwon catchment near the northern extent of the Murray-Darling Basin in New South Wales; (4) Moreton Bay, on the outskirts of Brisbane, a tall eucalyptus forest growing over regional sandstones along river terraces in south-eastern Queensland; (5) Fletcherview, a sheep grazing operation on basaltic soils and bedrock bordering the Burdekin River in Queensland, the source of most sediments and nutrients exported to the Great Barrier Reef.

Globally, the OzCZO sites offer an important point of reference for comparing the Northern Hemisphere CZs. The Australian CZs represented by these sites—in the main—include old, deeply weathered and (often) chemically depleted soils. For example, at the scale of the Australian continent, the mean depth of the regolith (i.e., depth to unweathered bedrock) is between 16 and 25 m below the land surface, extending to depths in excess of 100 m over extensive parts of the continent [134] (see Fig. 1).

The OzCZO sites function today in generally hot and semi-arid, Mediterranean to semi-humid climates (see Table 1), exposed to extremes of rainfall intensity, fire and drought [39, 47]. Australian ecosystems are Gondwanan [82, 115, 129] with often unique and diverse endemic biology [121, 125, 131] that remains in the stewardship of Indigenous people over large areas of the continent. Several OzCZO sites—notably Avon River and Calperum—sit in locations of unusually high agreement amongst climate change models as areas where protracted drying trends have been predicted and experienced in the past five decades [68, 88, 123]. Two OzCZO sites (Avon River, Moreton Bay) are in biodiversity hotspots or priority areas for conservation and restoration [117, 135]. Therefore, the OzCZO sites are valuable as diverse natural laboratories to explore global change impacts on CZs. OzCZOs act as end-member

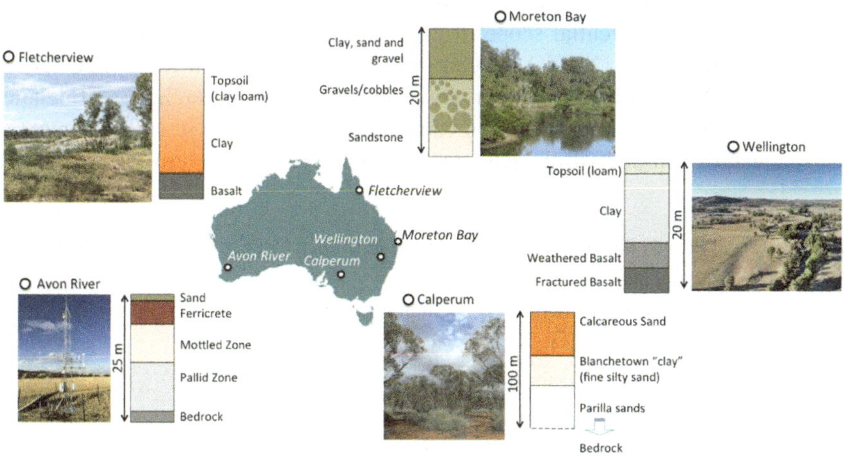

Fig. 2 Location of Critical Zone Observatories sites in Australia. Each site shows an indicative landscape photograph and, where it is known, the Critical Zone soil profile. The profile at Fletcherview is still under investigation

sites (Fig. 1), allowing robust tests of CZ theory developed in younger, cooler or wetter landscapes.

From an Australian perspective, the OzCZO sites offer an opportunity to develop local process knowledge and connect it to national environmental priorities, e.g. reducing terrestrial pressures on the health of the Great Barrier Reef [31, 42], understanding hydrological processes within Australia's major freshwater river system—the Murray-Darling [111, 132, 138], tracing the input and fate of carbon from ecosystems to soils to the groundwater and informing mitigation efforts such as 'carbon farming' [20, 128] and interrogating the different pathways by which disturbances such as drought, flood and fire are propagated, mitigated, amplified or otherwise interact with ecosystems and the CZ [39].

Locally, each OzCZO site creates opportunities to learn about CZ processes, linking them to Critical Zone Services [49, 102, 114] and drawing strategically on this understanding to inform environmental, agricultural, industrial and municipal actions.

Through the OzCZO programme, each site will have a standard set of equipment: an eddy covariance tower to measure energy, water and carbon fluxes, groundwater monitoring wells to measure water table dynamics and concentrations of solutes within and entering the groundwater, soil moisture sensors and a permanent electrical resistivity tomography (ERT) installation to observe and upscale soil moisture trends, and a vadose zone monitoring system to enable observation and sampling of water content, soil pore water and soil pore gases from undisturbed vertical profiles in the unsaturated zone, as illustrated in Fig. 3. Several sites (Avon River, Calperum, Fletcherview) have pre-existing ecosystem datasets, while Wellington has an existing detailed groundwater monitoring programme. Data from the sites will be made publicly available via the Terrestrial Ecosystem Research Network (TERN) Data Portal (https://portal.tern.org.au/#/f666a75d).

To illustrate the potential scope of the international, national and local opportunities created by the OzCZO network, we focus on the Avon River OzCZO site as a case study. At this site, we can draw on decades of research, recent detailed geophysical and soil characterisation campaigns, and the rich scientific context of the site in agricultural, hydrological, geological and ecosystem terms to:

Table 1 Location and climate characteristics of the Australian Critical Zone Observatories sites

Site	Latitude (degrees)	Longitude (degrees)	MAP (mm/yr)	Avg min T (°C)	Avg max T (°C)	Elev (m)
Avon River	−32.51	116.97	445	10.4	23.5	330
Calperum	−34.00	140.59	233	9.7	25.0	53
Wellington	−32.60	148.97	615	9.4	24.4	330
Moreton Bay	−27.27	152.98	1021	15.8	25.4	11
Fletcherview	−19.88	146.15	641	17.7	30.5	641

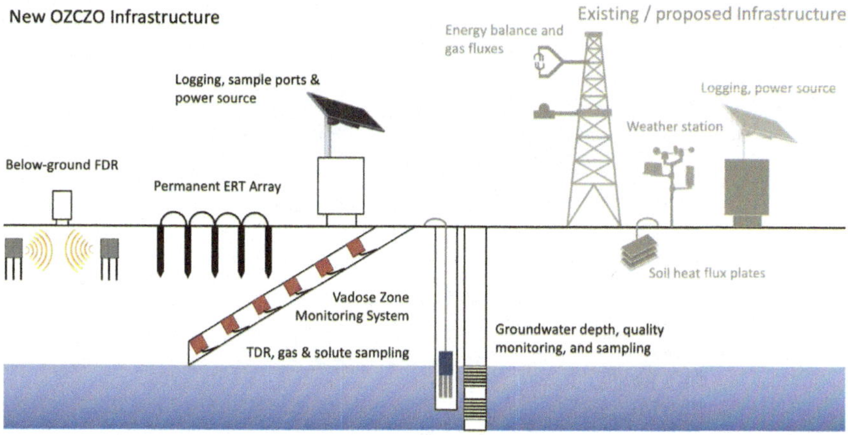

Fig. 3 Common infrastructure for all OzCZO sites

1. Illustrate how the contemporary CZ structure at the Avon River OzCZO site reflects and reveals a long history of landscape development and CZ formation.
2. Show how this history generated the features of today's CZ and its variation across the landscape.
3. Outline the profound implications of the contemporary CZ for sustaining agricultural practice, viable natural ecosystems and other CZ services in this landscape.

2 What Is the Avon River CZO?

The Avon River Critical Zone Observatory (AR-CZO) was founded in 2014 and added to the international CZO network (www.czen.org). The focal location for the AR-CZO is within The University of Western Australia (UWA)'s Ridgefield Farm (Fig. 4) (32.506° S, 116.991° E) near Pingelly in south-western Western Australia (WA). It now forms the westernmost node of the OzCZO.

The AR-CZO is characterised by a *ca.* 150-year-old recent history of European land clearing and agricultural land use, superimposed on *ca.* 40,000 years of Aboriginal land management, notably through fire management [130] and likely contributing to the extinction of megafaunal herbivores present in the south-west until approximately 45,000–43,100 years ago [2, 76]. The AR-CZO is in the south-western botanical province, a region known for its significant levels of floral endemism and biodiversity, and identified as a global biodiversity hotspot [56]. The regional topography is low relief, undulating hills of a dissected landscape, the climate Mediterranean, and the geological parent materials are granitic and extremely old with occasionally younger dolerite dykes. The site is on the Yilgarn Craton, for which the main intrusive emplacement of the protolith dates back 2.6 to 3 billion years B.P. [94]. It lies

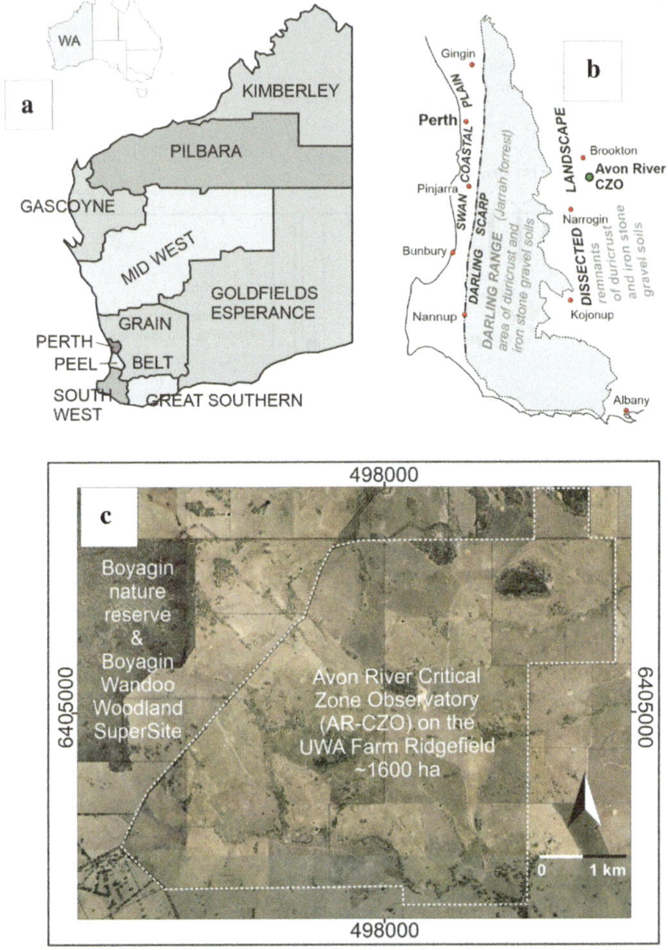

Fig. 4 **a** Landscape units of Western Australia (WA). The Avon River CZO (AR-CZO) is within the WA grain belt. **b** Major geological formations in south-western WA. The Darling Range is capped by iron-rich laterites that extended to the east until dissected and largely removed by fluvial erosion. **c** AR-CZO lies in the dissected landscape and is part of The University of Western Australia (UWA)'s Farm Ridgefield located close to the Boyagin nature reserve, which hosts the Boyagin Wandoo Woodland SuperSite as part of the Australian Terrestrial Ecosystem Research Network (TERN). The nature reserve was not cleared by European settlers and is a comparison site for the UWA Farm Ridgefield due to its contrasting land-use history

within the Avon River Basin, a 125,000 km^2 river basin extending approximately 400 km inland from the Indian Ocean [21]. The AR-CZO lies within the 'zone of mature drainage' in the Avon Basin [25, 69, 70, 97], a region bound by the 650 and 450 mm/yr rainfall contours [109], with an undulating landscape, broad valleys and moderate channel gradients (approximately 1 m/km, [21]). West of the AR-CZO landscape relief, channel incision and slopes increase with increasing annual rainfall, while east to below the 450 mm/yr rainfall contour, the landscape is low relief and poorly drained, with channel gradients as low as 0.2–0.3 m/km [21]. The eastern 40% of the basin is typically disconnected from the western region for years at a time, only contributing flow during periods of exceptional inland flooding [12]. Near the AR-CZO, today the channels are mostly ephemeral [12, 109], displaying extremely low specific discharge, e.g. annual drainage in several inland sub-catchments is 0.1–0.24 mm (relative to 350 mm precipitation) [66].

As illustrated in Fig. 5, these features mean that the AR-CZO lies towards the extremes of the key axes of topography, lithology, disturbance, time, climate and biology identified as determining CZ architecture and function [13], with implications for hydrology, ecosystems, agriculture and other CZ services.

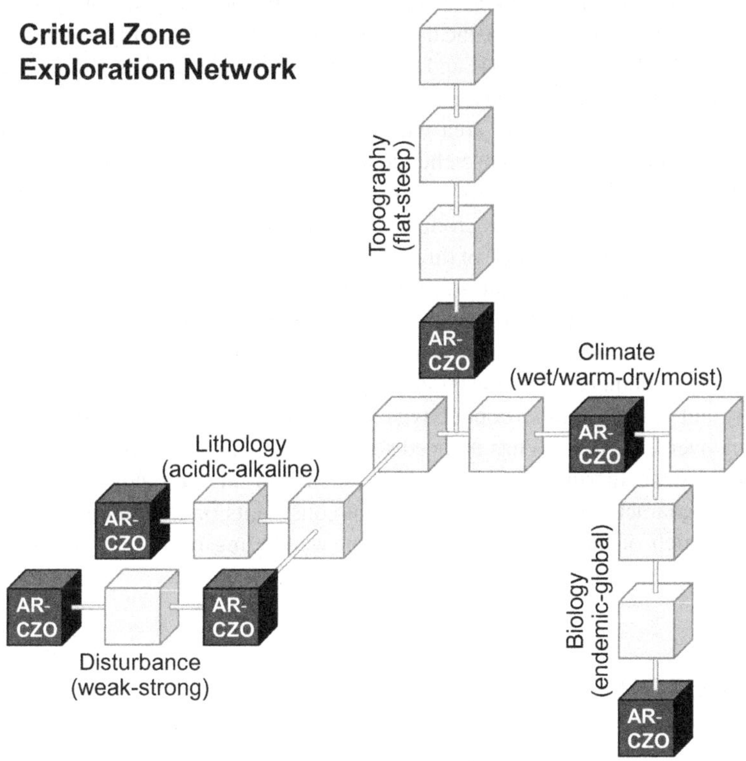

Fig. 5 Position of the Avon River Critical Zone Observatory (AR-CZO) in the global network. Its ancient weathering profiles, recent land-use history, pre-colonial aboriginal fire management, low topography and endemic biota often set the AR-CZO at the extremes of the global gradients as indicated

The central AR-CZO site is on the western edge of the agricultural region known as the WA grain belt (or wheatbelt). The varying topography, geology and weathering products lead to a diverse pattern of agricultural land use, including cropping and grazing paddocks, often in rotation, and remnant vegetation—enhanced at UWA's Farm Ridgefield by various experimental and revegetation land use treatments [58, 110].

Like much of south-western WA (Fig. 4), many parts of the AR-CZO is mantled by deep regolith, formed by intense subaerial weathering of the underlying granites since the mid-Proterozoic [7, 55, 78] without interruptions by marine transgressions. The underlying bedrock is the tectonically stable Yilgarn Craton, an Archaean igneous and metamorphic complex. Deep weathering of the Yilgarn results in an iron oxide-enriched, hard duricrust (ferricrete) near the surface overlying an intensively leached quartz- and kaolinite-rich horizon, known as the pallid zone (see Fig. 6). Additionally, pisolithic plinthisols [10], characterised by an abundance of ironstone gravels, are frequently found in areas where the duricrust was eroded or not fully formed. Anand and Paine [7], McCrea et al. [91] and Gilkes et al. [55] provide detailed descriptions of the nature and weathering layers of the regolith (Fig. 6).

The AR-CZO site is in an area where most of the original weathering profile has been eroded variably. Intense chemical weathering depleted most nutrients and other mobile elements, leading to kaolinisation and ferruginisation [55, 91]. Fluvial, aeolian and colluvial processes eroded parts of the weathering profile leaving truncated soils behind and a mixture of fluvial, colluvial and eolian processes re-deposited the eroded material along the slopes and valley floors [90]. These transported ancient materials are the parent materials of many contemporary soils, which inherit several legacies from this ancient parent material. The soils are highly depleted in nutrients and other elements. Kaolinisation has produced subsoils with significant microporosity but low hydraulic conductivity and high bulk density and acidity. Poor leaching over long periods has induced soil salinity from accumulated solutes from meteoric and weathering sources. Water repellency in surface sands is prevalent [3]. These soil characteristics present significant constraints on agricultural production and owe much of their origin to the ancient weathering profile and its formation processes [6, 90, 98].

3 AR-CZO Architecture

Critical Zone architecture comprises the physical, chemical, biological and anthropogenic structure of the CZ, while CZ processes highlight the input, storage, transportation and transformation of energy and matter (e.g. water, carbon, nutrients and solutes) within the CZ. Our understanding of CZ architecture and how it influences or is influenced by CZ processes at AR-CZO is still developing, but some key areas are outlined here.

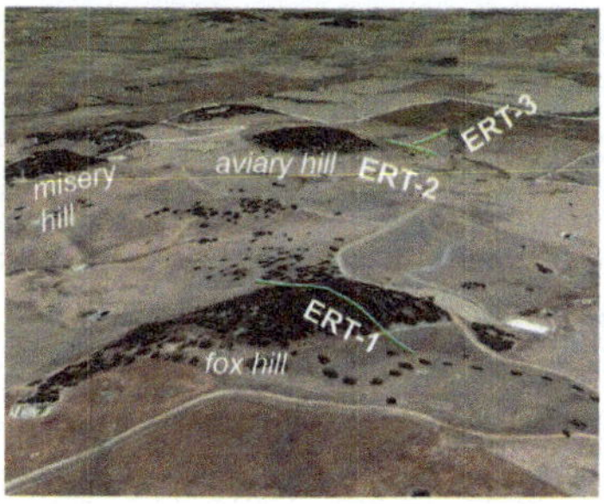

name	length [m]	electrode separation [m]	# of iterations	absolute error [%]
ERT-1	400	5	5	8.1
ERT-2	180	4	6	6.6
ERT-3	400	5	6	6.1

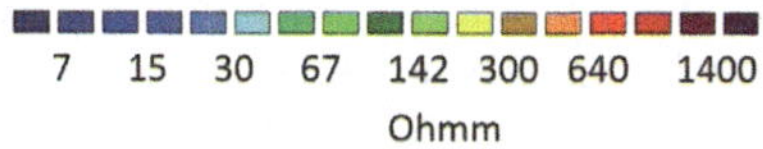

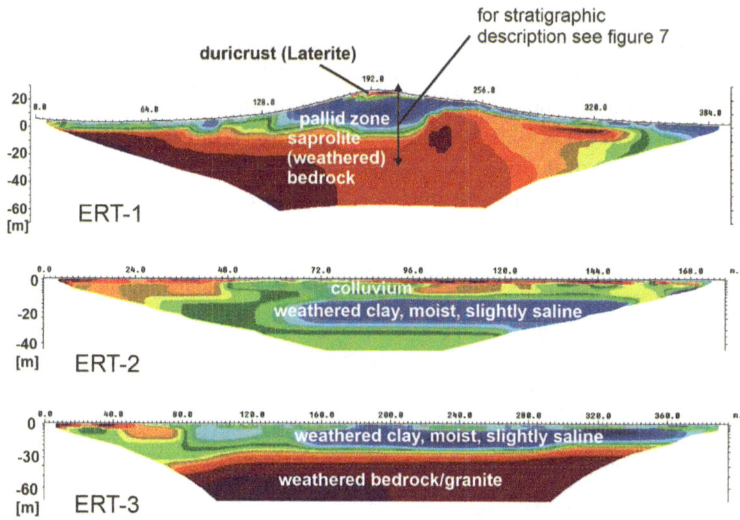

Fig. 6 Top left, 3D view of the remnant lateritic hills within the general dissected landscape of the western part of the Western Australian grain belt (white line represents farm boundary with paddock names for hills in italics). Indicated in green are the locations of three electrical resistivity tomography (ERT) lines, with ERT-1 crossing a lateritic hill with remnant native vegetation and ERT-2 and ERT-3 portraying subsurface information of accumulated and deeply weathered profiles currently used by agriculture. Top right, basic parameters of ERT lines and data inversions with legend for colour-values (Ohmm) for the three 2D-ERT lines on the bottom

3.1 Geophysical Architecture

Geophysical architecture at the AR-CZO has been investigated using ERT and shallow seismic refraction in two locations: (1) across one of the remaining lateritic hills with exposed duricrust; (2) in a cross-section in the toe-slope of such a hill. The measurements overview the general subsurface down to the bedrock (Fig. 6).

The critical zone architecture at the AR-CZO provides in-situ weathered laterite profiles that are well preserved by a hard, 1–2 m thick iron crust (duricrust) followed by deeply weathered and chemically depleted material (pallid zone/saprolite; Fig. 7). These materials can be delineated by ERT (ERT-1, Fig. 6) with the dry duricrust and its multiple voids resulting in high specific electric resistivities (up to 2000 Ohmm), whereas the clay-rich, partly salt-accumulating pallid zone displays very low specific resistivities of less than 20 Ohmm. The weathering front is clearly visible by a large gradient over small distance where the resistivity values increase from 20 to 1500 Ohmm and above, indicating a change in physical conditions that aligns with weathered granite bedrock before specific resistivity values further increase corresponding with bedrock. These lateritic hills are typically characterised by remnant vegetation, not used agriculturally due to the stored salinity in the pallid zone, and thus often fenced off to avoid erosion.

In contrast, transects ERT-2 and ERT-3 (Fig. 6) represent typical areas with agricultural land use. They are part of the dissected landscape where erosion and redistribution of the deep weathering profile materials lead to a complex mosaic of sediments forming the CZ. ERT-2 had a depth of exploration of roughly 40 m without detecting a signal from bedrock. ERT-3, which is 30 m downstream of ERT-2, seems to indicate bedrock at about 30 m depth at the start of the transect, decreasing to 20 m at the end of the transect. Undulating depths to bedrock are characteristic of the area, with

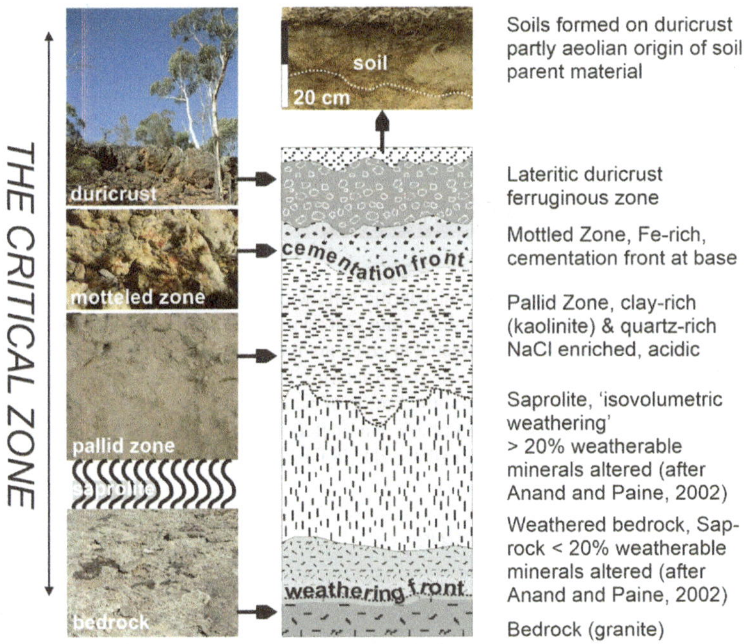

Fig. 7 Overview of the critical zone architecture at the Avon River Critical Zone Observatory (AR-CZO). Left, images of material found in outcrops and eroded sites along a remnant lateritic hill. Middle and right, conceptualised weathering zones at the AR-CZO with generalised descriptions following Anand and Paine [7]

bedrock outcrops being a frequent landscape feature. Substantial CZ architecture changes over short distances and the deeply weathered, chemically depleted profiles contrast with many other CZs in the Northern Hemisphere that are often only a few metres deep and more uniform. In addition, saline soils and groundwater frequently occur in the ancient weathering profiles of the AR-CZO, adding to the complexity and novelty that the OzCZO can provide to the global network.

3.2 Geochemical Architecture

The intense past weathering of the CZ depleted macronutrients (and many other elements) and enriched mainly Fe- and Al-oxides, altering the chemical properties through depletion and enrichment processes and the physical properties by cementing near-surface layers.

Absolute mass fluxes of elements [29], relative to bedrock, were calculated as negative (loss) for Si, Na, Mg, K, Ca and P for near-surface samples corresponding to exposed laterite, as seen in Fig. 8, corresponding to ongoing chemical depletion. Assuming conservative behaviour for Zr, several other elements (Al, Fe, Mn, Ti) varied in $\tau_{(Zr,\,j)}$ values with depth, with a tendency for Al and Fe to show net addition in near-surface regolith locations corresponding to the presence of ferruginous nodules or ferricrete.

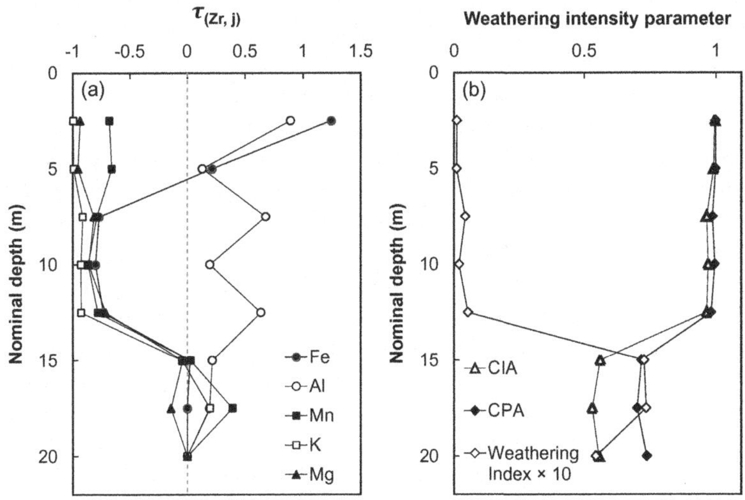

Fig. 8 **a** Absolute mass flux ($\tau_{(Zr,\,j)}$) estimates for Fe, Al, Mn, K and Mg generated from elemental analyses of near-surface samples collected at landscape elevations corresponding to original laterite profile horizons exposed by denudation. Values of $\tau_{(Zr,\,j)}$ for Ca and Na are similar to those for K, and thus omitted for clarity. **b** Estimates of weathering intensity: CIA = Chemical Index of Alteration [99], CPA = Chemical Proxy of Alteration [32] and Weathering Index [37]

This chemical architecture is similar to other well-known lateritic profiles in southwestern WA, e.g. the Jarrahdale Railway cutting site [30, 46] or the Boddington Gold mine [71] despite the AR-CZO's location east of the main laterite and ferruginous gravel soils (Fig. 2) of the region. McCrea et al. [91] pointed out that variability within the weathering zones will influence pedogenesis, as weathered materials are eroded and distributed in the landscape. The lateritic duricrust and the aggregation of ferruginous gravels are two examples of secondary alterations due to weathering, which significantly affect the physical properties of soils and their potential land uses.

Although the in-situ weathered and preserved laterite profiles provide insight into the history of weathering and landscape development at AR-CZO, they are limited in their extent in the present landscape because lateral erosion of these profiles produced the dissected contemporary landscape. The AR-CZO is thus characterised mostly by truncated laterite profiles, where subsequent eolian and colluvial processes deposited mainly sandy sediments that form the contemporary topsoils.

4 Biogeochemical Processes

The exposure of different lateritic profile horizons at the AR-CZO results in variations in local soil properties so that soil pH, salinity, water repellency, and texture vary across soils derived from different profile parts. These differences in soil properties are a major determinant of surface (0–10 cm) soil microbial community composition under remnant vegetation and managed permanent pasture [58, 105]. However, compositional differences in the microbial community do not always translate to functional differences. For example, although soils under agricultural land use had higher abundance of ammonia-oxidising bacteria and archaea than natural sites, the rates of microbial N cycling in response to the addition of N compounds were not distinguishable between sites [58]. This suggests that differences in community composition may be a mechanism to compensate for environmental conditions impairing microbial efficiency or functional capacity. Thus, functional capacities such as N turnover rate or CO_2 (respiration) evolution rate may appear static across different environmental conditions and community compositions. However, management practices will likely influence functional capacities and microbial community compositional differences. For example, N-rich fertilisers applied to crops and N-fixing pasture legumes may promote the proliferation of ammonia-oxidising bacteria and archaea and allow them to be relatively efficient in N use and cycling. A corollary of this hypothesis is that similar microbial community compositions might result in different functional properties due to the mediating effect of environmental conditions. The co-variation of soil properties and community composition highlights the importance of combining soil environment, microbiological compositional and functional studies to unravel the processes of element and nutrient cycling in the AR-CZO.

Within the CZ, soil as a habitat exhibits extreme spatial variability, and adjacent microhabitats can differ widely in their physico-chemical properties, being influenced by various factors, including mineralogy [58–60]. For the soils surrounding the AR-CZO, this influence of mineral chemistry is not restricted to oligotrophic rock surfaces [58–60] but seen in nutrient-depleted soils where mineralogy can exert a significant impact on the microbial community structure [36, 34]. The heterogeneous distribution of soil minerals may increase the variation between soil microhabitats and contribute to the wide spatial variation in microbial communities observed in these habitats—how this manifests within the AR-CZO is unclear; however, it is likely an important consideration given the nutrient status of these soils. Soil water and pore connectivity also play an important role in regulating the development of spatially distinct microbial populations and thus dictating the magnitude of soil microbial diversity. In other parts of the WA grain belt, decreasing soil water potential, resulting in drier soil, increased the diversity of complex bacterial communities, which was linked to a decrease in soil pore connectivity and thus microbial competition [35]. Additionally, bacterial diversity was low when the water potential was below field capacity, suggesting low pore connectivity increases bacterial diversity in most soils under field conditions [35]. How this plays out at the AR-CZO is unknown; however, it is likely important given that future climate changes will modify rainfall patterns, altering the frequency of periods of high pore connectivity that influence soil microbial diversity.

An open question remains on how patterns in microbiological community composition and function observed in surface soils at AR-CZO might translate across depths. The highly weathered CZ at the AR-CZO presents a paucity of minerals and organic matter from which soil microorganisms can extract large amounts of energy and carbon, particularly with minimal organic matter at depth. Layered onto the physical and chemical variations across soil horizons, we anticipate contrasting metabolic capacities and functions with depth at AR-CZO relative to younger sites that could sustain microbiological communities at depth with relatively high organic matter availability, high potential of other energy sources (e.g., chemical) in the profile, and overall weaker and shallower horizonation. For similar reasons, the AR-CZO might support a less abundant and diverse lithotrophic community due to the depletion of inorganic energy sources in the profile relative to younger sites. These highly weathered and nutrient-deficient soils are also likely to result in a dominance of ammonia-oxidising bacteria over ammonia-oxidising archaea, which could have implications for N cycling across the CZ [58, 75]. Thus, when comparing geologically younger sites (e.g., the Shale Hills CZO) or sites with landscape rejuvenation (e.g., IML CZO), the AR-CZO offers an excellent 'old' end-member for comparative soil microbiological studies that target understanding spatial and depth variations in community composition and function, and the subsequent consequences for soil formation, nutrient cycling and soil health.

Further investigation is required into how microbiological community composition and function influence greenhouse gas fluxes within the AR-CZO. Annual fluxes calculated from sub-daily measurements of nitrous oxide (N_2O) emissions from arable soils in other parts of the surrounding WA grain belt suggest relatively low emissions from N-fertilised cropping soils in the AR-CZO compared to global losses. For example, annual emissions ranged from 0.02–0.27 kg N/ha from cropping soils in the WA grain belt, representing less than 0.06% of N fertiliser applied [15–17]. The greatest fluxes occurred in response to short-lived but intense summer rainfall events rather than indirect responses to synthetic N fertiliser applications. Summer rainfall events in the WA grain belt, when agricultural soils are fallow, stimulate soil N mineralisation and mineral N availability to soil microorganisms responsible for producing N_2O emissions but with no active plant growth to use the mineralised N. Nitrous oxide emissions will likely vary within the AR-CZO's cropping soils in response to changes in soil organic C [15] but remain relatively low due to low soil organic C concentrations (compare [3, 118]). In some instances, annual N_2O emissions decreased by incorporating grain legumes in the cropping rotation (thus lowering the use of synthetic N fertiliser) or were offset by CH_4 oxidation (uptake) [19, 15]. Unlike temperate regions, manipulating current N fertiliser management strategies is not expected to influence N_2O emissions greatly as N fertiliser inputs are modest, and applications are 'split' during the growing season to align with expected crop N demand.

5 Hydrology and Anthropogenic Change

The woodland ecosystems endemic to the AR-CZO and WA grain belt are characterised by perennial sclerophyllous tree species, often having dimorphic root systems, with shallow roots that are dormant during dry periods but able to rapidly respond to shallow rain events and uptake nutrients from the surface profile and deep tap roots maximising the tree's ability to use deeply percolating winter rains. Together with the gentle topography and limited surface drainage of the Avon Basin it leads to conditions which produce a natural water balance between vapour fluxes (evaporation and transpiration) and annual rainfall input [51, 66]. Sea salt, aerosolised into the wet air masses over the ocean, was deposited into the CZ by rain as meteorological systems moved west–east across the landscape [67, 79]. The salt was flushed into the profile and concentrated by evaporation and transpiration below the woodland rooting zone [51, 66]. Salt concentrations in the CZ of the WA grain belt are high—in a separate catchment, estimated salt loads ranged from 300–1300 g/m^3 in the top 3 m to 9000–24,000 g/m^3 further below [119], comparable to seawater.

European colonisation of the WA grain belt from 1850 onwards led to the extensive clearing of the woodland and replacement with annual crops and pastures that are only present during a distinct autumn to early summer growing season, typically between April and November. The shallow root systems of the annual crops and pastures relative to perennial woodland species allowed more winter rainfall to percolate to depth. Fallowed fields, bare of all plant growth, did not transpire summer rainfall. Consequently, evaporation and especially transpiration decreased, and recharge increased. Rising groundwaters mobilised salt from deep in the regolith and transported it closer to the surface. Salinisation of soil and waterways is now threatening ecosystems and agriculture across the WA grain belt [5, 52], although dryland salinisation at AR-CZO remains mild.

The threat of salinisation is prompting agricultural and land management interventions relevant to CZ science, including exploring the productive use of saline land and water by evaluating new agricultural industries centred on perennial pastures, including saltland pastures [74, 96, 103, 113, 127], and efforts to integrate perennial vegetation, with a preference for local plant types [110], into productive agricultural systems [26]. Efforts to identify techniques to promote revegetation and understand the impact of revegetation on agroecosystem functioning are active at the AR-CZO site.

The presence of deep saline groundwater within the regolith also prompts questions about what constitutes the bottom boundary of the CZ in these systems—while the bedrock may remain the boundary by definition, the saline, anoxic, stable groundwater that overlies bedrock at AR-CZO and many parts of the Yilgarn are obstructing most forms of activity. Functionally, the transition to these groundwaters may mark the bottom of the dynamic CZ in these landscapes, which should be explored across the OzCZO.

6 Ecosystem and Agroecosystem

Ecosystem-scale measurements of plant–atmosphere interactions are ongoing in the AR-CZO catchment. These measurements, collected using the eddy covariance technique (OzFlux GCB paper), offer insight into climate- and land-management-driven variability in carbon, water and energy fluxes from ecosystems over time. Two ecosystem observatories exist in the AR-CZO at the UWA Ridgefield Farm site and the Australian TERN [39] Boyagin Wandoo Woodland SuperSite (Figs. 4 and 9). The UWA Ridgefield Farm has largely been cleared of native vegetation and consists of paddocks of annual crop and pasture systems (see Sect. 2), while Boyagin represents remnant *Eucalyptus wandoo* woodlands native to the area. Both sites are part of Australia's OzFlux network [24] and contribute open-access data to the global FLUXNET network [107].

The rationale for establishing these two observatories was to determine the balance of environmental demand for water, agricultural productivity, carbon storage and biodiversity from contrasting land uses within a catchment landscape. For example, over four years (2018–2021), there was a >70 mm difference between evapotranspiration at the two sites, with the Boyagin Wandoo Woodland SuperSite using 372 $\pm$ 29 mm/y and UWA Ridgefield Farm using 301 $\pm$ 29 mm/y (Fig. 9). Both sites did not use water in excess of mean annual precipitation (475 $\pm$ 75 mm), suggesting the excess remains available for replenishing groundwater stores or running into local and regional catchments and filling local dams for animal production. The establishment of CZ groundwater monitoring at these sites will determine differences in water table dynamics and the concentration of solutes between these two contrasting land uses.

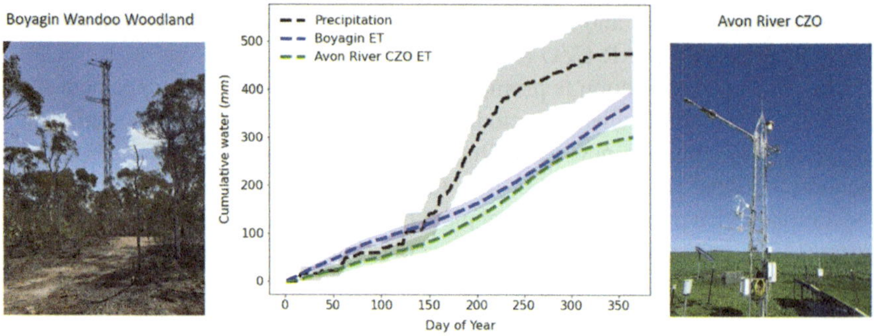

Fig. 9 Annual cumulative evapotranspiration (ET) and precipitation (middle panel) averaged over four years (2018–2021) at the Avon River Critical Zone Observatory and nearby (~4 km) Boyagin Wandoo Woodland SuperSite. ET was measured using the eddy covariance technique at each site, and data were processed using standardised routines described by Isaac et al. [73]

7 Agricultural Management

The dryland cropping systems in the WA grain belt, typical with those across southern Australia, are situated on land cleared of native perennial vegetation and resown mostly to exotic annual crop and pasture species with an autumn to early summer growing season when temperatures are low and rainfall is relatively high [81]. Some perennial pastures are also present. Many farms support mixed cropping and livestock enterprises, with paddocks often rotated between crops and pastures. Free-range livestock enterprises are based on ruminants, predominantly sheep, producing wool and meat. Livestock graze pastures over winter and graze dry stubbles or receive supplementary feed over summer and autumn when paddocks generally contain no living plants. Situated within the WA grain belt, the UWA Ridgefield Farm site hosts a mixed farming enterprise reasonably typical of farms in the region, with paddocks rotating between crops, mostly wheat (*Triticum aestivum*) and canola (*Brassica napus*), and self-regenerating pastures containing subterraneum clover (*Trifolium subterraneum*) and other grasses and broadleaf weeds, notably capeweed (*Arctotheca calendula*).

The low native soil nutrient status has shaped the development of WA farming systems. Soil nitrogen depletion in the original cereal-dominant systems catalysed the adoption of pasture legumes, particularly subterranean clover, and the development of a ley system where self-regenerating legume-based pastures were grown in short rotation with cereal crops [112]. The development of phosphorus fertilisers enabled the adoption of subterranean clover. However, across WA, the cropping frequency increased and the pasture frequency decreased due to shifts in enterprise profits, cheaper manufactured N fertilisers and the introduction of canola as a disease break in cereal rotations [81, 122], resulting in more intensive, cereal-dominated, higher-yielding systems, often with a reduced livestock enterprise. Crop and pasture legumes play a relatively small role in these systems [64], while N fertiliser application is routine [65]. Minimum tillage has been widely adopted [86] but has contributed to the development of herbicide-tolerant weeds (including on the UWA Farm Ridgefield) [104].

Even the best-managed WA winter wheat systems return modest yields by global standards, e.g. 2.39 t/ha projected for 2021–2022 [1] compared to global estimates of 3.5 t/ha for 2018 [116], reflecting soil and climatic constraints, notably the relatively low and variable rainfall. Achieving close to potential yields requires active management of the constraining topsoil and subsoil factors of the CZ. Improved water use efficiency is crucial [80] but is influenced by many factors, including summer fallow management, cultivar choice, crop agronomy, and soil physical and chemical constraints that require, for example, tailored application of lime (for pH control) and gypsum to mitigate sodic soils [63]. Crop nutrition programmes are based on soil and foliar testing and application of macronutrients, particularly nitrogen, potassium and phosphorus, and micronutrients, especially sulphur. Agricultural soils in the WA grain belt have become enriched in some nutrients due to fertiliser application, notably phosphorus, but the application of other nutrients often does not replace the nutrients removed in crop production [65]. At the AR-CZO site, soil nutrient levels and the soil microbes directly impacted by these, such as arbuscular mycorrhizal fungi [118], vary greatly among paddocks; soil pH is often low and liming may be required [118].

In the WA grain belt in general and AR-CZO specifically, other constraints exist, including various environmental challenges outlined in Sect. 2.6. A current high-profile issue is the need to understand the 'carbon intensity' of agricultural production. Encouragingly, these constraints continue to prompt experimentation with new approaches to land management, including improving agronomy, increasing complex technologies associated with precision agriculture, releasing better-adapted crop and pasture varieties and, most importantly, capturing the synergies between these by improving our understanding of the whole farming system [72, 96, 100]. More transformational ideas [23], include (1) revegetating agricultural land with native vegetation [137], (2) strategically including native ephemerals into summer fallows [101], perennial species to mitigate ruminant methane emissions [85] or benefit livestock health [83], or saltland pastures (see above), and (3) creating new agricultural industries based on woody crops [14, 22]. As drying climates place even greater constraints on agricultural systems, innovation in land management and land industries in the region will remain a high priority [120]. The AR-CZO will tackle these agricultural land use challenges via a multi-methodological and multi-disciplinary ecosystem approach.

8 CZ Services at AR-CZO

Critical Zone Services consider how the CZ constrains or enables ecosystems to provide the life-sustaining functions of ecosystem services [40, 49]. Compared to typical ecosystem services approaches, valuing CZ services emphasises belowground processes, slow processes and the coupling of biological systems to physical systems.

Adopting a CZ services perspective may help address some limitations of conventional ecosystem service frameworks that have struggled to capture the importance of some environmental features, such as soils and soil health (e.g. [4, 45]).

Considering the AR-CZO through a CZ services perspective strongly reinforces the importance of combining biotic, chemical and physical processes from the canopy to the groundwater. For example, the loss of hydrologic equilibrium following land clearing emerges as large changes in groundwater depth, expressed over ~30 m deep profiles. This provides a simple example of the whole CZ responding to a disturbance imposed at the surface and the propagation of the disturbance with changing water and energy balances with depth. The AR-CZO illuminates how deep-time histories produce legacies that emerge in terms of contemporary CZ services. The limitations and challenges of the CZ at AR-CZO arise from its ancient origins and long history of depletion and transformation. This history generated a contemporary CZ architecture that now constrains the microbiological community location, composition and function and thus soil creation, nutrient cycling and fertility, impacting agricultural land use. Importantly, the AR-CZO reflects its unique environmental, cultural and social context in an area where environmental threats, including salinisation and rapid climate change (heating and drying) can feel omnipresent. Restoration and innovation around land management in this context are valuable psychologically, socially, environmentally and economically [48, 124]. The CZO is on land with significant spiritual value to traditional owners, with much to learn by integrating Indigenous knowledge and values with scientific questions and European/migrant Australian perspectives [11, 93]. The deep-time history of the CZ can also be profoundly moving from a geomorphic point of view. To stand at the base of a contemporary duricrust capped hill and look at its surface 30 m higher, and appreciate that the top of the hill is the remnant of a former stream bed (so-called inverted topography) is a striking experience, one that brings abstract notions of landscape evolution over long timescales home in a visceral fashion. These features of AR-CZO offer exciting pathways for involving educators and researchers to enhance CZ service provision.

However, the AR-CZO also presents some challenges for CZ service provision. The properties of the AR-CZO CZ are challenging to ecosystems and to human use of the landscape. Ecosystem services are obtained from this landscape through finely tuned adaptations to its challenging properties (in the case of endemic ecosystems) or conscious manipulation and attempts to change those properties in the case of agriculture. The AR-CZO CZ is also not necessarily benign. Its CZ contains the ‘hidden menace’ [53] of salinity, increasing soil acidity and continuous soil carbon loss. In undisturbed conditions, the CZ at AR-CZO offered a ‘protective’ service that limited salinisation. However, this is not inevitable (and naturally occurring salt lakes and salt scalds cover much of the landscape in the ancient drainage zone). The Australian CZ can be obstructive to certain ecosystems. The potential for CZ disservice is not limited to salinity; chemical ‘menaces’ also lurk in many CZs worldwide, including serpentine soils [77], acid sulfate soils [27], the intriguing cases of ‘poisonous rocks’ [43]. These ‘menaces’ are widespread, e.g. ultramafic rocks generating serpentine soils cover some 1% of the global land surface [77], and acid sulfate soils represent some 3% of Australia's land surface [50]. Indeed, acid sulfate soils in the AR-CZO region are probably related to waterlogging/salinity and sulphate in the groundwater [44].

Like acid sulphate soils and salinity, these CZ features may restrict plant and microbial life only allowing heavily specialised organisms to survive, such that disturbing the co-evolved CZ and its ecosystem is profoundly destructive. These features may even preclude significant ecosystem formation [43] or generate hazards within the ecosystem; for example, plants becoming toxic due to heavy metal uptake [126]. Other forms of disservice, such as the propensity for land subsidence or landslides, should also be recognised as CZ properties and reconciled with the concept of CZ services (Table 2).

9 Conclusion

As part of a new network of Australian CZOs, the Avon River CZO illustrates how today's CZ can represent the outcome of many millions of years of landscape evolution. Formed by the transformation and transport of primary weathering materials, which are now present only in a few isolated hills on the landscape, the deeply weathered profile is poorly drained, highly depleted in nutrients and most minerals, and contains large stores of saline and acid weathering and deposition products. The history of the CZ emerges in the spatial structure of the landscape and the characteristics of the soil and the microbiological assemblages within it. It produces a template of agricultural fields of varying agricultural productivity, determining where different agricultural activities (e.g., water harvesting versus revegetation planting versus perennial pastures vs. rotation of annual crops and pastures) can occur. In equilibrium with the natural vegetation cover, the CZ sustains ecosystem services because the vegetation is adapted to its architecture and processes. In the absence of that vegetation cover, novel agroecosystems require significant management, the landscape hydrology forms a disequilibrium, and the threat of salinisation and acidification increases—mitigated in AR-CZO mostly by the relatively wet conditions that produced an undulating rather than topographically flat landscape. As outlined, AR-CZO presents exciting opportunities for research within the global community, such as (1) pedological and focused on laterite formation and weathering; (2) microbiological and focused on the deep vertical profiling of microbial communities, function and activity; (3) ecosystem and agroecosystem oriented, given the potential for paired-site comparisons between AR-CZO at the UWA Ridgefield Farm and Boyagin Wandoo Woodland SuperSite of the TERN network; (4) climatic, given the clear non-stationarities in the climate; (5) management-oriented and exploring the CZ implications of revegetation and sustainable agricultural practices. AR-CZO presents an end-member along key CZ axes. It provokes questions about the definition of a CZ and how the CZ extent relates to CZ dynamics at sites where low mobility, low reactivity groundwater occurs at higher elevations than bedrock. It also suggests critical examination of how to apply a CZ services framework at sites

Table 2 AR-CZO Critical Zone Services—processes involved and relevance of the AR-CZO architecture for those processes

	Processes	Relevance of AR-CZO architecture
Provisioning services		
CZ water storage (water supply and quality)	Surface soils partition seasonal runoff Surface runoff to dams/creeks provides summer water resources but can cause winter waterlogging Recharge to groundwater elevates (saline) water tables	Kaolinite-rich 'pallid zone' profile promotes seasonal perching and runoff Long-term vegetation change/climate change increase/decreases recharge
Food supply (food security)	Crop and livestock production are constrained by bulk density, sandy topsoils, low water retention, salinity, sodicity, acidity and nutrient depletion deep	Heavily weathered ancient CZ depleted in nutrients, low pH Low relief, highly-eroded landscape and clay subsoils inhibit drainage of saline waters
Habitat or supporting services		
Primary production (cereals, timber, pastures and livestock)	Maintain saline groundwater at depth Endemic vegetation adapted to depleted soil conditions Anthropogenic management (gypsum, lime, fertilising ripping)	Well-adapted vegetation efficiently uses water year-around Depletion (nutrients) and accumulation (acid-generating, saline) create a challenging profile for agriculture
Soil formation (water supply and quality, nutrient and carbon storage)	Microbiological activity is highest in surface soils and likely declines rapidly with depth Limited lithotrophic processes Decelerated pedogenesis	Soils formed by secondary weathering of primary weathering products Microbiological community structured (and possibly limited) by exposed horizons Fire disturbance likely contributes to weathering/erosion and soil formation

(continued)

Table 2 (continued)

	Processes	Relevance of AR-CZO architecture
Nutrient cycling (support of primary production, risks of eutrophication)	Slow and conservative (highly efficient) nutrient cycles to retain limited nutrients in undisturbed soils under natural vegetation Carbon and energy source availability likely limits nutrient cycling rates and magnitude of microbial contributions to bedrock weathering Oligotrophic aquatic ecosystems adapted to depleted soils Removal of nutrients in produce, especially grains	Deep soils limit surface carbon inputs (plants, fauna) from reaching bedrock weathering zones Likely sharp changes in microbial community composition (and potentially function) among horizons, given marked differences in horizon properties Adaptation to low nutrient conditions may cause high risk of excess nutrient losses → eutrophication
Regulating services		
Water quality regulation (water supply and quality)	Saline and acid groundwater poses a major water quality threat	Hydrological balance between evaporation and precipitation in poorly drained landscape
Water supply regulation (regulation of hydrological flows, creation of aquatic habitat)	Surface flooding is rare but can be persistent when occurring Winter flow regimes dominate—groundwater does not discharge into streams	Low relief, groundwater rise can produce long-lasting flooding in response to rare extremes Low waterholding capacity → flashy and ephemeral surface flows
Gas regulation (regulation of atmospheric chemical composition)	Expand shorter-term focus on plant and microbe responses to include longer time-scale constraints	Greenhouse gas regulation (e.g. mineral weathering rates)
Climate regulation (regulation of local temperature, precipitation)	Hypothesised alterations in climatology following tree clearing Hypothesised increases in surface temperatures following clearing	Phenology and root depth alter transpiration patterns; changing canopies alter roughness, energy partitioning, convection
Cultural services		

(continued)

Table 2 (continued)

	Processes	Relevance of AR-CZO architecture
Recreation (providing a platform for recreational activity)	Geotourism (e.g. Wave Rock) already takes place regionally Wildflowers/ecotourism could be linked explicitly to CZ constraints	Geologic aesthetic value—granite outcrops, laterite caps, dolerite dykes Rockhounding where there's mineralisation, pegmatites etc. Agritourism
Cognitive (opportunities for noncommercial activities, e.g., aesthetics, education and spiritual value)	Expand current perspectives related to biodiversity to include geological features, geological connections to Indigenous spirituality, and geological connections to deep-time	Educational and scientific value Spiritual connection to Indigenous owners Spiritual/aesthetic value of deep-time perspective

where the CZ places strong constraints on ecosystem productivity and where, if mismanaged or altered, the CZ function could become inimical to the provision of such services. Coupled with new state-of-the-art equipment and data portals expected to come online in 2023, we hope AR-CZO as part of the OzCZO also presents an attractive research partner and destination for the international CZ community. The authors are committed to collaborative and open science and look forward to sharing an ancient, deeply weathered AR-CZO with CZ scientists worldwide.

Acknowledgements Infrastructure for OzCZO sites is funded by the Australian Research Council (ARC) via LIEF grant LE210100035. Geophysical investigation of the CZ architecture at CZO-infrastructure installation sites was funded by AUSCOPE-opportunity funds.

References

1. ABARES (2022) Western Australia—Australian crop report: December edition from https://www.agriculture.gov.au/abares/research-topics/agricultural-outlook/australian-crop-report/western-australia
2. Abbott I (2003) Aborigines, settlers and native animals: a zoological history of the south-west Early Days. J R West Aust Hist Soc 12(3):231–249
3. Abdulkareem A, Leopold M, Murphy DV, McGrath G (2020) Thermal imaging of a hydrophobic soil's response to surfactant application at the Avon River Catchment Critical Zone Observatory. Geoderma 368:114309
4. Adhikari K, Hartemink AE (2016) Linking soils to ecosystem services—a global review. Geoderma 262:101–111
5. Ali R, Silberstein R, Byrne J, Hodgson G (2013) Drainage discharge impacts on hydrology and water quality of receiving streams in the wheatbelt of Western Australia. Environ Monit Assess 185(11):9619–9637
6. Allen and Jeffery (1990)

7. Anand RR, Paine M (2002) Regolith geology of the Yilgarn Craton, Western Australia: implications for exploration. Aust J Earth Sci 49(1):3–162
8. Anders AM, Bettis EA III, Grimley DA, Stumpf AJ, Kumar P (2018) Impacts of quaternary history on critical zone structure and processes: examples and a conceptual model from the intensively managed landscapes Critical Zone Observatory. Front Earth Sci 6:24
9. Anderson SP, Anderson RS, Tucker GE (2012) Landscape scale linkages in critical zone evolution. CR Geosci 344(11–12):586–596
10. Anjos L, Gaistardo C, Deckers J, Dondeyne S et al (2015) World reference base for soil resources. In: 2014 International soil classification system for naming soils and creating legends for soil maps. FAO, Rome, Italy, JRC91947. http://www.fao.org/3/a-i3794e.pdf
11. Austin BJ, Robinson CJ, Fitzsimons JA, Sandford M et al (2018) Integrated measures of Indigenous land and sea management effectiveness: challenges and opportunities for improved conservation partnerships in Australia. Conserv Soc 16(3):372–384
12. Avon Catchment Council (2005) Avon natural resource management strategy. The Regional Natural Resource Management Strategy for the Avon River Basin ACC, Northam
13. Banwart S, Bernasconi SM, Bloem J, Blum W et al (2011) Soil processes and functions in Critical Zone Observatories: hypotheses and experimental design. Vadose Zone J 10(3):974–987
14. Bartle J, Cooper D, Olsen G, Carslake J (2002) Acacia species as large-scale crop plants in the Western Australian wheatbelt. Conserv Sci West Aust 4(3)
15. Barton L, Hoyle FC, Stefanova KT, Murphy DV (2016) Incorporating organic matter alters soil greenhouse gas emissions and increases grain yield in a semi-arid climate. Agric Ecosyst Environ 231:320–330
16. Barton L, Kiese R, Gatter D, Butterbach-Bahl K, Buck R, Hinz C, Murphy DV (2008) Nitrous oxide emissions from a cropped soil in a semi-arid climate. Glob Chang Biol 14:177–192
17. Barton L, Murphy DV, Butterbach-Bahl K (2013) Influence of crop rotation and liming on greenhouse gas emissions from a semi-arid soil. Agric Ecosyst Environ 167:23–32
18. Barton L, Murphy DV, Kiese R, Butterbach-Bahl K (2010) Soil nitrous oxide and methane fluxes are low from a bioenergy crop (canola) grown in a semi-arid climate. Glob Chang Biol Bioenergy 2:1–15
19. Barton L, Thamo T, Engelbrecht D, Biswas WK (2014) Does growing grain legumes or applying lime cost effectively lower greenhouse gas emissions from wheat production in a semi-arid climate? J Clean Prod 83:194–203
20. Baumber A, Waters C, Cross R, Metternicht G, Simpson M (2020) Carbon farming for resilient rangelands: people, paddocks and policy. Rangeland J 42(5):293–307
21. Beard J (1999) Evolution of the river systems of the south-west drainage division, Western Australia. J R Soc West Aust 82:147
22. Bell S (2005) Constructing sustainable rural landscapes: oil mallees and the Western Australian wheatbelt. Geogr Res 43(2):194–208
23. Bell L, Moore A, Thomas D (2021) Diversified crop-livestock farms are risk-efficient in the face of price and production variability. Agric Syst 189:103050
24. Beringer et al (2022) Global change. Biology in final step before press, will be added soon
25. Bettenay E, Mulcahy M (1972) Soil and landscape studies in Western Aaustralia: (2) valley form and surface features of the south-west drainage division. J Geol Soc Aust 18(4):359–369
26. Blache D, Vercoe PE, Martin GB, Revell DK (2016) Integrated and innovative livestock production in drylands In: Farooq M, Siddique KHM (eds) Integrated and innovative livestock production in drylands. Innovations in dryland agriculture, pp 211–235
27. Bloomfleld C, Coulter J (1974) Genesis and management of acid sulfate soils. Adv Agron 25:265–326
28. Brantley SL, McDowell WH, Dietrich WE, White TS et al (2017) Designing a network of critical zone observatories to explore the living skin of the terrestrial earth. Earth Surf Dyn 5(4):841–860
29. Brimhall GH, Dietrich WE (1987) Constitutive mass balance relations between chemical composition, volume, density, porosity, and strain in metasomatic hydrochemical systems: results on weathering and pedogenesis. Geochim Cosmochim Acta 51:567–587

30. Brimhall GH, Lewis CJ, Ague JJ, Dietrich WE, Hampel J, Teague T, Rix P (1988) Metal enrichment in bauxites by deposition of chemically mature aeolian dust. Nature 333(6176):819–824
31. Brodie JE, Kroon F, Schaffelke B, Wolanski E et al (2012) Terrestrial pollutant runoff to the Great Barrier Reef: an update of issues, priorities and management responses. Mar Pollut Bull 65(4–9):81–100
32. Buggle B, Glaser B, Hambach U, Gerasimenko N, Marković S (2011) An evaluation of geochemical weathering indices in loess–paleosol studies. Quatern Int 240(1–2):12–21. https://doi.org/10.1016/j.quaint.2010.07.019
33. Buss HL, Brantley SL, Scatena F, Bazilievskaya E et al (2013) Probing the deep critical zone beneath the Luquillo Experimental Forest, Puerto Rico. Earth Surf Proc Land 38(10):1170–1186
34. Carson JK, Campbell L, Rooney DC, Clipson N, Gleeson DB (2009) Minerals in soil select distinct bacterial communities in their microhabitats. FEMS Microbiol Ecol 67:381–388
35. Carson JK, Gonzalez-Quiñones V, Murphy DV, Hinz C, Shaw JA, Gleeson DB (2010) Low pore connectivity increases bacterial diversity in soil. Appl Environ Microbiol 76:3936–3942
36. Carson J, Rooney DR, Gleeson DB, Clipson N (2007) Altering the mineral composition of soil causes a shift in microbial community structure. FEMS Microbiol Ecol 61(3):414–423. https://doi.org/10.1111/j.1574-6941.2007.00361.x
37. Chittleborough DJ (1991) Indices of weathering for soils and palaeosols formed on silicate rocks. Aust J Earth Sci 38(1):115–120. https://doi.org/10.1080/08120099108727959
38. Chorover J, Troch PA, Rasmussen C, Brooks PD et al (2011) How water, carbon, and energy drive critical zone evolution: the Jemez-Santa Catalina Critical Zone Observatory. Vadose Zone J 10(3):884–899
39. Cleverly J, Eamus D, Edwards W, Grant M et al (2019) TERN, Australia's land observatory: addressing the global challenge of forecasting ecosystem responses to climate variability and change. Environ Res Lett 14(9):095004
40. Costanza R, d'Arge R, De Groot R, Farber S et al (1997) The value of the world's ecosystem services and natural capital. Nature 387(6630):253–260
41. Coughlan MR, Nelson DR, Lonneman M, Block AE (2017) Historical land use dynamics in the highly degraded landscape of the Calhoun Critical Zone Observatory. Land 6(2):32
42. Creighton C, Waterhouse J, Day JC, Brodie J (2021) Criteria for effective regional scale catchment to reef management: a case study of Australia's Great Barrier Reef. Mar Pollut Bull 173:112882
43. Darmody R, Allen C, Thorn C, Dixon J (2001) The poisonous rocks of Kärkevagge. Geomorphology 41(1):53–62
44. Degens B, Shand P, Fitzpatrick R, George RG, Rogers S, Smith M (2008) Ch10 Avon Basin, WA Wheatbelt: an overview of geochemical risks of saline acidic discharge from deep drains used to manage dryland salinity in Western Australia. In: Fitzpatrick R, Shand P (eds) Inland acid sulfate soil systems across Australia. CRC LEME Open File Report No 249:169–175 CRC LEME, Perth, Australia
45. Dominati E, Patterson M, Mackay A (2010) A framework for classifying and quantifying the natural capital and ecosystem services of soils. Ecol Econ 69(9):1858–1868
46. Du X, Rate AW, Gee MM (2012) Redistribution and mobilization of titanium, zirconium and thorium in an intensely weathered lateritic profile in Western Australia. Chem Geol 330:101–115
47. Eamus D, Hatton T, Cook P, Colvin C (2006) Ecohydrology: vegetation function, water and resource management. CSIRO Publishing
48. Ellis NR, Albrecht GA (2017) Climate change threats to family farmers' sense of place and mental wellbeing: a case study from the Western Australian Wheatbelt. Soc Sci Med 175:161–168
49. Field JP, Breshears DD, Law DJ, Villegas JC et al (2015) Critical zone services: expanding context, constraints, and currency beyond ecosystem services. Vadose Zone J 14(1)

50. Fitzpatrick R, Powell B, Marvanek S (2008) Atlas of Australian acid sulfate soils. Inland Acid Sulfate Soil Syst Across Aust 249:75–89
51. George R, Clarke J, English P (2008) Modern and palaeogeographic trends in the salinisation of the Western Australian wheatbelt: a review. Soil Res 46(8):751–767
52. George R, Bennett D, Speed R (2004) Salinity management—the case for focussing on wheatbelt valleys. Salinity Solutions: Working with Science and Society
53. George R, and M Coleman (2001) Hidden menace or opportunity–groundwater hydrology, playas and commercial options for salinity in wheatbelt valleys. In: Wheatbelt valleys conference
54. Giardino JR, Houser C (2015) Introduction to the critical zone. Dev Earth Surf Process 19:1–13
55. Gilkes R, Scholz G, Dimmock G (1973) Lateritic deep weathering of granite. J Soil Sci 24(4):523–536
56. Gioia P, Hopper SD (2017) A new phytogeographic map for the Southwest Australian Floristic Region after an exceptional decade of collection and discovery. Bot J Linn Soc 184(1):1–15
57. Gleeson DB, Kennedy N, McDermott F, Melville K, Gadd GM, Clipson N (2005) Characterization of bacterial community structure on a weathered pegmatitic granite. Microb Ecol 51:526–534
58. Gleeson D, Mathes F, Farrell M, Leopold M (2016) Environmental drivers of soil microbial community structure and function at the Avon River Critical Zone Observatory. Sci Total Environ 571:1407–1418
59. Gleeson DB, Melville K, Gadd GM, McDermott FP, Clipson N (2006) Characterization of fungal community structure on a weathered pegmatitic granite. Microb Ecol 50:360–368
60. Gleeson DB, Melville K, McDermott F, Clipson N, Gadd GM (2010) Molecular characterization of fungal communities in sandstone. Geomicrobiol J (Spec Issue Eukaryotic Geomicrobiol) 27:559–571
61. Grant M, Thompson S (2021) The Australian CZO network from https://www.tern.org.au/critical-zone/
62. Hahm WJ, Dietrich WE Dawson TE, Lovill S, Rempe D (2016) Ecohydrological consequences of critical zone structure in the Franciscan Formation, Northern California Coast Ranges. AGU Fall meeting abstracts 2016, AGUFMEP41F07H
63. Hall D, Overhue T, Lemon J (2021) Managing dispersive (sodic) soils from https://www.agric.wa.gov.au/dispersive-and-sodic-soils/managing-dispersive-sodic-soils
64. Harries M, Anderson GC, Hüberli D (2015) Crop sequences in Western Australia: what are they and are they sustainable? Findings of a four-year survey. Crop Pasture Sci 66:634–647
65. Harries M, Flower KC, Scanlan CA (2021) Sustainability of nutrient management in grain production systems of south-west Australia. Crop Pasture Sci 72:197–212
66. Hatton T, Ruprecht J, George R (2003) Preclearing hydrology of the Western Australia wheatbelt: target for the future. Plant Soil 257(2):341–356
67. Hingston F, Gailitis V (1976) The geographic variation of salt precipitated over Western Australia. Soil Res 14(3):319–335
68. Hobbs TJ, Neumann CR, Meyer WS, Moon T, Bryan BA (2016) Models of reforestation productivity and carbon sequestration for land use and climate change adaptation planning in South Australia. J Environ Manag 181:279–288
69. Hodgkin E, Lenanton R (1979) Estuaries and coastal lagoons of south western Australia. In: Neilson BJ, Cronin LE (eds) Estuaries and nutrients. Contemporary Issues in Science and Society Humana Press
70. Hodgkin EP, Environmentalist E (1978) An environmental study of the Blackwood River Estuary, Western Australia: 1974–1975: a report to the Estuarine and Marine Advisory Committee of the Environmental Protection Authority Department of Conservation and Environment
71. Horbe AMC, Anand R (2011) Bauxite on igneous rocks from Amazonia and Southwestern of Australia: implication for weathering process. J Geochem Explor 111(1–2):1–12
72. Hunt JR, Kirkegaard JA, Harris FA, Porker KD et al (2021) Exploiting genotype × management interactions to increase rainfed crop production: a case study from south-eastern Australia. J Exp Bot 72:5189–5207

73. Isaac P, Cleverly J, McHugh I, van Gorsel E, Ewenz C, Beringer J (2017) OzFlux data: network integration from collection to curation. Biogeosciences 14:2903–2928
74. Jeffery RP, Ryan MH, Ayers NL, Nichols PGH (2021) Salinity tolerance and avoidance mechanisms at germination among messina (*Melilotus siculus*) accessions. Crop Pasture Sci 72:641–651
75. Jenkins SN, Murphy DV, Waite IS, Rushton SP, O'Donnel AG (2016) Ancient landscapes and the relationship with microbial nitrification. Sci Rep 6:30733
76. Van Der Kaars S, Miller GH, Turney CS et al (2017) Humans rather than climate the primary cause of Pleistocene megafaunal extinction in Australia. Nat Commun 8(1):1–7
77. Kazakou E, Dimitrakopoulos P, Baker A, Reeves R, Troumbis A (2008) Hypotheses, mechanisms and trade-offs of tolerance and adaptation to serpentine soils: from species to ecosystem level. Biol Rev 83(4):495–508
78. Kew GA, Mengler FC, Gilkes RJ (2007) Regolith strength, water retention, and implications for ripping and plant root growth in bauxite mine restoration. Restor Ecol 15:S54–S64
79. Keywood M, Chivas A, Fifield L, Cresswell R, Ayers G (1997) The accession of chloride to the western half of the Australian continent. Soil Res 35(5):1177–1190
80. Kirkegaard JA, Hunt JR, McBeath TM, Lilley JM et al (2014) Improving water productivity in the Australian grains industry—a nationally coordinated approach. Crop Pasture Sci 65:583–601
81. Kirkegaard JA, Peoples MB, Angus JF, Unkovich MF (2011) Diversity and evolution of rainfed farming systems in Southern Australia. In: Tow P, Cooper I, Partridge I, Birch C (eds) Rainfed farming systems. Springer, Netherlands, Dordrecht, pp 715–754
82. Kooyman RM, Watson J, Wilf P (2020) Protect Australia's gondwana rainforests. Science 367(6482):1083–1083
83. Kotze AC, O'Grady J, Emms J, Toofey AF et al (2009) Exploring the anthelmintic properties of Australian native shrubs with respect to their potential role in livestock grazing systems. Parasitology 136(9):1065–1080
84. Kumar P, Le PV, Papanicolaou AT, Rhoads BL et al (2018) Critical transition in critical zone of intensively managed landscapes. Anthropocene 22:10–19
85. Li X, Norman HC, Hendry JK et al (2018) The impact of supplementation with *Rhagodia preissii* and *Atriplex nummularia* on wool production, mineral balance and enteric methane emissions of Merino sheep. Grass Forage Sci 73:381–391
86. Llewellyn RS, D'Emden FH, Kuehne G (2012) Extensive use of no-tillage in grain growing regions of Australia. Field Crop Res 132:204–212
87. Lovill S, Hahm W, Dietrich WE (2018) Drainage from the critical zone: lithologic controls on the persistence and spatial extent of wetted channels during the summer dry season. Water Resour Res 54(8):5702–5726
88. Luo Q, Bellotti W, Williams M, Bryan B (2005) Potential impact of climate change on wheat yield in South Australia. Agric For Meteorol 132(3–4):273–285
89. Lybrand R, Rasmussen C, Jardine A, Troch P, Chorover J (2011) The effects of climate and landscape position on chemical denudation and mineral transformation in the Santa Catalina mountain critical zone observatory. Appl Geochem 26:S80–S84
90. McArthur WM, Johnston DAW, Snell LJ (2004) Reference soils of south-western Australia Perth, WA. Dept of Agriculture Western Australia on behalf of the Australian Society of Soil Science Inc
91. McCrea A, Anand R, Gilkes R (1990) Mineralogical and physical properties of lateritic pallid zone materials developed from granite and dolerite. Geoderma 47(1–2):33–57
92. Mix HT, Ibarra DE, Mulch A, Graham SA, Chamberlain CP (2016) A hot and high Eocene Sierra Nevada. Bulletin 128(3–4):531–542
93. Moggridge BJ, Thompson RM (2021) Cultural value of water and western water management: an Australian indigenous perspective. Australas J Water Resour 25(1):4–14
94. Mole DR, Fiorentini ML, Thebaud N, McCuaig TC et al (2012) Spatio-temporal constraints on lithospheric development in the southwest–central Yilgarn Craton, Western Australia. Aust J Earth Sci 59(5):625–656

95. Montanarella L, Panagos (2015) Policy relevance of critical zone science. Land Use Policy 49:86–91
96. Moore GA, Sanford P, Dolling PJ, Real D (2021) The challenges of developing resilient perennial pastures for a Mediterranean environment—a review for Western Australia. Crop Pasture Sci 72:613–633
97. Mulcahy M, Bettenay E (1972) Soil and landscape studies in Western Australia: (1) the major drainage divisions. J Geol Soc Aust 18(4):349–357
98. Mulchay MJ (1960) Laterites and lateritic soils in south-western Australia. J Soil Sci 11:206–225
99. Nesbitt HW, Young GM (1982) Early Proterozoic climates and plate motions inferred from major element chemistry of lutites. Nature 299:715–717
100. Nichols PGH, Loi A, Nutt BJ, Evans PM, Craig AD et al (2007) New annual and short-lived perennial pasture legumes for Australian agriculture—15 years of revolution. Field Crop Res 104(1–3):10–23
101. Nicol DL, Finlayson J, Colmer TD, Ryan MH (2013) Opportunistic Mediterranean agriculture—using ephemeral pasture legumes to utilize summer rainfall. Agric Syst 120:76–84
102. Nie W, Guo H, Banwart SA (2021) Economic valuation of earth's critical zone: framework, theory and methods. Environ Dev 40:100654
103. Norman H, Wilmot M, Dean T, Barrett-Lennard E, Masters D (2010) Sheep production, plant growth and nutritive value of a saltbush-based pasture system subject to rotational grazing or set-stocking. Small Rumin Res 91:103–109
104. Owen MJ, Martinez NJ, Powles SB (2014) Multiple herbicide-resistant *Lolium rigidum* (annual ryegrass) now dominates across the Western Australian grain belt. Weed Res 54(3):314–324
105. O'Brien FLM, Almaraz M, Foster MA, Hill AF et al (2019) Soil salinity and pH drive soil bacterial community composition and diversity along a lateritic slope in the Avon River Critical Zone Observatory, Western Australia. Front Microbiol 10
106. O'Geen AT, Safeeq M, Wagenbrenner J, Stacy E et al (2018) Southern Sierra Critical Zone Observatory and Kings River experimental watersheds: a synthesis of measurements, new insights, and future directions. Vadose Zone J 17
107. Pastorello G, Trotta C, Canfora E et al (2020) The FLUXNET2015 dataset and the ONEFlux processing pipeline for eddy covariance data. Sci Data 7:225
108. Pelletier JD, Broxton PD, Hazenberg P et al (2016) Global 1-km gridded thickness of soil, regolith, and sedimentary deposit layers. ORNL DAAC, Oak Ridge, Tennessee
109. Pen LJ, Hutchison J (1999) Managing our rivers : a guide to the nature and management of the streams of south-west Western Australia. Water and Rivers Commission East Perth, Western Austrailia
110. Perring MP, Standish RJ, Hulvey KB et al (2012) The Ridgefield multiple ecosystem services experiment: can restoration of former agricultural land achieve multiple outcomes? Agric Ecosyst Environ 163:14–27
111. Prosser IP, Chiew FH, Stafford Smith M (2021) Adapting water management to climate change in the Murray-Darling Basin, Australia. Water 13(18):2504
112. Puckridge DW, French RJ (1983) The annual legume pasture in cereal—Ley farming systems of southern Australia: a review. Agric Ecosyst Environ 9:229–267
113. Revell D, Norman H, Vercoe P, Phillips N et al (2013) Australian perennial shrub species add value to the feed base of grazing livestock in low-to medium-rainfall zones. Anim Prod Sci 53(11):1221–1230
114. Richardson M, Kumar P (2017) Critical zone services as environmental assessment criteria in intensively managed landscapes. Earth's Future 5(6):617–632
115. Rickards L (2016) Goodbye Gondwana? Questioning disaster triage and fire resilience in Australia. Aust Geogr 47(2):127–137
116. Ritchie H, Roser M (2018) Crop yields Online at: https://www.OurWorldInData.org
117. Rix MG, Edwards DL, Byrne M, Harvey MS, Joseph L, Roberts JD (2015) Biogeography and speciation of terrestrial fauna in the south-western Australian biodiversity hotspot. Biol Rev 90(3):762–793

118. Ryan MH, Kidd DR, Sandral GA, Yang Z et al (2016) High variation in the percentage of root length colonised by arbuscular mycorrhizal fungi among 139 lines representing the species subterranean clover (*Trifolium subterraneum*). Appl Soil Ecol 98:221–232
119. Salama R, Farrington P, Bartle G, Watson G (1993) Salinity trends in the wheatbelt of Western Australia: results of water and salt balance studies from Cuballing catchment. J Hydrol 145(1–2):41–63
120. Scanlon TT, Doncon G (2020) Rain, rain, gone away: decreased growing-season rainfall for the dryland cropping region of the south-west of Western Australia. Crop Pasture Sci 71(2):128–133
121. Schuster R, Germain RR, Bennett JR, Reo NJ, Arcese P (2019) Vertebrate biodiversity on indigenous-managed lands in Australia, Brazil, and Canada equals that in protected areas. Environ Sci Policy 101:1–6
122. Seymour M, Kirkegaard JA, Peoples MB, White PF, French RJ (2012) Break-crop benefits to wheat in Western Australia–insights from over three decades of research. Crop Pasture Sci 63(1):1–16
123. Silberstein R, Aryal S, Durrant J et al (2012) Climate change and runoff in south-western Australia. J Hydrol 475:441–455
124. Speldewinde PC, Cook A, Davies P, Weinstein P (2009) A relationship between environmental degradation and mental health in rural Western Australia. Health Place 15(3):880–887
125. Steffen W (2009) Australia's biodiversity and climate change. CSIRO Publishing
126. Trelease SF, Martin AL (1936) Plants made poisonous by selenium absorbed from the soil. Bot Rev 2(7):373–396
127. Turner NC, Ward PR (2002) The role of agroforestry and perennial pasture in mitigating water logging and secondary salinity: summary. Agric Water Manag 53(1–3):271–275
128. Vermeulen S, Bossio D, Lehmann J et al (2019) A global agenda for collective action on soil carbon. Nat Sustain 2(1):2–4
129. Wardell-Johnson G, Wardell-Johnson A, Bradby K, Robinson T et al (2016) Application of a Gondwanan perspective to restore ecological integrity in the south-western Australian global biodiversity hotspot. Restor Ecol 24(6):805–815
130. Wells G, Hopper SD, Dixon KW (2004) Fire regimes and biodiversity conservation: a brief review of scientific literature with particular emphasis on south-west Australian studies. Environmental Protection Authority, Perth, Austrailia
131. Westoby M (1994) Biodiversity in Australia compared with other continents. In: Ricklefs RE, Schluter D (eds) Species diversity in ecological communities: historical and geographical perspectives. University of Chicago Press, pp 170–177. http://www.press.uchicago.edu/ucp/books/book/chicago/S/bo3626513.html
132. Whetton P, Chiew F (2021) Climate change in the Murray-Darling Basin, Australia. Elsevier, pp 253–274
133. White TS, Brantley SL, Banwart SA, Chorover J et al (2015) The role of critical zone observatories in critical zone science. Dev Earth Surf Process 19:15–78
134. Wilford J, Searle R, Thomas M, Pagendam D, Grundy M (2016) A regolith depth map of the Australian continent. Geoderma 266:1–13
135. Williams KJ, Ford A, Rosauer DF, Silva ND, Mittermeier R, Bruce C, Larsen FW, Margules C (2011) Forests of East Australia: the 35th biodiversity hotspot. In: Biodiversity hotspots, pp 295–310
136. Wilson CG, Abban B, Keefer LL, Wacha K et al (2018) The intensively managed landscape critical zone observatory: a scientific testbed for understanding critical zone processes in agroecosystems. Vadose Zone J 17(1):1–21
137. Yates CJ, Hobbs RJ (1997) Woodland restoration in the Western Australian wheatbelt: a conceptual framework using a state and transition model. Restor Ecol 5(1):28–35
138. Zheng H, Teng J, Chiew F, Post D, Zhang L (2020) Plausible hydroclimate futures for the Murray-Darling Basin. A report for the Murray–Darling Basin Authority. CSIRO, EP201750. https://doi.org/10.25919/samv-ey93